ENGINEERING MECHANICS

ENGINEERING MECHANICS

Revised and Updated

R.K. Singal
Professor
Educationist and Technologist
rk.singal@gmail.com

Mridul Singal
B.Tech. (IIT, Kanpur), M.Tech. (IISc, Bangalore),
Leading Architect, IBM India Pvt. Ltd.

Rishi Singal
B.Tech. (PEC Chandigarh), PMP (PMI, USA)
Principal Consultant, Oracle India Pvt. Ltd.

I.K. International Publishing House Pvt. Ltd.

NEW DELHI • BANGALORE

Published by

I.K. International Publishing House Pvt. Ltd.
S-25, Green Park Extension
Uphaar Cinema Market
New Delhi – 110 016 (India)
E-mail: info@ikinternational.com
Website: www.ikbooks.com

ISBN 978-81-907462-1-2

Reprint 2013

Published by Krishan Makhijani for I.K. International Publishing House Pvt. Ltd., S-25, Green Park Extension, Uphaar Cinema Market, New Delhi – 110 016 ·nd Printed by Rekha Printers Pvt. Ltd., Okhla Industrial Area, Phase II, Delhi – 110 020.

Preface

"Engineering Mechanics" is meant for the students of all branches of Engineering. The book will also be helpful to engineers preparing for various competitive examinations.

The primary attractions of the book are the presentation style of the text, its simple and clear language, and use of comparison tables and illustrative drawings. The difficulties usually faced by new engineering students have been taken care of while preparing the book.

In addition to exhaustive theory, a large number of numerical problems have been selected from university examination papers and question banks, graded properly, solved and arranged in various chapters. Exhaustive question banks on theory problems have also been added at the end of every chapter. The students need supplementary material for the subject, the same has been provided as annexures in the book.

The authors are grateful to the publisher for bringing out this book in a very short time. Although the authors are constantly in touch with the students while teaching and understanding their requirements and difficulties, any constructive criticism will be welcome with gratitude.

Authors

Contents

Unit IV: Dynamics of Rigid Body

Unit V: Stress and Strain Analysis

UNIT I: TWO-DIMENSIONAL FORCE SYSTEM

CHAPTER 1

Force and Equilibrium

1.1 INTRODUCTION

Mechanics deals with dynamic behaviour of rigid bodies under the action of forces. Engineering mechanics deals with the laws and principles of mechanics and their applications to engineering problems. Engineering mechanics is also called structural mechanics or rigid-body mechanics and the body is idealized as a continuum (continuous distribution of matter) that undergoes theoretically no deformation whatsoever. In actual practice, no solid body is perfectly rigid and every body undergoes some deformation (change of shape and size) under the action of forces. However, the deformation may by very small and may not be considered.

In the study of dynamic behaviour of a body, the latter may be at rest and in equilibrium or it may be in motion. Statics deals with the behaviour of bodies at rest and dynamics deals with the bodies in motion under the action of forces.

Force is an agent or disturbance which changes or tends to change the state of rest or of uniform motion of a body in a straight line. Force may cause mechanical disturbance of a body on another body in direct contact or without any contact where force may be created by gravitational action.

1.2 LAWS OF MECHANICS

The laws of motion and gravity were established by Isaac Newton (1642–1727) which are the foundations of engineering mechanics. The Newton's laws of motion form a complete definition of force. The following basic laws and principles are considered to be the foundation of mechanics:

1. Newton's first and second laws of motion
2. Newton's third law
3. Newton's law of gravitation
4. The parallelogram law of forces
5. Principle of transmissibility of forces.

1.2.1 Newton's Laws of Motion

Newton's laws of motion establish relationship between force and motion. There are three laws of motion.

1. *Newton's First Law of Motion*

   It states, "Every body continues in its state of rest or of uniform motion in a straight line unless it is acted upon by some external force to change that state".

As per Newton's first law of motion the action of a force is to change the velocity of a body, i.e., to produce an acceleration on the body on which it acts. This is the functional definition of force.

The first law of motion is also called the *law of inertia.* Inertia is the property of a body to remain in its present state of rest or of uniform motion.

2. *Newton's Second Law of Motion*

   It states, "The rate of change of momentum of a body is directly proportional to the impressed force and takes place in the direction of the straight line in which the force acts."

The quantity of motion possessed by a body moving in a straight line is known as *momentum*. It is the product of mass of a body and its velocity.

If m = mass of a body, kg

u = initial velocity, m/s

v = final velocity, m/s

a = acceleration, m/s^2

t = time required to change the velocity of a body from u to v.

F = force required to change the above velocity [N]

Initial momentum $= mu$

Final momentum $= mv$

Change of momentum $= mv - mu$

Rate of change of momentum

$$= \frac{mv - mu}{t} = \frac{m(v-u)}{t}$$

$$= ma$$

As per Newton's second law of motion, force is the product of mass and acceleration produced on the mass by the force. Alternatively, the acceleration of a body is inversely proportional to its mass for a given force.

$$a \propto \frac{1}{m}$$

or, $$a = \frac{F}{m}$$

or, $$F = ma$$

where, F is the constant of proportionality and is called force.

Unit of Force

The unit of force in SI system is *newton*. This is the force which acting on a mass of kilogram produces an acceleration of one metre per second, per second.

$$1\text{ N} = 1\text{ kg} \times 1\text{ m/s}^2$$

$$= \frac{1\text{ kg m}}{\text{s}^2}$$

$$[\text{F}] = [\text{M}]\frac{[\text{L}]}{[\text{T}]^2}$$

The second law of motion is the *quantitative* and *dimensional* definition of force.

1.2.2 Newton's Third Law of Motion

It states, "To every action there is an equal and opposite reaction".

Action is the force exerted by a body on another body and reaction is the equal and opposite force exerted by another body on the first body.

Example: A body C of weight W is resting on a plane surface and is exerting a force equal to mg (= W) downwards. As per Newton's third law, a reaction R_A equal and opposite to W will be acting upwards on the body C at the point A.

Similarly, a force produces motion of the body on which it acts. If the motion of the body is stopped by another body, as per Newton's third law, there will be equal and opposite reaction by the second body on the first.

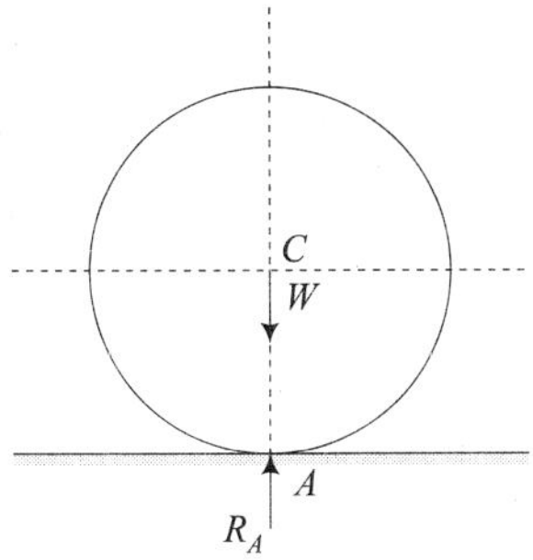

Fig. 1.1 Reaction

Therefore, as per Newton's third law of motion, force is an interaction and that is the nature of the force.

The Newton's three laws of motion form a complete definition of force.

The principle of reaction is used in ensuring equilibrium of bodies at rest, propulsion of jet aeroplanes and rockets, power production by steam, gas and water turbines, propulsion of ship, etc.

1.2.3 Newton's Law of Gravitation

Newton's law of gravitation states, "Every particle attracts any other particle with a *gravitational force* whose magnitude is given by the following equation."

$$F = G\frac{m_1 m_2}{r^2}$$

where, m_1 and m_2 = mass of particles, [kg]

r = distance between particles [m]

G = gravitational constant

$= 6.67 \times 10^{-11}$ N.m²/kg

$= 6.67 \times 10^{-11}$ m³/kg.s²

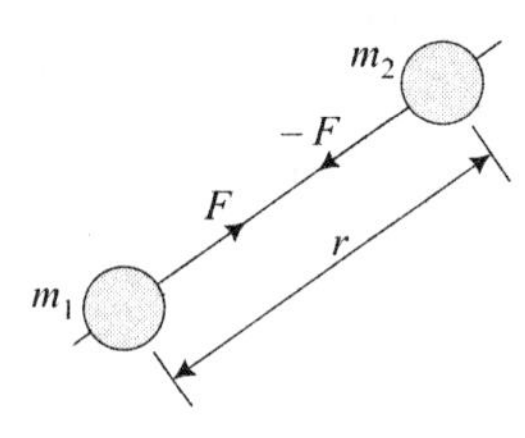

Fig. 1.2 Gravitation

Gravitational force near earth's surface is given by

$$F = G\frac{Mm}{r^2}$$

where, M = mass of the earth

From Newton's second law,

$$F = ma$$
$$= mg$$

$$\therefore \quad g = \frac{GM}{r^2}$$
$$= 9.81 \text{ m/s}^2$$

g = Acceleration due to gravity

The force can be applied directly or it can be a load of mass of body due to gravitational attraction called weight of the body. The weight of the body is found out from Newton's law of gravitation.

In SI system of units, a kilogram is the amount of mass that will accelerate 1 m/s^2 under the action of 1 newton. One kilogram of force is the weight of 1 kilogram of mass at earth's surface, where acceleration of gravity is 9.81 m/s^2.

The weight,

$$W = mg \text{ [N]}$$

where, m = mass of body [kg]

g = acceleration due to gravity [m/s^2]

= 9.81 m/s^2.

1.2.4 Principle of Transmissibility

It states, "If a force acts at a point or a rigid body, it is assumed to act at any other point on the line of action of a force within the same body."

The effect of a force depends upon its following *characteristics:*

1. Magnitude
2. Direction or line of action with respect to a coordinate system
3. Sense of nature (push or pull)
4. Point of applications

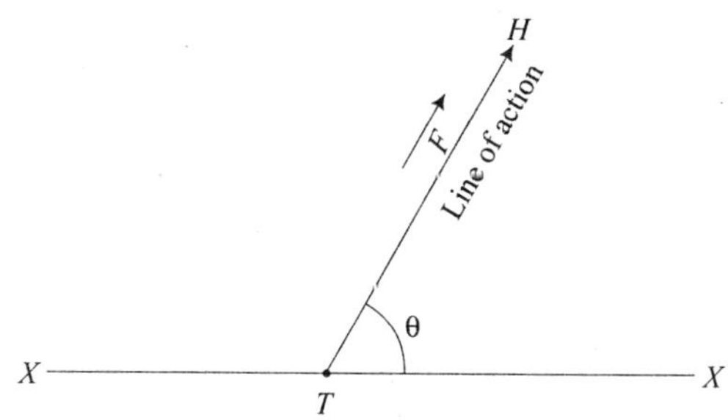

Fig. 1.3 Representation of a force

Force is a vector quantity as it possesses magnitude and direction. It can be represented as follows:

1. The *magnitude F* is indicated by the length of line *TH*, where *T* is the tail and *H* is the head of the line.
2. The *directon* is defined by angle θ with reference to axis *X-X*.
3. The *sense* is represented by arrowhead.

The sliding of a force along its line of action without altering the external static effect of the rigid body is called the principle of transmissibility.

Example:

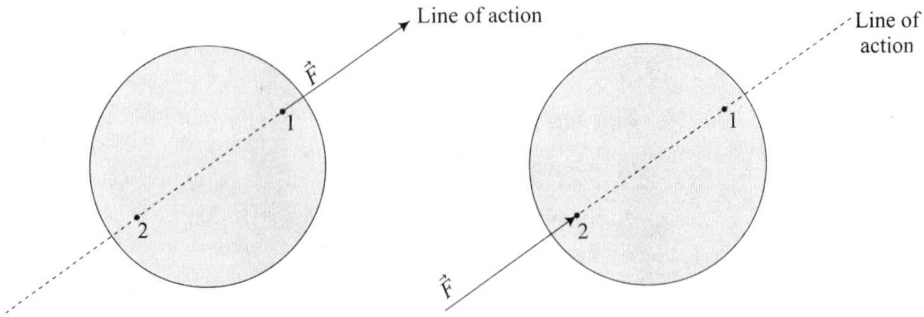

Fig. 1.4 Principle of transmissibility

The force $\vec{F}$ applied at point 1 of the rigid body can be slided along its line of action and applied at point 2. The magnitude, direction and line of action of the force do not change and, therefore, equilibrium conditions of rigid body remain unchanged.

Two forces having the same magnitude, direction and line of action but acting at different points 1 and 2 and producing the same external static effect on the rigid body are called *equivalent* forces.

1.3 SYSTEM OF FORCES

When two or more forces act on a body, they are called to form a system of forces.

1.3.1 Classification of Forces

The forces are classified according to the relative positions of their lines of action.

(a) Coplanar forces

A system of forces in which lines of action of all forces lie in one plane. The coplanar forces can be further classified as:

(i) *Collinear forces*: The lines of action lie on the same straight line.

(ii) *Concurrent forces*: The lines of action of all forces of the system meet at one point. The direction of forces are different and all forces lie in one plane.

(iii) *Non-concurrent forces*: The lines of action of all forces of the system lie in one plane but do not pass through one point. The forces may be parallel or non-parallel.

Fig. 1.5 Collinear forces

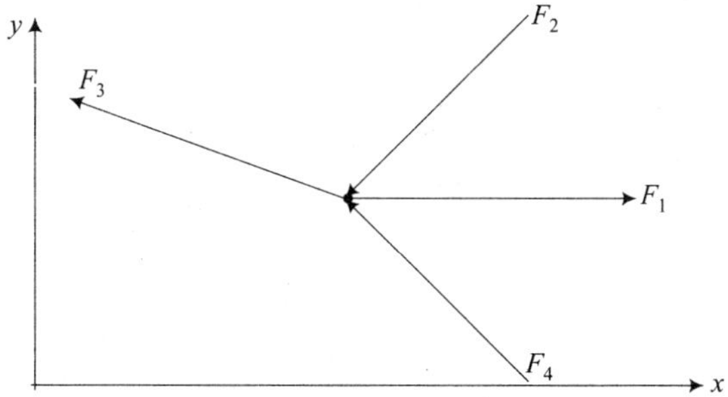

Fig. 1.6 Concurrent forces

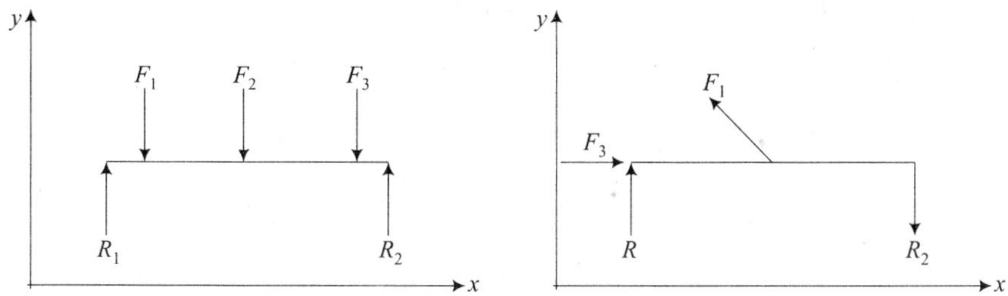

(a) Non-concurrent parallel forces

(b) Non-concurrent non-parallel forces

Fig. 1.7 Non-concurrent forces

(b) Non-coplanar forces

The lines of action of forces do not lie in one plane. The non-coplanar forces can be further classified as:

(i) *Concurrent forces*: The lines of action pass through a common point. For example, forces in the legs of tripod support of a camera.

(ii) *Non-concurrent forces*: The lines of action of all forces do not lie in the same plane nor do they meet in one point. See Fig. 1.8.

1.3.2 Basic Operations

In solving engineering problems, some basic operations have to be carried out with the system of forces, i.e., addition of forces, subtraction of forces, composition of forces, resolution of forces, etc.

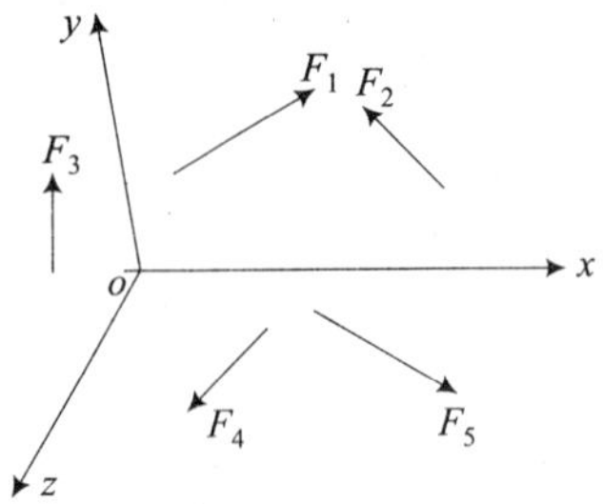

Fig. 1.8 Non-coplanar and non-concurrent forces

(i) *Composition of forces*: Combination of two or more forces to obtain a single force which produces the same effect as the original system of forces is called composition of forces. This single equivalent force is called the *resultant* of the original system of forces.

(ii) *Resolution of a force*: The process of splitting up a given force into a number of components, without changing its effect on the body is called resolution of a force. A force is generally resolved into rectangular components.

These operations can be carried out:

- Graphically by the use of polygon law of forces.
- Analytically by the use of trigonometry or vector algebra.

1.4 COMPOSITION OF FORCES

1.4.1 Polygon Law of Forces (Graphical Method)

The resultant of several forces acting at a point (concurrent forces) is found out graphically by the use of polygon law of forces. "If a number of coplanar forces are acting at a point such that they can be represented in magnitude and direction by the sides of a polygon taken in the same order, then their resultant is represented in magnitude and direction by the closing side of the polygon taken in the opposite direction."

Example:

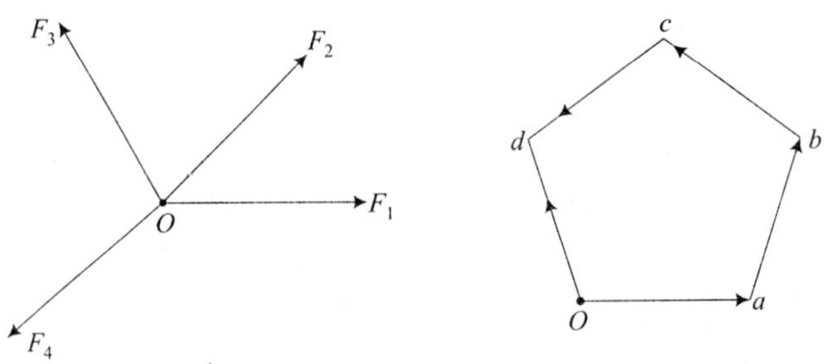

Fig. 1.9 Polygon law of forces

If the forces F_1, F_2, F_3, and F_4 acting simultaneously on a particle O be represented in magnitude and direction by the sides Oa, ab, bc and cd of a polygon respectively, their resultant is represented in magnitude and direction by the side Od drawn in the opposite direction.

1.4.2 Triangle Law of Forces

If two forces acting simultaneously on a body are represented in magnitude and direction by the two sides of a triangle taken in order, then their resultant may be represented in magnitude and direction by the third side taken in opposite direction.

Example: If F_1 and F_2 are two coplanar concurrent forces and are represented by sides Oa and ab of a triangle, their resultant is represented in magnitude and direction by the side Ob drawn in opposite direction.

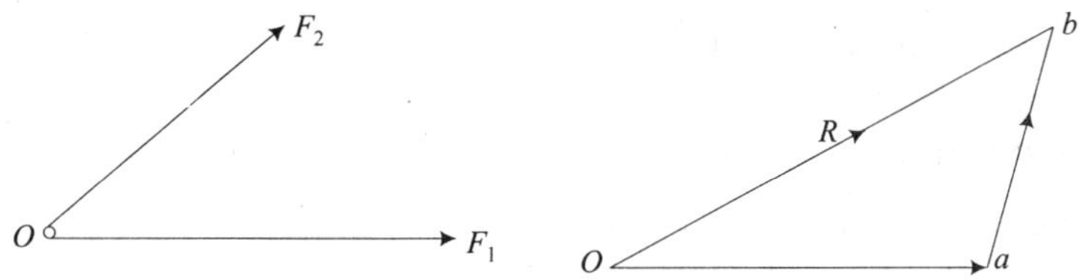

Fig. 1.10 Triangle law of forces

Triangle law of forces is useful in determining the condition of equilibrium.

"If three forces acting at a point be represented in magnitude and direction by the three sides of a triangle, taken in order, they will be in equilibrium."

Example: There are three forces F_1, F_2 and F_3. Draw Oa, ab and bO to represent the forces F_1, F_2 and F_3 in magnitude and direction. If the triangle Oab is closed, the forces are in equilibrium.

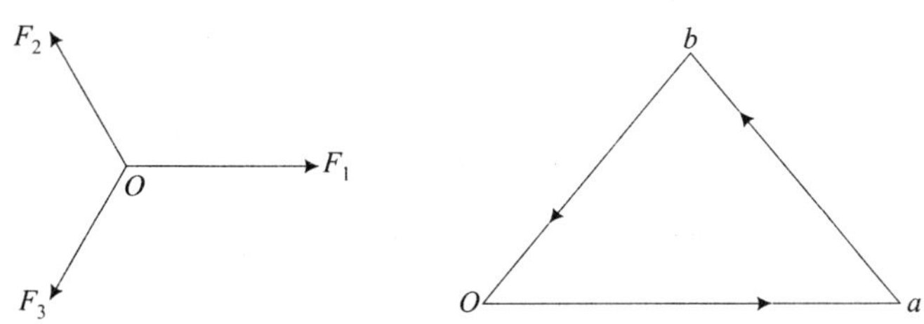

Fig. 1.11 Forces in equilibrium

1.4.3 Parallelogram Law of Forces

This law is also used to find the resultant of two concurrent forces.

"If two forces, acting at a point be represented in magnitude and direction by the two adjacent sides of a parallelogram, then their resultant is represented in

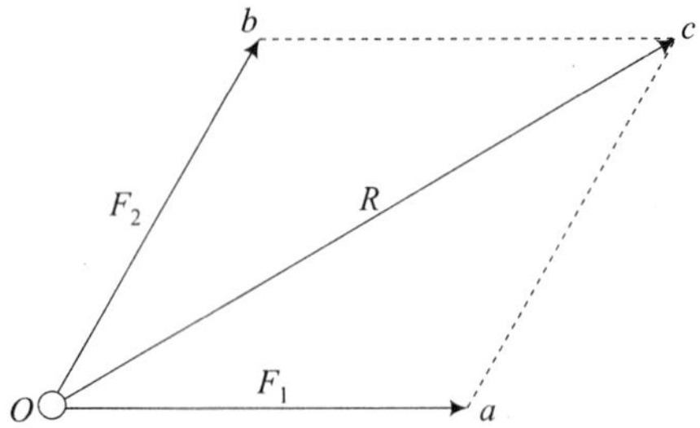

Fig. 1.12 Parallelogram law of forces

magnitude and direction by the diagonal of the parallelogram passing through that point".

Example: Two forces F_1 and F_2 are acting at point O of a body. Oa and Ob are the adjacent sides of parallelogram $Oacb$. The diagonal Oc represents the resultant R in magnitude and direction.

1.4.4 Cosine Law of Forces

This is the analytical method of parallelogram law of forces. If two forces F_1 and F_2 are acting at a point O of a body making an angle α between them, then the magnitude of resultant R is given by:

$$R = \sqrt{F_1^2 + F_2^2 + 2F_1F_2 \cos\alpha}$$

The direction of resultant R with force F_1 is given by:

$$\theta = \tan^{-1}\left[\frac{F_2 \sin\alpha}{F_1 + F_2 \cos\alpha}\right]$$

Case 1. If $\alpha = 0$, forces F_1 and F_2 are collinear. Put $\alpha = 0$

$$\therefore \quad R = F_1 + F_2 \qquad (\because \cos\theta° = 1)$$

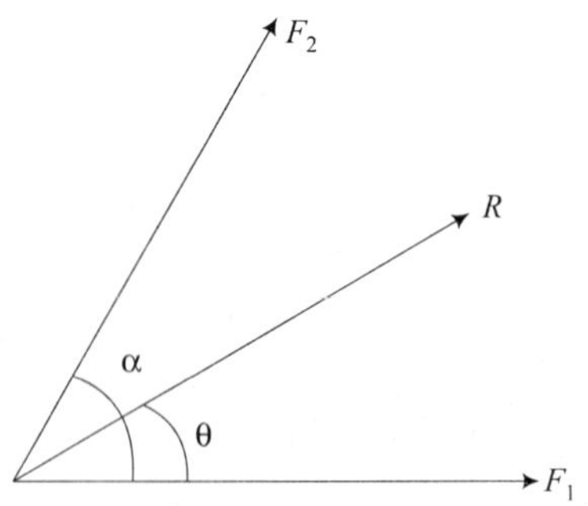

Fig. 1.13 Cosine law

Case 2. If $\alpha = 90°$, forces F_1 and F_2 are acting at right angles to each other.

$$\text{Cos } 90° = 0$$

$$\therefore \quad R = \sqrt{F_1^2 + F_2^2}$$

Case 3. If $\alpha = 180°$, the forces F_1 and F_2 are collinear but are acting in opposite direction to each other.

$$R = F_1 - F_2 \qquad (\because \cos 180° = -1)$$

1.4.5 Sine Law of Forces (Lami's Theorem)

"If three forces acting at a point are in equilibrium, each force will be proportional to the sine of the angle between the other two forces." This is also called *Lami's Theorem*.

$$\frac{F_1}{\sin\beta} = \frac{F_2}{\sin\gamma} = \frac{F_3}{\sin\alpha}$$

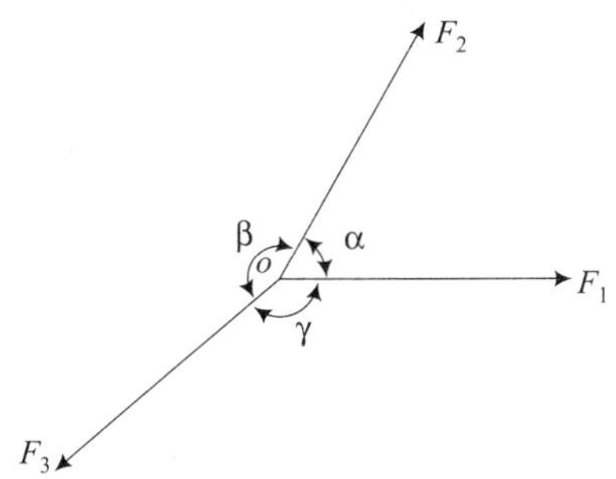

Fig. 1.14 Lami's theorem

1.5 RESOLUTION OF A FORCE

Resolution of a force is the reverse operation of composition of forces. In the operation of composition of forces, several forces acting at a particle are replaced by a resultant force which produces the same effect as original forces. In resolution of forces, a single force can be replaced by two forces in magnitude and direction different from the single force but producing the same effect as the given force. The given force is called the *resolved force* and parts are called *component forces* or the *resolutes*. A force is generally resolved into rectangular components.

The operation of resolution of forces can be carried out by the use of trigonometry or vector algebra.

A force F represented by a vector Oc inclined at an angle α with the horizontal axis has the following components.

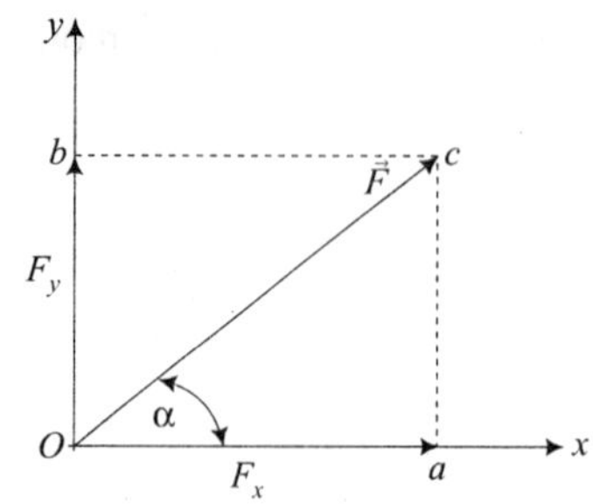

Fig. 1.15 Components of a force

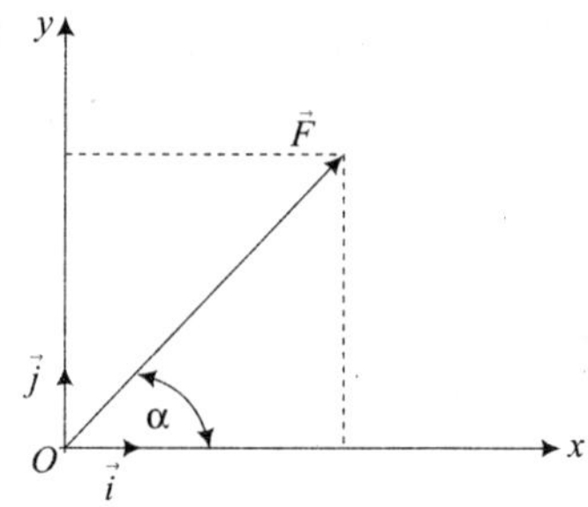

Fig. 1.16 Vector components of a force

Along x-axis: vector O-a

$$F_x = F \cos \alpha$$

Along y-axis: vector O-b

$$F_y = F \sin \alpha$$

The component forces satisfy the condition:

$$F = \sqrt{F_x^2 + F_y^2}$$

$$= \sqrt{(F \cos \alpha)^2 + (F \sin \alpha)^2}$$

$$= F\sqrt{\cos^2 \alpha + \sin^2 \alpha}$$

$$= F$$

Also, $$\vec{F} = F \times \vec{i} + F_{yx} \vec{j}$$

where, $\vec{i}$ is unit vector in x-direction

$\vec{j}$ is the unit vector in y-direction.

1.5.1 Principle of Resolved Components

If a large number of forces are acting, the principle of resolved components states: "The sum of resolved components of forces acting at a point in any direction is equal to the resolved components of the resultant of forces in that direction".

If three forces F_1, F_2 and F_3 are acting at a point O making angles α_1, α_2 and α_3 with the x-axis respectively, the resolved components are:

On x-axis:

$F_1 \cos \alpha_1$, $F_2 \cos \alpha_2$ and $F_3 \cos \alpha_3$

On y-axis:

$F_1 \sin \alpha_1$, $F_2 \sin \alpha_2$ and $F_3 \sin \alpha_3$

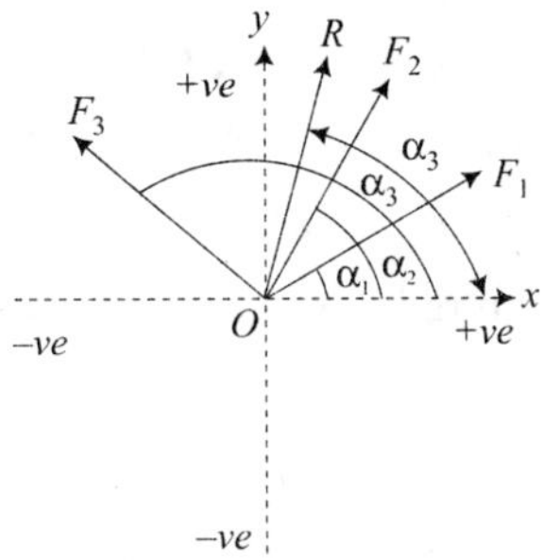

Fig. 1.17 Principle of resolved components

If R is resultant force making an angle θ with x-axis, its components are $R\cos\theta_1$ and $R\sin\theta_1$. According to the principle of resolved components:

$$R\cos\theta_1 = F_1\cos\alpha_1 + F_2\cos\alpha_2 + F_3\cos\alpha_3$$
$$= \Sigma F_x$$
$$R\sin\theta_1 = F_1\sin\alpha_1 + F_2\sin\alpha_2 + F_3\sin\alpha_3$$
$$= \Sigma F_y$$
$$R = \sqrt{(\Sigma F_x)^2 + (\Sigma F_y)^2}$$

and

$$\theta = \tan^{-1}\frac{\Sigma F_y}{\Sigma F_x}$$

1.5.2 Resultant of Coplanar Concurrent Forces

The application of laws and principles of mechanics to the solution of engineering problems is the main role of engineering mechanics. The engineering problems involve the analysis of static and dynamic stability of structures and machines against the action of forces. The analytical tools available are:

(i) Trigonometry
(ii) Vector algebra
(iii) Graphics

These tools have relative merits and demerits and can be used with advantage depending upon particular problem. The resultant of coplanar concurrent forces can best be solved by the use of principle of resolved components as discussed above.

Example 1.1: Following forces act at a point:

(i) 20 N inclined at 20° towards North of East.
(ii) 25 N towards North.

(iii) 30 N towards North-West.

(iv) 35 N inclined at 40° towards South of West.

Find the resultant and direction of resultant forces.

Solution:

1. Draw the force diagram reducing directions w.r.t. x-axis.

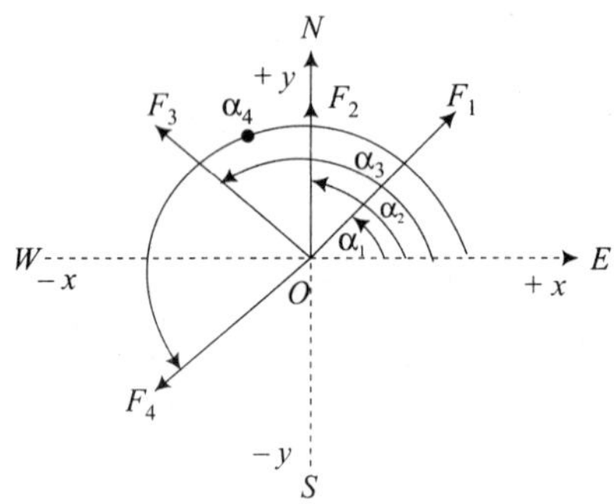

2. Find the component of forces along x-axis and y-axis.
3. Find the resultant forces in magnitude and direction employing the principle of resolved components.

$F_1 = 20$ N $\qquad \alpha_1 = 20°$ (North of East)

$F_2 = 25$ N $\qquad \alpha_2 = 90°$ (North)

$F_3 = 30$ N $\qquad \alpha_3 = 135°$ (North-West)

$F_4 = 35$ N $\qquad \alpha_4 = 180° + 40° = 220°$ (40° South of West)

$$\Sigma F_x = 20\cos 20° + 25\cos 90° + 30\cos 135° + 35\cos 220°$$
$$= 18.79 + 0 - 21.2 - 26.81 = -29.22 \text{ N}$$

$$\Sigma F_y = 20\sin 20° + 25\sin 90° + 30\sin 135° + 35\sin 220°$$
$$= 6.84 + 25 + 21.2 - 22.49 = 30.55 \text{ N}$$

The resultant, $\quad R = \sqrt{(\Sigma F_x)^2 + (\Sigma F_y)^2}$

$$= \sqrt{(-29.22)^2 + (30.55)^2}$$
$$= 42.27 \text{ N}$$

The angle of resultant with x-axis (East)

$$\theta = \tan^{-1}\frac{\Sigma F_y}{\Sigma F_x} = \tan^{-1}\frac{30.55}{-29.22}$$
$$= -42.27°$$

Example 1.2: A small block of weight 100 N is placed on an inclined plane which makes an angle, $\theta = 30°$ with the horizontal. What is the component of this weight parallel to inclined plane and perpendicular to inclind plane?

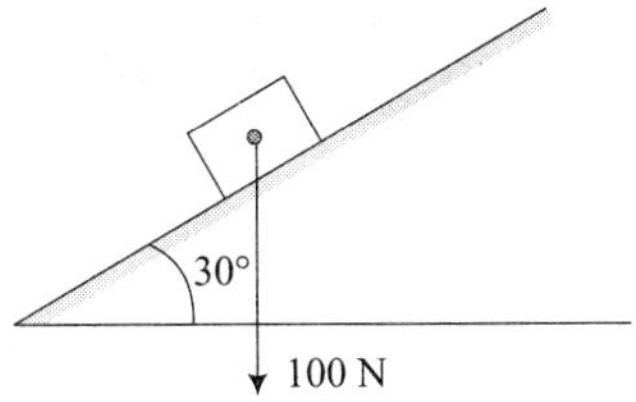

Solution:

1. Select the axis:

   x-axis parallel to inclined plane

   y-axis perpendicular to inclined plane

2. Draw the force diagram,

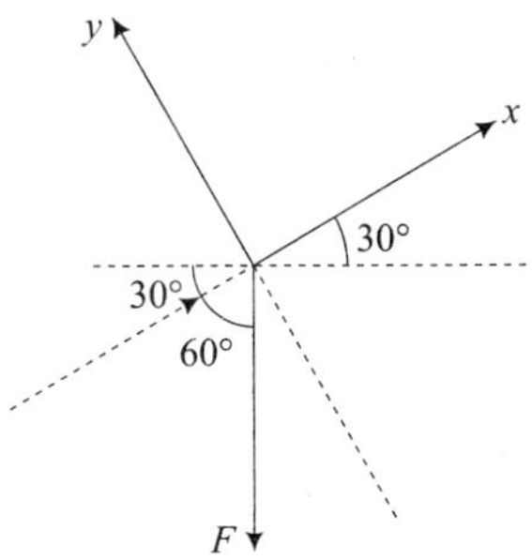

3. Find the force components,

$$F - 100 \text{ N}$$

$$\alpha = 60°$$

$$F_x = F\cos\alpha = 100\cos 60° = 50 \text{ N}$$

$$F_y = F\sin\alpha = 100\sin 60° = 86.6 \text{ N}$$

Example 1.3: The angle between two forces of magnitude 20 kN and 15 kN is 60°, the force of 20 N being horizontal. Determine the resultant in magnitude and direction, if

(i) The forces are pulls.

(ii) The 15 kN force is a push and 20 kN force is a pull.

Solution:

Case 1. Draw the force diagram,

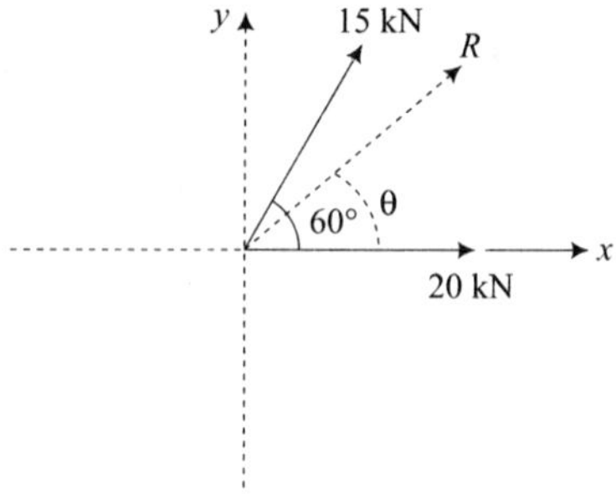

$$F_1 = 20 \text{ kN} \quad \alpha_1 = 0$$
$$F_2 = 15 \text{ kN} \quad \alpha_2 = 60°$$
$$\Sigma F_x = F_1 \cos \alpha_1 + F_2 \cos \alpha_2$$
$$= 20 \cos 0° + 15 \cos 60°$$
$$= 20 + 7.5 = 27.5 \text{ kN}$$
$$\Sigma F_y = F_1 \sin \alpha_1 + F_2 \sin \alpha_2$$
$$= 20 \sin 0° + 15 \sin 60°$$
$$= 0 + 12.99 = 12.99 \text{ N}$$

The resultant, $R = \sqrt{(\Sigma F_x)^2 + (\Sigma F_y)^2}$

$$= \sqrt{(27.5)^2 + (12.99)^2}$$

$= \mathbf{30.4\ kN}$ **Ans.**

The direction of resultant with x-axis,

$$\theta = \tan^{-1} \frac{\Sigma F_y}{\Sigma F_x} = \tan^{-1} \frac{12.99}{27.5} = \mathbf{25.28°} \quad \textbf{Ans.}$$

Case 2. Draw the force diagram,

$$\Sigma F_x = F_1 \cos \alpha_1 + F_2 \cos \alpha_2$$
$$= 20 \cos 0° + 15 \cos 240°$$
$$= 20 - 7.5 = 12.5 \text{ kN}$$
$$\Sigma F_y = F_1 \sin \alpha_1 + F_2 \sin \alpha_2$$
$$= 20 \sin 0° + 15 \sin 240°$$
$$= 0 - 12.99 = -12.99 \text{ kN}$$

The resultant, $R = \sqrt{(\Sigma F_x)^2 + (\Sigma F_y)^2}$

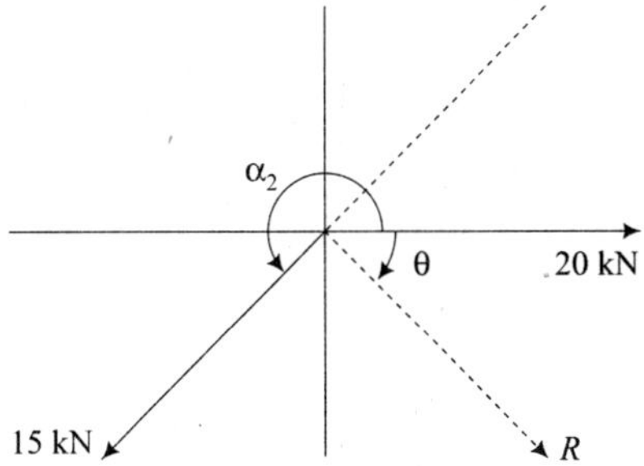

$$= \sqrt{(12.5)^2 + (-12.99)^2}$$

$$= \mathbf{18\ kN} \quad \textbf{Ans.}$$

The angle of resultant with x-axis,

$$\theta = \tan^{-1}\frac{\Sigma F_y}{\Sigma F_x} = \tan^{-1}\frac{-12.99}{12.5} = \mathbf{-46°} \quad \textbf{Ans.}$$

1.5.3 Equilibrium of Concurrent Forces

The principle of equilibrium states, "A stationary body subjected to a system of concurrent forces will be in equilibrium if the resultant of the forces is zero".

The forces are generally resolved into components along Cartesian coordinates. Therefore,

$$\Sigma F_x = 0$$

$$\Sigma F_y = 0.$$

Example 1.4: The five forces F_1, F_2, F_3, F_4 and F_5 are acting at a point on a body as shown and the body is in equilibrium. If $F_1 = 18$ N, $F_2 = 22.5$ N, $F_3 = 15$ N and $F_4 = 30$ N, determine the force F_5 in magnitude and direction.

[U.P.T.U. I Sem., 2004–05]

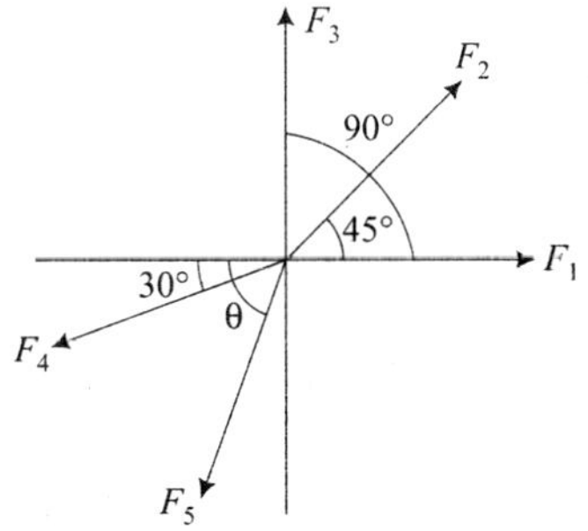

Solution:

1. Draw the force diagram taking angles of forces w.r.t. *x*-axis.

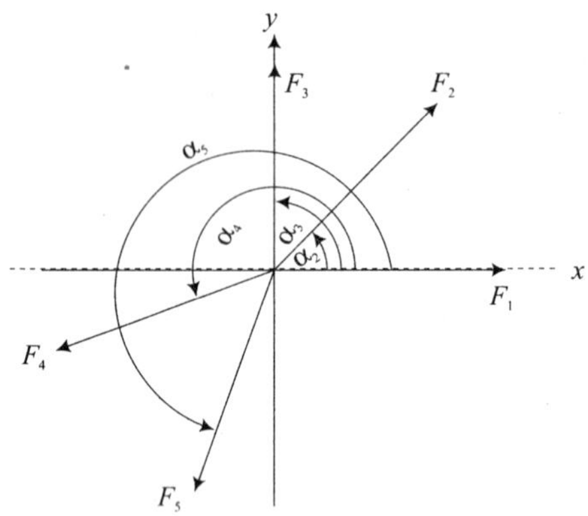

$$
\begin{array}{ll}
F_1 = 18 \text{ N} & \alpha_1 = 0^\circ \\
F_2 = 22.5 \text{ N} & \alpha_2 = 45^\circ \\
F_3 = 15 \text{ N} & \alpha_3 = 90^\circ \\
F_4 = 30 \text{ N} & \alpha_4 = 210^\circ \\
F_5 = ? & \alpha_5 = (180^\circ + \theta)
\end{array}
$$

2. Calculate the components of forces

$$
\begin{aligned}
\Sigma F_x &= F_1 \cos \alpha_1 + F_2 \cos \alpha_2 + F_3 \cos \alpha_3 + F_4 \cos \alpha_4 + F_5 \cos \alpha_5 \\
&= 18\cos 0^\circ + 22.5\cos 45^\circ + 15\cos 90^\circ + 30\cos 210^\circ + F_5\cos (180^\circ + \theta) \\
&= 18 + 15.90 + 0 - 25.98 - F_5 \cos \theta \\
&= 7.92 - F_5 \cos \theta^\circ \\
&= 0
\end{aligned}
$$

$\therefore \quad F_5 \cos \theta = 7.92$

$$
\begin{aligned}
\Sigma F_y &= F_1 \sin \alpha_1 + F_2 \sin \alpha_2 + F_3 \sin \alpha_3 + F_4 \sin \alpha_4 + F_5 \sin \alpha_5 \\
&= 18 \sin 0^\circ + 22.5 \cos 45^\circ + 15 \sin 90^\circ + 30 \sin 210^\circ + F_5 \sin (180^\circ + \theta) \\
&= 0 + 15.90 + 15 - 15 - F_5 \sin \theta \\
&= 15.90 - F_5 \sin \theta \\
&= 0
\end{aligned}
$$

$\therefore \quad F_5 \sin \theta = 15.90^\circ$

$$\tan \theta = \frac{F_5 \sin \theta}{F_5 \cos \theta} = \frac{15.90}{7.92} = 2$$

$\therefore \quad \boldsymbol{\theta = 63.5^\circ}$ **Ans.**

$F_5 \cos 63.5^\circ = 7.92$

$$\therefore \quad F_5 = \frac{7.92}{\cos 63.5^\circ} = \mathbf{17.76 \text{ N}} \quad \textbf{Ans.}$$

Example 1.5: Determine the value of F and θ so that particle A is in equilibrium.

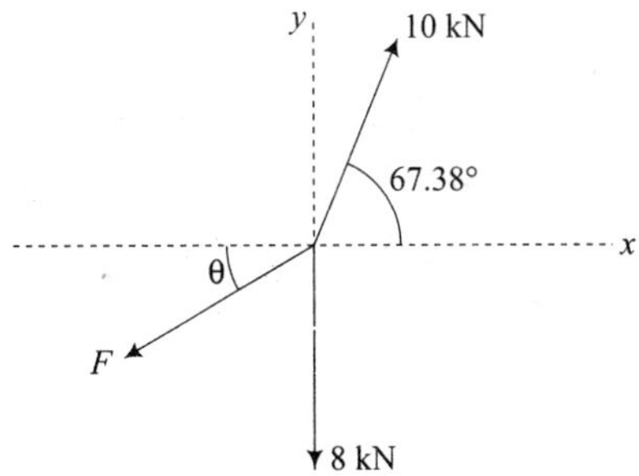

Solution:

1. $$\Sigma F_x = 10 \cos 67.38° + F \cos (180° + \theta) + 8 \cos 270°$$
$$= 3.846 - F \cos \theta - 0$$
$$= 0$$

$\therefore$ $$F \cos \theta = 3.846$$

2. $$\Sigma F_y = 10 \sin 67.38° + F \sin (180° + \theta) + 8 \sin 270°$$
$$= 9.23 - F \sin \theta - 0$$
$$= 0$$

$\therefore$ $$F \sin \theta = 1.238$$

3. $$\tan \theta = \frac{F \sin\theta}{F \cos\theta} = \frac{1.23}{3.846} = 0.3198$$

$\therefore$ $$\theta = 17.73°$$
$$F \cos \theta = 3.846$$

$\therefore$ $$F = \frac{3.846}{\cos 17.73°} = 4 \text{ kN} \quad \textbf{Ans.}$$

Example 1.6: Four coplanar forces are acting at a point. Three forces have magnitude of 20 N, 50 N and 20 N at angles of 45°, 200° and 270° respectively. Fourth force is unknown. Resultant force has a magnitude of 50 N and acts along x-axis at an angle of 0°. Determine the unknown force and its direction from x-axis.

Solutions:

1. Draw the force diagram.
2. Component forces,

$$\Sigma F_x = 20 \cos 45° + 50 \cos 200° + 20 \cos 270° + F \cos \theta$$
$$= 14.14 - 46.98 + 0 + F \cos \theta$$
$$= F \cos \theta - 32.84$$

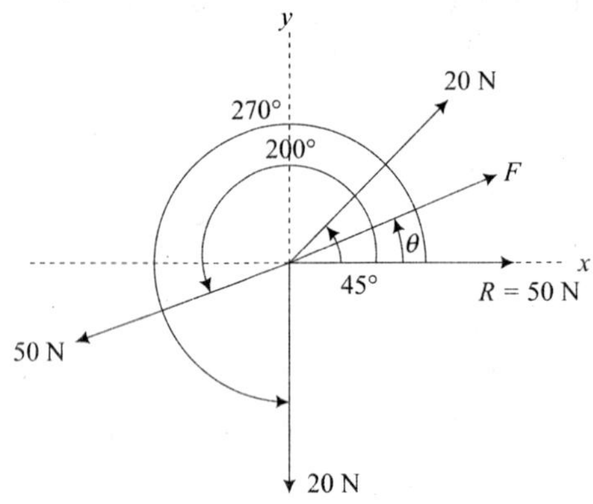

$$\Sigma F_y = 20 \sin 45° + 50 \sin 200° + 20 \sin 270° + F \sin \theta$$
$$= 14.14 - 17.1 - 20 + F \sin \theta$$
$$= - 22.96 + F \cos \theta$$

3. Components of resultant,

$$R_x = R = 50°$$
$$= \Sigma F_x = F \cos \theta - 32.84$$

$\therefore$ $$F \cos \theta = 50 + 32.84 = 82.84$$

$$R_y = 0$$
$$= \Sigma F_y = F \sin \theta - 22.96$$

$\therefore$ $$F \sin \theta = 22.96$$

$$\tan \theta = \frac{F \sin\theta}{F \cos\theta} = \frac{22.96}{82.84} = 0.277$$

$\therefore$ $$\theta = 15.49°$$

$$F \sin \theta = 22.96$$

$\therefore$ $$F = \frac{22.96}{\sin 15.49°} = \mathbf{85.96\ N} \quad \textbf{Ans.}$$

Example 1.7: A lamp of mass 1 kg is hung from the ceiling by a chain and is pulled aside by a horizontal chord until the chain makes an angle of 60° with the ceiling. Find the tension in the chain and chord.

Solution:

1. Draw force diagram.
2. Apply equilibrium conditions.

(i) $$\Sigma F_x = 0$$

$$F_2 \cos 0° + F_1 \cos 120° + 9.8 \cos 270° = 0$$

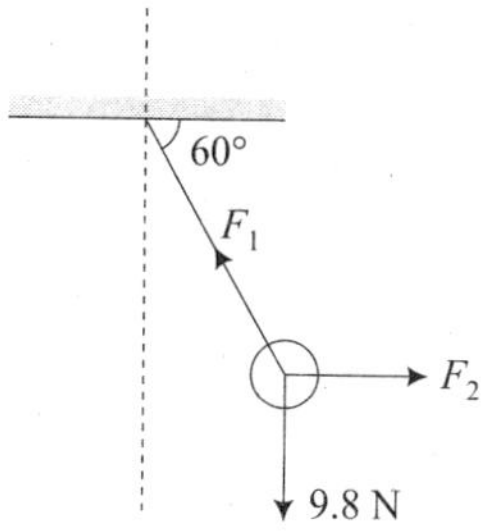

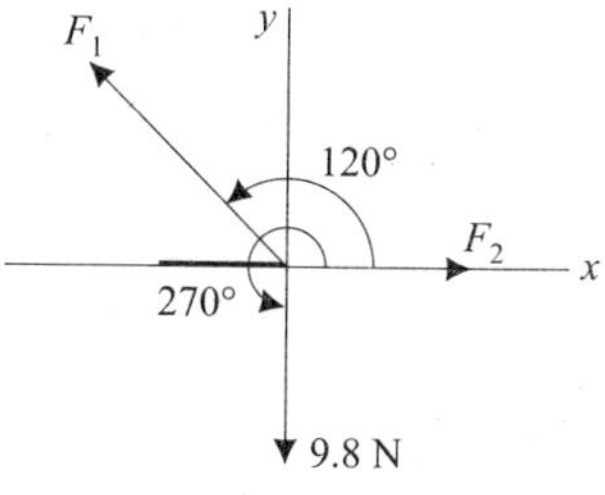

Force Diagram

$$\therefore \quad F_2 - 0.5\,F_1 + 0 = 0$$

$$\text{or} \quad F_1 = 2F_2$$

(ii) $$\Sigma F_y = 0$$

$$F_2 \sin 0° + F_1 \sin 120° + 9.8 \sin 270° = 0$$

$$\therefore \quad 0 + 0.866\,F_1 - 9.8 = 0$$

$$\therefore \quad F_1 = \frac{9.8}{0.866} = 11.3 \text{ N}$$

$$\therefore \quad F_2 = \frac{F_1}{2} = \frac{11.3}{2} = \mathbf{5.658\ N} \quad \textbf{Ans.}$$

Example 1.7 (a): Forces 7, 1, 1 and 3 kN act at one of the angular points of a regular pentagon towards four other angular points taken in order obtain the resultant of this force system. What is its direction?

Solution: The force diagram is shown.

For a pentagon, total included angle = 3 × 180° = 540°

$$\alpha_4 = \frac{540}{5} = 108°$$

$$\alpha_2 = \frac{108}{3} = 36°$$

$$\alpha_3 = 2\,\alpha_2 = 72°$$

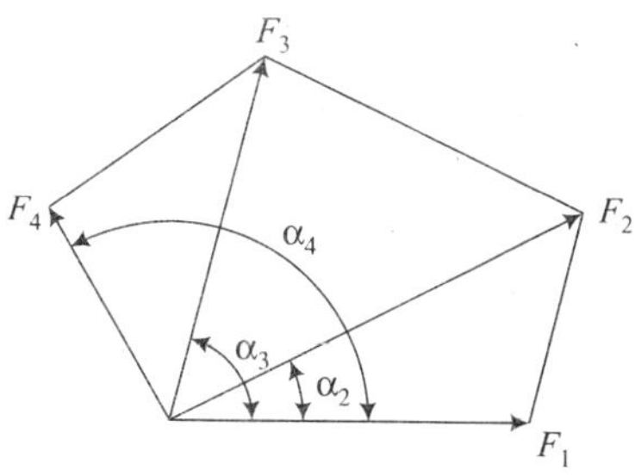

$F_1 = 7$kN; $\alpha_1 = 0$

$F_2 = 1$ kN;` $\alpha_2 = 36°$

$F_3 = 1$ kN; $\alpha_3 = 72°$

$F_4 = 3$ kN; $\alpha_4 = 108°$

Using principle of resolved components,

$$\Sigma F_x = F_1 \cos \alpha_1 + F_2 \cos \alpha_2 + F_3 \cos \alpha_3 + F_4 \cos \alpha_4$$
$$= 7 (\cos 0°) + 1 (\cos 36°) + 1 (\cos 72°) + 3 (\cos 108°)$$
$$= 7 + 0.8 + 0.3 - 0.927 = 7.173 \text{ kN.}$$
$$\Sigma F_y = F_1 \sin \alpha_1 + F_2 \sin \alpha_2 + F_3 \sin \alpha_3 + F_4 \sin \alpha_4$$
$$= 7 (\sin 0°) + 1 (\sin 36°) + 1 (\sin 72°) + 3 (\sin 108°)$$
$$= 0 + 0.58778 + 0.95 + 2.853 = 4.39 \text{ kN.}$$

Resultant,
$$R = \sqrt{(\Sigma Fx)^2 + (\Sigma Fy)^2}$$
$$= \sqrt{(7.173)^2 + (4.39)^2} = \mathbf{8.4\ kN} \quad \textbf{Ans.}$$

Direction with x-axis,
$$\theta = \tan^{-1} \frac{\Sigma Fy}{\Sigma Fx} = \tan^{-1} \frac{4.39}{7.173} = \mathbf{31.46°} \quad \textbf{Ans.}$$

1.6 MOMENT OF A FORCE

A force tends to move a body in the direction of its applications. It also tends to rotate a body about an axis. The rotational tendency of the force is called moment. The magnitude of the moment is given by the product of the force and the perpendicular distance of the axis of rotation and line of action of force. The unit of moment is N-m. The moment can be clockwise or anticlockwise. The moment is, therefore, a vector quantity as it has magnitude as well as direction.

Example: A force $\vec{F}$ is acting at point A of a rigid body making an angle θ with the horizontal axis passing through the axis O of the body. The perpendicular distance between line of action and point O is given by OB.

$$OB = d = r \sin \theta$$

where, $r = OA$.

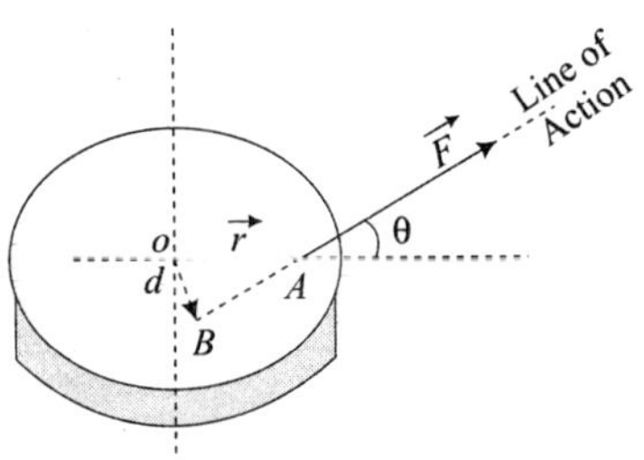

Fig. 1.18 Moment

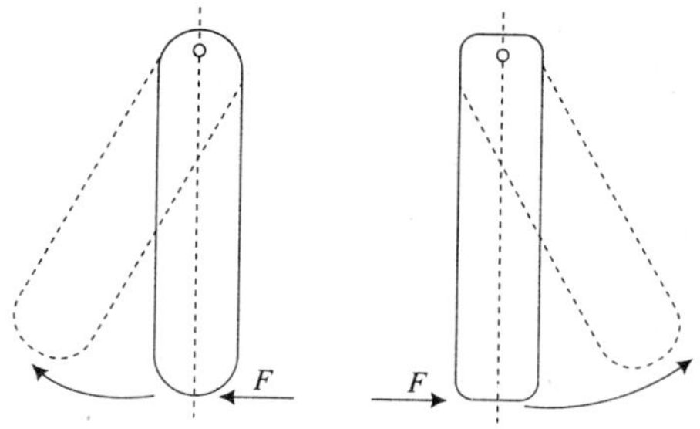

Fig. 1.19 Direction of moment

The moment of force $\vec{F}$ about point O can be expressed as

$$\vec{M}_o = \vec{r} \times \vec{F} \quad \text{(Vector equation)}$$

$\vec{r}$ = displacement vector of point A w.r.t. point O.

$$M_o = rF \sin \theta$$

$$= F \times d \quad \text{(Scalar equation)}$$

If a force tries to rotate a body in the clockwise direction, the moment is called clockwise moment.

Similarly, a body is rotated in anticlockwise direction by an anticlockwise moment.

1.6.1 Principle of Moments

The principle of moments may be stated as "When a body acted upon by several forces is in rotational equilibrium, the sum of the clockwise moments of the forces about any point is equal to the sum of the anticlockwise moments of the forces about the same point".

There are four coplanar forces F_1, F_2, F_3 and F_4 acting on a body keeping the body in rotational equilibrium. The respective perpendicular distances from z-axis (axis of rotation) are d_1, d_2, d_3 and d_4. The moment of the forces about point O are:

$$M_1 = F_1 \times d_1 \text{ (clockwise)}, -\text{ve}$$

$$M_2 = F_2 \times d_2 \text{ (anticlockwise)}, +\text{ve}$$

$$M_3 = F_3 \times d_3 \text{ (clockwise)}, -\text{ve}$$

$$M_4 = F_4 \times d_4 \text{ (anticlockwise)}, +\text{ve}$$

The resultant moment $= -F_1d_1 + F_2d_2 - F_3d_3 + F_4d_4 = 0$

$$\therefore \quad F_1d_1 + F_3d_3 = F_2d_2 + F_4d_4$$

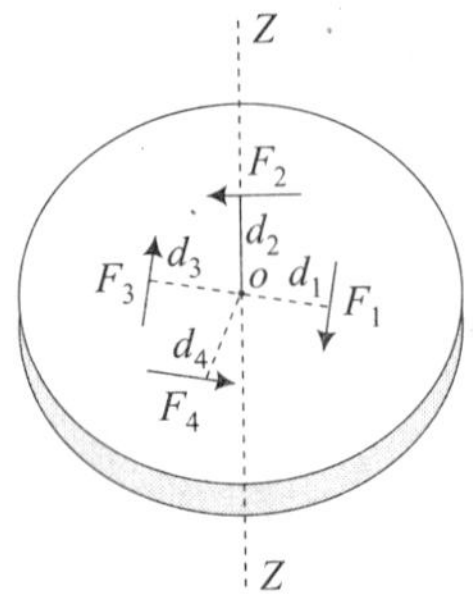

Fig. 1.20 Principle of moments

Sum of clockwise moments = sum of anticlockwise moment.

Thus principle of moments is verified.

1.6.2 Varignon's Theorem

Varignon's theorem states, "The moment of the resultant of several concurrent forces about a given point is equal to the sum of the moments of individual forces about the same point".

The theorem can be best expressed in vector algebra.

$\vec{F_1}$, $\vec{F_2}$ and $\vec{F_3}$ are three concurrent forces acting at point O. The resultant,

$$\vec{R} = \vec{F_1} + \vec{F_2} + \vec{F_3} + \ldots$$

The moment of resultant

$$= \vec{r} \times \vec{R}$$

The sum of moment of original forces

$$= \vec{r}\,\vec{F_1} + \vec{r}\,\vec{F_2} + \vec{r}\,\vec{F_3} + \ldots$$

$$\vec{r} \times (\vec{R}) = \vec{r} \times (\vec{F_1} + \vec{F_2} + \vec{F_3} + \ldots)$$

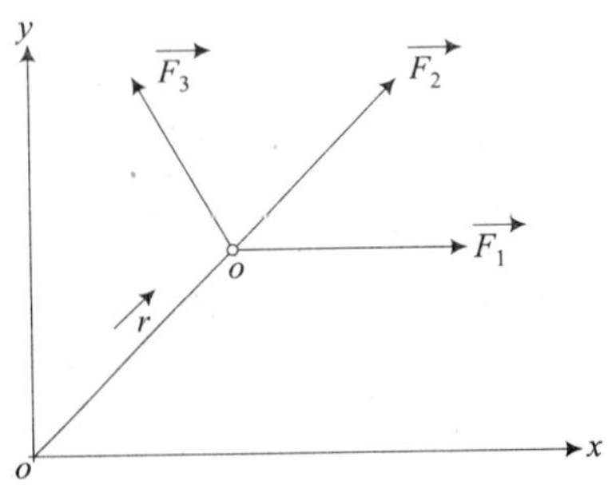

Fig. 1.21 Varignon's theorem

Using the distributive property of vector products,

$$\vec{r} \times (\vec{F_1} + \vec{F_2} + \vec{F_3} + ...) = \vec{r}\,\vec{F_1} + \vec{r}\,\vec{F_2} + \vec{r}\,\vec{F_3} + ...$$

1.7 PRINCIPLE OF EQUILIBRIUM

A body is acted upon by active forces including external applied forces (F) and weight (W) due to gravitational attraction which cause motion of the body. It is also acted upon by reactive forces including support reaction (R), frictional force (μR) and inertia force (ma) which oppose the motion of the body. A body will be in equilibrium if there is no linear motion due to combined action of active and reactive forces and also there is no rotation due to moments of these forces.

The principle of equilibrium states, "A stationary body subjected to a system of forces will be in equilibrium if the resultant of the forces and resultant moment are zero".

If F includes all active and reactive forces

$$\Sigma F = 0$$

$$\Sigma M = 0$$

The forces are generally resolved into components along Cartesian coordinates. Therefore,

$$\Sigma F_x = 0 \quad \text{and} \quad \Sigma M_x = 0$$

$$\Sigma F_y = 0 \quad \text{and} \quad \Sigma M_y = 0$$

$$\Sigma F_z = 0 \quad \text{and} \quad \Sigma M_z = 0$$

For coplanar forces, the principle of equilibrium can be simplified as:

$$\Sigma F_x = 0, \qquad \Sigma F_y = 0 \quad \text{and} \quad \Sigma M_o = 0$$

where 0 is any point.

1.7.1 Force Law of Equilibrium

The force law of equilibrium can be applied to different force systems.

(a) Two-force system A body subjected to two forces will be in equilibrium if the two forces are collinear, equal and opposite.

(b) Three-force system A body subjected to three forces will be in equilibrium if the resultant of any two forces will be equal and opposite to the third force.

(c) Multi-force system A body subjected to four or more forces will be in equilibrium if it satisfies both the force law of equilibrium as well as the moment law of equilibrium.

$$\Sigma F_x = 0, \quad \Sigma F_y = 0 \quad \text{and} \quad \Sigma M_o = 0.$$

1.7.2 Coplanar Concurrent Forces

When several concurrent forces act on a body, it will be in equilibrium if there is no unbalanced force acting on it. The resultant of all forces should be zero. The following equilibrium conditions should be satisfied.

(a) Analytical conditions

The sum of all the components in the x-direction as well as y-direction must be zero,

$$\Sigma F_x = 0 \quad \text{and} \quad \Sigma F_y = 0$$

(b) Graphical condition

The force polygon must close.

1.7.3 Coplanar Non-concurrent Forces

When a body is under the action of coplanar non-concurrent force system, it may move linearly in horizontal or vertical direction due to unbalanced force components in respective directions. It may also rotate due to resultant moment of force system. The body will be in equilibrium if the algebraic sum of all external forces and their moments about any point in their plane are nil.

The conditions of equilibrium may be expressed mathematically as follows:

1. Sum of all component forces in x-direction must be zero,

$$\Sigma F_x = 0$$

2. Sum of all component forces in y-direction must be zero,

$$\Sigma F_y = 0$$

3. Sum of moments of all forces about any point O must be zero,

$$\Sigma M_o = 0$$

The *resultant* of coplanar non-concurrent non-parallel force system can be found out as follows:

(a) Magnitude

The magnitude of the resultant can be found out analytically as follows:

The resultant, $$R = \sqrt{(\Sigma F_x)^2 + (\Sigma F_y)^2}$$

where, ΣF_x = algebraic sum of component forces in x-direction

ΣF_y – algebraic sum of component forces in y-direction

(b) Direction

The direction of the resultant with respect to x-axis can be found out as follows:

$$\theta = \tan^{-1} \frac{\Sigma F_y}{\Sigma F_x}$$

(c) Position

The position of resultant along x-axis can be found out by taking the moment of component force of resultant in y-direction about a point O and equating the same to the sum of moments of force components in y-direction about point O.

The sum of moments of component forces in y-direction about point O,

$$= \Sigma F_y . x_i$$

where, x_i = perpendicular distance of component forces in x-direction.

The moment of y-component of resultant about point O,

$$= Ry.x$$

where, x = perpendicular distance of R_y from point O along x-direction.

$$\therefore \quad Ry.x = \Sigma Fy.x_i$$

$$\therefore \quad x = \frac{\Sigma F_y . x_i}{R_y}$$

1.7.4 Coplanar Parallel Forces

There is no component force say in x-direction. The equilibrium conditions, therefore, reduce to the following:

1. The sum of parallel forces in y-direction must be zero,

$$\Sigma F_y = 0$$

2. The sum of moment of parallel forces about any point O must be zero,

$$\Sigma M_o = 0$$

The *resultant* of parallel forces be found out as follows:

(a) Magnitude

The magnitude of resultant will be the algebraic sum of parallel forces,

$$R = \Sigma F_y$$

(b) Direction

The direction of resultant will be parallel to the original forces.

(c) Position

The position of the resultant from a reference point O will be

$$x = \frac{\Sigma F_y . x_i}{R}$$

where, $\Sigma F_y x_i$ = algebraic sum of moment of parallel forces about reference point

R = resultant of parallel forces.

1.8 EQUIVALENT SYSTEMS

1.8.1 Shifting of a Force to Parallel Position

A force acting on a rigid body has two effects:

1. Tendency to push or pull the body in its direction.
2. Tendency to rotate the body about any axis which does not intersect the line of force.

The above dual effect of a force can be replaced by the following equivalent system.

1. An equal parallel force acting at point O, and
2. A couple to compensate for the change in moment of force.

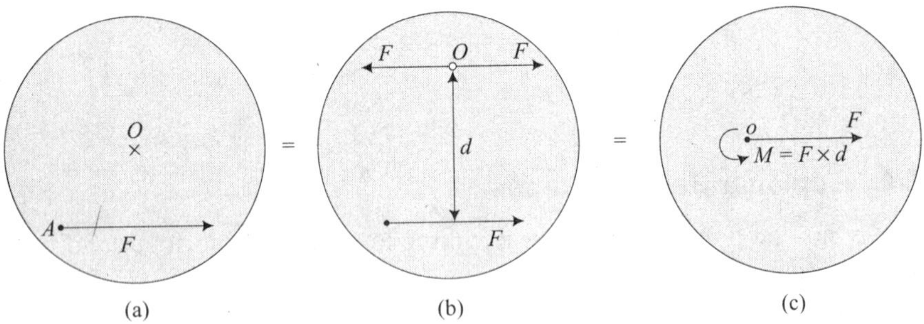

Fig. 1.22 Equivalent system

Example: A force F is acting on a rigid body at point A tending to move the body towards right and rotate the body in anticlockwise direction about point O which is at a distance d from the line of action of force. The force can be shifted to point O by applying two equal and opposite forces F. The new system is equal to a direct force F acting towards right and a couple with a moment $F \times d$.

Therefore, the equivalent system shown in Fig. 1.21(c) is a direct force F acting at point O and an anticlockwise moment,

$$M = F \times d$$

This method of movement of a force to parallel position is a powerful tool used in solving many problems in engineering mechanics.

1.8.2 Resultant of Force System

(a) Two-dimensional concurrent force system

Forces F_1, F_2, F_3, F_4, ... are acting at point O of a rigid body making angles α_1, α_2, α_3, α_4, ... with $+x$ direction.

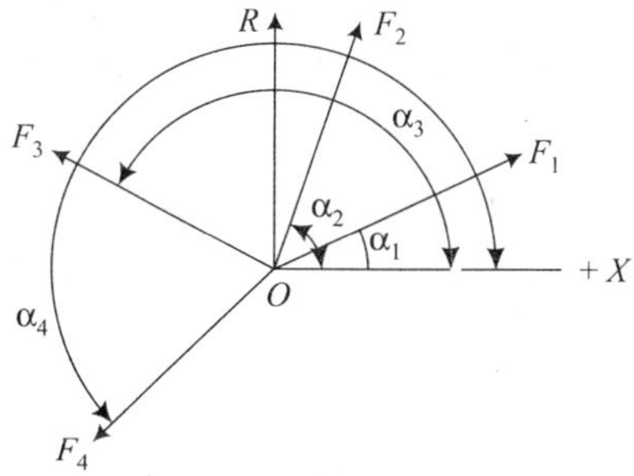

Fig. 1.22 (a) Resultant of concurrent forces

The x component of resultant force

$$R_y = \Sigma F_x = F_1 \cos \alpha_1 + F_2 \cos \alpha_2 + F_3 \cos \alpha_3 + F_4 \cos \alpha_4 + ...$$

The y component of resultant force,

$$R_x = \Sigma F_y = F_1 \sin \alpha_1 + F_2 \sin \alpha_2 + F_3 \sin \alpha_3 + F_4 \sin \alpha_4 + ...$$

The resultant

$$R = \sqrt{R_x^2 + R_y^2} = \sqrt{(\Sigma Fx)^2 + (\Sigma Fy)^2}$$

The direction of resultant with $+x$ direction,

$$\theta = \tan^{1} \frac{R_y}{R_x} = \tan^{1} \frac{\Sigma F_y}{\Sigma F_x}$$

(b) Two-dimensional non-concurrent force system

Forces F_1, F_2, F_3, F_4 are acting at the corners of a block $ABCD$ making angles α_1, α_2, α_3, α_4 with $+x$ direction. The block side is l and weight W.

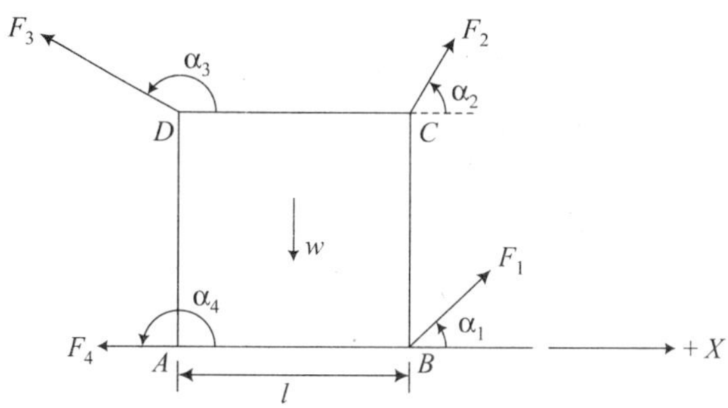

Fig. 1.22 (b) Resultant of non-concurrent forces

The x component of resultant force,

$$R_x = \Sigma F_x = F_1 \cos \alpha_1 + F_2 \cos \alpha_2 + F_3 \cos \alpha_3 + F_4 \cos \alpha_4$$

The y component of resultant force,

$$R_y = \Sigma F_y = W + F_1 \sin \alpha_1 + F_2 \sin \alpha_2 + F_3 \sin \alpha_3 + F_4 \sin \alpha_4$$

The magnitude of resultant force,

$$R = \sqrt{R_x^2 + R_y^2} = \sqrt{(\Sigma Fx)^2 + (\Sigma Fy)^2}$$

The direction of resultant with $+x$ direction

$$\theta = \tan^{-1} \frac{R_y}{R_x} = \tan^{-1} \frac{\Sigma F_y}{\Sigma F_x}$$

The position of line of action of resultant on horizontal line from corner A,

$$x = \frac{\Sigma F_y\, x_i + \dfrac{Wl}{2}}{R_y}$$

where x_i and x are measured from corner A.

(c) Two-dimensional parallel force system

Vertical forces F_1, F_2, F_3, F_4 are acting on a beam at horizontal distances of x_1, x_2, x_3, x_4 from a reference point O. The resultant force is vertical and is algebraic sum of all the forces.

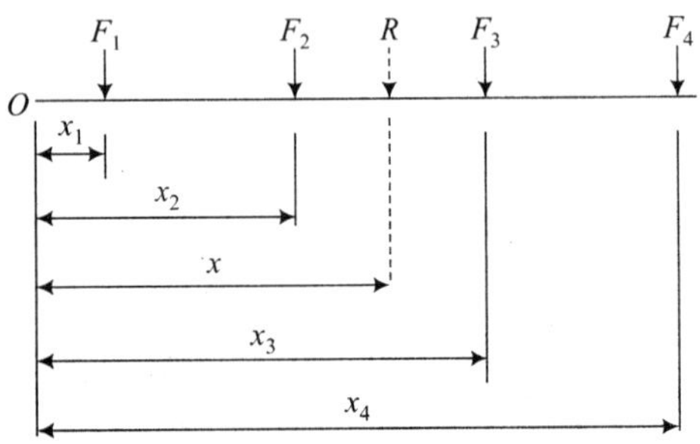

Fig. 1.22 (c) Resultant of Parallel Forces

$$R = F_1 + F_2 + F_3 + F_4$$

The position of resultant force from point O,

$$x = \frac{F_1x_1 + F_2x_2 + F_3x_3 + F_4x_4}{R}$$

1.8.3. Distribution Force System

(a) Uniformly distributed load (UDL)

A beam supporting a wall or roof or slab or a vertical wall subjected to wind pressure or a water pipe supported over a span are examples of uniformly distributed load (UDL). The loading is represented by a rectangle.

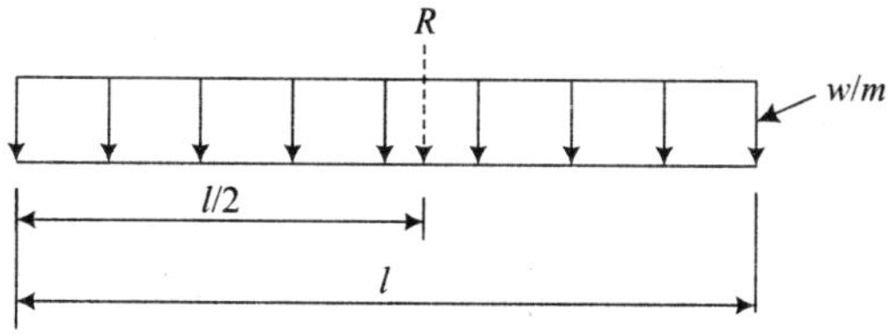

Fig. 1.22 (d) Resultant of UDL

The intensity of UDL = w N/m

The span of beam = l m

The resultant force will be area of loading.

$$R = wl\ (N)$$

The position of resultant will be the centre of gravity of the loading. For a rectangle c.g. acts at a distance of $l/2$ from either end.

(b) Uniformly varying load (UVL)

A beam AB of span l is subjected to uniformly varying load from w/m at A to zero at B. The wall of a container or a dam is subjected to UVL with zero load at the top.

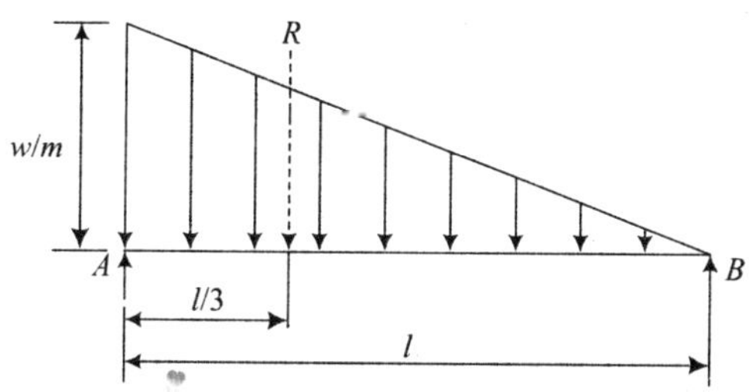

Fig. 1.22 (e) Resultant of UVL

The resultant force will be area of the loading, i.e., triangle.

$$R = \frac{wl}{2}$$

The position of resultant force passes through c.g. of triangle, i.e., $l/3$ from A.

(c) Trapezoidal Loading

A beam *AB* of span l is subjected to trapezoidal uniformly varying load of w_2 at *A* to w_1 at *B*. Divide the loading into a rectangle of height w_1 and a triangle of height $(w_2 - w_1)$.

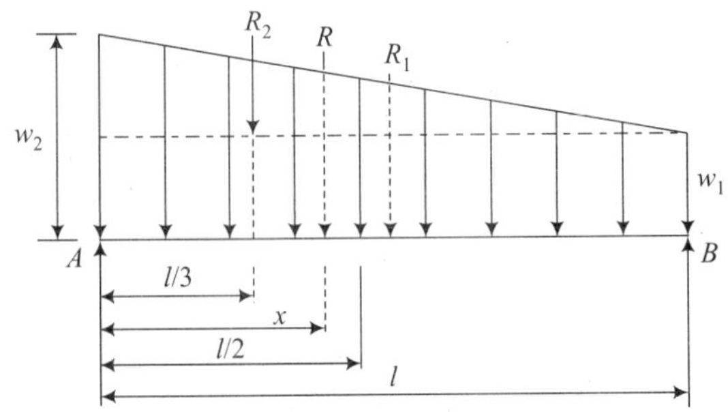

Fig. 1.22 (f) Resultant of Trapezoidal loading

$$R_1 = w_1 l \text{ at a distance } \frac{l}{2} \text{ from A}$$

$$R_2 = \frac{(w_2 - w_1)l}{2} \text{ at a distance } \frac{l}{3} \text{ from A}$$

Total resultant force,

$$R = R_1 + R_2 = w_1 l + \frac{(w_2 - w_1)l}{2}$$

The position of resultant from A,

$$x = \frac{\frac{R_1 l}{2} + \frac{R_2 l}{3}}{R}$$

1.9 SUPPORT REACTIONS

A rigid body to be in equilibrium requires both the balance of forces to prevent translation and balance of moments to prevent rotation. The balancing forces are provided by the support reactions as per Newton's third law of motion. Different types of supports used in mechanics and structures and their reactions are shown in Table 1.1.

Table 1.1 Support reactions

Sl. No.	Type of support/connections		Type of reaction offered		
	Description	Sketch	Description	Sketch	No. of unknown
1.	Roller	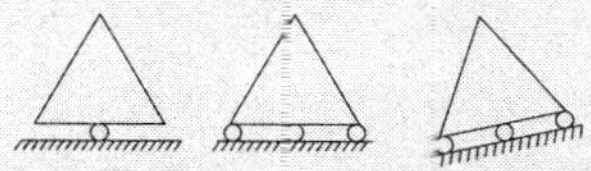	Force with line of action perpendicular to contact point	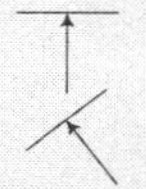	1 1
2.	Rocker	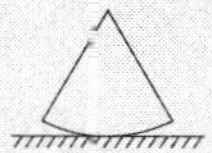	-do-		1
3.	Pin or Hinge support	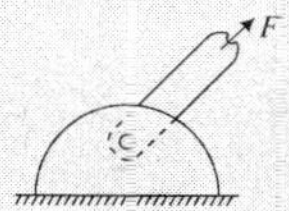	Reactions are two components of force	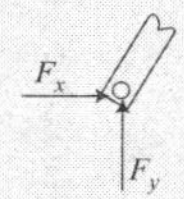	2
4.	Fixed support	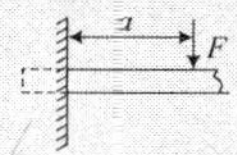	Reactions are couple moments and force components	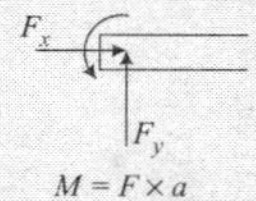	3
5.	Cable	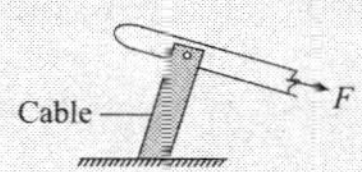	Tension acting in the direction of cable	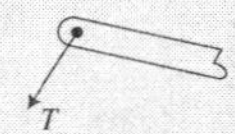	1
6.	Link	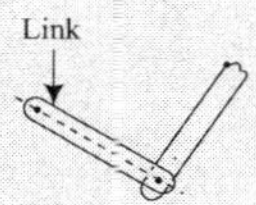	Force acting along the axis of link	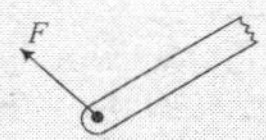	1

1.10 FREE BODY DIAGRAM

A free body diagram is a sketch of the particle/body which represents it being isolated from its surrounding/system. All the forces that act on it are represented on the sketch. The forces include active forces which cause motion of the particle/body and reactive forces arising from supports or constraints. The reactions always act normal to the plane of contact as shown in Table 1.1.

Problems in static can be solved easily from free body diagram. Force system acting on bodies in contact can be shown in the sketch. The solution of problem becomes simple.

Example 1: A sphere resting on a horizontal surface,

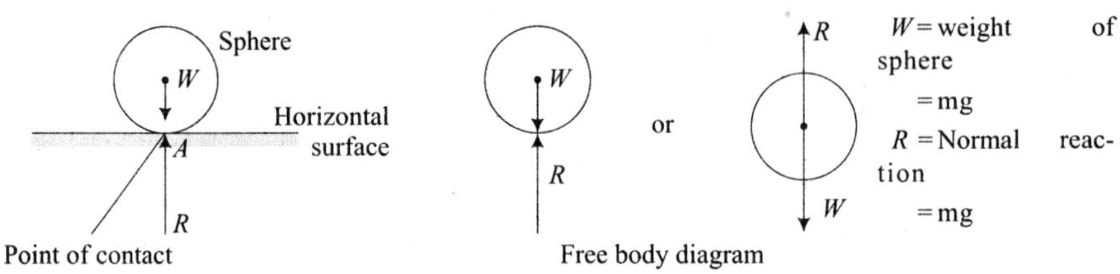

Fig. 1.23 FBD of a sphere on horizontal surface

Example 2: A sphere resting in a *V*-notch,

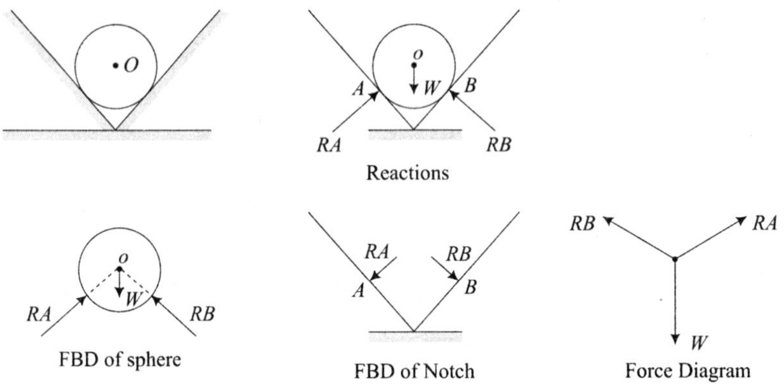

Fig. 1.24 FBD of sphere and a notch

Example 1.8: Determine the resultant *R* of five vertical forces acting on the beam as shown.

Solution:

1. Draw force diagram.
2. Apply equilibrium conditions.

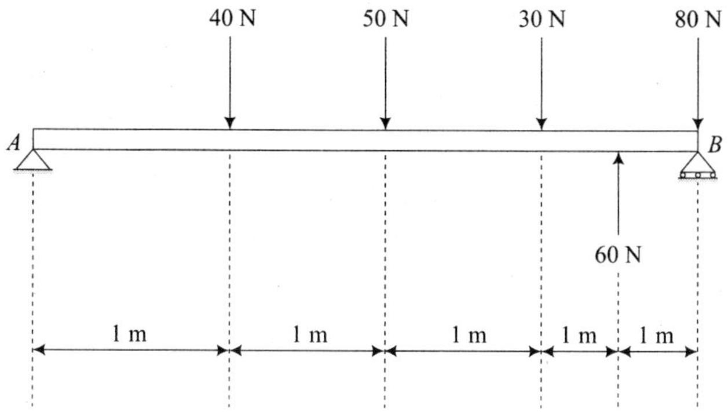

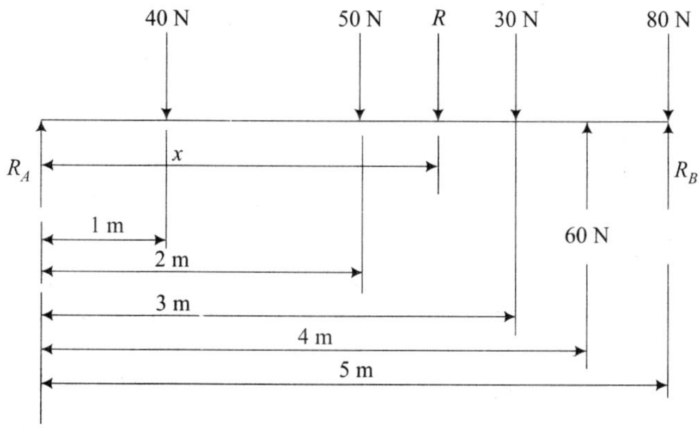

Force Diagram

(i) $$\Sigma F_y = 0$$

$$R = 40 + 50 + 30 + 80 - 60$$

$$= 1140 \text{ N downwards}$$

(ii) $$\Sigma M_A = 0$$

$$R(x) = 40 \times 1 + 50 \times 2 + 30 \times 3 \quad 60 \times 4 + 80 \times 5$$

$\therefore$ $$x = \frac{40 + 100 + 90 - 240 + 400}{140}$$

$$= \frac{390}{140} = \mathbf{2.786 \text{ m}} \quad \textbf{Ans.}$$

Example 1.9: A horizontal beam *PQRS* is 12 m long, where *PQ* = *QR* = *RS* = 4 m. Forces of 1000 N, 1500 N, 1000 N and 500 N act at point *P*, *Q*, *R* and *S* respectively with the downward direction. The lines of action of these forces make angles 90°, 60°, 45° and 30° respectively with *PS*. Find the magnitude, direction and position of resultant.

Solution:

1. Draw force diagram.

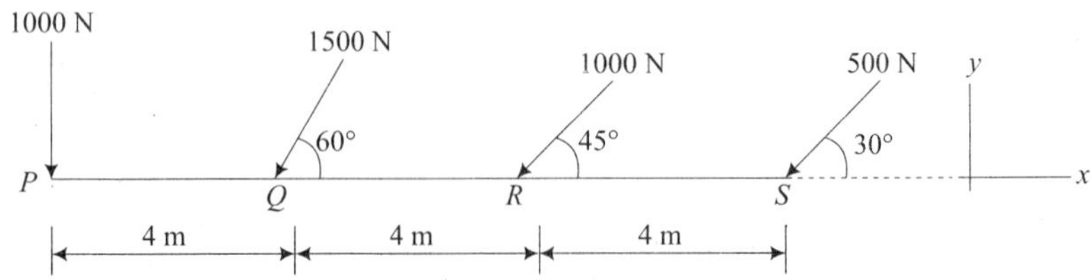

Force Diagram

2. Find the Σx and Σy

(i) $\Sigma F_x = 1000 \cos 90° + 1500 \cos 60° + 1000 \cos 45° + 500 \cos 30°$

$= 1000 \times 0 + 1500 \times 0.5 + 1000 \times 0.707 + 500 \times 0.866$

$= 1890$ N

(ii) $\Sigma F_y = 1000 \sin 90° + 1500 \sin 60° + 1000 \sin 45° + 500 \sin 30°$

$= 1000 \times 1 + 1500 \times 0.866 + 1000 \times 0.707 + 500 \times 0.5$

$= 3256$ N

(iii) The resultant,

$$R = \sqrt{(\Sigma F_x)^2 + (\Sigma F_y)^2}$$

$$= \sqrt{(1890)^2 + (3256)^2}$$

$$= 3760 \text{ N}$$

$$\tan\theta = \frac{\Sigma F_y}{\Sigma F_x} = \frac{3256}{1890} = 1.865$$

$\therefore$ $\theta = 61°48'$

(iv) $\Sigma M_p = 1000 \times 0 + 1500 \sin 60° \times 4 + 1000 \sin 45° \times 8$
$+ 500 \sin 30° \times 12$

$= 0 + 5196.15 + 5656.85 + 3000 = 13853$

$= R_y \times x$

$\therefore$ $x = \dfrac{13853}{R \sin 61°48'} = \dfrac{13853}{3760 \sin 61°48'}$ = **4.18 m from *P*** **Ans.**

Example 1.10: Determine the moment of the force acting on the railway sign post about point O.

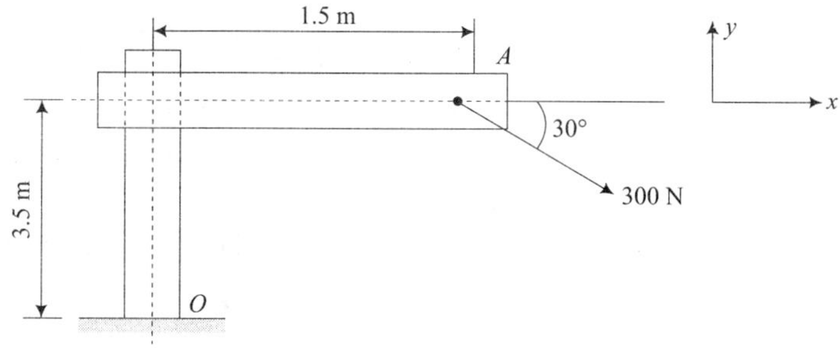

Solution:

1. The rectangular components of the force at A

$$F_x = F\cos\theta$$
$$= 300\cos 30° = 259.8 \text{ N}$$
$$F_y = F\sin\theta = 300\sin 30°$$
$$= 150 \text{ N}$$

2. Applying Varignon's theorem that the moment of resultant is equal to the sum of moments of component forces about the same point.

$$M_o = F_x\,(3.5) + F_y\,(1.5)$$
$$= 259.8\,(3.5) + 150\,(1.5)$$
$$= 1134.34 \text{ N-m (clockwise)}$$

Example 1.11: Replace the 20 kN force acting on steel column by a force-couple system at O.

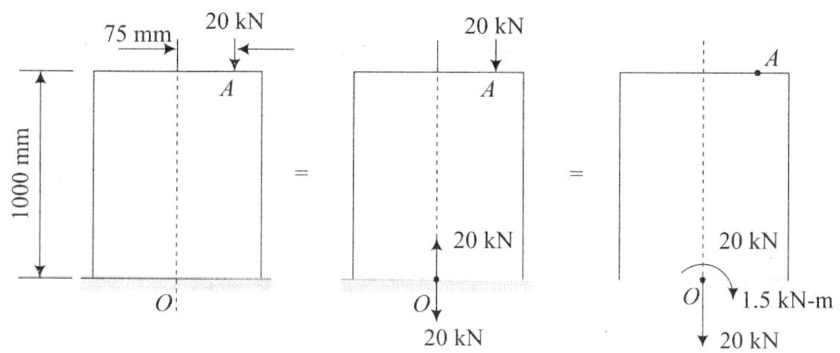

Solution: Step 1. Apply two concurrent forces of 20 kN each at O acting in opposite direction.

Step 2. Remove force of 20 kN at *A* and 20 kN upward force at *O* cancelling equal and opposite force.

Step 3. Apply a moment of couple of 20 kN acting at a distance of 75 mm clockwise.

$$M_o = 20 \times 0.075 = 1.5 \text{ kN-m}$$

Therefore, equivalent force system at *O* will consist of a vertical force of 20 kN acting downwards and a moment of 1.5 kN-m acting clockwise.

Example 1.12: What force moment is transmitted to the supporting wall at *A* in given cantilever beam? [U.P.T.U. II Sem., 2002–03]

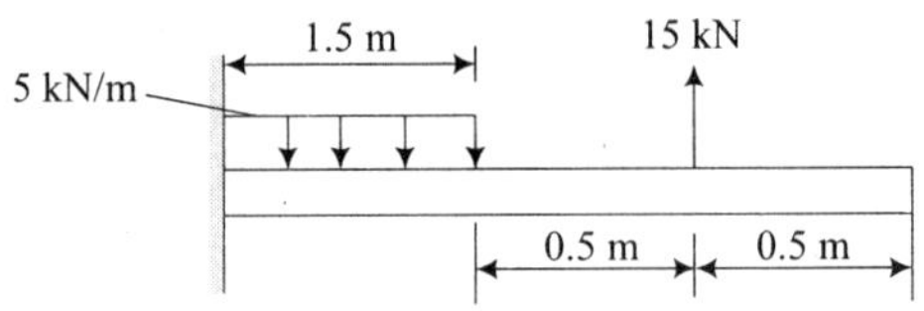

Solution:

1. Draw the force diagram,

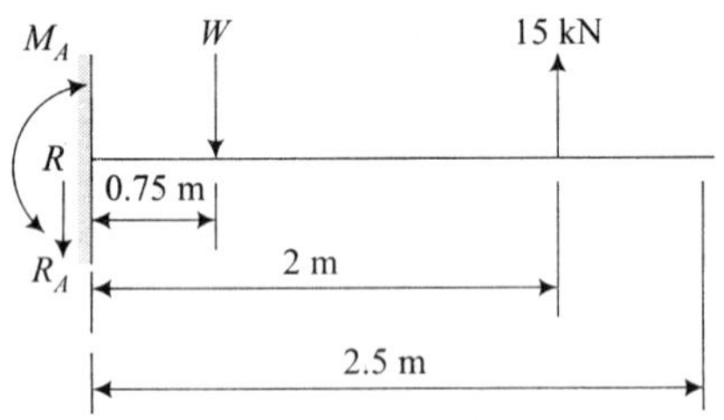

Force Diagram

The equivalent force *W* is equal to the area under the load curve and will act at c.g. of load curve.

$W = wL = 5 \times 1.5 = 7.5$ kN downwards at a distance of 0.75 m from *A*.

2. Apply equilibrium condition to find the resultant reaction.

$$R_A = 15 - 7.5 = 7.5 \text{ kN} \quad \text{downwards}$$

$$M_A = 15 \times 2 - 7.5 \times 0.75 = 30 - 5.625$$

$$= 24.375 \text{ kN-m} \quad \text{Anticlockwise}$$

Example 1.13: Four forces having magnitude of 20 N, 40 N, 60 N and 80 N respectively are acting along the fours sides (1 m each) of equate *ABCD*, taken in order, as shown. Determine the magnitude and direction of the resultant force.

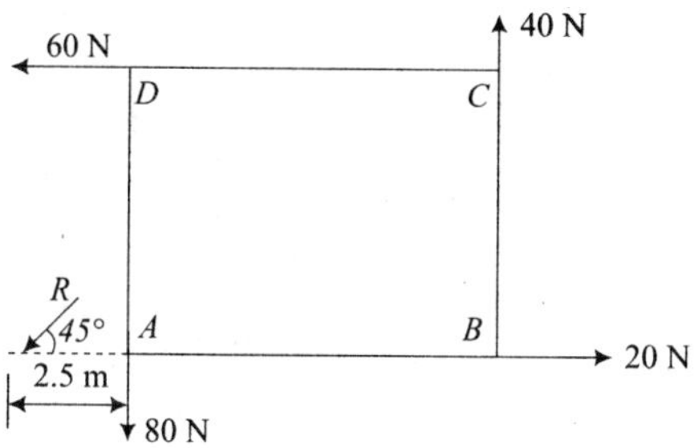

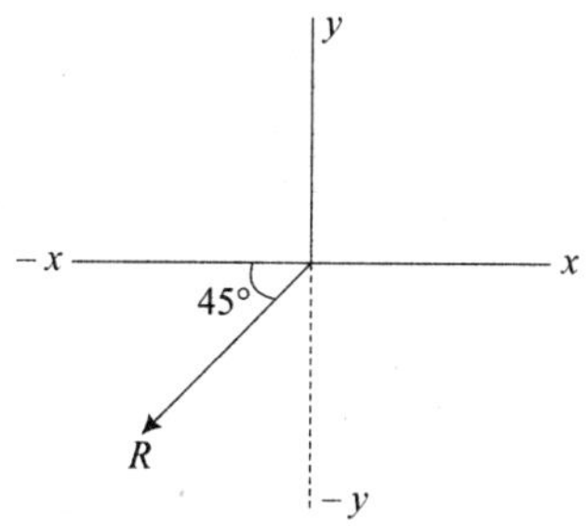

Solution:

1. Resolve the forces along x-axis and y-axis

$$\Sigma F_x = 20 - 60 = -40 \text{ N}$$

$$\Sigma F_y = 40 - 80 = -40 \text{ N}$$

2. The resultant

$$R = \sqrt{(\Sigma F_x)^2 + (\Sigma F_y)^2} = \sqrt{(-40)^2 + (-40)^2}$$

$$= 56.57 \text{ N}$$

3. The direction of resultant with x-axis

$$\tan\theta = \frac{\Sigma F_y}{\Sigma F_x} = \frac{-40}{-40} = 1$$

$\therefore$ $$\theta = 45°$$

As ΣF_x and ΣF_y both are negative, the resultant will make an angle of 45° with –ve x-axis.

4. Position of resultant

If x is the horizontal distance of resultant R from point A.

$$\Sigma M_A = 20 \times 0 + 40 \times 1 + 60 \times 1 + 80 \times 0$$

$$= 100 \text{ N-m}$$

$\therefore$ $$x = \frac{\Sigma M_A}{R_y} = \frac{100}{-40} = -2.5 \text{ m}$$

Resultant will be acting at a distance of 2.5 m to left of *A*. Ans.

Example 1.14: A square block of wood of mass M is hinged at A and rests on roller at B. It is pulled by means of a string attached at D and inclined at an angle 30° with the horizontal. Determine the force P which should be applied to the string to just lift the block off the roller. [U.P.T.U. I Sem., 2004–05]

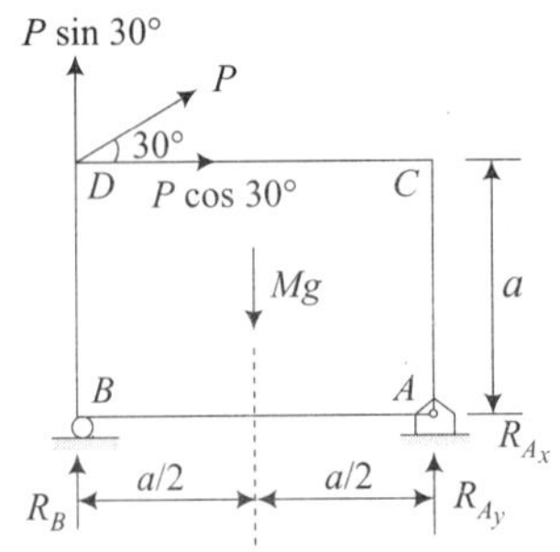

Solution:

$$R_B = 0$$

Apply equilibrium conditions,

(i) $$\Sigma F_x = 0$$

$\therefore$ $$P\cos 30° - R_{Ax} = 0$$

$\therefore$ $$R_{Ax} = P\cos 30°$$

(ii) $$\Sigma F_y = 0$$

$\therefore$ $$P\sin 30° - Mg + R_{Ay} = 0$$

(iii) $$\Sigma M_A = 0$$

$$P\cos 30° \times a - Mg \times \frac{a}{2} + P\sin 30° \times a = 0$$

$\therefore$ $$Pa(\cos 30° + \sin 30°) = Mg\frac{a}{2}.$$

$\therefore$ $$P = \frac{Mg\dfrac{a}{2}}{a(0.866 + 0.5)} = \mathbf{0.366\ Mg} \quad \textbf{Ans.}$$

Example 1.15: A man weighing 75 N stands on the middle rung of a 25 N ladder resting on smooth floor and against a wall. The ladder is prevented from slipping by a string *OD*. Find the tension in the string and reactions at *A* and *B*.

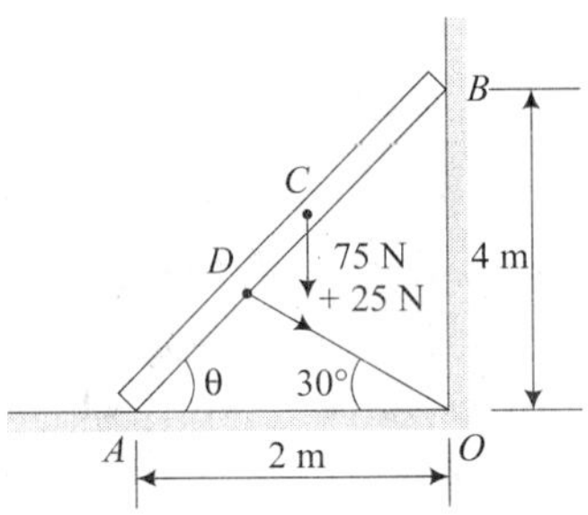

Solution:

1. Draw force diagram or free body diagram.

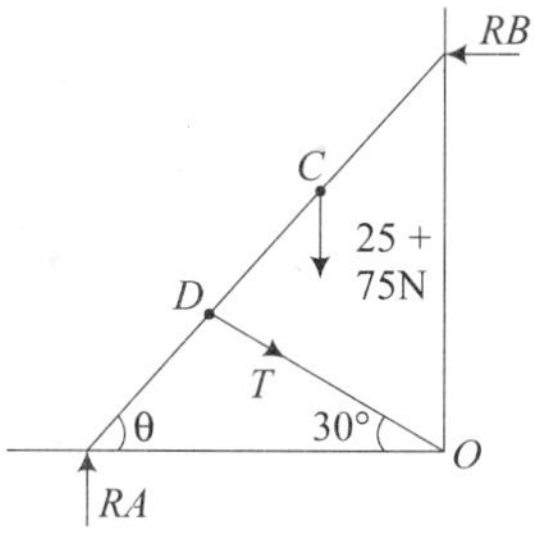

Force diagram

$$\tan\theta = \frac{OB}{OA} = \frac{4}{2} = 2$$

$\therefore$ $\theta = 63.43°$

2. Apply equilibrium conditions.

(i) $\Sigma F_x = 0$

$$T\cos 30° - R_B = 0$$

$$R_B = T\cos 30°$$

(ii) $\Sigma F_y = 0$

$$R_A - (75 + 25) - T\sin 30° = 0$$

$$T\sin 30° = R_A - 100$$

(iii) $\Sigma M_o = 0$

$$R_B(4) + 100(1) - R_A(2) = 0$$

$$2R_A - 4R_B = 100$$

$\therefore$ $2(T\sin 30° + 100) - 4T\cos 30° = 100$

$$2T \times 0.5 - 3.464T = 100 - 200$$

$\therefore$ $T = \dfrac{-100}{-2.464} = 40.56 \text{ N}$

$\therefore$ $R_A = 40.56 \sin 30° + 100 = 120.28 \text{ N}$

$R_B = 40.56 \cos 30° =$ **35.13 N Ans.**

Example 1.16: A man weighing 100 N stands on the middle rung of a ladder whose weight can be neglected. The end A rests on the ground against a stop and the end B rests on the corner of a wall as shown. Find the reactions at A and B. Neglect friction between the ladder and the ground and the ladder and the wall.

Solution:

1. Draw the force diagram or free body diagram.

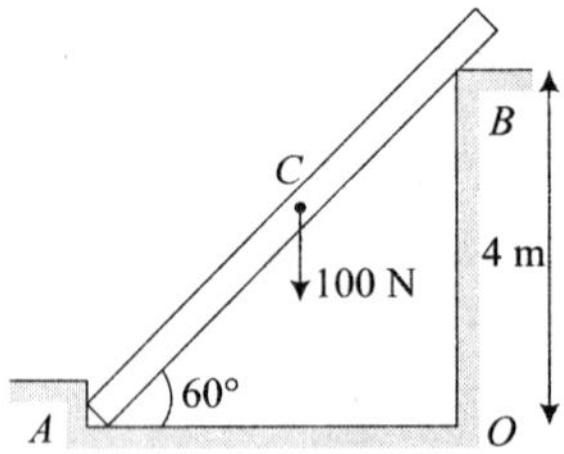

$$\sin 60° = \frac{OA}{AB}$$

$$\therefore \quad AB = \frac{4}{\sin 60°} = 4.62 \text{ m}$$

$$AD = AC \cos 60° = \frac{4.62}{2} \cos 60° = 1.155 \text{ m}$$

2. Apply equilibrium conditions.

(i) $\Sigma F_x = 0$

$$R_{Ax} - R_B \sin 60° = 0$$

(ii) $\Sigma F_y = 0$

$$R_{Ay} - W - R_B \cos 60° = 0$$

(iii) $\Sigma M_A = 0$

$$R_B(AB) - 100(AD) = 0$$

$$R_B = \frac{100(AD)}{AB}$$

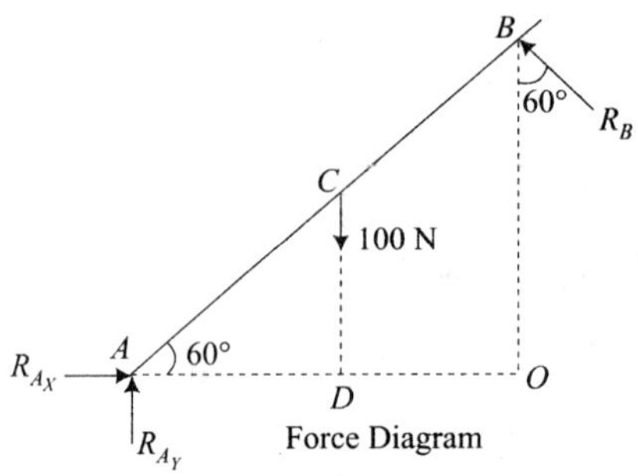

Force Diagram

$$= \frac{100 \times 1.155}{4.62} = 25 \text{ N}$$

$$R_{Ax} = R_B \sin 60° = 25 \sin 60° = 21.65 \text{ N}$$

$$R_{Ay} = R_B \cos 60° + W = 25 \cos 60° + 100 = 112.5 \text{ N}$$

$$R = \sqrt{R_{A_X}^2 + R_{A_Y}^2} = \sqrt{(21.65)^2 + (112.5)^2} = 114.56 \text{ N}$$

Example 1.17: If the end B of the above ladder rests against a wall, find the reactions at A and B.

Solution:

1. Draw the force diagram.

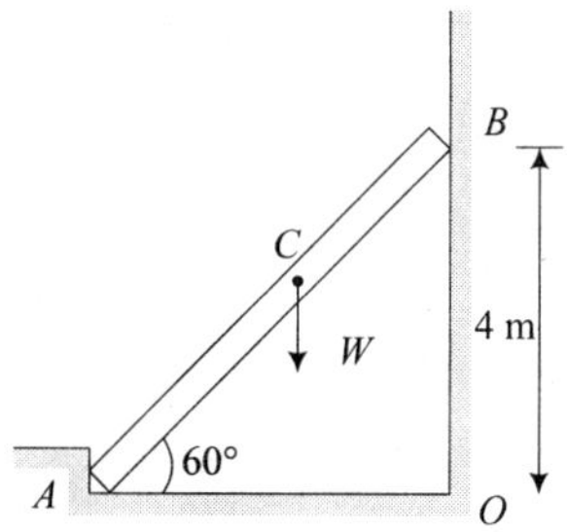

2. Apply equilibrium condition.

(i) $\Sigma F_x = 0$

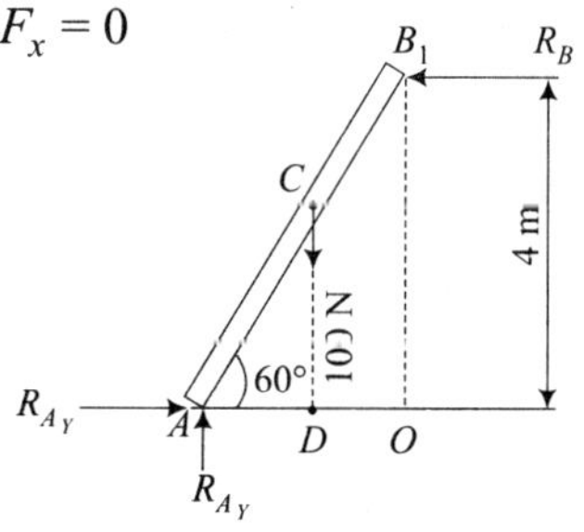

$$R_{Ax} - R_B = 0$$

(ii) $\Sigma F_y = 0$

$$R_{Ay} - 100 = 0$$

(iii) $\Sigma M_A = 0$

$$100(AD) - R_B(OB) = 0$$

$\therefore$ $$R_B = \frac{100(AB)}{OB}$$

$$R_B = \frac{100(1.155)}{4.62 \cos 60°} = 50 \text{ N}$$

$$R_{Ax} = R_B = 50 \text{ N}$$

$$R_{Ay} = 100 \text{ N}$$

$$\therefore \qquad R = \sqrt{R_{AX}^2 + R_{AY}^2} = \sqrt{(50)^2 + (100)^2} = 111.8 \text{ N}$$

Example 1.18: Three cylinders are piled up in a rectangular channel as shown. Determine the reaction R_6 between cylinder A and vertical wall of the channel.

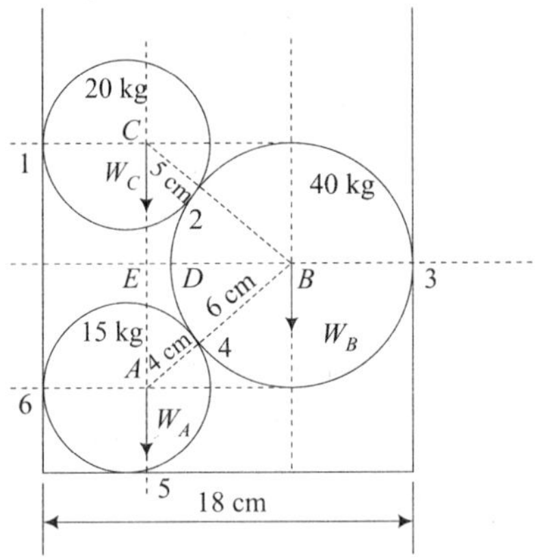

Solution:

1. Draw free body diagram of the three cylinders.

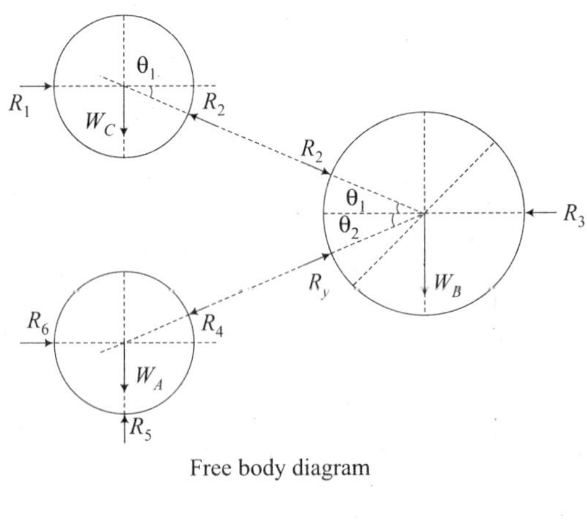

Free body diagram

2. Find the angles,

$$\cos\theta_1 = \frac{BD}{BC} = \frac{18-5-6}{6+5} = \frac{7}{11}$$

$\therefore$ $$\theta_1 = 50.5°$$

$$\cos\theta_2 = \frac{BE}{AC} = \frac{18-4-6}{4+6} = \frac{8}{10}$$

$\therefore$ $$\theta_2 = 36.87°$$

3. Apply equilibrium conditions to roller *c*,

(i) $$\Sigma F_x = 0$$

$$R_1 = R_2\cos\theta_1$$

(ii) $$\Sigma F_y = 0$$

$$R_2\sin\theta_1 = W_C$$

$\therefore$ $$R_2 = \frac{W_C}{\sin\theta_1} = \frac{20\times 9.81}{\sin 50.5°} = 254.5 \text{ N}$$

$$R_1 = R_2\cos\theta_1 = 254.5\cos 50.5° = 161.88 \text{ N}$$

4. Apply equilibrium condition to roller *B*.

(i) $$\Sigma F_x = 0$$

$$R_3 = R_4\cos\theta_2$$

(ii) $$\Sigma F_y = 0$$

$$R_4\sin\theta_2 - R_2\sin\theta_1 - W_B = 0$$

$\therefore$ $$R_4 = \frac{R_2\sin\theta_1 + W_B}{\sin\theta_2} = \frac{254.5\sin 50.5° + 40\times 9.81}{\sin 36.87°}$$

$$= \frac{196.378 + 392.4}{0.60} = 980.8 \text{ N}$$

$$R_3 = R_4\cos\theta_2 = 980.8\cos 36.87° = 784.64 \text{ N}$$

5. Apply equilibrium conditions to roller *A*,

(i) $$\Sigma F_x = 0$$

$$R_6 = R_4\cos\theta_2$$

$$R_6 = 980.8\cos 36.87° = \mathbf{784.6\ N} \quad \textbf{Ans.}$$

Example 1.19: A smooth cylinder of weight W and radius r rests in a V-shaped groove whose sides are inclined at angle α and β with horizontal. Find the reactions R_A and R_B at points of contact. Given $\alpha = 25°$, $\beta = 65°$, $W = 500$ N.

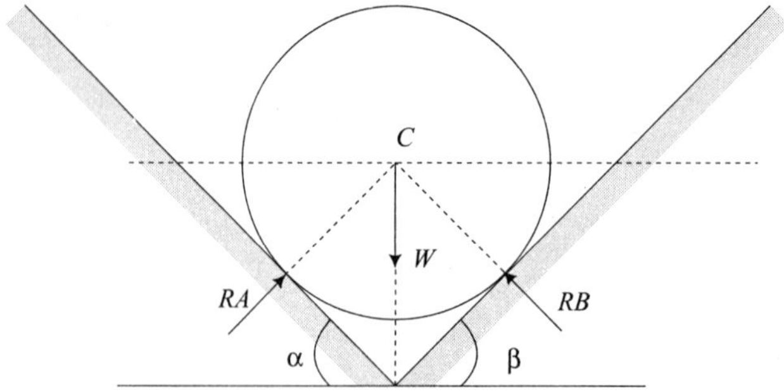

Solution:

1. Draw force diagram or free body diagram.

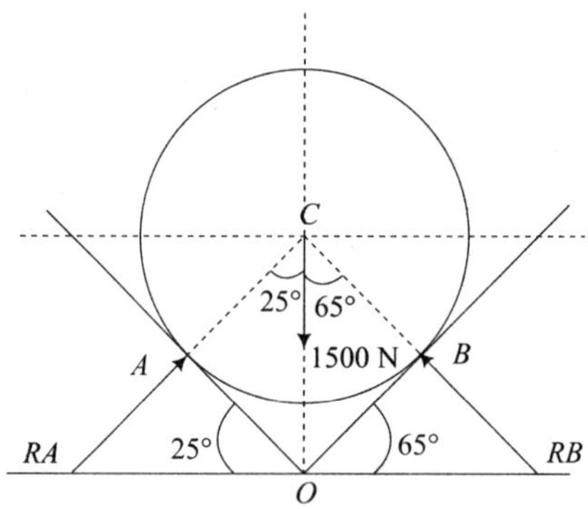

Free Body Diagram

Angle $ACO = \alpha = 25°$

Angle $BCO = \beta = 65°$

2. Apply equilibrium conditions to roller C.

(i) $\Sigma F_x = 0$

$$R_A \sin\alpha - R_B \sin\beta = 0 \quad ...(1)$$

(ii) $\Sigma F_y = 0$

$$R_A \cos\alpha + R_B \cos\beta - W = 0 \quad ...(2)$$

From equation (1)

$$R_A = \frac{R_B \sin\beta}{\sin\alpha} = \frac{R_B \times 0.9063}{0.4226} = 2.144\ R_B$$

From equation (2)

$2.144\ R_B \cos\alpha + R_B \cos\beta - W = 0$

$$\therefore \quad R_B = \frac{500}{2.144 \cos\alpha + \cos\beta} = \frac{500}{1.963 + 0.4226} = 211.36 \text{ N}$$

$$\therefore \quad R_A = 2.144\ R_B$$

$$= 2.144 \times 211.36 = \mathbf{453.2\ N} \quad \textbf{Ans.}$$

Example 1.20: Two similar spheres P and Q each of weight W rest inside a hollow cylinder which is resting on a horizontal plane. Draw the free-body diagram of:

(a) Both the spheres taken together
(b) The sphere P
(c) The sphere Q
(d) The cylinder.

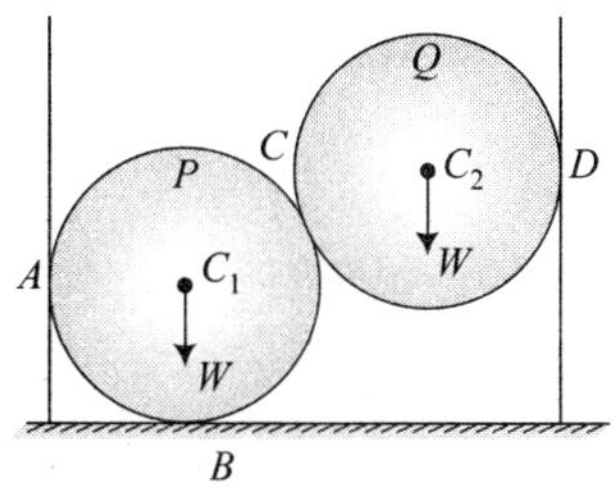

Fig. 1

Solution:

(a) Free body diagram of spheres P and Q (Fig. 2)
(b) Free body diagram of sphere P (Fig. 3)
(c) Free body diagram of sphere Q (Fig. 4)
(d) Free body diagram of cylinder (Fig. 5)

$$R_C(P) < R_C(P)$$

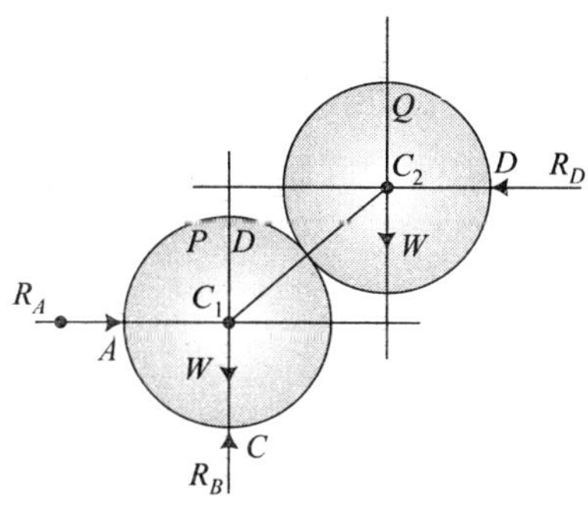

Fig. 2

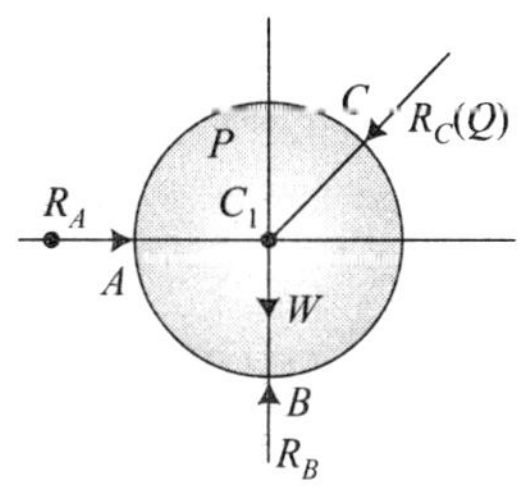

Fig. 3

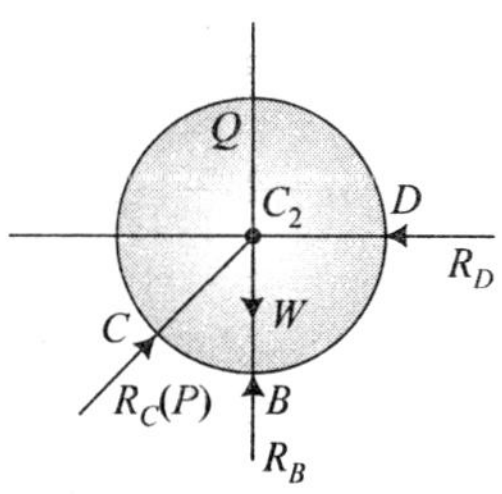

Fig. 4

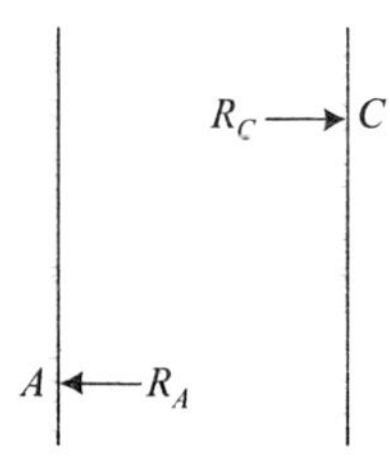

Fig. 5

Example 1.21: The figure shows a sphere resting in a smooth *V*-shaped groove and is subjected to a spring force. The spring is compressed to a length of 100 mm from its free length of 150 mm. If the stiffness of spring is 2 N/mm, determine the contact reactions at *A* and *B*. [U.P.T.U., May, 2003]

Solution:

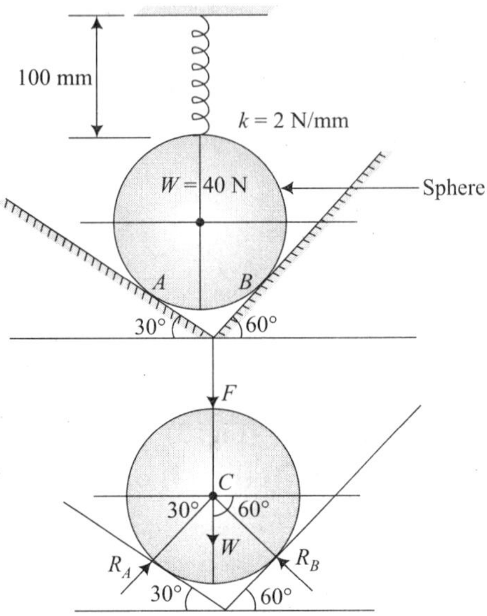

1. Draw the free-body diagram of sphere.

   Spring force, $F = k \times \delta = 2 \times (150 - 100) = 100$ N

2. Equilibrium of sphere *C*.

$$\Sigma F_x = 0; \quad R_A \sin 30° - R_B \sin 60°$$

$$\therefore \qquad R_A = \frac{\sin 60°}{\sin 30°} - R_B = 1.732 R_B$$

$$\Sigma F_y = 0; \quad R_A \cos 30° + R_B \cos 60° - W - F = 0$$

$$1.732\ R_B \cos 30° + R_B \cos 60° = 40 + 100 = 140$$

$$\therefore \qquad R_B = \frac{140}{1.732 \cos 30° + \cos 60°} = \mathbf{70\ N} \quad \textbf{Ans.}$$

$$R_A = 1.732 \times 70$$

$$= \mathbf{1121.2\ N} \quad \textbf{Ans.}$$

Example 1.22: Two cylinders A and B rest in a horizontal channel as shown. The cylinder A has a weight of 1000 N and radius of 9 cm. The cylinder B has a weigth of 400 N and a radius of 5 cm. The channel is 18 cm wide at the bottom, with one side vertical. The other side is inclined at 60° with the horizontal. Find the reactions.

Solution:

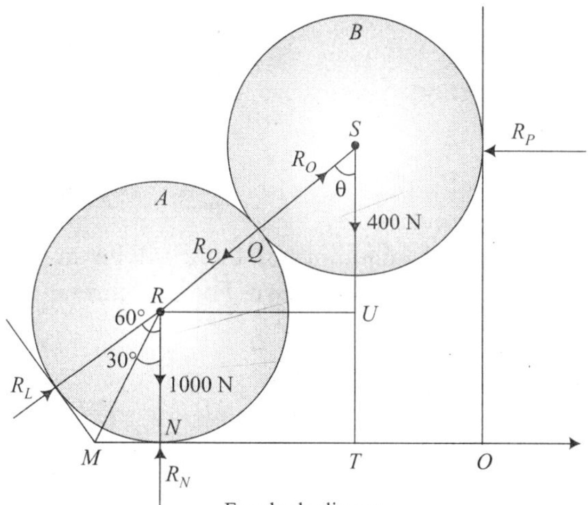

Free body diagram

1. Free body diagram is drawn.

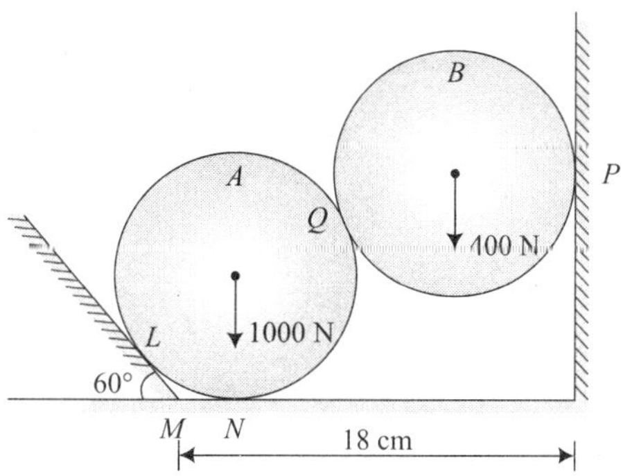

2. $$\sin\theta = \frac{RU}{RS} = \frac{18 - MN - OT}{9+5}$$

$$= \frac{18 - RN\tan 30° - 5}{14}$$

$$= \frac{18 - 9\tan 30° - 5}{14}$$

$$\theta = 33.86°$$

3. Equilibrium of cylinder B.

$$\Sigma F_x = 0, \quad R_Q \sin(33° - 86°) - R_P = 0$$

$$\Sigma F_y = 0, \quad R_Q \cos 33.86° - 400 = 0$$

$$\therefore \quad R_Q = 481.9 \text{ N}$$

$$R_P = 481.9 \sin 33.86 = 268.5 \text{ N.}$$

4. Equilibrium of cylinder A.

$$\Sigma F_x = 0; \quad R_L \sin 60° - R_Q \sin 33.86° = 0$$

$$\therefore \qquad R_L = \frac{481.9 \sin 33.86°}{\sin 60°} = 310 \text{ N}$$

$$\Sigma F_y = 0, \quad R_N - 1000 - R_Q \cos 33.86° + R_L \cos 60° = 0$$

$$\therefore \qquad R_N = 1248.2 \text{ N.}$$

Example 1.23: A roller of mass 150 kg has to be started rolling over the block A as shown. Find out the force T required. [U.P.T.U. 2008-09]

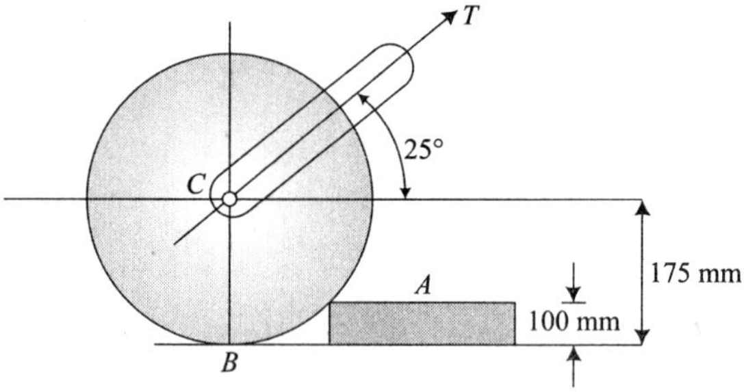

Solution: When the roller is about to be started, it will lose contact with ground at point B, $\therefore$ $R_B = 0$. As the roller is in equilibrium, there will be three concurrent forces acting at C. The free body diagram is drawn. For equilibrium of roller,

$$\Sigma F_x = 0, \quad T \cos 25° - R_A \cos \theta = 0$$

$$\Sigma F_y = 0, \quad T \sin 25° + R_A \sin \theta - W = 0$$

$$\theta = \tan^{-1} \frac{100}{175} = 53°$$

$$\therefore \qquad R_A = \frac{T \cos 25°}{\cos 53°} = \frac{0.906T}{0.602} = 1.5T$$

$$\therefore \qquad T \sin 25° + 1.5T \sin 53° = 150 \times 9.81$$

$$\therefore \qquad T = \frac{150 \times 9.81}{\sin 25° + 1.5 \sin 53°}$$

$$= \frac{1471.5}{0.4226 + 1.19795} = \textbf{908 N} \quad \textbf{Ans.}$$

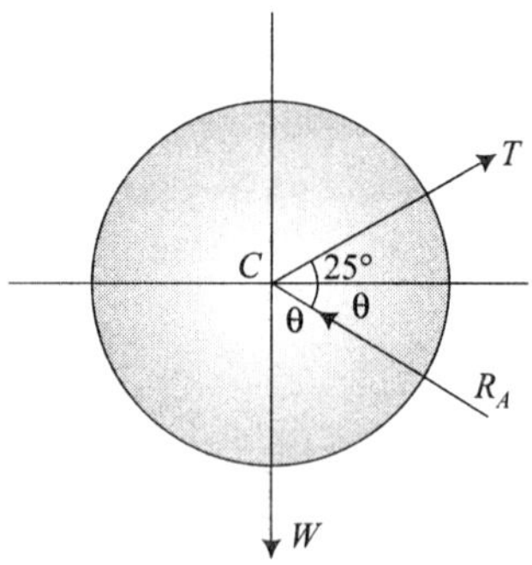

Free body diagram

Example 1.24: A right circular roller of weight 5000 N rests on a smooth inclined plane and is held in position by a cord AC as shown. Find the tension in the cord if there is a horizontal force of magnitude 1000 N acting at C.

[U.P.T.U. May, 2003]

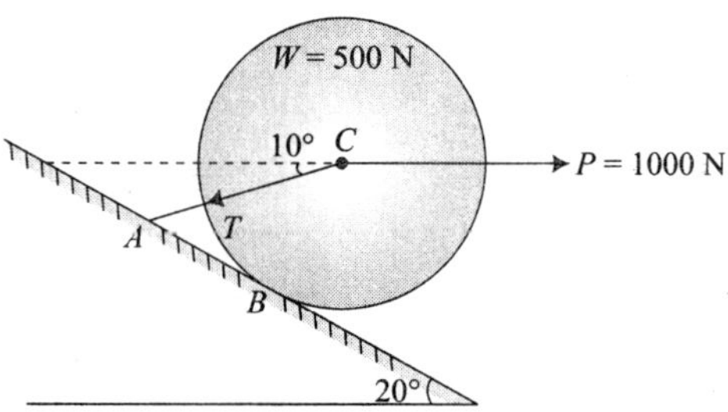

Solution:

1. Draw free-body diagram of the roller.
2. For equilibrium

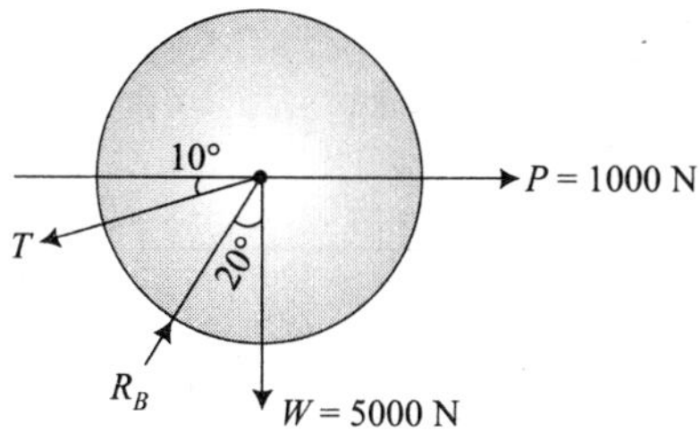

Free body diagram

$$\Sigma F_x = 0$$

$$P - T\cos 10° + R_B \sin 20° = 0 \quad ...(1)$$

$$\Sigma F_y = 0$$

$$W - R_B \cos 20° + T \sin 10° = 0 \quad ...(2)$$

From equation (1)

$$1000 - 0.9848T + 0.342R_B = 0$$

$$5000 - 0.93969R_B + 0.173648T = 0$$

$$\therefore \; R_B = \frac{0.9848T - 1000}{0.342} = 2.8795T - 2923.9766$$

Substituting the value of R_B into equation (2).

$$5000 - 0.9369(2.8795T - 2923.9766) + 0.173648T = 0$$

$$5000 - 2.7T + 2747.63 + 0.173648T = 0$$

$$\therefore \quad T = \frac{5000 + 2747.63}{2.7 - 0.173648} = \frac{7747.63}{2.52635} = \mathbf{3066.73\ N}$$ **Ans.**

Example 1.25: Three cylinders *A*, *B* and *C* each weighing 100 N and diameter 80 mm are placed in a channel of 180 mm width as shown. Determine the pressure exerted by the cylinders *A* and *B* at the point of contact.

[U.P.T.U. 2009-10]

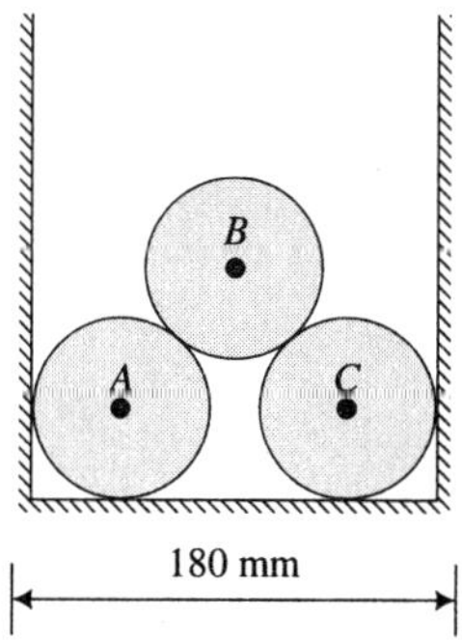

Solution:

Draw FBD of cylinder *B*

$$\cos\alpha = \frac{AD}{AB} = \frac{(180-80)\times\frac{1}{2}}{80}$$

$$= \frac{50}{80} = \frac{5}{8},$$

$$\alpha = 51.3°$$

$$\beta = 90° - 51.3° = 38.7°.$$

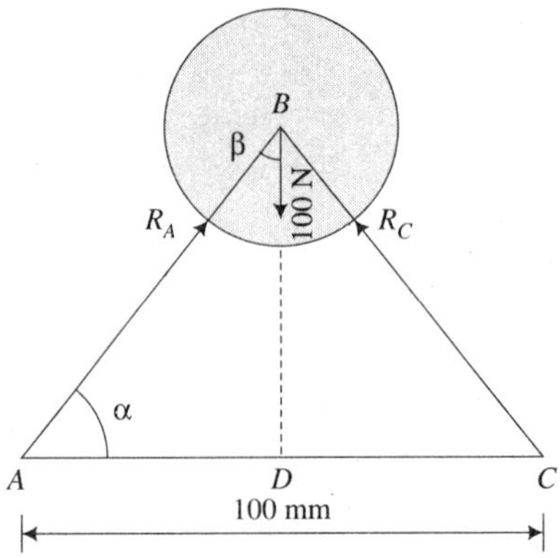

For equilibrium of cylinder B,

$$\Sigma F_y = 0$$

$$R_A \cos\alpha + R_C \cos\alpha = W$$

or $$2 R_A \cos\alpha = W \qquad [\because R_A = R_C]$$

$\therefore$ $$R_A = \frac{W}{2\cos\alpha} = \frac{100}{2 \times \cos 51.3°} = 64 \text{ N}$$ **Ans.**

Example 1.26: Two cylinders A and B of diameter 3 cm and 6 cm weighing 20 kN and 80 kN respectively are placed as shown. Assuming all the contact surfaces to be smoth, find the reactions at the wall. [U.P.T.U. 2008-09]

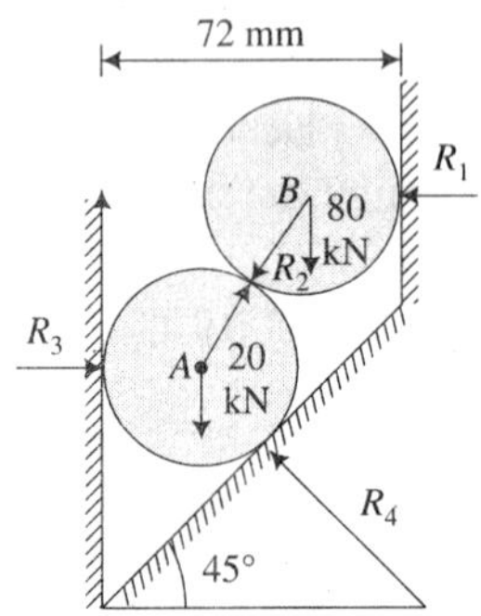

Solution:

(i) Draw FBD of cylinder B.

$$AB = \frac{D_A + D_B}{2} = \frac{3+6}{2} = 4.5 \text{ cm}$$

$$AC = 7.2 - \frac{D_A + B_B}{2} = 7.2 - 4.5$$

$$= 2.7 \text{ cm.}$$

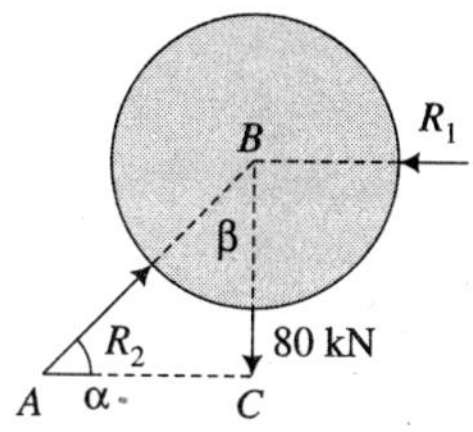

$$\sin\beta = \frac{AC}{AB} = \frac{2.7}{4.5} = 0.6$$

$$\beta = 36.87°$$

$$\alpha = 90° - 36.87° = 53.13°$$

Consider the equilibrium of cylinder B,

$$\Sigma F_y = 0$$

$$R_2 \cos\alpha = W_B$$

$$\therefore \quad R_2 = \frac{W_B}{\cos\alpha} = \frac{80}{\cos 36.87°} = 100 \text{ kN}$$

$$\Sigma F_x = 0$$

$$R_1 = R_2 \sin\alpha = 100 \sin 36.87° = 60 \text{ kN}.$$

(ii) Draw FBD of cylinder A

$$\Sigma F_x = 0,$$

$$R_3 = R_2 \cos\alpha + R_4 \cos 45°$$

$$= 100 \cos 36.87 + R_4 \cos 45°$$

$$= 80 + 0.7\, R_4 \qquad (1)$$

$$\Sigma F_y = 0.$$

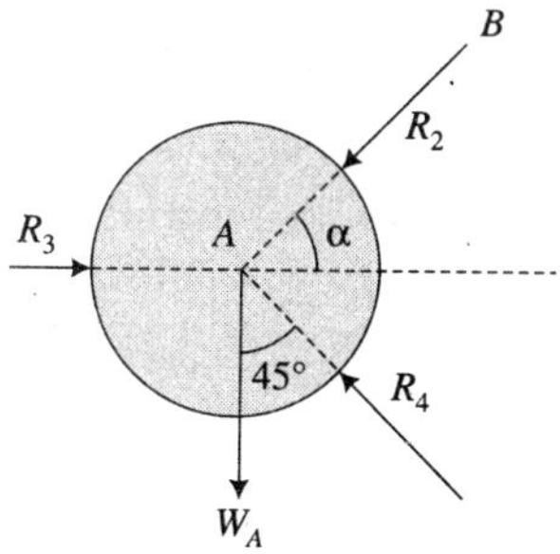

$$R_2 \sin\alpha - R_4 \sin 45° = W_A.$$

$$100 \sin 36.87° - R_4 \sin 45° = 20$$

$$60 - 0.7\, R_4 = 20.$$

$$\therefore \qquad R_4 = \frac{60-20}{0.7} = 57 \text{ kN.}$$

From equation 1,

$$R_3 = 80 + 0.7 \times 57 = 120 \text{ kN.}$$

QUESTION BANK NO. 1

1. Enumerate different laws of motion, discussing the significance of each of them.
2. Discuss various laws and principles of engineering mechanics.
3. Explain various force systems.
4. What do you understand by the resultant of a force system? Discuss various methods used for determination of the resultant of coplanar force systems.
5. Explain the principle of transmissibility of a force.
6. What are the necessary and sufficient conditions of equilibrium of a system of coplanar force system?
7. Explain the polygon law of forces.
8. Define moment of force about a point. Also explain Varignon's theorem of moments.
9. What do you understand by transfer of force to a parallel position?
10. Explain free-body diagram with suitable examples.

QUICK REVIEW INFORMATION NO. 1

1. Statics is the study of forces in rigid bodies.
2. Dynamics is the study of forces in moving bodies.
3. Kinematics is the study of displacement, velocity and acceleration.
4. Kinetics is the study of forces that cause motion.
5. Lami's theorem deals with equilibrium of three concurrent forces.
6. Varignon's theorem deals with the principle of moments.
7. D'Alembert's principle deals with dynamic equilibrium of a particle.
8. Maxwell theorem deals with force analysis of trusses.
9. Two forces can be in equilibrium only if they are equal in magnitude, opposite in direction and collinear in action.
10. In order to determine the effects of a force acting on a body, we must know its magnitude, its direction and line of action.
11. In coplanar force system, the line of action of all forces lies in the same plane.
12. In concurrent forces, line of action of all forces passes through a common point.

13. In collinear forces, line of action of all forces lies along the same line.
14. A system of coplanar forces will be in equilibrium if $\Sigma F_x = \Sigma F_y = \Sigma M_0 = 0$.
15. A system of coplanar concurrent forces will be in equilibrium if $\Sigma F_x = \Sigma F_y = 0$.
16. If a number of forces act simultaneously on a particle, it is possible to replace them by a single force.
17. When non-concurrent coplanar forces act on a body, it is possible that the resultant force is zero if the body is not in equilibrium.
18. The magnitude of the moment does not change if the point of force is transmitted along its line of action.
19. The algebraic sum of moments of two concurrent forces with respect to any moment centre in their plane of action is equal to the moment of their resultant with respect to the same centre.

CHAPTER 2

Friction

2.1 INTRODUCTION

Whenever the surfaces of two bodies are in contact, there is some resistance to sliding between them. The opposing force to the movement is called friction or force of friction. It is due to interlocking of surfaces as a result of the presence of some roughness and irregularities at the contact surfaces. The resisting force acts in the direction opposite to the movement. A force of friction comes into play whenever there is a relative motion between two parts. Some energy is wasted in order to overcome the resistance due to force of friction.

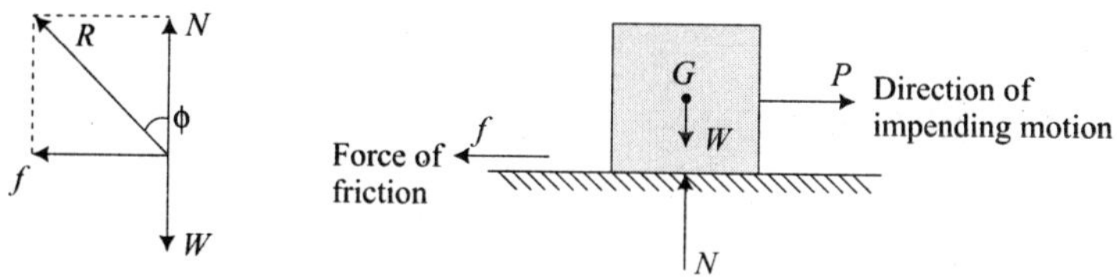

Fig. 2.1 Force of friction

Force of friction or frictional force is the opposing force to the movement of one body over the surface of another body.

In Fig. 2.1,

W = Weight of the body (mg)
N = Normal reaction
f = Friction force
P = Force applied to the body
R = Total reaction
ϕ = Angle of friction

2.1.1 Engineering Applications of Friction

Friction is both desirable and undesirable in engineering applications.

1. *Friction devices:* There are devices in engineering which work due to friction. Examples are lifting machines such as screw jack, power trans-

mission by belts and ropes, friction clutches, brakes, nuts and bolts, jib and cotter joints, etc. The friction should be utilized by the use of proper materials and shape of components.
2. The vehicles can run on roads and inclined surfaces due to the friction only.
3. A man cannot walk if there is no friction.

However, there is resistance and wastage of energy in the bearings and sliding surfaces.

1. Friction causes loss of energy in moving parts of machines.
2. Friction causes heating of bodies and may affect the life of the machine.
3. The efficiency of machine decreases due to presence of friction.
   Solid, liquid and gaseous lubricants are used to reduce friction by separating the moving surfaces.

The *characteristics* of the friction force are:

1. It always acts in a direction opposite to the motion and opposes the tractive effort.
2. It is a reactive or passive force and exists as long as tractive effort is present.
3. The friction force adjusts itself depending upon the magnitude of tractive effort or applied force.

2.2 TYPES OF FRICTION

1. *Static friction:* A body at rest tending to move encounters static friction.
2. *Dynamic friction:* The force of friction encountered by a body in motion is called dynamic or kinetic friction. The friction can be sliding friction on flat surfaces or rolling friction when one body rolls over other or surfaces are separated by balls or rollers. The rolling friction is always less than sliding friction when surfaces slide one over the other. The friction that exists when no lubricant is used, is called *dry friction* or *solid friction.* Laws of static and dynamic friction are based on experiments conducted by Prof. Coulomb and are called laws of *Coulomb Friction*.

2.3 COULOMB LAWS OF FRICTION

2.3.1 Laws of Static Friction

The laws of static friction are as follows:

1. The force of friction always acts opposite to the direction of impending motion.
2. The magnitude of force of friction is equal to applied force tending to move the body.

$$f = P$$

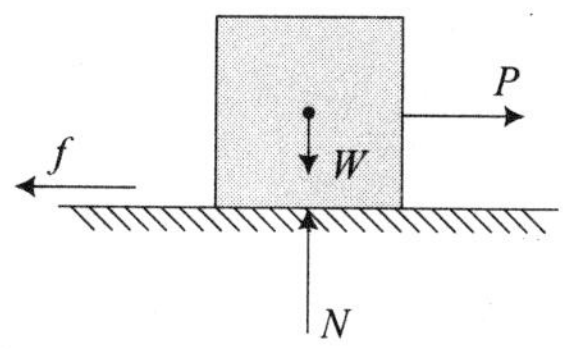

Fig. 2.2 Static friction

3. The force of friction is directly proportional to the normal reaction between the surfaces.

$$f \propto N$$

or

$$\frac{f}{N} = \text{constant} = \mu$$

μ is called coefficient of friction and depends upon the nature of contact surfaces.

$$\therefore \qquad f = \mu N$$

4. The force of friction is independent of the area of contact of the two surfaces.
5. The force of friction depends upon the roughness of the surface.

2.3.2 Laws of Dynamic Friction

Limiting Friction

The friction force, f increases to oppose P until a maximum value of f_m is reached. If P is further increased, the body starts sliding as force of friction cannot balance the force P. The value of f is lowered from f_m, which is the *limiting* and maximum friction force that is exerted at the time when motion is about to begin.

The laws of dynamic friction are as follows:

1. The force of friction always opposes the motion of the body and acts in the direction opposite to the motion.
2. The frictional force is directly proportional to the normal reaction between the contacting surfaces in relative motion.
3. The force of friction although directly proportional to normal reaction but is slightly less than for static friction

$$f = \mu N$$

The coefficient of friction μ is less for dynamic friction than for limiting friction.

4. For low velocities, total amount of friction force that can be developed is practically independent of velocity.

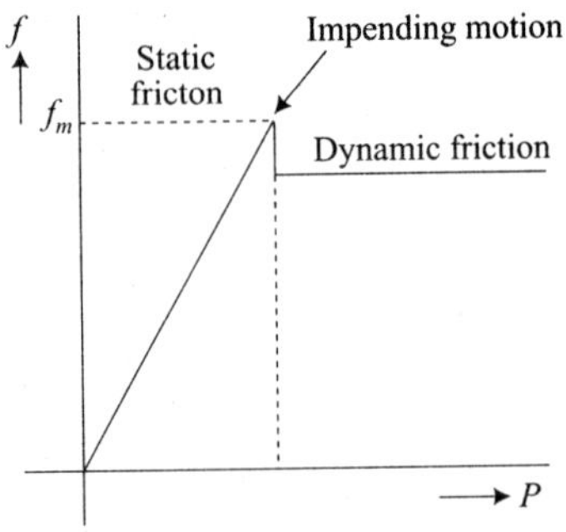

Fig. 2.3 Limiting friction

2.3.3 Important Definitions

(a) Angle of Friction

It is the angle made by the resultant (R) of the normal reaction (N) and limiting force of friction (f) and made with the direction of normal reaction.

R is the resultant of normal reaction N and force of friction f.

$$R = \sqrt{N^2 + f^2}$$

ϕ is the angle of friction

$$\tan \phi = \frac{f}{N}$$

or $$\phi = \tan^{-1} \frac{f}{N}$$

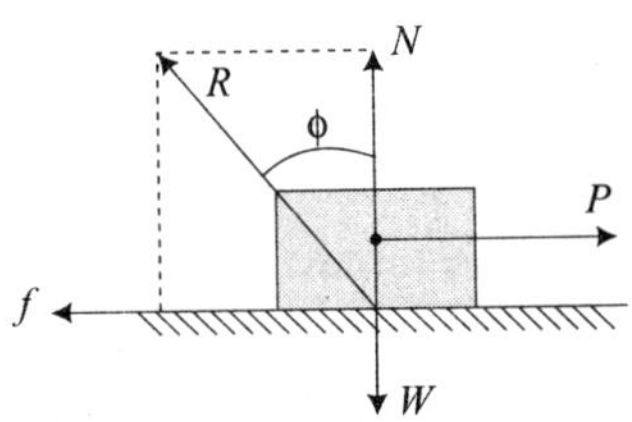

Fig. 2.4 Angle of friction

(b) Coefficient of friction

It is ratio of limiting frictional force and the normal reaction.

The coefficient of friction,

$$\mu = \frac{f}{N} = \tan \phi$$

$\therefore$ $$f = \mu N$$

μ depends upon the nature of contacting surface. Its value is very low for lubricated surfaces and high for dry friction.

(c) Angle of Repose

A body of weight W is lying on a rough plane inclined at an α with the horizontal. The body is in equilibrium under the action of following forces:

(i) *Weight of the body W*. It has two components: $W \sin \alpha$ parallel to inclined plane and $W \cos \alpha$ normal to the plane.

(ii) *Normal reaction*, N acting in a direction normal to inclined plane.

(iii) *Friction force*, f acting in a direction opposite to the motion.

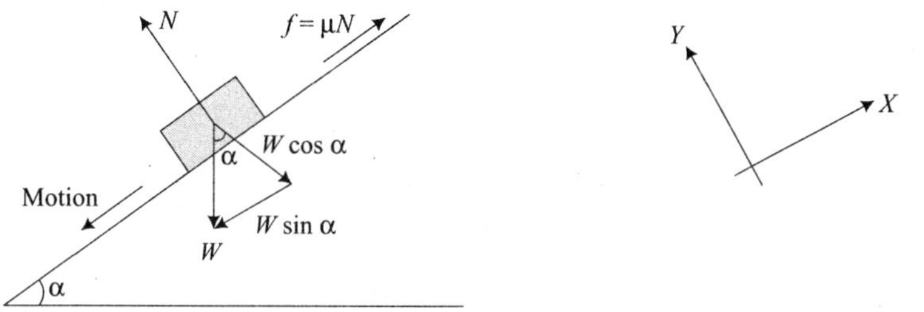

Fig. 2.5 Angle of repose

When the body tends to slide down the plane, the frictional force must act up the plane and when the body is being pulled up, the force of friction acts downwards to oppose the motion.

Selecting the reference coordinate system with X-axis in the direction of inclined plane and Y-axis perpendicular to inclined plane,

Applying equilibrium conditions,

$$\Sigma F_X = 0$$

$$f = W \sin \alpha$$

$$\Sigma F_Y = 0$$

$$N = W \cos \alpha$$

$$\therefore \quad \frac{f}{N} = \frac{W \sin \alpha}{W \cos \alpha} = \tan \alpha$$

But, $$\frac{f}{N} = \mu = \tan \phi$$

where φ is called the angle of friction.

If angle α of inclined plane is altered so that the body is in the condition of limiting equilibrium, i.e., on the verge of sliding down, it is equal to angle of friction ϕ. This value of α is called *angle of repose*. Therefore, angle of repose may be defined as the angle of inclined plane on which body is in the condition of limiting equilibrium. The angle of repose is equal to angle of friction.

(d) Cone of Friction

In Fig. 2.6,

$$OA = N = \text{Normal reaction}$$
$$OC = f_{max} = \text{Limiting frictional force}$$
$$OR = \text{Resultant force}$$
$$\angle AOB = \phi = \text{Maximum angle of friction}$$

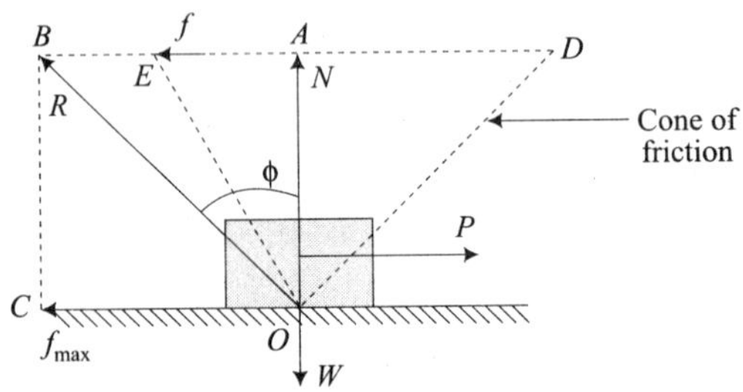

Fig. 2.6 Cone of friction

If line OB is revolved about vertical axis OA, the cone generated OBD is called cone of friction.

If the force of friction f (AE) and resultant OC fall within the cone of friction, there will be no motion of the body. This principle is used in self-locking mechanism of lifting machines (i.e., screw jack) and taper pins.

2.4 EQUILIBRIUM OF BODIES INVOLVING DRY FRICTION

The force of dry friction is involved in the equilibrium and motion of bodies under following cases:

1. Rough Horizontal Plane
2. Rough Inclined Plane
3. Ladders
4. Wedges
5. Belt Drive
6. Lifting Machines

In drawing the free body diagrams, in addition to active and reactive forces, the force of friction is applied in a direction opposite to the motion. The forces are resolved into rectilinear components. The unknown forces are found from the equations of equilibrium.

2.4.1 Rough Horizontal Plane

There can be following cases of motion of body on rough horizontal plane. The equilibrium conditions for each case are discussed.

(a) No moving force

$$P = 0$$

∴ Friction force, $f = 0$

$$W = N$$

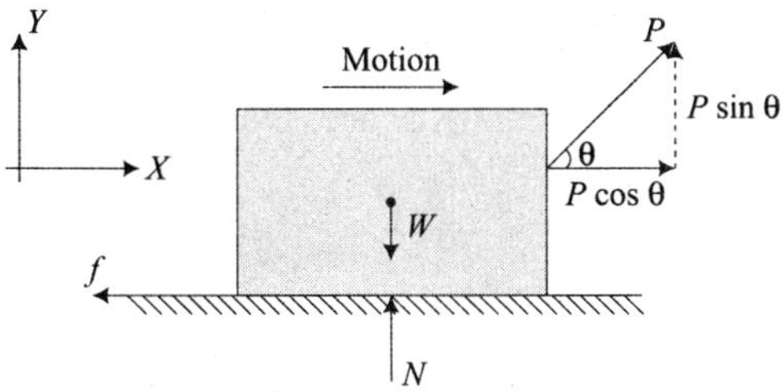

Fig. 2.7 Rough horizontal plane

(b) Body moving under pull or push

A force P is applied to the body. The force body diagram is shown. Considering the equilibrium of the body.

$$\Sigma F_X = 0$$

$$P \cos \theta = f$$

But by definition, $f = \mu N$

$$\therefore \quad P \cos \theta = \mu N \qquad ...(1)$$

$$\Sigma F_Y = 0$$

$$\therefore \quad N = W \pm P \sin \theta$$

$$\therefore \quad W = N + P \sin \theta$$

$$\text{or} \quad N = W - P \sin \theta \qquad ...(2)$$

From equations (1) and (2)

$$P \cos \theta = \mu(W - P \sin \theta)$$

$$P \cos \theta + \mu P \sin \theta = \mu W$$

$$P\left(\cos \theta + \frac{\sin \phi}{\cos \phi} \sin \theta\right) = \frac{\sin \phi}{\cos \phi} W \qquad \left[\because \mu = \tan \phi = \frac{\sin \phi}{\cos \phi}\right]$$

$$p(\cos\theta\cos\phi + \sin\theta\sin\phi) = W\sin\phi$$

$$P\cos(\theta - \phi) = W\sin\phi$$

$$P = \frac{W\sin\phi}{\cos(\theta - \phi)}$$

For P to be minimum, $\cos(\theta - \phi)$ should be maximum,

$\therefore$ $$\cos(\theta - \phi) = 1$$

$$\theta = \phi$$

The angle of inclination of force P should be equal to the angle of friction, ϕ.

Example 2.1: The force required to pull the body of weight 50 N on a rough horizontal surface is 20 N where it is applied at an angle of 25° with the horizontal as shown. Determine the coefficient of friction and magnitude of reaction N between the body and the horizontal surface. Does the reaction pass through the centre of gravity of the body? [U.P.T.U. I Sem., 2002–03]

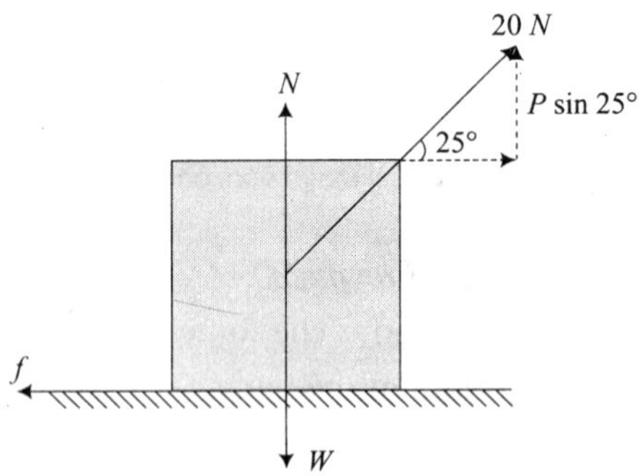

Solution: The body is in equilibrium

$\therefore$ $$\Sigma F_y = 0$$

$$N = W - P\sin 25°$$

$$= 50 - 20 \times 0.42$$

$$= 41.55\ N$$

$$\Sigma F_x = 0$$

$\therefore$ $$f = 20\cos 25° = 18.13\ N$$

Now, $$\mu = \frac{f}{N} = \frac{18.13}{41.55} = \mathbf{0.436} \quad \textbf{Ans.}$$

The reaction passes through the centre of gravity of the body as it is equal and opposite to weight of body W.

Example 2.2: A block of weight 5 kN is pulled by a force P as shown. The coefficient of friction between the contact surface is 0.35. Find the direction θ for which P is minimum and find the corresponding value of P.

[U.P.T.U. II Sem., 2003–04]

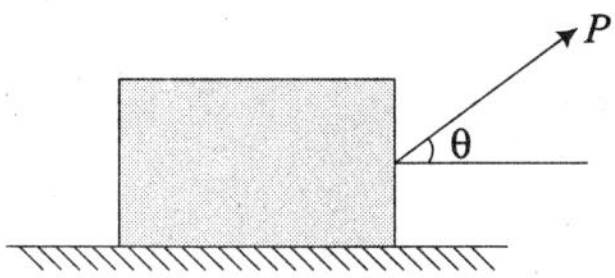

Solution:

(i) Draw free body diagram of the block.

(ii) Apply equilibrium conditions.

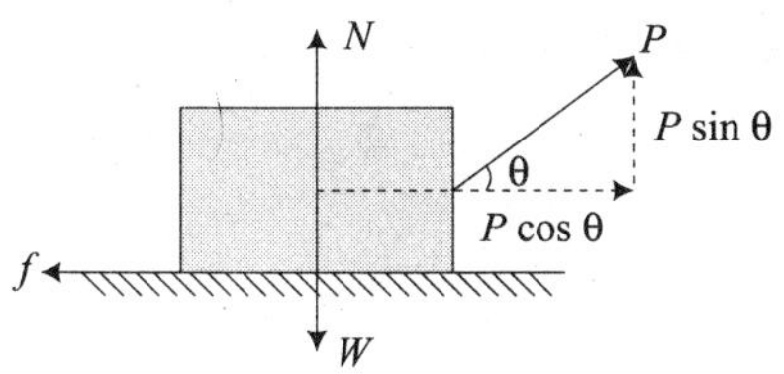

$$\Sigma F_x = 0$$

$$f = P \cos \theta$$

$$\Sigma F_y = 0$$

$$W = N + P \sin \theta$$

(iii) Thc anglc of inclination of forcc P will bc cqual to thc anglc of friction for minimum value of P.

$\therefore$ $$\theta - \phi - \tan^{-1} \mu - \tan^{-1} 0.35$$

$$= 20.48°$$

$$P = \frac{W \sin \phi}{\cos (\theta - \phi)}$$

For P to be minimum, $\cos (\theta - \phi) = 1$

$\therefore$ $P = W \sin \phi = 5 \sin 20.48° =$ **1.75 kN** **Ans.**

Examsple 2.3: Obtain the expression for minimum force required to drag a body on a rough horizontal plane.

Solution:

(i) Draw free body diagram.

(ii) Apply equilibrium conditions.

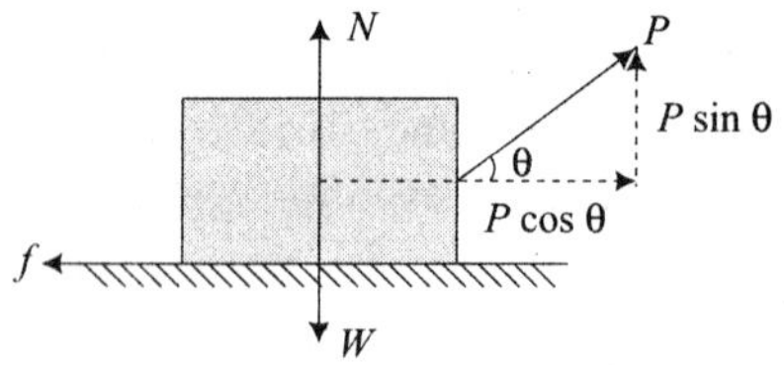

$$\Sigma F_x = 0$$

$$\therefore \quad f = P\cos\theta = \mu N \qquad \text{...(1)}$$

$$\Sigma F_y = 0$$

$$N = W + P\cos\theta \qquad \text{...(2)}$$

From equations (1) and (2)

$$P\cos\theta = \mu N$$

$$= \mu(W - P\sin\theta)$$

$$P\cos\theta + \mu P\sin\theta = \mu W$$

Now, $\mu = \tan\phi = \dfrac{\sin\phi}{\cos\phi}$

$$P\left(\cos\theta + \frac{\sin\phi}{\cos\phi}\sin\theta\right) = \frac{\sin\phi}{\cos\phi}.W$$

$$P(\cos\theta\cos\phi + \sin\theta\sin\phi) = W\sin\phi$$

$$P\cos(\theta - \phi) = W\sin\phi$$

$$P = \frac{w\sin\phi}{\cos(\theta - \phi)}$$

For P to be minimum, $\cos(\theta - \phi)$ should be maximum

$$\therefore \quad \cos(\theta - \phi) = 1$$

$$\theta = \phi$$

Minimum value of P = $\boldsymbol{W \sin\phi}$

or $= \boldsymbol{W\sin\theta}$ **Ans.**

2.4.2 Rough Inclined Plane

(a) Equilibrium Condition for Different Angle of Inclination

(i) Angle of inclination less than angle of friction

1. Draw free body diagram.
2. Apply equilibrium conditions.

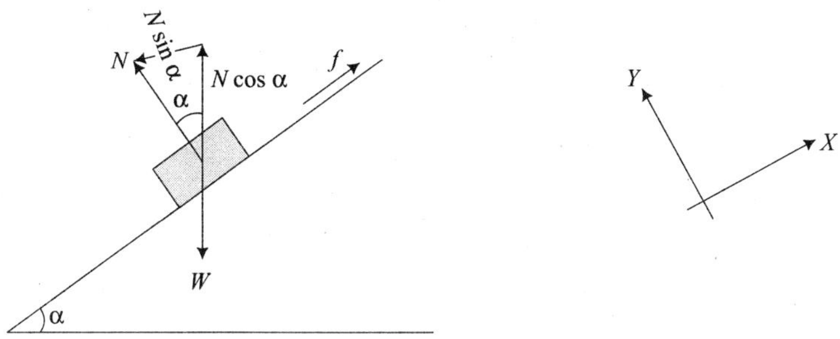

Fig. 2.8 Rough inclined plane ($\alpha < \phi$)

$$\Sigma F_x = 0$$

$$f = \mu N = N \sin \alpha$$

$$\Sigma F_y = 0$$

$$W = N \cos \alpha$$

The body is in equilibrium.

(i) Angle of inclination more than angle of friction.

The body will slide down and an upward force P is required to restrict the body from moving down. The restricting force can be applied in different ways.

1. *Along the inclined plane*

   Draw free body diagram.

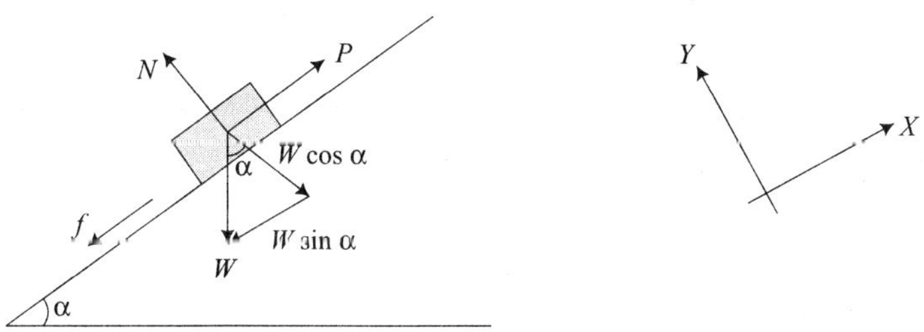

Fig. 2.9 Force parallel to inclined plane

Equilibrium conditions.

$$\Sigma F_x = 0$$

$$f = P - W \sin \alpha$$

$$\Sigma F_y = 0$$

$$N = W \cos \alpha$$

2. *The restricting force applied horizontally*

   Draw free body diagram.

   Apply equilibrium condition.

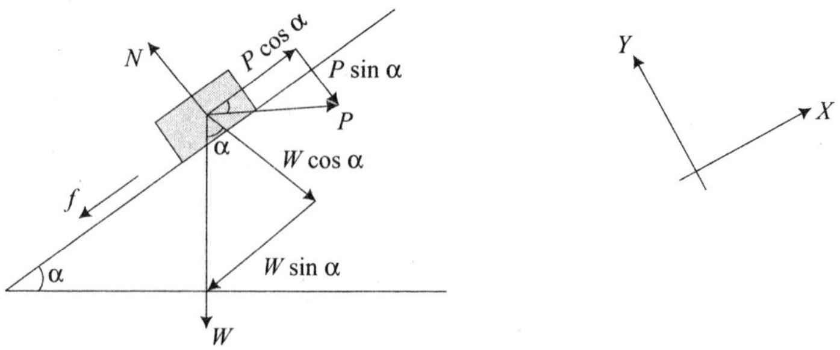

Fig. 2.10 Force applied horizontally

$$\Sigma F_x = 0$$
$$f = P \cos \alpha - W \sin \alpha$$
$$\Sigma F_y = 0$$
$$N = W \cos \alpha + P \sin \alpha$$

3. *Force applied at an angle θ with the inclined plane*
   Draw free body diagram.
   Apply equilibrium conditions.

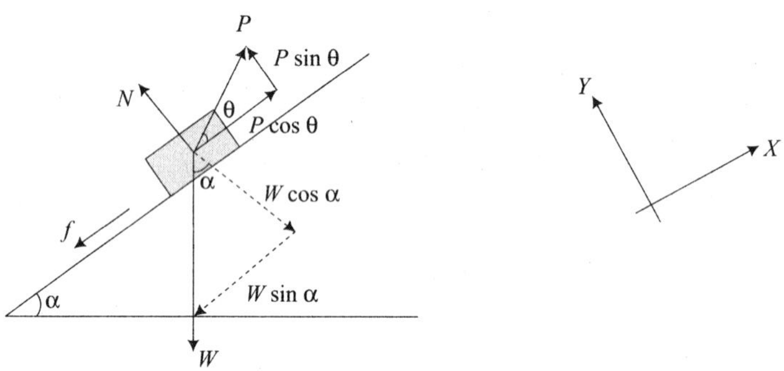

Fig. 2.11 Force applied at an angle θ with plane

$$\Sigma F_x = 0$$
$$f = P \cos \theta - W \sin \alpha$$
$$\Sigma F_y = 0$$
$$N = W \cos \alpha - P \sin \theta$$

(b) Body sliding downwards

A body of weight W is on the verge of sliding downwards. A minimum force P is required to restrict the motion.

(i) Draw free body diagram.
(ii) Apply equilibrium conditions.

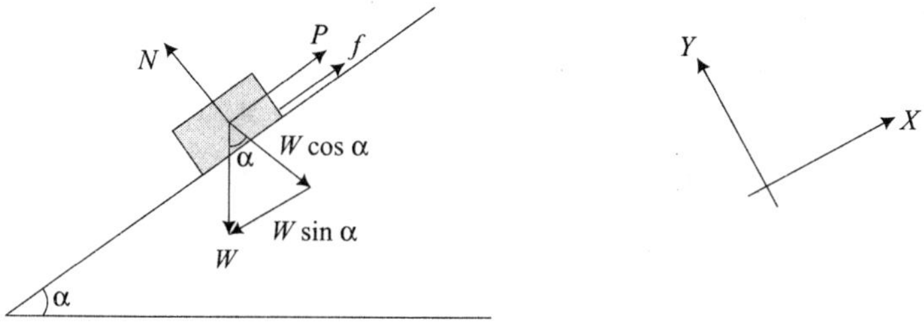

Fig. 2.12 Body sliding downwards

$$\Sigma F_x = 0$$

$$f = \mu N = W \sin \alpha - P$$

$\therefore$ $$P = W \sin \alpha - \mu N \qquad ...(1)$$

$$\Sigma F_y = 0$$

$$N = W \cos \alpha \qquad ...(2)$$

From equations (1) and (2)

$$P = W \sin \alpha - \mu(W \cos \alpha)$$

$$= W(\sin \alpha - \mu \cos \alpha)$$

$$= W\left(\sin \alpha - \frac{\sin \phi}{\sin \theta} \cos \alpha\right)$$

$\because$ $$\mu = \tan \phi = \frac{\sin \phi}{\cos \phi}$$

$$P \cos \phi = W(\sin \alpha \cos \phi - \cos \alpha \cos \phi)$$

$\therefore$ $$P = \frac{W \sin(\alpha - \phi)}{\cos \phi}$$

(c) Body upwards an inclined plane

Minimum force required to keep the body in equilibrium.

(i) Draw free body diagram.

(ii) Apply equilibrium conditions.

$$\Sigma F_x = 0$$

$$f = P - W \sin \alpha$$

$\therefore$ $$P = W \sin \alpha + f = W \sin \alpha + \mu N \qquad ...(1)$$

$$\Sigma F_y = 0$$

$$N = W \cos \alpha \qquad ...(2)$$

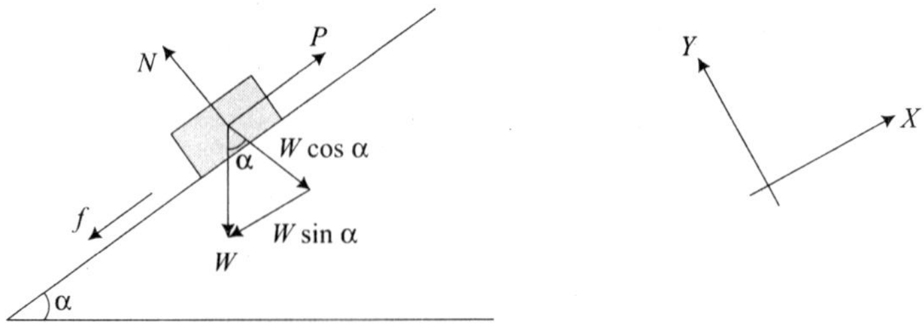

Fig. 2.13 Body upwards an inclined plane

From equations (1) and (2)

$$P = W \sin \alpha + \mu W \cos \alpha$$

$$= W \sin \alpha + \frac{\sin \phi}{\cos \phi} W \cos \alpha \qquad \left[\mu = \tan \phi = \frac{\sin \phi}{\cos \phi}\right]$$

$$P \cos \phi = W \sin \alpha \cos \phi + W \sin \cos \alpha$$

$$= W \sin (\alpha + \phi)$$

$$\therefore \qquad P = \frac{W \sin (a + \phi)}{\cos \phi}$$

Example 2.4: A body of weight 500 N is pulled up along an inclined plane having an inclination of 30° with the horizontal. If the coefficient of friction between the body and the plane is 0.3 and the force is applied parallel to inclined plane, determine the force required.[U.P.T.U. II Sem., 2005–06]

Solution:

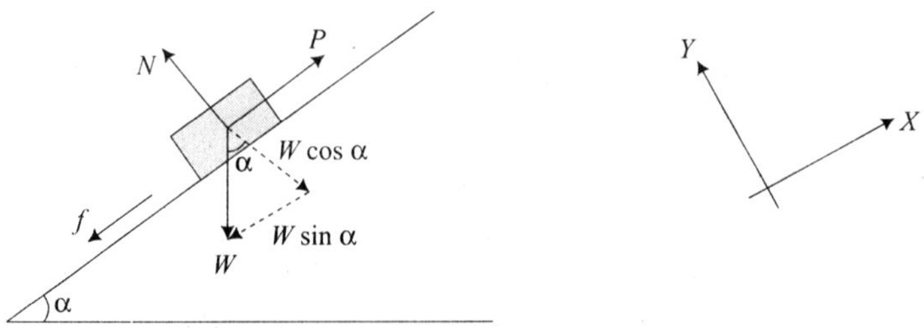

$$W = 500 \text{ N}$$

$$\alpha = 30°$$

$$\mu = 0.3$$

Angle of friction,

$$\phi = \tan^{-1} \mu = \tan^{-1} 0.3 = 16.7°$$

The force of friction will be acting downwards as the body is tending to move upwards.

Draw free body diagram and apply equilibrium conditions.

$$\Sigma F_x = 0$$

$$P - \sin\alpha = f = \mu N$$

$\because$
$$P = W \sin\alpha + \mu N \qquad ...(1)$$

$$\Sigma F_y = 0$$

$$N = W \cos\alpha \qquad ...(2)$$

From equations (1) and (2)

$$P = W \sin\alpha + \mu W \cos\alpha$$

$$= 500 \sin 30° + 0.3 \times 500 \cos 30°$$

$$= 250 \times 0.5 + 0.3 \times 500 \times 0.866$$

$$= 250 + 129.9 = \mathbf{379.9\ N} \quad \textbf{Ans.}$$

Example 2.5: A body of weight W is placed on a rough inclined plane having inclination α to the horizontal. The force P is applied horizontal to drag the body. If the body is on the point of motion up the plane, prove that P is given by:

$$P = W \tan(\alpha + \phi)$$

where, ϕ = angle of friction.

Solution: Draw free body diagram and apply equilibrium conditions.

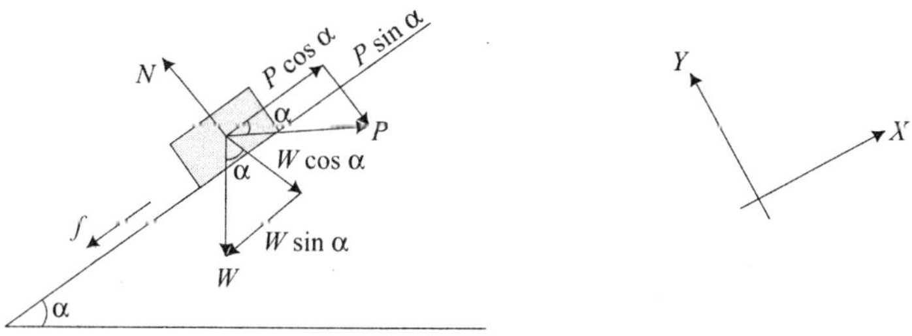

$$\Sigma F_x = 0$$

$$P \cos\alpha = W \sin\alpha + f = W \sin\alpha + \mu N \qquad ...(1)$$

$$\Sigma F_y = 0$$

$$N = W \cos\alpha + P \sin\alpha \qquad ...(2)$$

From equations (1) and (2)

$$P \cos\alpha = W \sin\alpha + \mu(W \cos\alpha + P \sin\alpha)$$

$$= W \sin\alpha + \mu W \cos\alpha + \mu P \sin\alpha$$

$$P \cos\alpha - \mu P \sin\alpha = W \sin\alpha + \mu W \cos\alpha$$

$$\therefore \qquad P = W \frac{\sin \alpha + \mu \cos \alpha}{\cos \alpha - \mu \cos \alpha}$$

But,
$$\mu = \tan \phi \quad \text{(By definition)}$$

$$\therefore \qquad P = W \frac{\sin \alpha + \tan \phi \cos \alpha}{\cos \alpha - \tan \phi \sin \alpha}$$

$$= W \frac{\sin \alpha \cos \phi + \sin \phi \cos \alpha}{\cos \alpha \cos \phi - \sin \alpha \sin \phi}$$

$$= W \frac{\sin (\alpha + \phi)}{\cos (\alpha + \phi)}$$

$$\therefore \qquad P = W \tan (\alpha + \phi) \quad \textbf{Proved.}$$

Example 2.6: A four wheel drive car as shown has a mass of 2000 kg with passengers. The roadway is inclined at an angle θ with the horizontal. If the coefficient of friction between tyers and road is 0.3, what is the maximum inclination θ that can be climbed?

Solution: Draw free body diagram as shown. The wheels are rotating clockwise to climb the roadway, the forces of friction on the front wheel and rear wheel f_F and f_R will be acting upwards to oppose the direction of motion of the wheel.

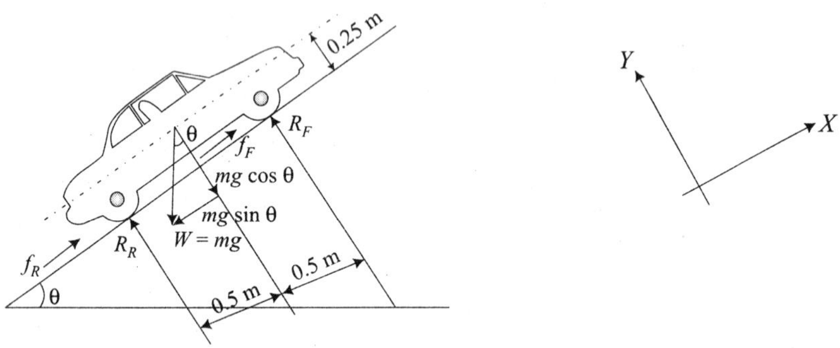

Apply equilibrium conditions,

$$\Sigma F_x = 0$$

$$\therefore \qquad f_R + f_F = mg \sin \theta$$

$$= 2000 + 9.81 \sin \theta$$

$$= 19620 \sin \theta \; (N)$$

$$\Sigma F_y = 0$$

$$R_R + R_F = mg \cos \theta$$

$$= 2000 \times 9.81 \cos \theta$$

$$= 19620 \cos \theta \ (N)$$

But, $$f_R = \mu R_R$$

and $$f_F = \mu R_F$$

$\therefore$ $$f_R + f_F = \mu (R_R + R_F)$$

$$19620 \sin \theta = \mu (R_R + R_F)$$

$\therefore$ $$R_F + R_R = \frac{19620 \sin \theta}{0.3} = 65400 \sin \theta$$

$$= 19620 \cos \theta$$

$\therefore$ $$\frac{\sin \theta}{\cos \theta} = \frac{19620}{65400} = 0.3$$

$\therefore$ $$\theta = \mathbf{16.7°}$$

2.4.3 Ladder Friction

A ladder is placed against a rough wall and a rough floor with coefficient of friction μ_2 and μ_1 respectively.

Example: A ladder, *AB* of length *L* and of weight *W* is placed against a rough wall and rough floor. The reactions R_A and R_B will be acting perpendicular to the point of supports at floor and wall. The ladder has tendency to slide down. Therefore, the forces of friction f_A and f_B will be acting to oppose the motion.

Draw free body diagram of ladder and apply conditions of equilibrium.

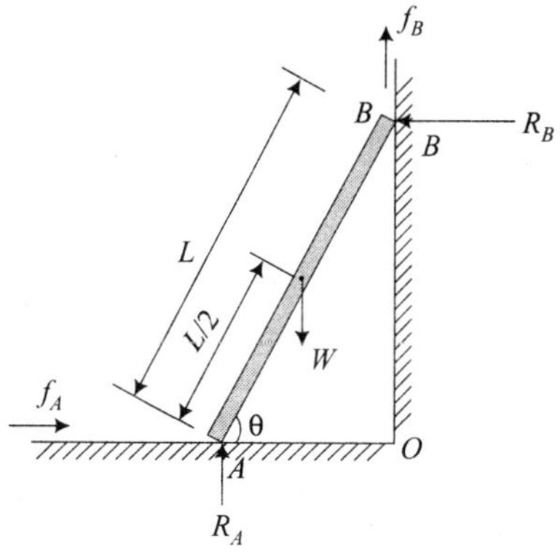

Fig. 2.14 Ladder friction

$$\Sigma F_x = 0 \quad \therefore f_A = R_B$$
$$\Sigma F_y = 0 \qquad W = R_A + f_B$$

The number of unknowns are four. Therefore, take moments about point O.

$$\Sigma M_O = 0$$

Example 2.7: A uniform ladder of weight 300 N rests against a smooth vertical wall a rough horizontal floor making an angle 60° with the horizontal. Find the force of friction at floor. [U.P.T.U. I/II Sem., 2001–02]

Solution: Draw free body diagram of ladder AB and apply condition of equilibrium.

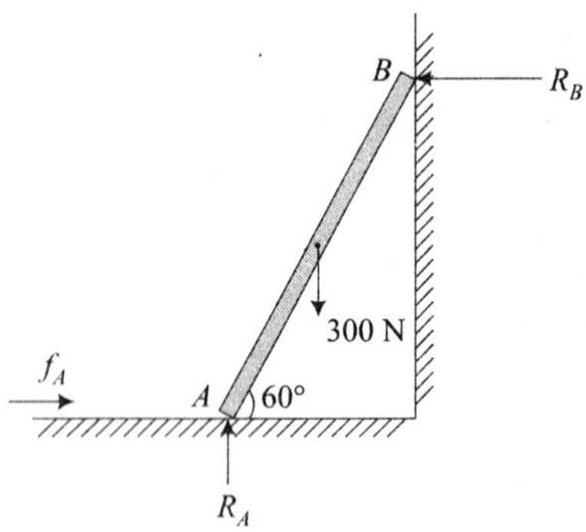

As the wall is smooth, $f_B = 0$.

$$\Sigma F_x = 0$$

$$\therefore \quad f_A = R_B$$
$$\mu R_A = R_B$$

$$\Sigma F_y = 0$$

$$R_A = W = 300 \text{ N}$$
$$\therefore \quad f_A = \mu R_A = 300\mu$$

Assume $\mu = 0.3$ between ladder and floor.

The force of friction,

$$f_A = 0.3 \times 300 = \mathbf{90\ N} \quad \textbf{Ans.}$$

Example 2.8: A uniform ladder of length 10 m and weighing 20 N is placed against a smooth vertical wall with its lower end 8 m from the wall. In this position the ladder is just to slip. Determine:

(i) Coefficient of friction between ladder and floor.

(ii) Frictional force acting on the ladder at the point of contact between the ladder and floor. [U.P.T.U. I Sem., 2004–05]

Solution: Draw free body diagram of ladder.

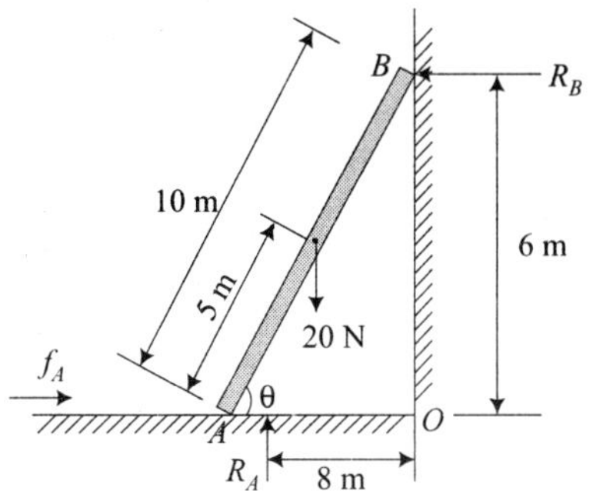

Angle of inclination,

$$\cos\theta = \frac{8}{10} = 0.8$$

$$\therefore \quad \theta = 36.87°$$

$$\sin\theta = 0.6$$

Applying equilibrium conditions,

$$\Sigma F_x = 0$$

$$\therefore \quad f_A = R_B$$

$$\Sigma F_y = 0$$

$$R_A = W = 20 \text{ N} \quad (\because f_B = 0)$$

$$\Sigma M_O = 0$$

$$R_A \times 8 = 20 \times 4 + R_B \times 6$$

$$R_B = \frac{R_A \times 8 - 20 \times 4}{6} = \frac{20 \times 8 - 20 \times 4}{8}$$

$$= 10 \text{ N}$$

Force of friction, $f_A = R_B = 10$ N

$$\mu = \frac{R_B}{N} = \frac{10}{20} = \mathbf{0.5} \quad \textbf{Ans.}$$

Example 2.9: A ladder of length '*l*' rests against a wall, the angle of inclination being 45°. If the coefficient of friction between the ladder and the ground and that between the ladder and the wall is 0.5 each, what will be the maximum distance on ladder to which a man whose weight is 1.5 times the weight of the ladder may ascend before the ladder begins to slide? [U.P.T.U. I Sem., 2005–06]

Solution: Draw free body diagram as shown and apply equilibrium conditions.

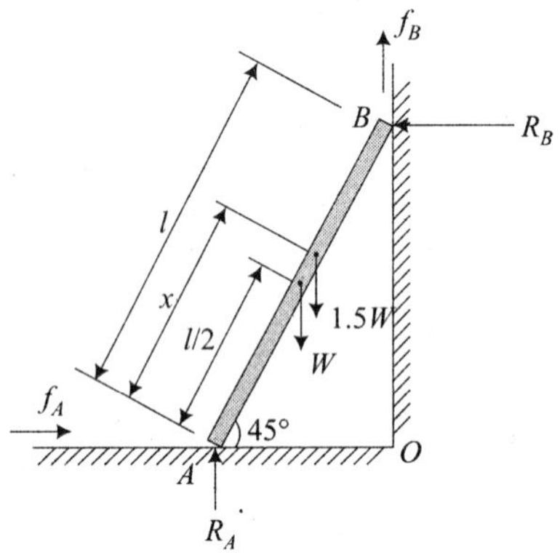

(i) $\Sigma F_x = 0$

$$f_A = R_B = \mu N = \mu R_A$$
$$= 0.5 R_A$$

$\therefore$ $$R_A = 2R_B$$

(ii) $\Sigma F_y = 0$

$$R_A + f_B = W + 1.5W = 2W$$
$$R_A = 2R_B$$
$$f_B = \mu R_B = 0.5\, R_B$$

$\therefore$ $$2R_B + 0.5\, R_B = 2W$$

$$R_B = \frac{2}{2.5}\, W = 0.8\, W$$

$\therefore$ $$f_B = 0.5 R_B = 0.4\, W$$

(iii) $\Sigma M_A = 0$

$$W \times \frac{l}{2} \cos 45° + 1.5\, W\, x \cos 45° = R_B \times l \sin 45° + f_B\, l \cos 45°$$
$$= 0.8\, W\, l \sin 45° + 0.4\, W\, l \cos 45°$$
$$0.353\; Wl + 1.06\; Wx = 0.5656\, Wl + 0.2828\, Wl$$
$$1.06x = (0.5656 + 0.2828 - 0.353)l = 0.49544\, l$$

$\therefore$ $$\frac{x}{l} = \frac{0.49544}{1.06} = 0.467$$

The man can ascend 46.7% of ladder length.

Example 2.10: A uniform ladder of weight 800 N and length 7 m rests on a horizontal ground and leans against a smooth vertical wall. The angle made by the ladder with the horizontal is 60°. When a man of weight 600 N stands on the ladder 4 m from the top of the ladder, the ladder is at the point of slipping. Determine the coefficient of friction between the ladder and the floor.

Solution: Draw free body diagram of the ladder and apply equilibrium conditions.

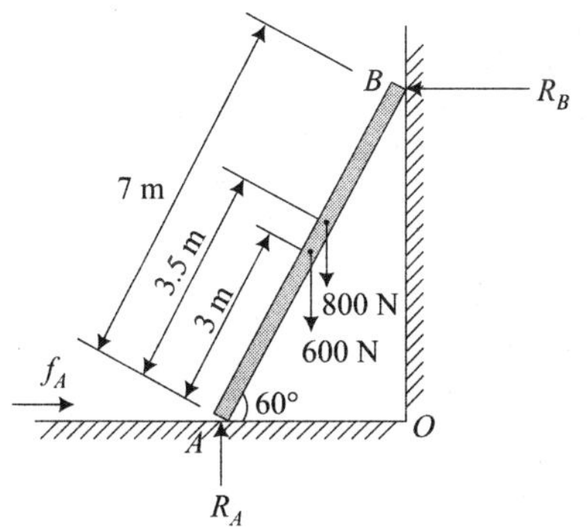

(i) $\Sigma F_x = 0$

$$f_A = R_B$$

(ii) $\Sigma F_y = 0$

$$R_A = 800 + 600 = 1400 \text{ N}$$

$\therefore$ $$f_A = R_B = \mu N = \mu R_A$$
$$= 1400\mu$$

(iii) $\Sigma M_B = 0$

$$R_A \times 7 \cos 60° = f_A \times 7 \sin 60° + 800 \times 3.5 \cos 60° + 600 \times 3 \cos 60°$$
$$1400 \times 7 \cos 60° = 1400\mu \times 7 \sin 60° + (800 \times 3.5 + 600 \times 3) \cos 60°$$

$\therefore$ $$\mu = \frac{1400 \times 7 \cos 60° - 4600 \cos 60°}{1400 \times 7 \sin 60°}$$

$$= \frac{4900 - 2300}{8487} = \mathbf{0.3} \quad \textbf{Ans.}$$

2.4.4 Wedge Friction

A wedge is used to lift loads vertically by applying horizontal force. These are made in triangular or trapezoidal shape. It works on the principle of inclined plane. A wedge of angle α is pushed under a block of weight W by applying a force P horizontally. The solution consists of finding the value P.

1. Draw free body diagram of block (Fig. 2.16). The frictional force f_1, between block and wall will act downwards to oppose the motion of block moving upwards. The frictional force f_2, between block and wedge will act downwards of inclined plane of block to oppose the motion. The reactions N_1 and N_2 will act normal to the supporting surface. R_1 and R_2 are the resultants of N_1, f_1 and N_2, f_2 respectively. ϕ is the angle of limiting friction.

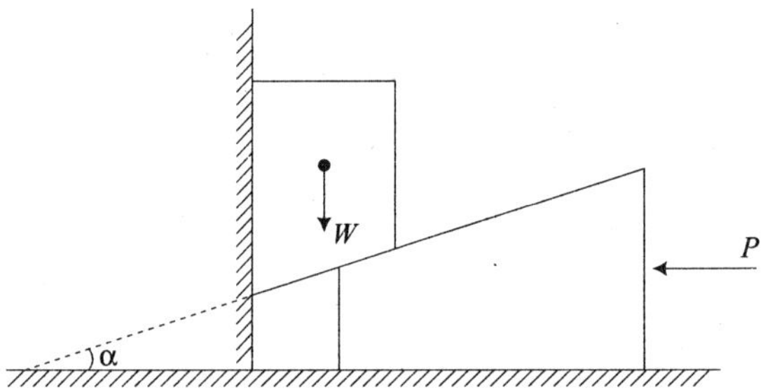

Fig. 2.15 Wedge friction

2. Apply equilibrium conditions.

(i) $\Sigma F_x = 0$

$$R_1 \cos \phi - R_2 \sin (\alpha + \phi) = 0 \quad ...(1)$$

(ii) $\Sigma F_y = 0$

$$R_1 \sin \phi + W - R_2 \cos (\alpha + \phi) = 0 \quad ...(2)$$

Calculate R_2 from equations (1) and (2).

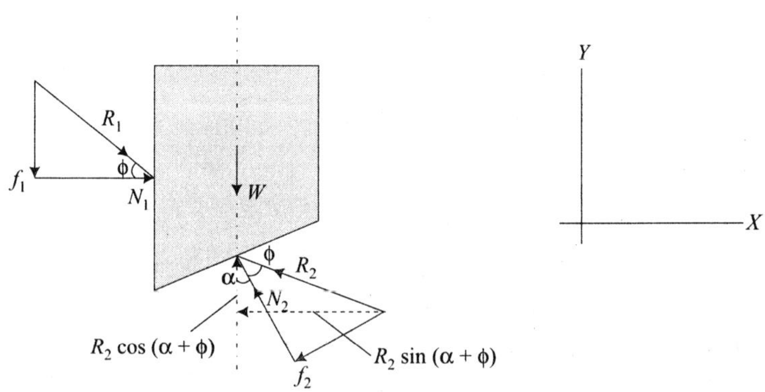

Fig. 2.16 FBD of block

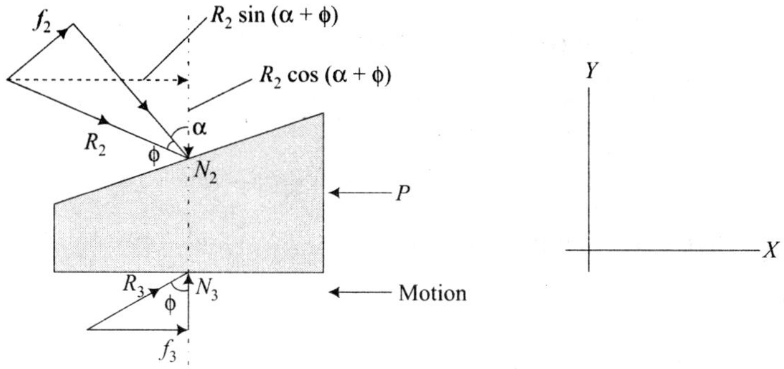

Fig. 2.17 FBD of wedge

3. Draw free body diagram of wedge (Fig. 2.17). The reactions $N_2' = -N_2$ and frictional force f_2 will be equal to frictional force acting on block but in opposite direction. Reactions N_2' and N_3 will be acting normal to supporting surfaces.
4. Apply equilibrium conditions to the wedge.

   (i) $\Sigma F_x = 0$

   $$R_2 \sin(\alpha + \phi) + R_3 \sin\phi - P = 0 \qquad \text{...(3)}$$

   (ii) $\Sigma F_y = 0$

   $$R_3 \cos\phi - R_2 \cos(\alpha + \phi) = 0 \qquad \text{...(4)}$$

   Solve equations (3) and (4) simultaneously to find P.

Example 2.11: A block weighing 1500 N, over laying a 10° wedge on a horizontal floor leaning against a vertical wall, is to be raised by applying a horizontal force to the wedge. Assuming the coefficient of friction between all surfaces in contact to be 0.3, determine the minimum horizontal force.

Solution:

1. Draw free body diagram for block (Fig. 2.16), and apply equilibrium conditions.

   (i) $\Sigma F_x = 0$

   $$R_1 \cos\phi - R_2 \sin(\alpha + \phi) = 0$$

   $$W = 1500 \text{ N}$$

   $$\alpha = 10°$$

   $$\mu = \tan\phi = 0.3$$

   $$\phi = 16° 42'$$

   $$\therefore \quad R_1 \cos 16°42' - R_2 \sin(10° + 16°42') = 0$$

   $$\therefore \quad R_2 = 2.132 R_1$$

(ii) $\Sigma F_y = 0$

$$-R_1 \sin \phi - W + R_2 \cos (\alpha + \phi) = 0$$

$$-R_1 \sin 16°42' - 1500 + 2.13\text{R}^1 \cos (10° + 16°42') = 0$$

$\therefore$ $$R_1 = 927.3 \text{ N}$$

$$R_2 = 2.132 \text{ R}_1 = 1997 \text{ N}$$

2. Draw free body diagram of wedge (Fig. 2.17) and apply equilibrium conditions.

   (i) $\Sigma F_x = 0$

$$R_2 \sin (\alpha + \phi) + R_3 \sin \phi - P = 0$$

   (ii) $\Sigma F_y = 0$

$$R_3 \cos \phi - R_2 \cos (\alpha + \phi) = 0$$

$$R_3 \cos 16°42' - 1977 \cos (10° + 16°42') = 0$$

$\therefore$ $$R_3 = 1844 \text{ N}$$

$\therefore$ $$P = R_2 \sin (\alpha + \phi) + R_3 \sin \phi$$

$$= 1977 \sin (10° + 16°42') + 1844 \sin 16°42'$$

$$= \mathbf{1418.3 \text{ N}} \quad \textbf{Ans.}$$

2.4.5 Belt Friction

Power is transmitted from one shaft to another by friction between the belt and the pulleys on which it passes. The driving pulley pulls the belt from one side and delivers the same to the other. The tension T_1 on tight side is more than T_2 on slack side.

If T_1 = Tension in belt on tight side, [N]
T_2 = Tension in belt on slack side, [N]
v = Velocity of the belt, [m/s]
r = Radius of the pulley, [m]

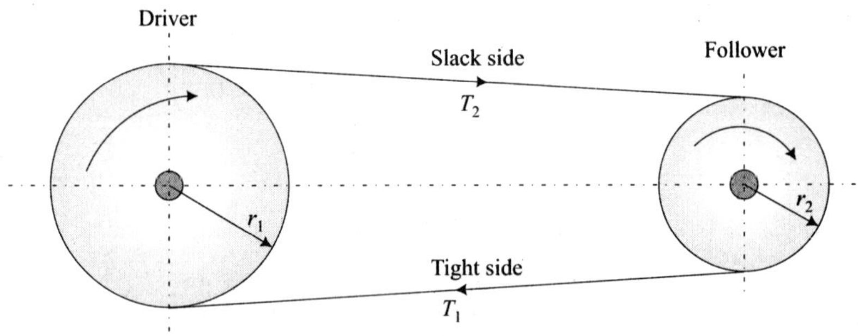

Fig. 2.18 Belt drive

Torque transmitted $= (T_1 - T_2)r$ [N – m]

Work done = Force × Distance $= (T_1 - T_2)v$ [Nm/s]

Power transmitted $= (T_1 - T_2)v$ [W]

Example 2.12: Find expression for the following due to belt friction is a pulley drive.

(i) Tension in the belt on tight and slack side.

(ii) Torque transmitted.

(iii) Reaction on the bearing of the pulley.

Solution: The belt and pulley system is shown in Fig. 2.19. The pulley is rotating in clockwise direction with angular velocity ω.

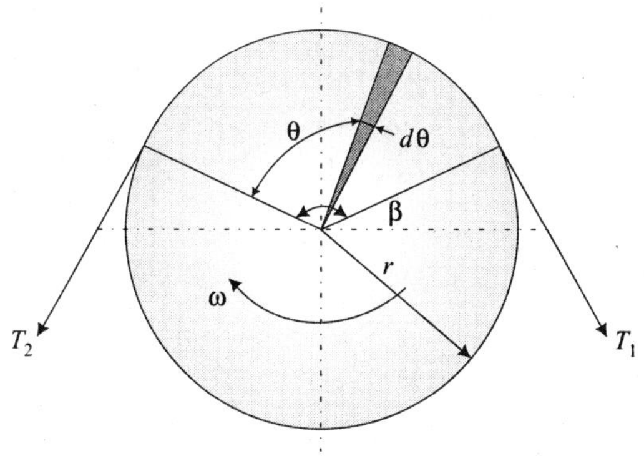

Fig. 2.19 Belt-pulley system

$T_1 > T_2$

r = Radius of pulley [m]

ω = Angular velocity [rad/s]

β = Angle of wrap of the belt over pulley.

Consider a small segment of belt of subtended angle $d\theta$ at an angle θ.

Draw free body diagram of segment of belt as shown in Fig. 2.20.

T = Tension in the belt segment on slack side

$T + dT$ = Tension in belt segment on tight side

dN = Normal reaction at the pulley rim.

$d\theta$ = Angle of embrace.

f = Frictional force

$= \mu\, dN$

μ = Coefficient of friction between belt and pulley.

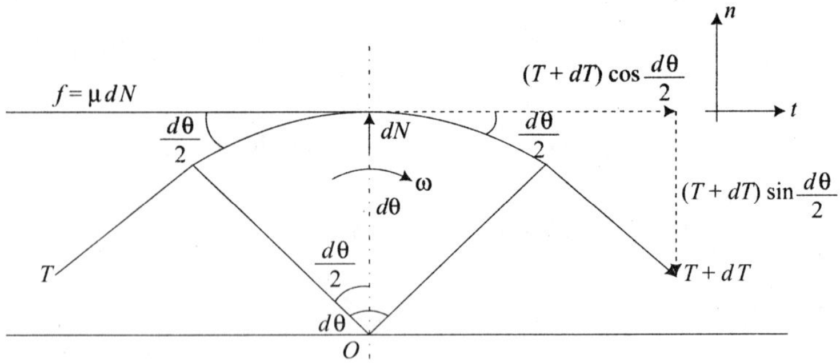

Fig. 2.20 Free body diagram

Apply equilibrium conditions,

(i) $\Sigma F_t = 0$

$$(T + dT)\cos\frac{d\theta}{2} - T\cos\frac{d\theta}{2} - \mu dN = 0$$

$$\therefore \qquad dT\cos\frac{d\theta}{2} = \mu dN \qquad ...(1)$$

(ii) $\Sigma F_n = 0$

$$dN - (T + dT)\sin\frac{d\theta}{2} - T\sin\frac{d\theta}{2} = 0$$

$$dN - 2T\sin\frac{d\theta}{2} - dT\sin\frac{d\theta}{2} = 0$$

If $d\theta$ is very small, $\sin\frac{d\theta}{2} = \frac{d\theta}{2}$

and $$\cos\frac{d\theta}{2} = \cos 0° = 1$$

$$\therefore \qquad dN - Td\theta - dT\frac{d\theta}{2} = 0 \qquad ...(2)$$

From equation (1)

$$dT = \mu dN$$

$$dN = \frac{dT}{\mu}$$

From equation (2),

$$dN = -\,Td\theta$$

$\left[\because\ dT.\dfrac{d\theta}{2} \approx 0\right.$, product of two small quantities may be neglected]

$$\therefore \qquad Td\theta = -\,dN = -\,\frac{dT}{\mu}$$

$$\therefore \qquad \frac{dT}{T} = -\,\mu d\theta$$

Integrating from $\theta = 0$ to $\theta = \beta$

$$\int_{T_2}^{T_1} \frac{dT}{T} = -\int_{\beta}^{0} \mu d\theta = \int_{0}^{\beta} \mu d\theta$$

$$\therefore \qquad \ln\frac{T_1}{T_2} = \mu\beta$$

$$\therefore \qquad \frac{T_1}{T_2} = e^{\mu\beta}$$

Torque transmitted = $\boldsymbol{(T_1 - T_2)r}$ **[N-m]** **Ans.**

Reaction on pulley bearing = $\boldsymbol{(T_1 + T_2)}$ **[N]** **Ans.**

Example 2.13: A flat belt is used to transmit torque from pulley *A* to pulley *B*. The radius of each pulley is 50 cm and coefficient of friction is 0.3. Determine the largest torque that can be transmitted if allowable belt tension is 3 kN.

Solution: The maximum allowable tension will be T_1

$$\therefore \qquad T_1 = 3 \text{ kN}$$

$$\mu = 0.3$$

$\beta = 180° = \pi$ rad, as the pulley diameters are equal.

$$r = 50 \text{ cm} = 0.5 \text{ m}$$

The belt tension ratio,

$$\frac{T_1}{T_2} = e^{\mu\beta}$$

$$\ln\frac{T_1}{T_2} = \mu\beta = 0.3\pi$$

$$\therefore \qquad \frac{T_1}{T_2} = 2.566$$

$$\therefore \qquad T_2 = \frac{T_1}{2.566} = \frac{3}{2.566} = 1.169 \text{ N}$$

$$\text{Torque} = (T_1 - T_2)r$$
$$= (3 - 1.169)0.5 = \mathbf{0.9155\ N\text{-}m} \quad \textbf{Ans.}$$

Example 2.14: A belt drives a pulley of 200 mm diameter such that the ratio of tension in tight side and slack side is 1.2. If the maximum tension in the belt is not to exceed 240 kN. Find the safe power transmitted by the pulley at 60 rpm.

Solution: The maximum tension in the belt, T_1= 240 kN.

$$\frac{T_1}{T_2} = 1.2$$

$$\therefore \qquad T_2 = \frac{240}{1.2} = 200 \text{ kN}$$

Speed of belt, $$V = \frac{\pi d N}{60} = \pi \times \left(\frac{200}{1000}\right) \times \frac{60}{60} = 0.628 \text{ m/s}$$

$\therefore$ Power transmitted, $$P = (T_1 - T_2)V = (240 - 200)0.628$$
$$= \mathbf{25\ kW} \quad \textbf{Ans.}$$

Example 2.15: A belt 100 mm wide and 8.0 mm thick is transmitting power at a belt speed of 160 m/min. The angle of lap of smaller pulley is 165° and coefficient of friction is 0.3. The maximum permissible stress in the belt is 2 MN/m^2 and mass of belt is 0.9 kg/m. Find the power transmitted and the initial tension in the belt.

Solution: The data given is:

$$b = 100 \text{ mm} = 0.1 \text{ m}$$
$$t = 8 \text{ mm} = 0.008 \text{ m}$$
$$V = 160 \text{ m/min} = \frac{160}{60} = 2.667 \text{ m/sec}$$
$$\beta = 165° \times \frac{\pi}{180} = 2.88 \text{ rad}$$
$$\mu = 0.3$$
$$m = 0.9 \text{ kg/m}$$

$$\sigma = 2 \text{ MN/m}^2 = 2 \text{ N/mm}^2$$

$$T_{max} = \sigma bt = 2 \times 100 \times 8 = 1600 \text{ N}$$

Centrifugal tension, $T_C = mv^2 = 0.9\,(2.667)^2 = 6.4$ N

$$T_1 = T_{max} - T_C = 1600 - 6.4$$

$$= 1593.6 \text{ N}$$

$$\frac{T_1}{T_2} = e^{\mu\beta} = e^{0.3 \times 2.88} = 2.3726$$

$$\therefore \quad T_2 = \frac{1593.6}{2.3726} = 671.66 \text{ N}$$

Power transmitted,

$$P = (T_1 - T_2)V = (1593.6 - 671.66) \times 2.667 = 2.46 \text{ kW}$$

Initial tension,

$$T_o = \frac{T_1 + T_2}{2} + T_C = \frac{1593.6 + 671.66}{2} + 6.4 = \mathbf{1139\ N} \quad \textbf{Ans.}$$

Example 2.16: A belt is stretched over two identical pulleys of diameter D metre. The initial tension in the belt throughout is 2.4 kN, when the pulleys are at rest. In using these pulleys and belt to transmit torque, it is found that the increase in tension on one side is equal to the decrease on the other side.

Find the maximum torque that can be transmitted by the belt drive, given that the coefficient of friction between belt and pulley is 0.30.

[U.P.T.U. I Sem., 2002–03]

Solution:

$\beta = \pi$ as the pulleys are identical

$\mu = 0.30$

$$\frac{T_1}{T_2} = e^{\mu\beta} = e^{0.3\pi} = 2.566$$

$$T_o = \text{Initial tension}$$

$$= \frac{T_1 + T_2}{2} \qquad [\because T_1 - T_0 = T_0 - T_2]$$

$$= 2.4 \text{ kN}$$

$$T_1 + T_2 = 4.8$$

$$\therefore \quad T_1 + \frac{T_1}{2.566} = 4.8$$

$$T_1 = 3.454 \text{ kN}$$
$$T_2 = 4.8 - 3.454 = 1.346 \text{ kN}$$

$$\text{Max. torque} = (T_1 - T_2)\frac{D}{2} = (3.454 - 1.346)\frac{D}{2} = \mathbf{1.054}\boldsymbol{D}\ \mathbf{kN.m} \quad \textbf{Ans.}$$

Example 2.17: A horizontal drum of a belt drive carries the belt over a semi-circle around it. It is rotated anticlockwise to transmit a torque of 300 N-m. If coefficient of friction between the belt and drum is 0.3, calculate the tension in the limbs 1 and 2 of the belt shown, and the reaction on the bearings. The drum has a mass of 20 kg and the belt is assumed to be massless.

[U.P.T.U. I Sem., 2001–02]

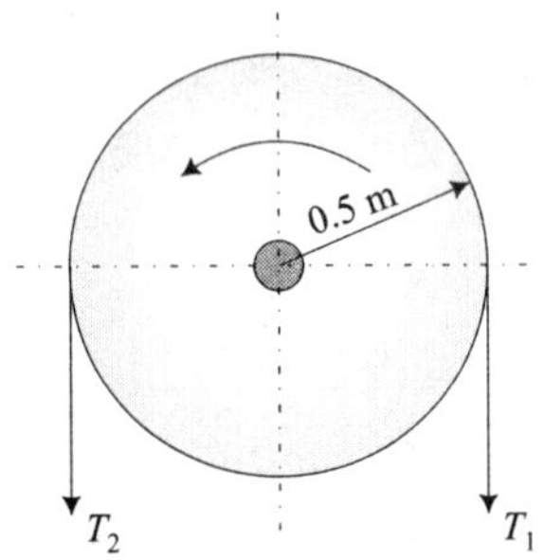

Solution:

$$T_2 > T_1 \text{ given}$$
$$\beta = \pi$$
$$\mu = 0.3$$

$$\therefore \quad \frac{T_2}{T_1} = e^{\mu\beta} = e^{0.3\pi} = 2.566$$

$$T_2 = 2.566\ T_1$$

$$\text{Torque} = (T_2 - T_1)r$$

$$300 = (2.566\ T_1 - T_1) \times 0.5$$

$$T_1 = 383 \text{ N}$$

$$T_2 = 2.566\ T_1 = 983 \text{ N}$$

$$\text{Reaction on the bearing} = T_1 + T_2 + mg = 383 + 983 + 20 \times 9.81$$

$$= \mathbf{1562.2\ N} \quad \textbf{Ans.}$$

Example 2.17(a): Find the power transmitted by a belt running over a pulley of 600 mm diameter at 200 rpm. The coefficient of friction between the belt and pulley is 0.25, angle of lap 160° and maximum tension in the belt is 2.5 kW.

[U.P.T.U. 2009-10]

Solution:

The power transmitted,

$$P = \frac{2\,\pi\, NT}{60 \times 1000}\ [\text{kW}].$$

where $N = 200$ rpm.

$$T = (T_1 - T_2)R$$

$$T_1 = 2.5 \text{ kN}$$

$$\frac{T_1}{T_2} = e^{\mu\theta} = e^{0.25} \times \frac{160}{180}\pi = 2.01$$

$$T_2 = \frac{T_1}{2.01} = \frac{2.5}{2.01} = 1.244 \text{ kN.}$$

$$\therefore \quad P = \frac{2\,\pi \times 200 \times (2.5 - 1.244)0.3}{60} = \mathbf{7.89\ kW.}\quad \textbf{Ans.}$$

Example 2.17(b): Rod *AB* of weight 500 N is supported by a cable wrapped around a semicircle having coefficient of friction of 0.2 A weight *c* weighing 100 N can slide without friction on rod *AB*. What is the maximum range *x* from centre line the mass *c* can be placed without causing slippage?

[U.P.T.U. 2010-11]

Solution:

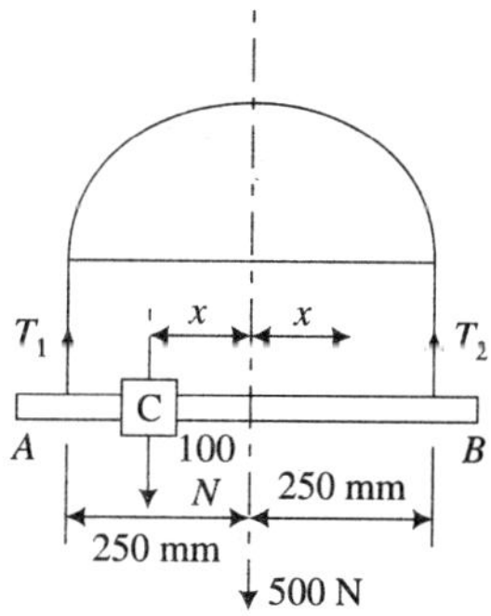

$$T_1 + T_2 = 500 + 100 = 600 \text{ N.}$$

$$\frac{T_1}{T_2} = e^{\mu\theta} = e^{0.2\pi} = 1.874$$

$$T_1 = 1.874\, T_2$$

$$\therefore \quad 1.874\, T_2 + T_2 = 600$$

$$T_2 = 208.768 \text{ N}$$

$$T_1 = 391.2 \text{ N}$$

For equilibrium of the system

$$(T_1 - T_2)r = 100(2x)$$

$$(391.2 - 208.768)250 = 200\,x$$

$$\therefore \quad x = \frac{(391.2 - 208.768) \times 250}{200} = \mathbf{228\ mm.}\quad \textbf{Ans.}$$

Example 2.18: A 100 N weight is on the verge of sliding downwards. It is prevented by a weight *W* as shown. The coefficient of friction between the rope and the fixed pulley, and between other surfaces of contact $\mu = 0.3$. Determine the minimum weight *W* to prevent the downward motion of 100 N body.

[U.P.T.U. I Sem., 2001–02]

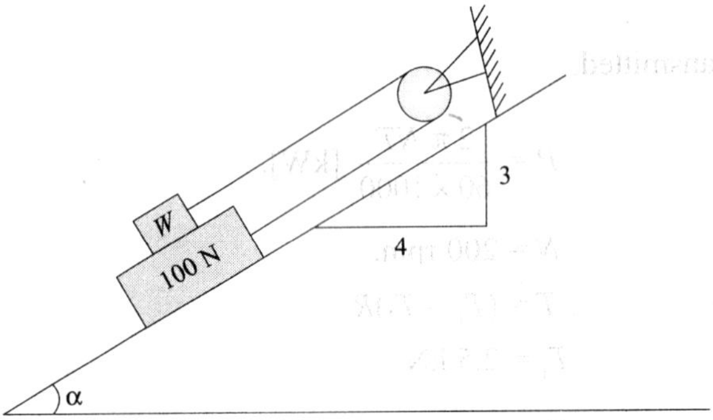

Solution:

$$\tan\alpha = \frac{3}{4}$$

$$\alpha = 36.87°$$

$$\cos\alpha = 0.8$$

$$\sin\alpha = 0.6$$

1. Draw free body diagram of weight W and apply conditions of equilibrium.

(i) $\Sigma F_y = 0$

$$N_1 = W\cos\alpha = 0.8W$$

$$f_1 = \mu N_1 = 0.3 \times 0.8W$$

$$= 0.24W$$

(ii) $\Sigma F_x = 0$

$$T_1 = f_1 + W\sin\alpha$$

$$= 0.24W + 0.6W = 0.84W$$

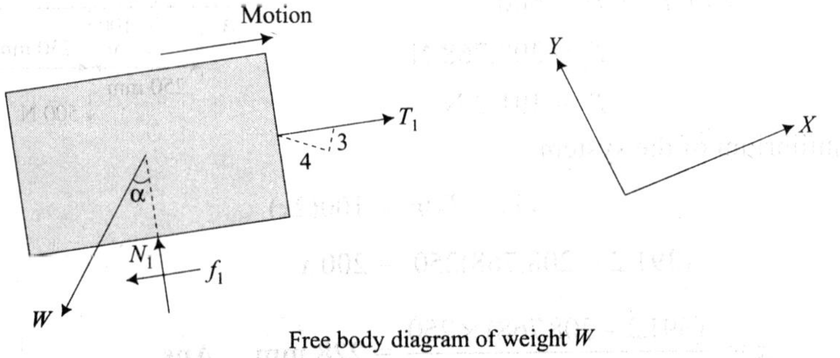

Free body diagram of weight W

2. Draw free body diagram of pulley and apply conditions of equilibrium. The pulley will rotate clockwise as weight 100 N is on the verge of sliding downwards.

$$\frac{T_2}{T_1} = e^{\mu\beta}$$

$$\therefore \quad T_2 = T_1\, e^{0.3\pi} = 1.874\, T_2$$

$$= 1.874\,(0.84\, W)$$

$$= 1.575\, W$$

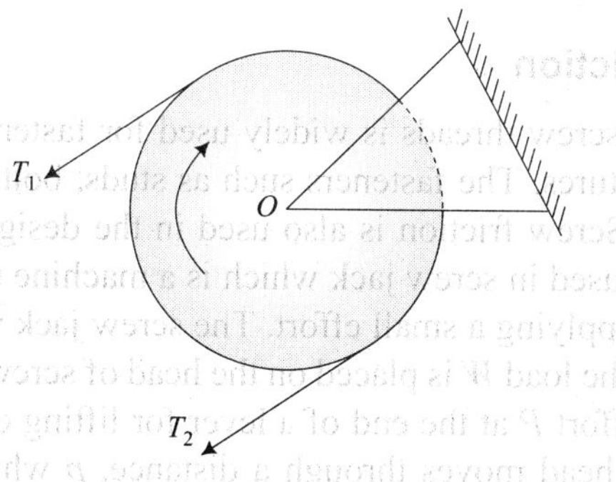

FBD of pulley

3. Draw free body diagram of body of 100 N and apply conditions of equilibrium.

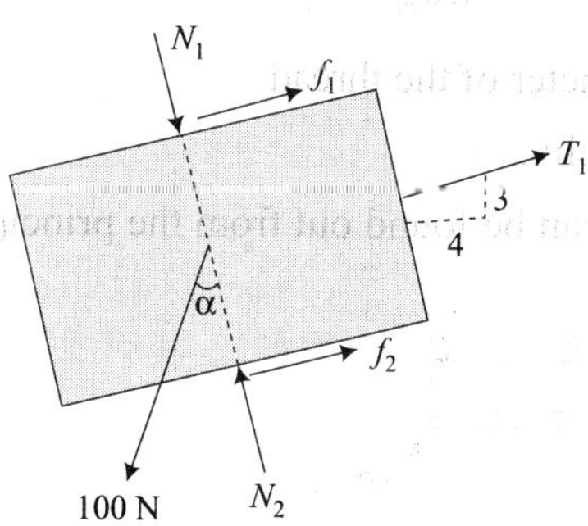

FBD of body of 100 N

The normal reactions and directions of force of friction f_1 or f_2 are shown to oppose the motion.

(i) $\Sigma F_y = 0$

$$N_2 - N_1 - 100 \cos \alpha = 0$$

$$\therefore \qquad N_2 = N_1 + 100 \cos \alpha = 0.8W + 80$$

$$f_2 = \mu N_2 = 0.3(0.8W + 80)$$

$$= 0.24W + 24$$

(ii) $\Sigma F_x = 0$

$$f_1 + f_2 - 100 \sin \alpha + T_2 = 0$$

$$0.24W + (0.24W + 24) - 100 \times 0.6 + 1.575W = 0$$

$$2.55W = 36$$

$$\therefore \qquad W = \mathbf{17.5\ N} \quad \textbf{Ans.}$$

2.4.6 Screw Friction

The friction in the screw threads is widely used for fastening of components in machines and structures. The fasteners such as studs, bolts, nuts and screws are very widely used. Screw friction is also used in the design of lifting machines. Square threads are used in screw jack which is a machine used for raising/lifting of heavy loads by applying a small effort. The screw jack works on the principle of inclined plane. The load W is placed on the head of screw jack which is rotated by application of effort P at the end of a lever for lifting or lowering of load. In one revolution the head moves through a distance, p which is the pitch of the screw. In case of multi-start screw, the distance moved is np where n is the number of starts. The inclination of the thread, α can be found out as:

$$\tan \alpha = \frac{p}{\pi d_m}$$

where, d_m = mean diameter of the thread

α = angle of helix.

The effort required can be found out from the principle of inclined plane.

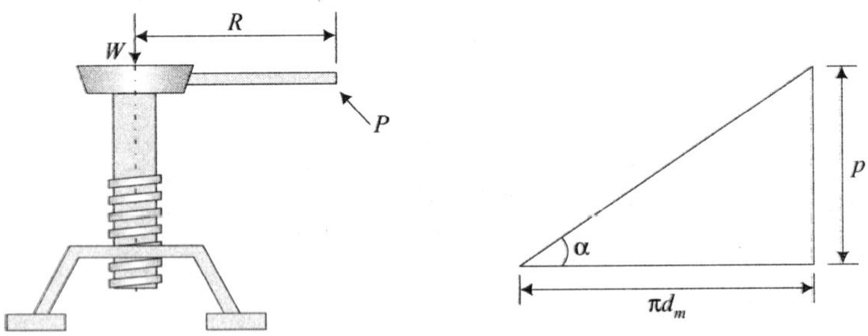

Fig. 2.21 Screw jack

The effort required to lift the load,

$$P = W \tan(\alpha + \phi)$$

where, ϕ = angle of friction.

The effort required to lever the load,

$$P = W \tan(\alpha - \phi) \qquad \text{(when } \alpha > \phi\text{)}$$

$$P = W \tan(\phi - \alpha) \qquad \text{(where } \alpha < \phi\text{)}$$

For ideal conditions, when there is no friction in the screw and nut,

$$P_o = W \tan \alpha$$

where, P_o = ideal effort.

Example 2.19: Show that efficiency of a screw jack for raising a load W is given by,

$$\eta = \frac{\tan \alpha}{\tan(\alpha + \phi)}$$

where, α = angle of screw

ϕ = angle of friction

What is the condition for maximum efficiency?

Solution: The efficiency of a screw jack is defined as the ratio of ideal effort and actual effort.

The effort required to raise the load,

$$P = W \tan(\alpha + \phi)$$

The ideal effort required when there is no screw friction,

$$P_o = W \tan \alpha \qquad [\because \phi = 0]$$

$$\therefore \qquad \eta = \frac{\text{Ideal effort}}{\text{Actual effort}} = \frac{P_o}{P} = \frac{W \tan \alpha}{W \tan(\alpha + \phi)}$$

$$= \frac{\tan \alpha}{\tan(\alpha + \phi)}$$

$$1 - \eta = 1 - \frac{\tan \alpha}{\tan(\alpha + \phi)} = 1 - \frac{\sin \alpha \cos(\alpha + \phi)}{\cos \alpha \sin(\alpha + \phi)}$$

$$= \frac{\cos \alpha \sin(\alpha + \phi) - \sin \alpha \cos(\alpha + \phi)}{\cos \alpha \sin(\alpha + \phi)}$$

Let, $\alpha + \phi = A$ and $\alpha = \beta$

$$\therefore \qquad \sin(A - B) = \sin A \cos B - \cos A \sin B$$

$$1-\eta = \frac{\sin(\alpha+\phi-\alpha)}{\cos\alpha\sin(\alpha+\phi)} = \frac{\sin\phi}{\cos\alpha\sin(\alpha+\phi)}$$

$$= \frac{2\sin\phi}{2\cos\alpha\sin(\alpha+\phi)}$$

But, $\quad 2\cos A\sin B = \sin(A+B)+\sin(A-B)$

$$\therefore \quad 1-\eta = \frac{2\sin\phi}{\sin(2\alpha+\phi)+\sin\phi}$$

For efficiency to be maximum, $1-\eta$ should be minimum.

$\therefore$ $2\sin(2\alpha+\phi)$ should be maximum.

$$\therefore \quad (2\alpha+\phi) = 90°$$

$$2\alpha = 90° - \phi$$

$$\alpha = \left(45° - \frac{\phi}{2}\right) \quad \textbf{Ans.}$$

This is the condition for maximum efficiency of screw jack.

Example 2.20: Show that maximum efficiency of a screw jack is given by

$$\eta_{max} = \frac{1-\sin\phi}{1+\sin\phi} \text{ at } \alpha = \left(45° - \frac{\phi}{2}\right)$$

where, α is the helix angle and ϕ is the angle of friction.

Solution:

$$\eta_{max} = \frac{P_o}{P} = \frac{\tan\alpha}{\tan(\alpha+\phi)}$$

For maximum efficiency,

$$\alpha = 45° - \frac{\phi}{2}$$

$$\therefore \quad \eta_{max} = \frac{\tan\left(45° - \frac{\phi}{2}\right)}{\tan\left(45° - \frac{\phi}{2} + \frac{\phi}{2}\right)} = \frac{\tan\left(45° - \frac{\phi}{2}\right)}{\tan\left(45° + \frac{\phi}{2}\right)}$$

$$= \frac{\dfrac{\tan 45° - \tan\phi/2}{1+\tan 45°\tan\phi/2}}{\dfrac{\tan 45° + \tan\phi/2}{1-\tan 45°\tan\phi/2}} = \frac{\dfrac{1-\tan\phi/2}{1+\tan\phi/2}}{\dfrac{1+\tan\phi/2}{1-\tan\phi/2}}$$

$$\eta_{max} = \frac{(1 - \tan \phi/2)^2}{(1 + \tan \phi/2)^2} = \frac{\left[1 - \dfrac{\sin \phi/2}{\cos \phi/2}\right]^2}{\left(1 + \dfrac{\sin \phi/2}{\cos \phi/2}\right)^2}$$

$$\therefore \qquad \eta_{max} = \frac{\left(\cos\dfrac{\phi}{2} - \sin\dfrac{\phi}{2}\right)^2}{\left(\cos\dfrac{\phi}{2} + \sin\dfrac{\phi}{2}\right)^2}$$

$$= \frac{\cos^2\dfrac{\phi}{2} + \sin^2\dfrac{\phi}{2} - 2\cos\dfrac{\phi}{2}\sin\dfrac{\phi}{2}}{\cos^2\dfrac{\phi}{2} + \sin^2\dfrac{\phi}{2} + 2\cos\dfrac{\phi}{2}\sin\dfrac{\phi}{2}}$$

$$\frac{1 - 2\cos\dfrac{\phi}{2}\sin\dfrac{\phi}{2}}{1 + 2\cos\dfrac{\phi}{2}\sin\dfrac{\phi}{2}} = \frac{1 - \sin\phi}{1 + \sin\phi}$$

Hence, shown.

Example 2.21: A screw thread of a screw jack has a mean diameter of 10 cm and a pitch of 1.25 cm. The coefficient of friction between the screw and its nut housing is 0.25 cm. Determine the force F that must be applied at the end of a 50 cm lever arm to raise a mass of 5000 kg. Is the device self-locking? Also determine its efficiency. [U.P.T.U. I Sem., 2001–02]

Solution:

Load, $W = 5000 \times 9.81 = 49050$ N

Pitch, $p = 12.5$ mm

Lever arm, $R = 50 \text{ cm} = 0.5$ m

Mean screw radius, $r = \dfrac{10 \text{ cm}}{2} = 50$ mm

$\mu = 0.25$

$\phi = \tan^{-1} \mu = \tan^{-1} 0.25 = 14°$

$$\tan \alpha = \frac{p}{2\pi r} = \frac{12.5}{2\pi \times 50} = 0.0397887$$

$$\therefore \qquad \alpha = 2.2785°$$

Effort required to raise the load,

$$p = W \tan(\alpha + \phi)$$
$$= 49050 \tan(2.2785° + 14°)$$
$$= 14323.25 \text{ N}$$

Torque,
$$T = Pr = 14323.25 \times 50 \times 10^{-3} \text{ N-m}$$
$$= 716.16 \text{ N-m}$$

Force required at the end of the lever,

$$F = \frac{T}{R} = \frac{716.16}{0.5} = 1432.32 \text{ N}$$

Efficiency,
$$\eta = \frac{\tan\alpha}{\tan(\alpha+\phi)} = \frac{0.0397887}{0.2920} = 0.136$$

$$\therefore \quad \eta = \mathbf{13.625\%} \quad \textbf{Ans.}$$

Because η is less than 50%, it is self-locking device.

Example 2.22: A screw jack has square threads of mean diameter of 10 cm and a pitch of 1.25 cm. Determine the force required at the end of 50 cm lever.

(i) to raise the load

(ii) to lower the load of 50 kN.

Find the efficiency of the jack. Is it self-locking?

Assume $\mu = 0.20$.

Solution:

$$\mu = 0.20$$
$$\therefore \quad \phi = \tan^{-1}\mu = \tan^{-1} 0.20 = 11.31°$$

$$\tan\alpha = \frac{p}{2\pi r} = \frac{12.5}{2\pi \times \frac{100}{2}} = 0.0397887$$

$$\therefore \quad \alpha = 2.279°$$

(i) *Torque required to raise the load*

Effort,
$$P = W \tan(\alpha + \phi)$$
$$- 50 \tan(2.279° + 11.31°) - 12.1 \text{ kN}$$

Torque,
$$T = P \times r = 12.1 \times \frac{50}{1000} = 0.604 \text{ kN-m}$$

Force to be applied at the end of the lever,

$$F \times R = T$$

$$\therefore \qquad F = \frac{T}{R} = \frac{0.604}{0.50} = 1.208 \text{ kN}$$

(ii) *Torque required to lower the load*

Effort, $P = W \tan(\phi - \alpha)$

$= 50 \tan(11.31 - 2.279)$

$= 7.947$ kN

Torque required, $T = P \times r = 7.947 \dfrac{50}{1000} = 0.397$ kN

Force to be applied at the end of the lever,

$$\therefore \qquad F = \frac{T}{R} = \frac{0.0397}{0.50} = 1.7947 \text{ kN}$$

(iii) Efficiency, $\eta = \dfrac{\tan\phi}{\tan(\alpha + \phi)} = \dfrac{\tan 2.279°}{\tan(11.31° + 2.279°)} = 0.1646$

$= \mathbf{16.46\%}$ **Ans.**

The efficiency is less than 5%. Therefore, it is self-locking.

QUESTION BANK NO. 2

1. What is friction? Give some useful applications of friction.
2. Explain laws of static and dynamic friction.
3. What are the characteristics of friction force? Explain the concept of equilibrium of bodies involving dry friction.
4. Write notes on:
   (i) Cone of friction
   (ii) Coefficient of friction
   (iii) Angle of friction
   (iv) Angle of repose
   (v) Limiting friction

5(a). Find expressions for the following due to belt friction in a pulley drive:
   (i) Tension in the belt on tight and slack side
   (ii) Torque transmitted
   (iii) Reaction on the bearing of pulley.

5(b). Derive an expression for the ratio of belt tensions in a flat belt drive.

6. Draw free-body diagram of the following systems:
   (i) A block pulled along a rough horizontal plane.
   (ii) A body sliding down a rough inclined plane.
   (iii) A body being pushed up a rough inclined plane.
   (iv) A wedge lifting a load.
   (v) A ladder resting on a rough floor and a wall.
7. Explain briefly different types of friction.

QUICK REVIEW INFORMATION NO. 2

1. If a weight 100 N is resting on a rough horizontal table, the friction force acting on it is zero.
2. When a body slides down an inclined plane (making an angle θ with the horizontal) the acceleration of the body is $g \sin \theta$.

UNIT II:
BEAMS AND TRUSSES

CHAPTER 3

Beams

3.1 INTRODUCTION

A beam is a structural member to support external loads. Normally a beam support transverse loads acting at right angle to its axis. The length of the beam is usually large as compared to its cross-section.

The load applied on the beam gets transferred to its supports. Any section of the beam experiences shearing off by the load and bending by a moment.

The beam supports different types of loads such as point loads, uniformly distributed loads, uniformly varying loads and the combination of the loads. The beams can be supported on roller supports, hinged supports or fixed supports.

3.1.1 Types of Loads

The various types of loads can be classified as:

(a) Concentrated loads

A concentrated load or a point load is considered to act at a point P_1, P_2 and P_3 are concentrated loads.

(b) Uniformly distributed loads (UDL)

A uniformly distributed load is considered distributed or spread over the length of the beam. In UDL, the intensity of load w is uniform over the length and is expressed as N/m or kN/m.

(c) Uniformly varying loads

If the spread or intensity of the load is uniformly varying from one end to another, it is called uniformly varying load. The triangular or trapezoidal distributed load is considered in this category. In trapezoidal distribution, intensity of load may vary linearly from w_1/m from left to w_2/m on right hand.

(d) Externally applied moments

External moments may also be applied to the beam. *A* moment M is applied to the beam at point C.

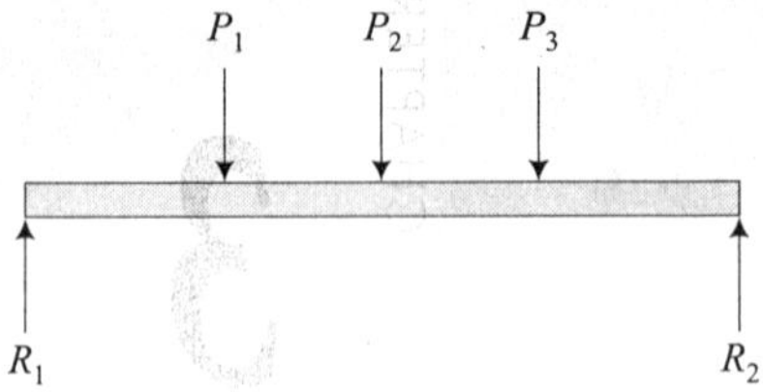

Fig. 3.1 Concentrated load

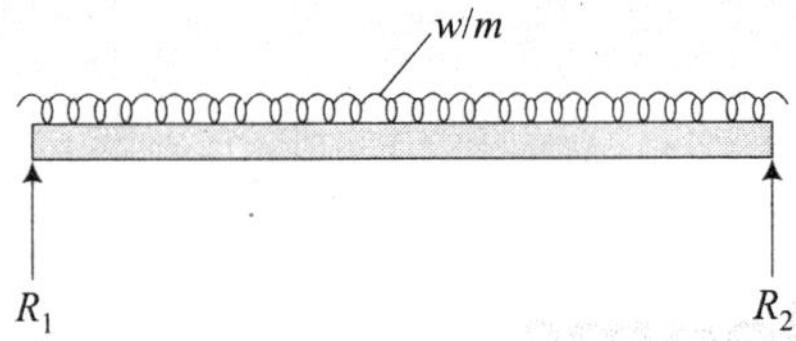

Fig. 3.2 Uniformly distributed load

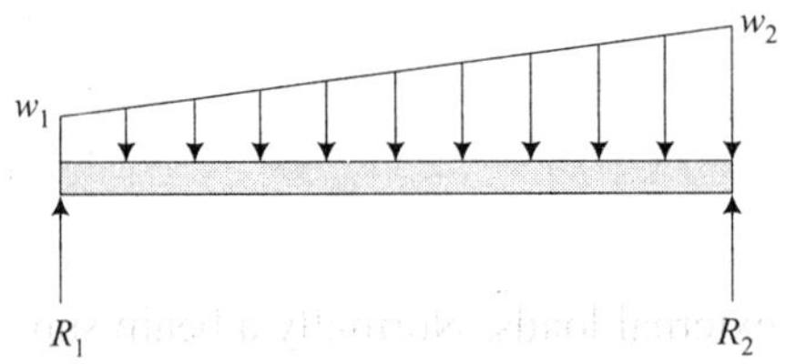

Fig. 3.3 Uniformly varying load

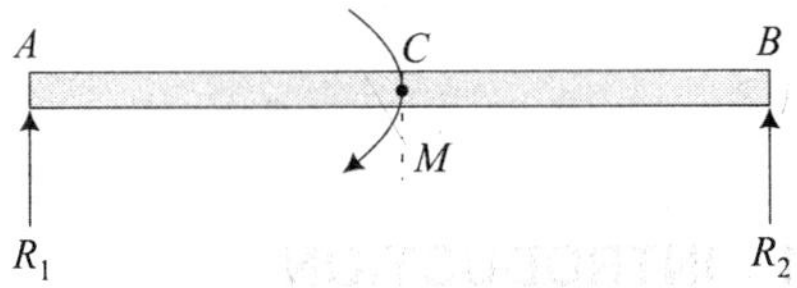

Fig. 3.4 External moment

3.1.2 Types of Supports

The beams are supported on columns or walls by the following methods:

(a) Roller support or free support

The beam is supported on a ball support to allow for thermal expansion or thermal contraction and thus allow unhindered stretching or shrinking of the beam in axial direction. The reaction R at the roller support of the beam will act perpendicular to the roller base.

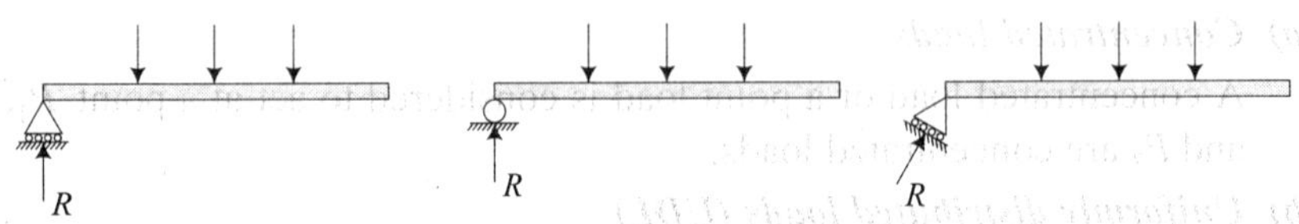

Fig. 3.5 Roller supports

(b) Pin support or hinged support

The beams with pin support are also called simply supported beams. The support does not allow vertical deflection or movement but rotation of beam to a slope angle is possible. The direction of reaction, R will depend upon the direction of loads on the beams.

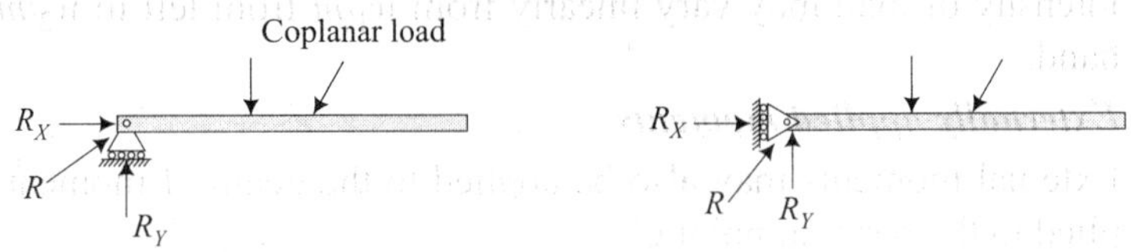

Fig. 3.6 Pin support

(c) Fixed support

The support does not allow the beam either to deflect vertically or to rotate. In addition to reaction R_2, there is a fixing moment M_2 also.

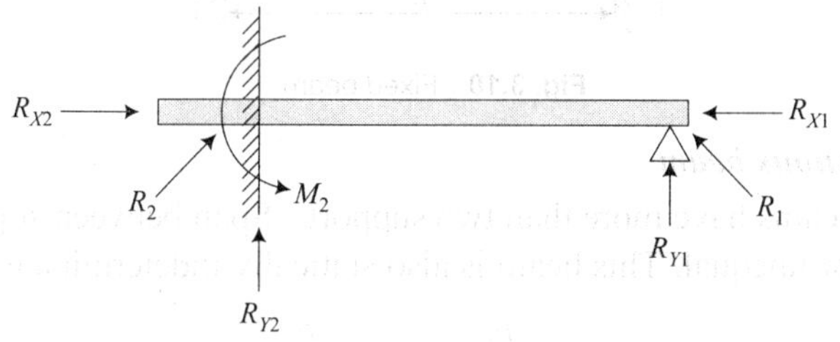

Fig. 3.7 Fixed support

3.2 TYPES OF BEAMS

The beams are classified as per type of support. Various types of beams are:

(a) Cantilever beam

It is fixed at one end and free at the other. In addition to reaction R, there is a fixing moment at the support.

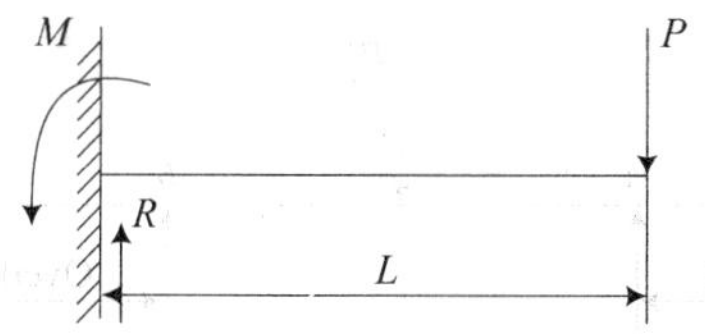

Fig. 3.8 Cantilever beam

(b) Simply supported beam

The ends of a simply supported beam are supported on a wall or column. The reactions at the support are vertical reactions.

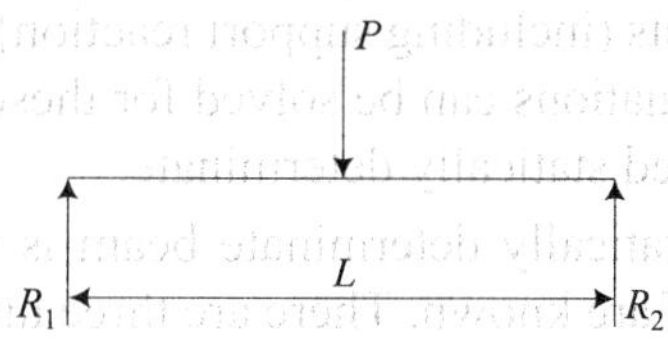

Fig. 3.9 Simply supported beam

(c) Fixed beam

It has fixed supports at both ends. It is also called *built-in* beam or *encastre* beam. The ends are fixed into walls or columns. In addition to support reactions R_1 and R_2, there are fixing moments M_1 and M_2. This beam is statically indeterminate beam.

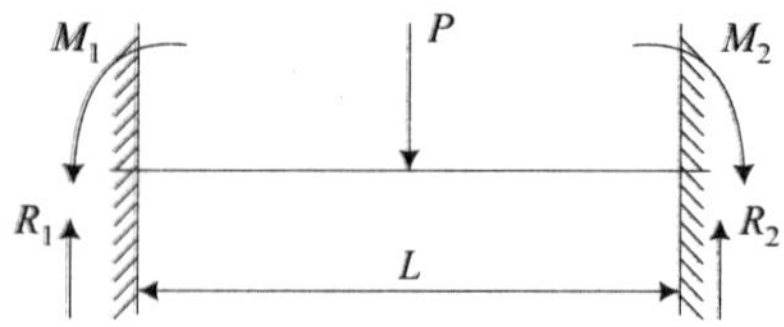

Fig. 3.10 Fixed beam

(d) Continuous beam

Such beams have more than two supports. Span between supports may be equal or unequal. This beam is also statically indeterminate beam.

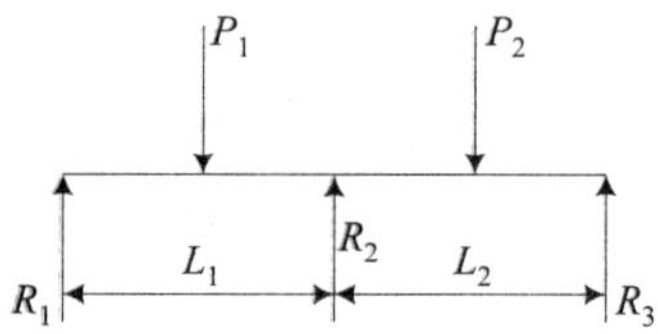

Fig. 3.11 Continuous beam

(e) Overhanging beam

The beam has one or both ends projecting beyond the supports.

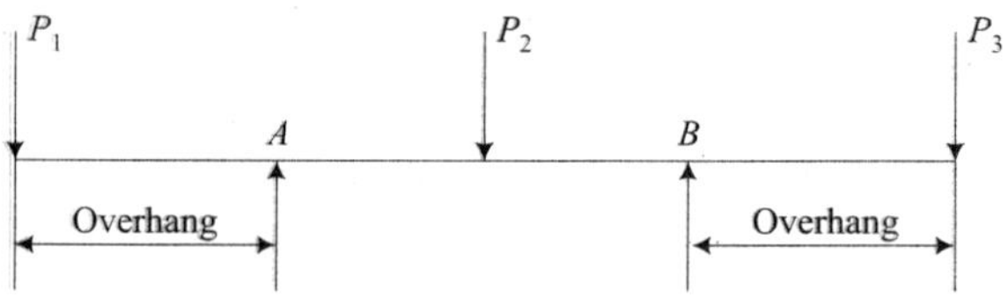

Fig. 3.12 Overhanging beam

3.3 STATICALLY DETERMINATE BEAM

If for a beam or a truss or any rigid body there are as many unknown supporting forces and couple-moments (including support reaction) as there are equations of equilibrium, and if the equations can be solved for these unknowns, the beam or truss or rigid body in called statically determinate.

1. The example of statically determinate beam is shown in Fig. 3.13. The loads F_1, F_2 and W are known. There are three unknown reactions R_A, R_{BY}

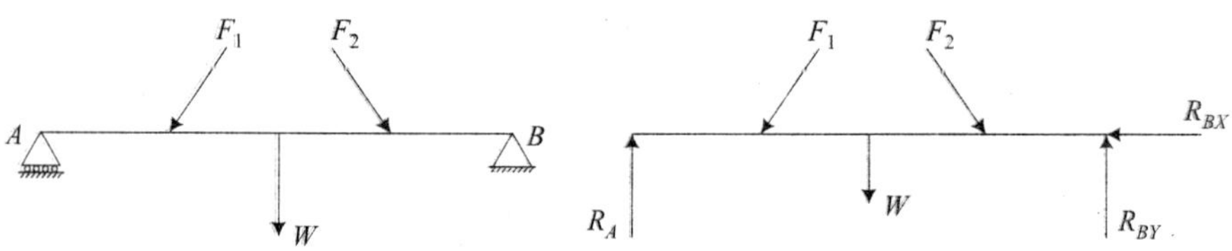

(a) Beam AB (a) FBD of beam

Fig. 3.13 Statically determinate beam

and R_{BX} which can be determined from the three equations of equilibrium $\Sigma F_x = 0$; $\Sigma F_y = 0$; $\Sigma M_o = 0$. Therefore, the problem can the solved from statics alone, the beam is called statically determinate.

2. The example of statically indeterminate beam is shown in Fig. 3.14. The beam is supported on three supports A, B and C. There are four unknown, i.e., R_A, R_C, R_{BY} and R_{BX}. The beam becomes statically indeterminate as there are insufficient number of equilibrium equations (three) to calculate the four support reactions. In order to solve such problems, use of deflection of beam at various points has to be used.

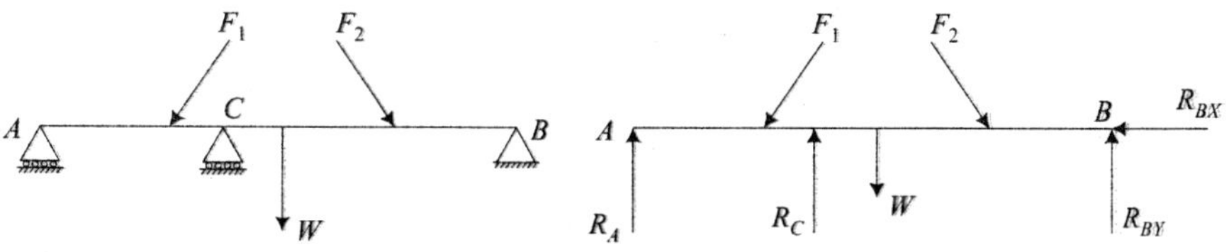

(a) Beam *ABC* (b) FBD of beam

Fig. 3.14 Statically indeterminate beam

3.4 SUPPORT REACTIONS

The solution of problems on beams starts with finding the support reactions. The solution consists of the following steps.

1. Load diagram/Free body diagram.

   The following principles are used to draw the load diagrams.

2. Use the following equations of equilibrium to find out the reactions.

$$\Sigma F_x = 0; \quad \Sigma F_y = 0; \quad \Sigma M_o = 0$$

where, 0 is any point (A or B).

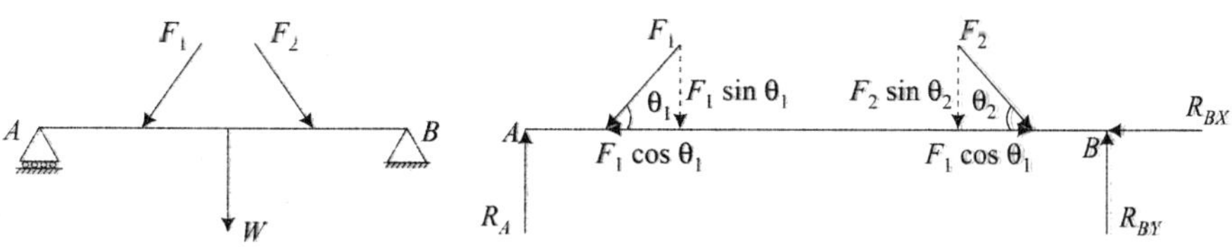

(a) Beam *AB* with inclined loads (b) Load diagram

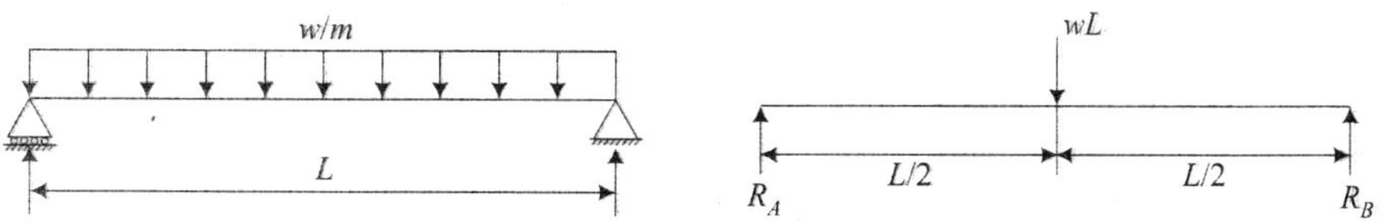

(a) Beam *AB* with UDL (b) Load diagram

(Fig. 3.15 Contd.)

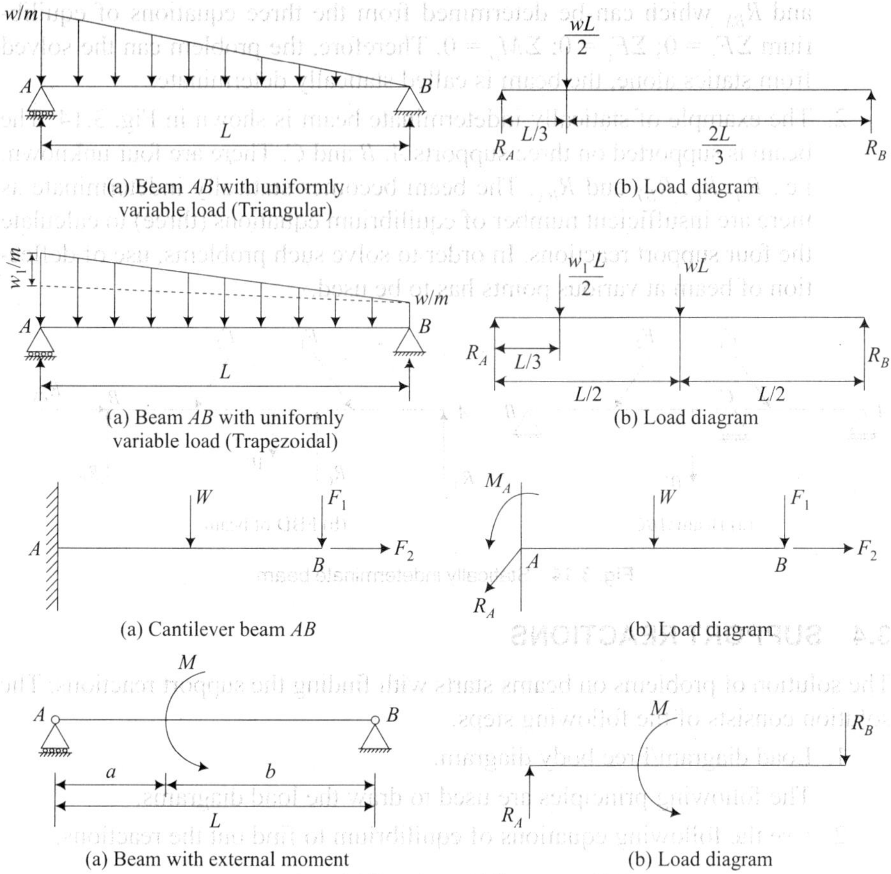

(a) Beam AB with uniformly variable load (Triangular) (b) Load diagram

(a) Beam AB with uniformly variable load (Trapezoidal) (b) Load diagram

(a) Cantilever beam AB (b) Load diagram

(a) Beam with external moment (b) Load diagram

Fig. 3.15 Load diagrams

Example 3.1: Find the reactions of the beam ABC.

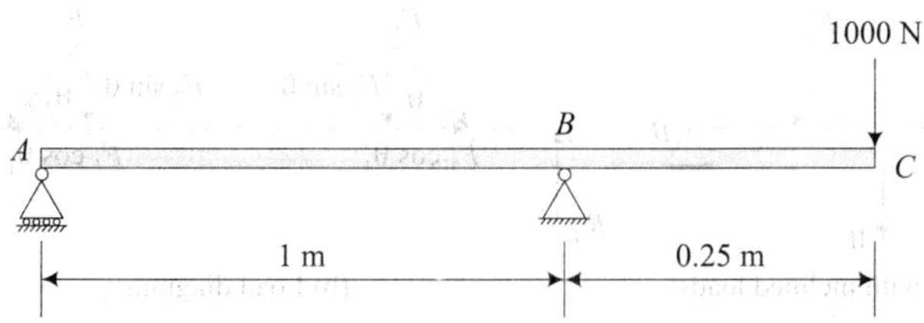

Solution:

1. Draw load diagram (or FBD).

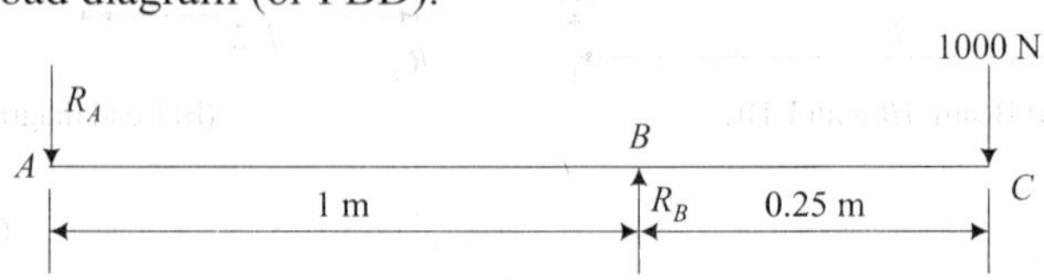

2. Apply equilibrium conditions.

(i) $\Sigma F_y = 0$

$$R_B - R_A - 1000 = 0$$

(ii) $\Sigma M_A = 0$

$\therefore$ $R_B(1) - 1000(1.25) = 0$

$$\therefore \quad R_B = \frac{1000 \times 1.25}{1} = 1250 \text{ N (upwards)}$$

$$R_A = R_B - 1000 = 1250 - 1000$$

$$= 250 \text{ N (downwards)}$$

Example 3.2: *A* beam supports a distributed load. Determine the resultant of this distributed load and reactions at supports *A* and *B*.

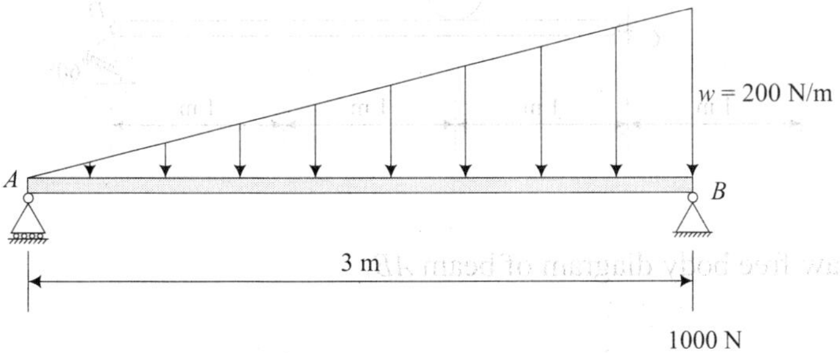

Solution:

1. The resultant of the variable load is area under the load curve.

$$W = \frac{wL}{2} = \frac{2000 \times 3}{2} = 3000 \text{ N}$$

$$AC = \frac{2}{3} L = \frac{2 \times 3}{3} - 2 \text{ m}$$

2. Draw load diagram or free body diagram.

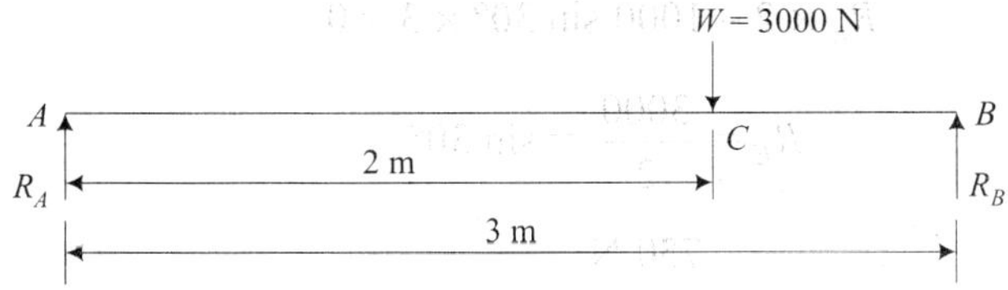

Load diagram

3. Apply equilibrium conditions

(i) $\Sigma F_y = 0$

$\therefore$ $R_A + R_B = W$

(ii) $\Sigma M_A = 0$

$\therefore$ $R_B \times 3 - W \times 2 = 0$

$$R_B = \frac{3000 \times 2}{3} = 2000 \text{ N.}$$

$$R_A = W - R_B = 3000 - 2000 = 1000 \text{ N.}$$

Example 3.3: Two beams *AB* and *CD* are arranged and supported as shown. Find the reaction at *D* due to a force of 1000 N acting at *B*.

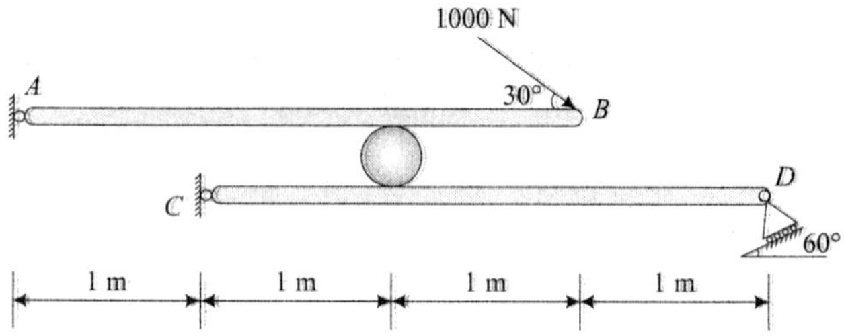

Solution:

1. Draw free body diagram of beam *AB*.

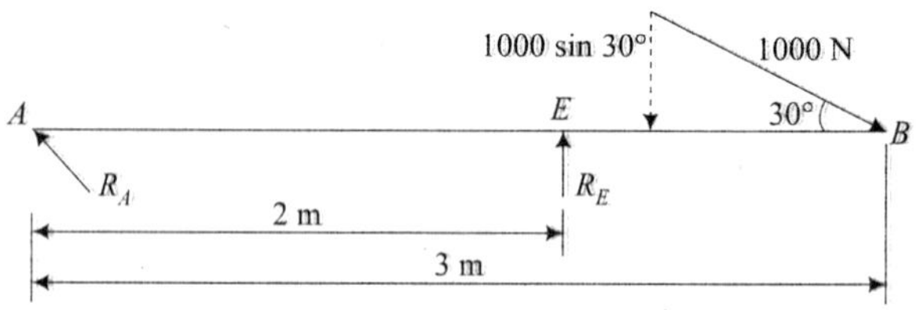

FBD of beam *AB*

2. Apply equilibrium conditions.

$\Sigma M_A = 0$

$\therefore$ $R_E \times 2 - 1000 \sin 30° \times 3 = 0$

$\therefore$ $R_E = \dfrac{3000}{2} = \sin 30°$

$= 750$ N

3. Draw free body diagram of beam *CD*.

4. Apply equilibrium condition

$$\Sigma M_C = 0$$

$$R_D \sin 30° \times 3 - R_E \times 1 = 0$$

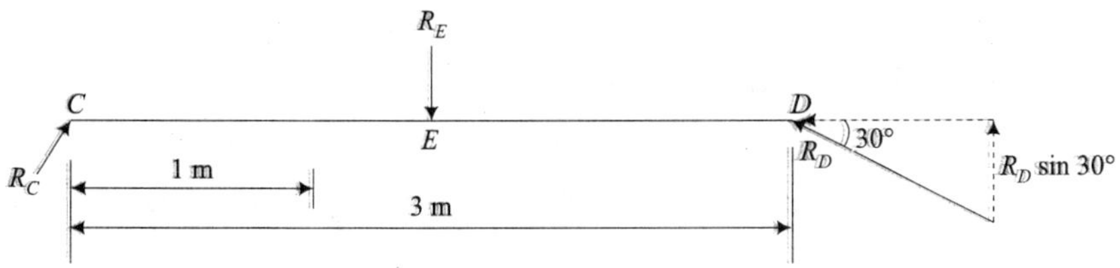

FBD of beam *CD*

$$\therefore \qquad R_D = \frac{R_E \times 1}{3 \sin 30^\circ} = \frac{75^\circ \times 1}{3 \times 0.5}$$

$$= \mathbf{500\ N}$$ **Ans.**

Example 3.4: A beam 8 m long is hinged at *A* and supported on rollers over a smooth surface inclined at 30° to the horizontal at *B*. The beam is loaded as shown. Determine the support reactions. [U.P.T.U. II Sem., 2002–03]

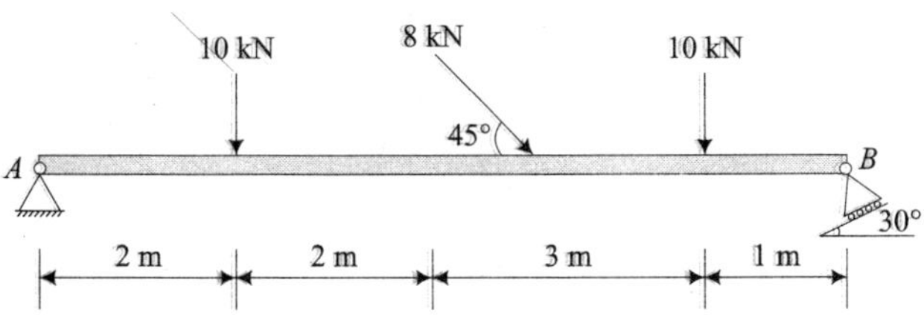

Solution:

1. Draw load diagram or free-body diagram.

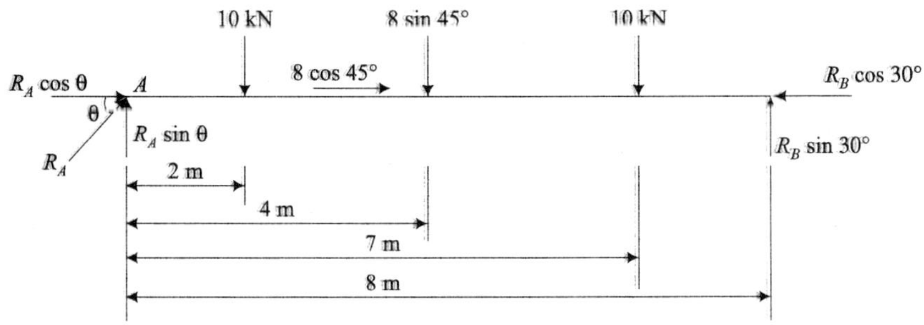

FBD of beam *AB*

2. Apply equilibrium condition

   (i) $\Sigma F_y = 0$

   $R_A \sin\theta - 10 - 8 \sin 45^\circ - 10 + R_B \sin 30^\circ = 0.$

(ii) $\Sigma F_x = 0$

$$R_A \cos\theta + 8\cos 45° - R_B \cos 30° = 0.$$

(iii) $\Sigma M_A = 0$

$$10\,(2) + 80 \sin 45°\,(4) + 10\,(7) - R_B \sin 30°\,(8) = 0$$

$$\therefore \quad R_B = \frac{20 + 22.63 + 70}{8 \times 0.5} = 28.16 \text{ kN}$$

$$R_A \sin\theta = 10 + 8\sin 45° + 10 - 28.16 \sin 30°$$
$$= 11.66 \text{ kN}$$
$$R_A \cos\theta = R_B = \cos 30° - 8\cos 45°$$
$$= 28.16 \cos 30° - 8\cos 45°$$
$$= 18.73 \text{ kN}$$
$$R_A = \sqrt{(11.66)^2 + (18.73)^2}$$
$$= 22 \text{ kN}$$
$$\theta = \tan^{-1}\frac{11.66}{18.73} = 31.9°$$

Example 3.5: Compute the simplest resultant force for the loads shown acting on the cantilever. What force and moment is transmitted by this force to supporting wall at *A*? [U.P.T.U. II Sem., 2004–05]

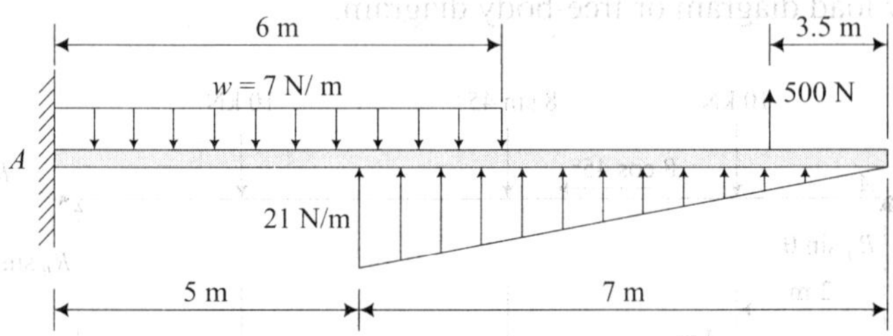

Solution:

1. Draw force diagram/free body diagram.
2. Find resultant.

(i) *Magnitude*

$$R = \Sigma F_y$$
$$= -42 + 73.5 + 500$$
$$= 531.5\ N$$

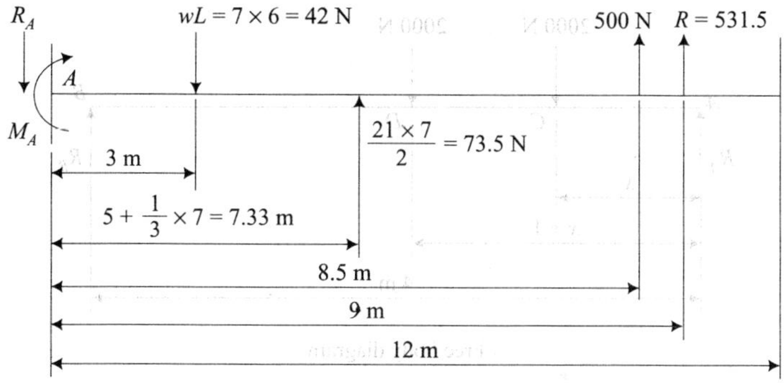

Free body diagram

(ii) *Direction*

Vertically upwards

(iii) *Position*

$$\Sigma M_A = 0$$

$$42 \times 3 - 73.5 \times 7.33 - 500 \times 8.5 - R_x = 0$$

$$\therefore \quad x = \frac{-126 + 538.75 + 4250}{531.5} = 8.77 \text{ m from } A$$

3. Wall reactions

$$R_A = -R = 531.5 \text{ N downwards}$$

$$M_A = Rx = 531.5 \times 8.77 = 4661.25 \text{ N clockwise.}$$

Example 3.6: Two weights $C = 2000$ N and $D = 1000$ N are located on a horizontal beam AB as shown. Find the distance of weight C from support A, i.e. x so that the support reaction at A is twice that at B. [U.P.T.U. II Sem., 2000-01].

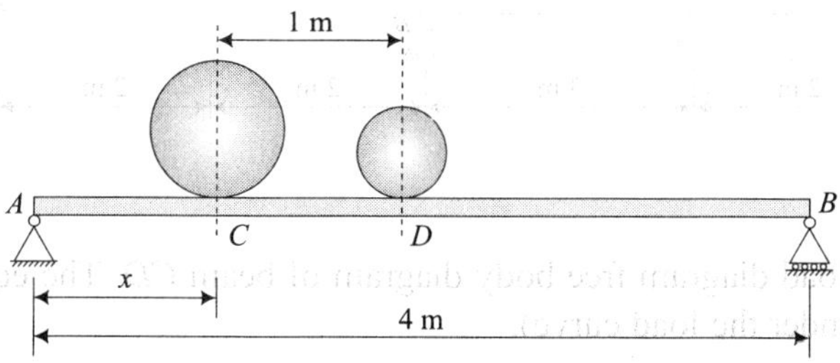

Solution:

1. Draw load diagram/free-body diagram.

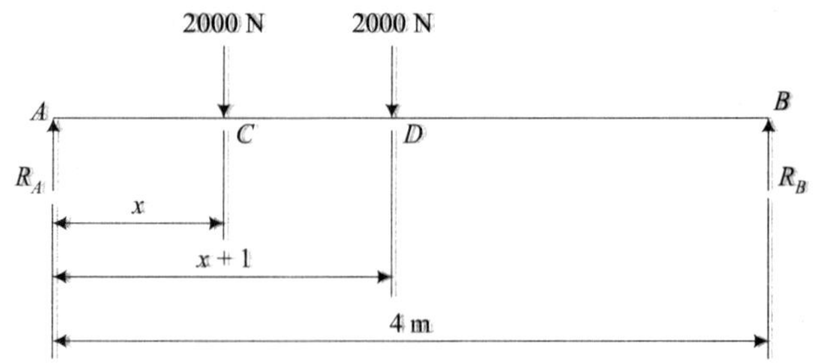

Free body diagram

2. Apply equilibrium conditions

 (i) $\Sigma F_y = 0$

$$R_A + R_B = 2000 + 1000 = 3000 \text{ N}$$
$$R_A = 2\ R_B$$

∴ $$2R_B + R_B = 3000 \text{ N}$$
$$R_B = 1000 \text{ N}$$
$$R_A = 2000 \text{ N}$$

 (ii) $\Sigma M_A = 0$

$$2000x + 1000(x + 1) - R_B(4) = 0$$
$$2000x + 1000x + 1000 = 1000(4)$$

∴ $$x = \frac{4000 - 1000}{3000} = \mathbf{1\ m} \quad \textbf{Ans.}$$

Example 3.7: Determine the reactions *A*, *B* and *D* of the system.

[U.P.T.U. I Sem., 2001–02]

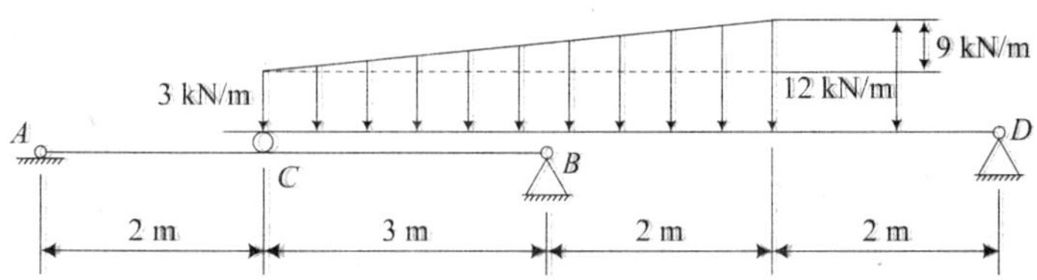

Solution:

1. Draw load diagram/free body diagram of beam *CD*. The equivalent load (area under the load curve),

$$W_1 = 3 \times 5 = 15 \text{ kN}$$
$$x_1 = \text{distance of load from } C$$
$$= \frac{5}{2} = 2.5 \text{ m}$$

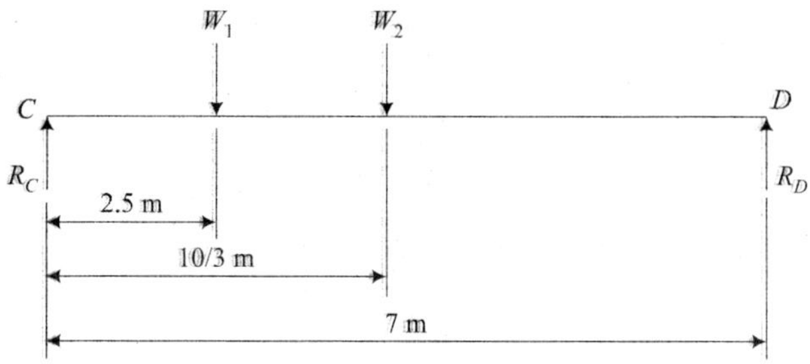

FBD of beam CD

$$W_2 = \frac{9 \times 5}{2} = 22.5 \text{ kN}$$

$$x_2 = \text{distance of load } W_2 \text{ from } C$$

$$= \frac{2L}{3} = \frac{2 \times 5}{3} = \frac{10}{3} \text{ m}$$

2. Apply equilibrium conditions.

(i) $\Sigma M_c = 0$

$$15 \times 2.5 + 22.5 \times \frac{10}{3} = R_D \times 7$$

$\therefore$ $R_D = 16$ kN

$\Sigma F_y = 0,$

(ii) $R_c = W_1 + W_2 - R_D = 15 + 22.5 - 16 = 21.5$ kN

3. Draw free body diagram of beam *AB*.

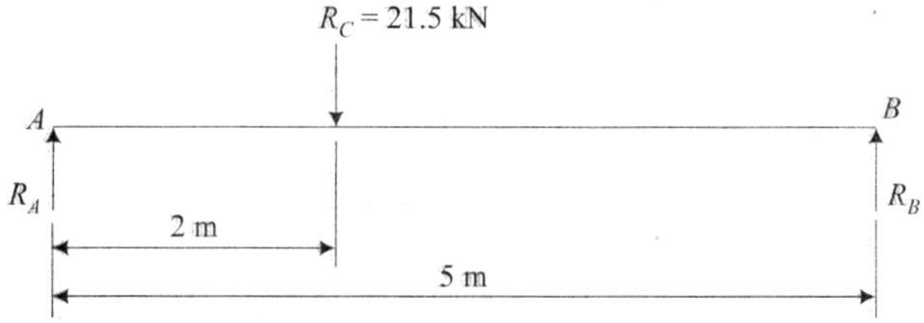

4. Apply equilibrium conditions.

(i) $\Sigma M_A = 0$

$$21.5\ (2) = R_B\ (5)$$

$\therefore$ $$R_B = \frac{21.5 \times 2}{5} = 8.6 = \text{kN}$$

(ii) $\Sigma F_Y = 0$

$\therefore$ $R_A = 21.5 - R_B = 21.5 - 8.6 = 12.9$ kN.

Example 3.8: Find the support reactions in the beam shown below.

[U.P.T.U. I Sem., 2000–01]

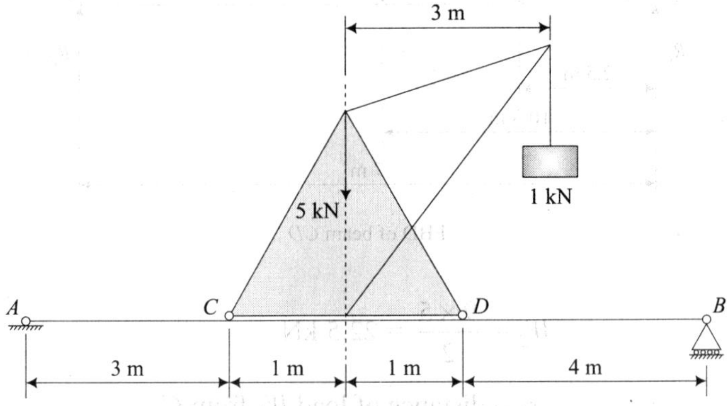

Solution:

1. Draw free body diagram of crane.

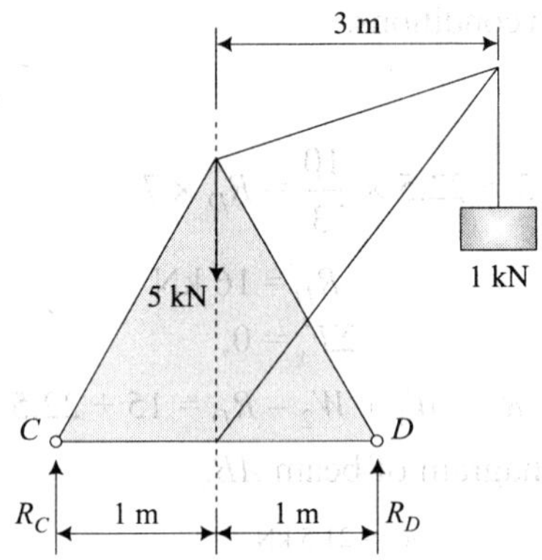

FBD of crane

2. Apply equilibrium conditions.

(i) $\Sigma F_y = 0$

$$R_C + R_D = 5 + 1 = 6 \text{ kN}$$

(ii) $\Sigma M_C = 0$

$$R_D \times 2 = 5 \times 1 + 1 \times 4$$

$$R_D = 4.5 \text{ kN}$$

$$R_C = 6 - 4.5 = 1.5 \text{ kN}$$

3. Draw free body diagram of beam *AB*.
4. Apply equilibrium conditions.

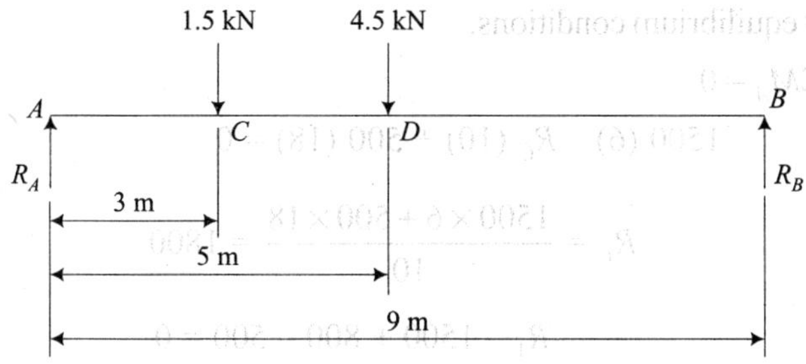

FBD of beam *AB*

(i) $\Sigma M_A = 0$

$1.5(3) + 4.5\,(5) = R_B\,(9)$

$$\therefore \quad R_B = \frac{4.5 + 22.5}{9} = 3 \text{ kN}$$

(ii) $\Sigma F_y = 0$

$$R_A + R_B = 1.5 + 4.5 = 6 \text{ kN}$$

$$R_A = 6 - R_B = 6 - 3 = 3 \text{ kN.}$$

Example 3.9: Find the reactions R_1, R_2 and R_3 in case of two beams placed one over the other as shown in the figure. [U.P.T.U. II Sem., 2000–01]

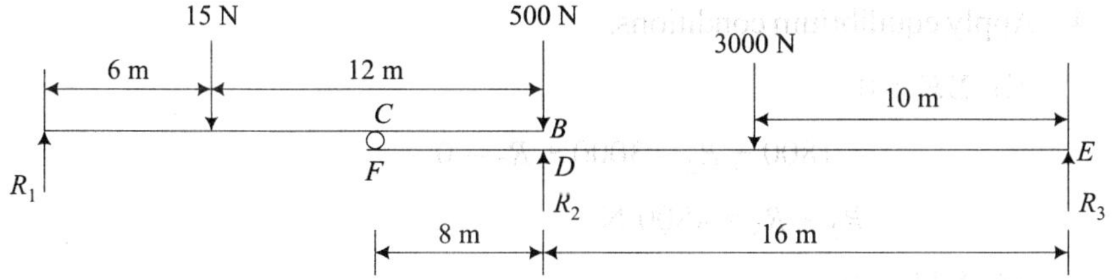

Solution:

1. Draw free body diagram of beam *AB*.

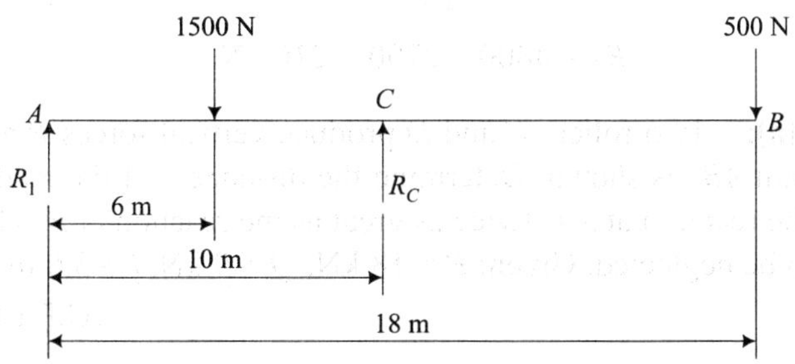

FBD of beam *AB*

2. Apply equilibrium conditions.

(i) $\Sigma M_A = 0$

$$1500\ (6) - R_C\ (10) + 500\ (18) = 0$$

$$\therefore \quad R_C = \frac{1500 \times 6 + 500 \times 18}{10} = 1800$$

$$R_1 - 1500 + 800 - 500 = 0$$

$$R_1 = 1500 + 800 - 500 = 0$$

$$R_1 = 200 \text{ N}$$

3. Draw free-body diagram of beam *FE*.

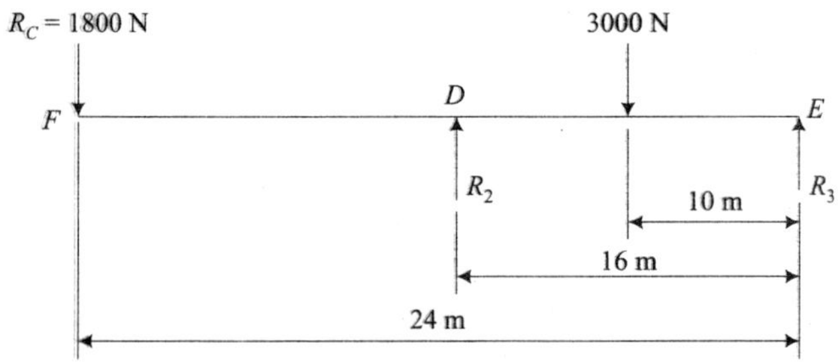

FBD of beam *FE*

4. Apply equilibrium conditions.

(i) $\Sigma F_y = 0$

$$-1800 + R_2 - 3000 + R_3 = 0$$

$$R_2 + R_3 = 4800 \text{ N}$$

(ii) $\Sigma M_E = 0$

$$1800\ (24) - R_2(16) + 3000(10) = 0$$

$$\therefore \quad R_2 = \frac{1800 \times 24 \times 10}{16} = 2700 \text{ N}$$

$$\therefore \quad R_3 = 4800 - 2700 = 2100 \text{ N}$$

Example 3.9(a): Two rollers *C* and *D* produce vertical forces *P* and *Q* on the horizontal beam *AB*, as shown. Determine the distance *x* of the load *P* from the support *A* if the reaction at *A* is twice as great as the reaction at *B*. The weight of the beam is to be neglected. Given: $P = 18$ kN, $Q = 9$ kN, $l = 3.6$ m, $c = 0.9$ m.

[U.P.T.U. 2008-09]

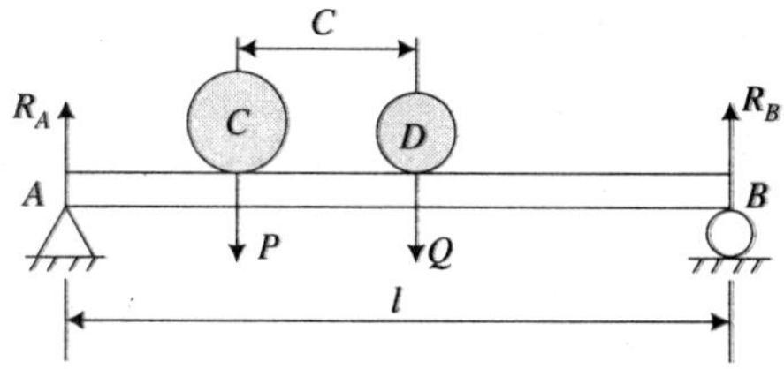

Solution:

Draw the free-body diagram of the beam.

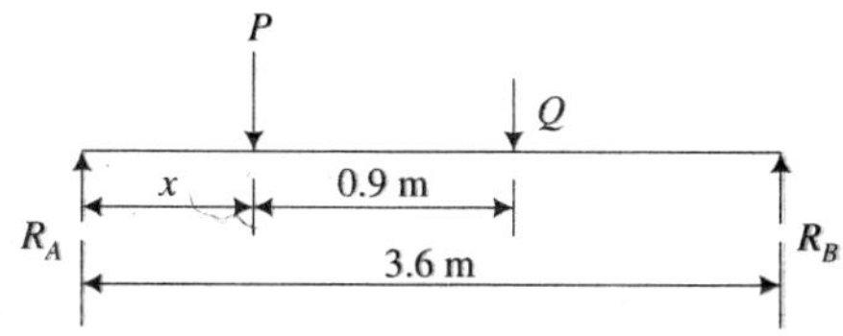

$$\Sigma F_y = 0.$$

$$R_A + R_B = P + Q = 18 + 9 = 27 \text{ kN}.$$

$$R_A = 2\,R_B.$$

$$\therefore \quad R_B = \frac{27}{3} = 9 \text{ kN}.$$

$$R_A = 18 \text{ kN}.$$

$$\Sigma M_A = 0$$

$$\therefore \quad R_B \times 3.6 = Q \times (0.9 + x) + Px$$

$$9 \times 3.6 = 9(0.9 + x) + 18x.$$

$$27x = 9 \times 3.6 - 9 \times 0.9 = 2.7 \times 9$$

$$\therefore \quad x = \frac{2.7 \times 9}{27} = \mathbf{0.9\ m} \quad \textbf{Ans.}$$

Example 3.9(b): A system of levers is supporting a load of 500 N. Determine the reactions at the supports *A* and *B*. [U.P.T.U. 2009-10]

Solution:

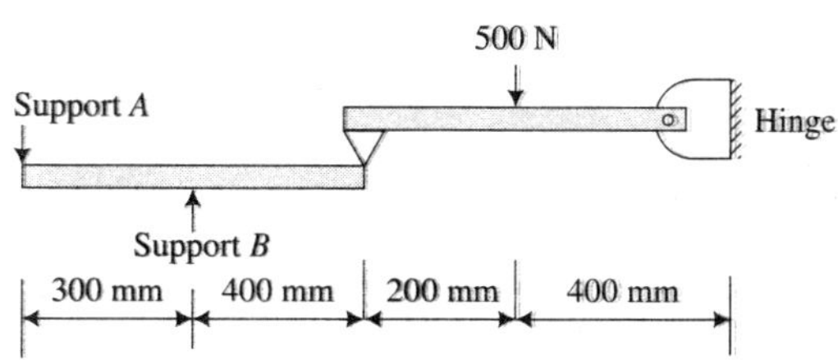

(i) Draw FBD of top lever.

$$\Sigma F_y = 0.$$

$$R_C + R_D = 500 \text{ N}.$$

$$\Sigma M_D = 0.$$

$$R_C \times 600 = 500 \times 400$$

$$\therefore \quad R_C = \frac{500 \times 400}{600} = 333.3 \text{ N}$$

(ii) Draw FBD of bottom lever.

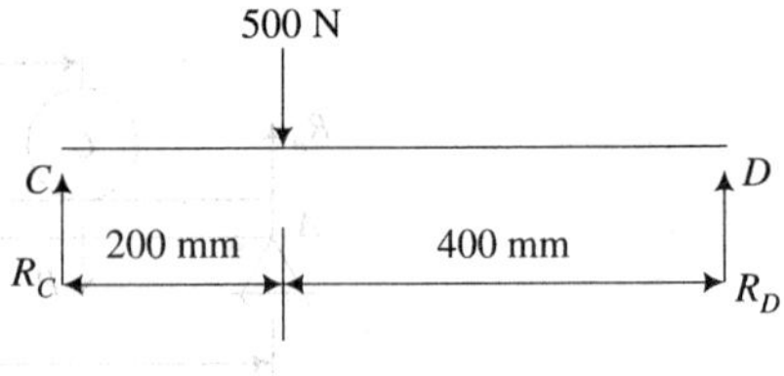

$$\Sigma F_y = 0$$

$$R_B = R_A + R_C$$

$$\Sigma M_A = 0.$$

$$R_B \times 300 = R_C \times 700$$

$$\therefore \quad R_B = \frac{333.3 \times 700}{300} = 777.7 \text{ N.}$$

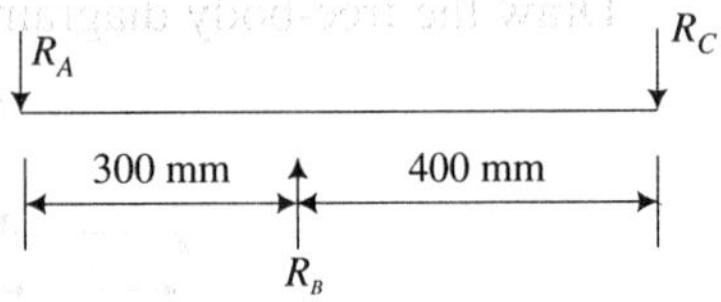

$$R_A = R_B - R_C = 777.7 - 333.3$$

$$= \mathbf{444.4\ N} \quad \textbf{Ans}$$

Example 3.9(c): Determine the reactions at A and E. [U.P.T.U. 2008-09]

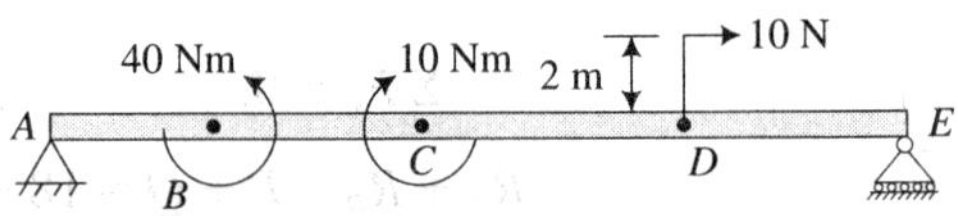

Solution:

Draw FBD of the beam.

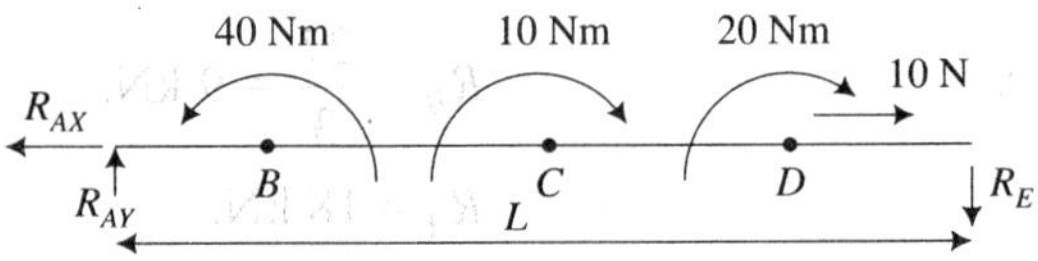

$$\Sigma F_x = 0$$

$$\therefore \quad R_{AX} = 10 \text{ N}$$

$$\Sigma F_y = 0$$

$$R_{AY} + R_E = 0$$

$$\Sigma M_A = 0, R_E \times L + 40 - 10 - 20 = 0.$$

$$\therefore \quad R_E = -\frac{10}{L} \text{ (Downward)}$$

$$\therefore \quad R_{AY} = +\frac{10}{L} \text{ (Upward)}$$

Example 3.9(d): A system of forces acting on a cantilever beam is shown. Reduce this system to a single-force system and find the point of application of this force on the beam.

[U.P.T.U. 2010-11]

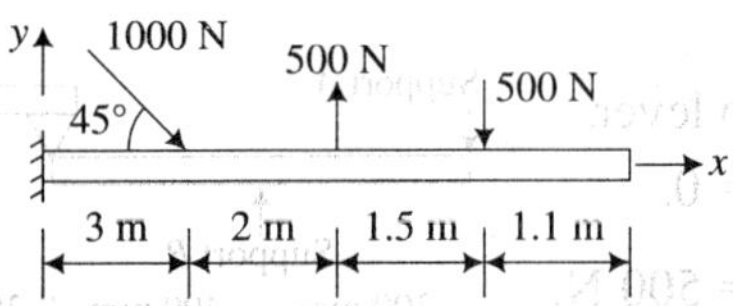

Solution:

Draw FBD of the cantilever.

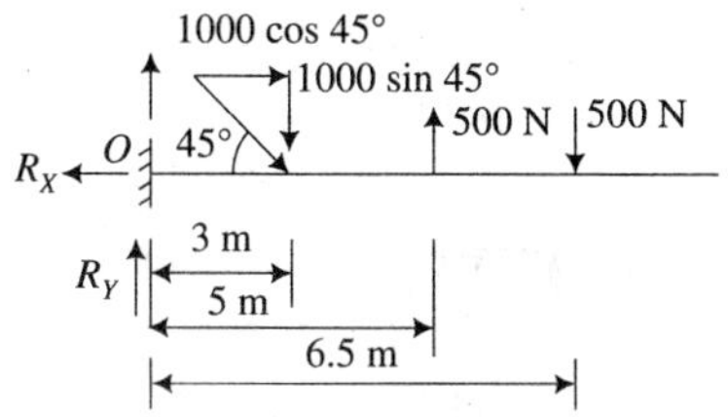

$$\Sigma F_x = 0 \quad R_x = 1000 \cos 45° = 700 \text{ N}$$

$$\Sigma F_y = 0 \quad R_y = 1000 \sin 45° - 500 + 500$$

$$= 700 \text{ N}$$

The resultant force,

$$R = \sqrt{R_x^2 + R_y^2} = \sqrt{(700)^2 + (700)^2} = 990 \text{ N.}$$

$$\Sigma M_o = 0$$

$$700 \times 3 + 500 \times 6.5 - 500 \times 5 = R \times x$$

$$\therefore \quad x = \frac{2100 + 3250 - 2500}{990} = 1.36 \text{ m from O}$$

$$\tan \theta = \frac{700}{700} = 1$$

$$\therefore \quad \theta = 45°$$

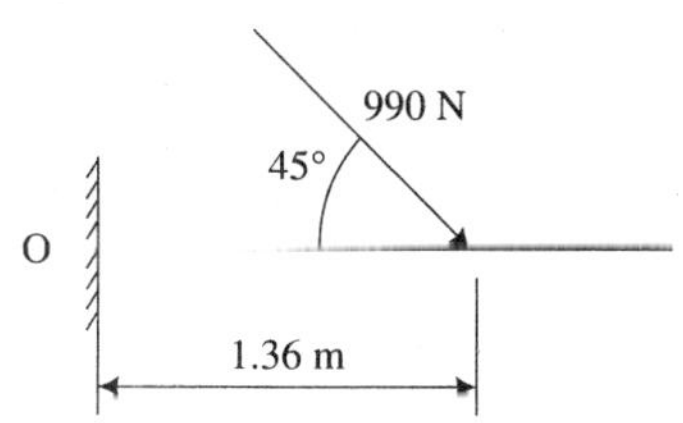

The single equivalent force is 990 N inclined at 45° at a distance 1.36 m from support.

3.5 SHEAR FORCE AND BENDING MOMENT

The load applied on a beam gets transferred to its supports. Any section of the beam will experience shearing off by vertical forces and bending by moments.

3.5.1 Shear Force

A beam *AB* is supporting vertical loads W_1, W_2 and W_3 and the beam is supported at *A* and *B*. The supports *A* and *B* are providing reactions R_1 and R_2. The beam is in equilibrium under five forces R_1, W_1, W_2, W_3 and R_2. If the beam is cut at section

X-X, a force F_X must he applied at the section to maintain equilibrium of left portion and right. This force F_X is called shear force. This force was supported by left portion to the right portion. Similarly, right portion was supported by left portion. Hence, there will be a shearing force at the action.

Shear force is defined as the force that tries to shear off the section of the beam and is numerically equal to the algebraic sum of all the forces acting normal to the axis of the beam either to the left or to the right of the section. In the given example,

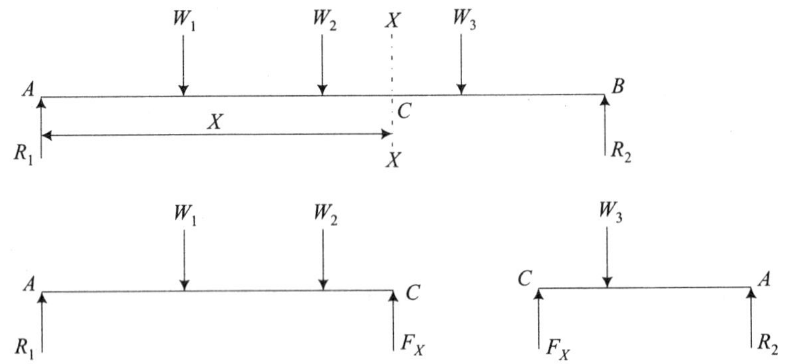

Fig. 3.16 Shear force

$$F_x = R_1 - W_1 - W_2$$

or

$$F_x = R_2 - W_{3.}$$

The shear force that tends to move the left portion upwards relative to right portion is taken as positive shear force.

The sign convention of shear force is illustrated in Fig. 4.17.

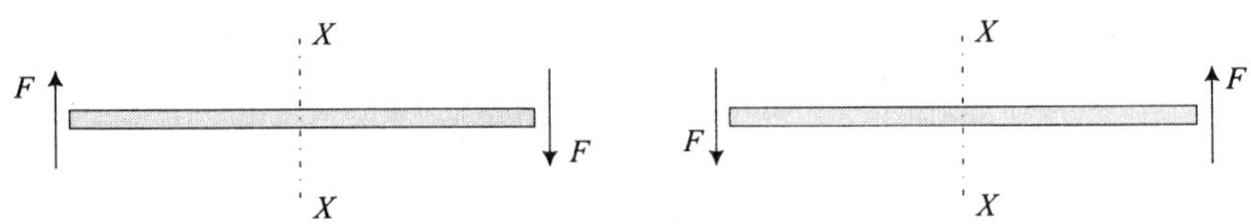

Fig. 3.17 Sign convention of shear force

3.5.2 Bending Moment

Bending moment at any section is the moment of forces which tries to bend it.

A beam *AB* is in equilibrium and there is no resultant moment. If the beam is cut at *X-X* there should be a balancing moment M_X at point *C* to keep separately the right portion and left protion in equilibrium. This moment is provided by the internal forces acting in the beam at section *X–X*. M_X is called the bending moment at the section.

The bending moment is numerically equal to the algebraic sum of moments of all the forces about the section, either acting to the left or to the right of the section. In the given example,

$$M_X = R_1\,(x) - W_1\,(x - l_1) - W_2\,(x - l_2)$$

or

$$M_X = R_2\,(l - x) - W_3\,(x - l_3)$$

The bending moment which tends to sag the beam is taken as positive. The bending moment which tries to hog the beam is taken as negative.

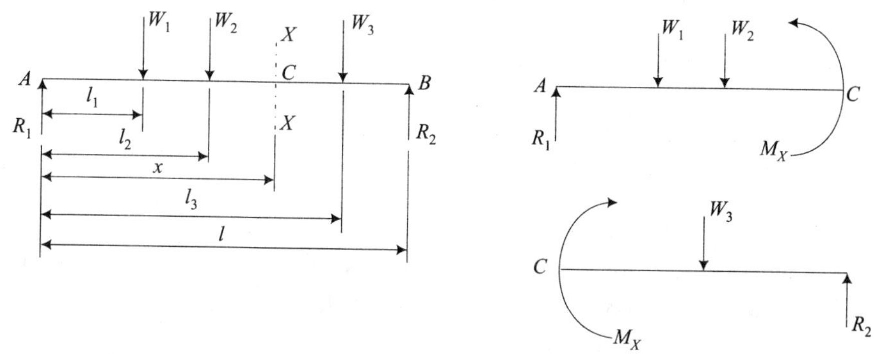

Fig. 3.18 Bending moment

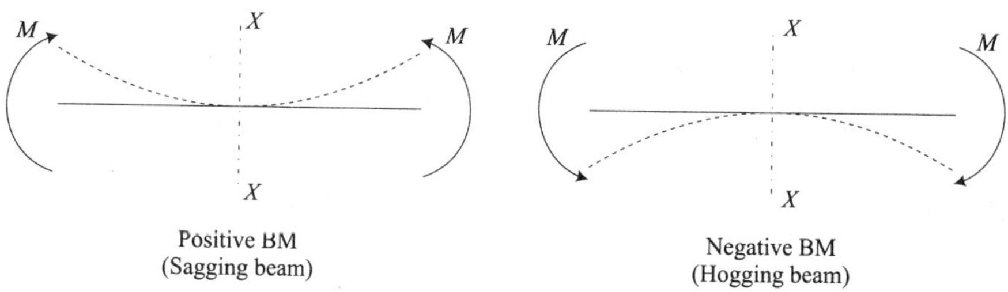

Fig. 3.19 Sign convention of bending moment

3.5.3 Differential Equations for Equilibrium

There is a relationship between load on the beam, the shearing force and bending moment.

Consider the equilibrium of a small element of length δx in a beam of span l and under *udl* of w/m. The values of shear force and bending moment at left section be F and M which change to $F + \delta F$ and $M + \delta M$ over the elemental length δx. Following sign convention for bending moment and shear force, the free body diagram of elemental length of beam δx is drawn.

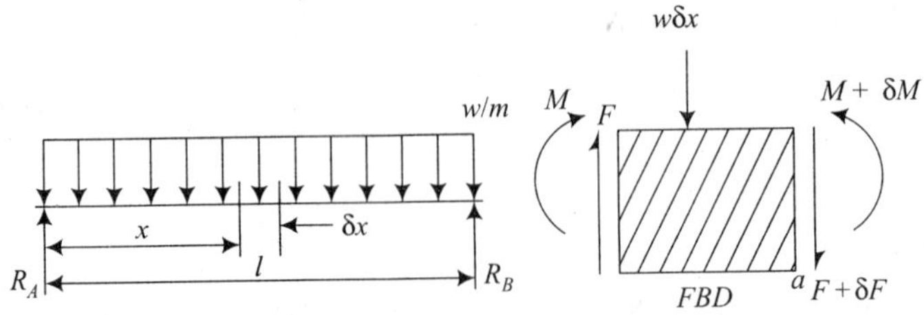

Fig. 3.20 Differential equation for equilibrium

(i) $\Sigma F_y = 0$

$$F - w\delta x = F + \delta F$$

$$\frac{\delta F}{\delta X} = -w \qquad \text{...(1)}$$

The rate of change of shear force is equal to intensity of loading on the beam.

(ii) $\Sigma M_a = 0$ Taking moments about corner a of element

$$M + F\delta x - w\delta x\left(\frac{\delta X}{2}\right) = M + \delta M$$

$$F\delta x + \frac{w}{2}(\delta x)^2 = \delta M$$

δx^2 can be neglected being very small.

$\therefore$ $$\delta M = F\delta x$$

$$\frac{\delta M}{\delta x} = F \qquad \text{...(2)}$$

The rate of change of bending moment is equal to shear force.

Integrating equations (1) and (2)

$$F_2 - F_1 = -\int_1^2 w dx$$

$\therefore$ $$F_2 = F_1 - \int_1^2 w dx \qquad \text{...(3)}$$

$$M_2 - M_1 = \int_1^2 F\delta x$$

$$M_2 = M_1 + \int_1^2 F\delta x \qquad \text{...(4)}$$

Equations (1) and (2) are the differential equations for equilibrium of beam.

For maximization of bending moment

$$\frac{\delta M}{\delta X} = 0$$

$$\therefore \qquad F = 0$$

The bending moment is maximum or minimum where shear force is zero.

3.6 SHEAR FORCE AND BENDING MOMENT DIAGRAMS

The shear force and bending moment in a beam vary from section to section along its length.

3.6.1 Shear Force Diagram (SFD)

The variation of shear force along the length of the beam can be represented graphically in which ordinate represents the shear force. Such a graph is called Shear Force Diagram (SFD). Positive shear force is represented above the beam line and negative force below the beam. Salient values of shear force are written on the diagram.

The section at which shear force changes its sign is the point or section where maximum bending moment occurs.

3.6.2 Bending Moment Diagram (BMD)

The variation of bending moment along the length of the beam can be represented graphically. Such a graph is called Bending Moment Diagram (BMD). Positive bending moment is represented above the beam line and negative bending moment below the line. Salient values of bending moment are written on the diagram.

The bending moment changes its sign at the point of contraflexure. The bending moment at that section is obviously zero.

3.6.3 Some Rules for SFD and BMD

(i) Whenever a concentrated load is acting (including support reactions) the shear force will suddenly change. The change is numerically equal to the load and in the direction of load.

(ii) If there is no loading between any two sections, the magnitude of shear force will not change between the given vertical loads and SFD will be a horizontal line.

(iii) The variation of shear force will be a horizontal line for vertical loads, it is an inclined straight line for uniformly distributed load and a parabola for uniformly varying load between the sections.

(iv) Bending moment is zero at supports of simply supported beam.

(v) Bending moment is zero at the free end of a cantilever or overhanging beam.

(vi) The bending moment is maximum or minimum where shear force is zero.

(vii) The bending moment curve will be one degree higher than that of shear force and two degree higher than load curve.

(viii) Since $\frac{dF}{dX} = P$ and $\frac{dM}{dX} = -F$, nature of variation of shear force and bending moment diagram with load will be as follows.

	Load	*SFD*	*BMD*
1.	Concentrated or no load.	Horizontal line	Inclined linear line
2.	UDL (horizontal line)	Inclined linear line	Parabolic
3.	Uniformly varying load (inclined line)	Parabolic	Cubic

(ix) If an external moment is acting at a point or a section of a beam, there will be sudden change in the value of bending moment, the change being numerically equal to external moment.

While proceeding from left, it will increase in value, if the moment is clockwise. There will be drop in bending moment value if the external moment is anticlockwise. Exactly opposite phenomenon will be observed while proceeding from right to left.

The external moment will affect the magnitude of support sections. However, there is no sudden change in the value of shear force at the section where external moment is acting.

(x) Inflection or contraflexure point occurs where bending moment is zero.

(xi) The location of points of maximum bending moment and contraflexure must be shown in the bending moment diagram.

3.7 SFD AND BMD FOR STANDARD CASES

The load diagram, S.F. diagrams and B.M. diagram for some standard cases are tabulated in Table 3.1 (See Page 3.24).

Example 3.10: Draw shear force and bending moment diagram for a cantilever of length L carrying a concentrated load W at the free end.

Solution:

1. *Draw the load diagram.*

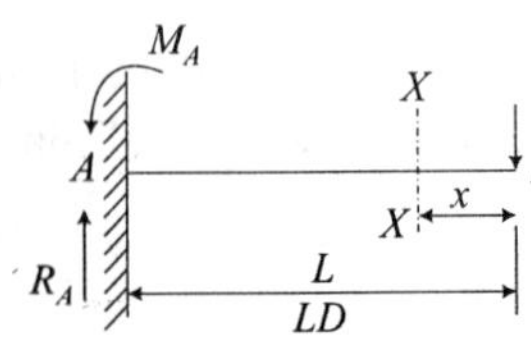

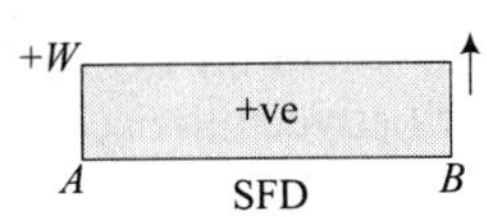

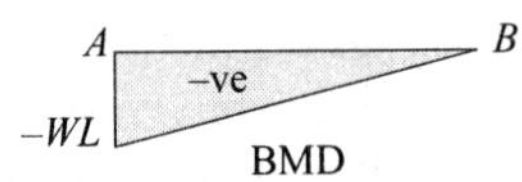

Table 3.1 SFD and BMD for standard case

Sl. No.	Type of beam	Load diagram	S.F. diagram	B.M. diagram
1.	Cantilever	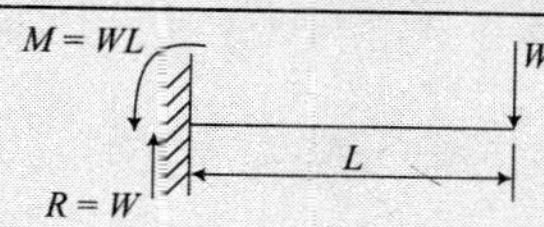	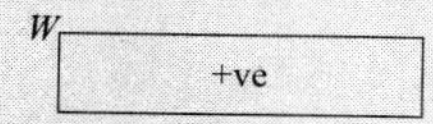	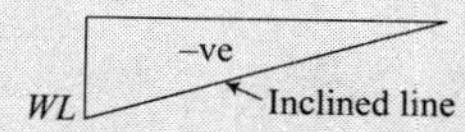
		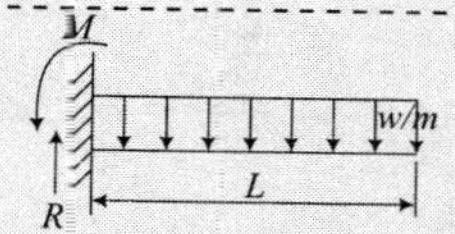	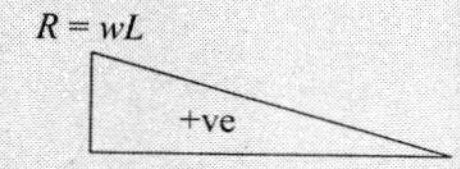	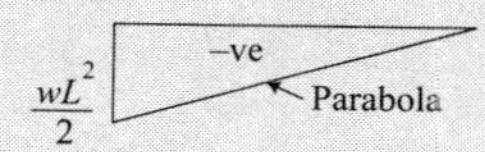
2.	Simply supported beam	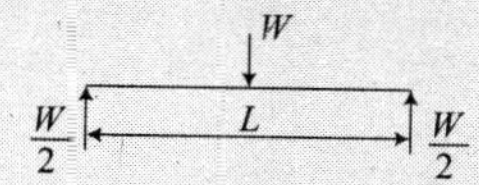	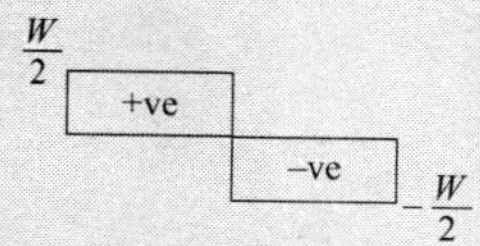	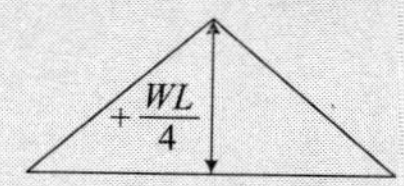
		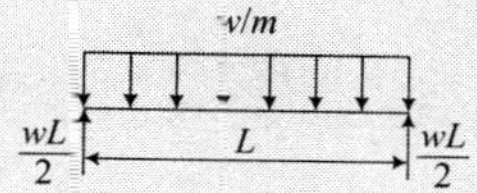	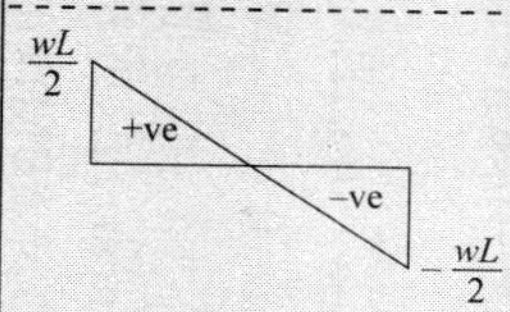	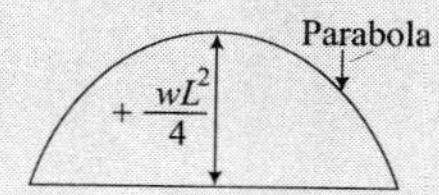
		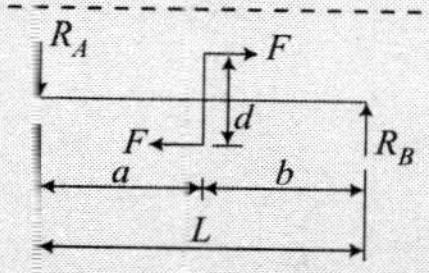	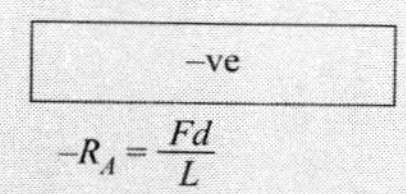	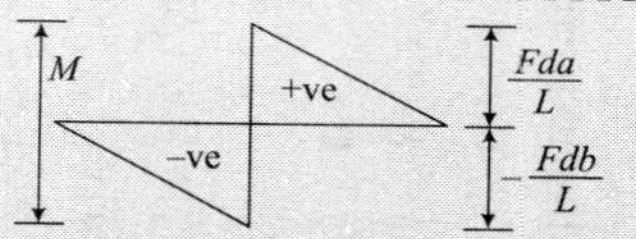
		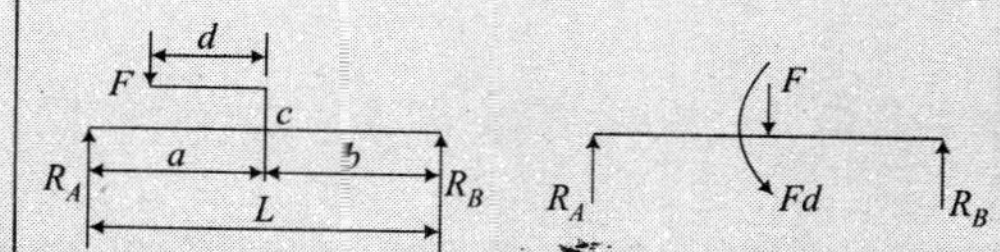	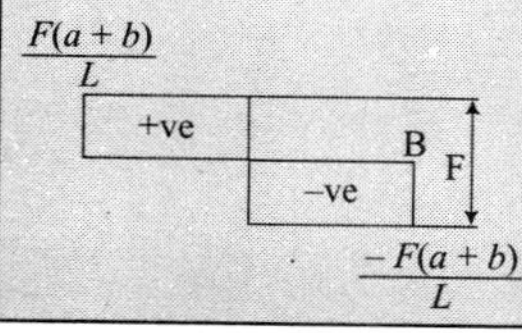	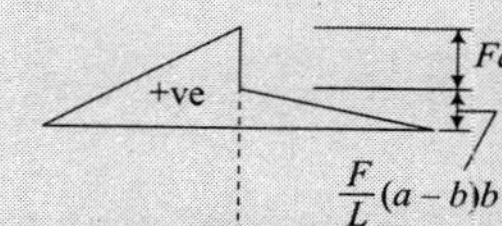

2. *Support reaction*

Applying equations of equilibrium, find the support reactions

(i) $\Sigma F_y = 0$

$\therefore \quad R_A = W$

(ii) $\Sigma M_A = 0$

$\therefore \quad -M_A = +WL = 0$

$M_A = WL$ (Anticlockwise)

3. *Shear Force*

Cut a section *X-X* at a distance *X* from end *B*.

$F_X = +W$ which is constant over the length

4. *Bending Moment*

$M_X = -Wx$ (Hogging)

when, $X = 0 \qquad M = 0$

when, $X = L \qquad M = WL$

The B.M. changes linearly from *B* to *A*.

Example 3.11: Draw SFD and BMD for the cantilever shown carrying several concentrated loads.

Solution:

1. Draw the load diagram.

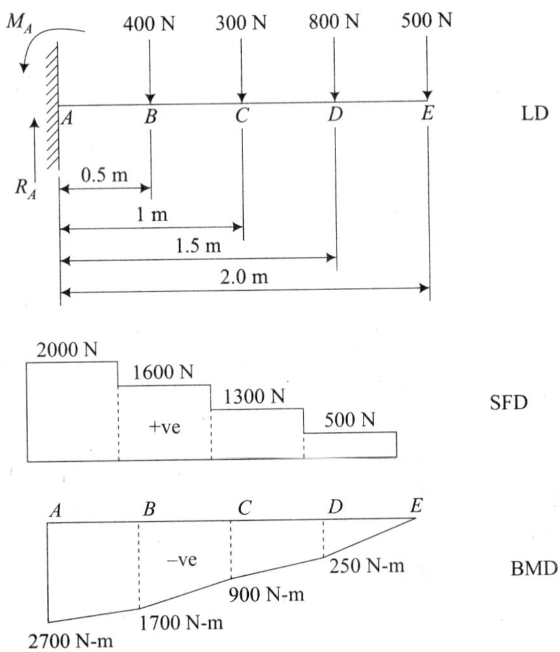

2. *Support Reactions,*

Apply equilibrium conditions

(i) $\Sigma F_y = 0$

$R_A = \Sigma W$

$500 + 800 + 300 + 400 = 2000$ N

(ii) $\Sigma M_A = 0$

$M_A = \Sigma Wx$

$= 500 \times 2 + 800 \times 1.5 + 300 \times 1 + 400 \times 0.5$

$= 2700$ N-m

3. *Shear Force*

$F_A = R_A = 2000$ N

$F_B = F_A - W_B = 2000 - 400 = 1600$ N

$F_C = F_B - W_C = 1600 - 300 = 1300$ N

$F_D = F_C - W_D = 1300 - 800 = 500$ N

$F_E = F_D - W_E = 500 - 500 = 0$ N

Draw SFD using horizontal lines between the loads.

4. *Bending Moment*

$M_E = 0$ (Free end)

$M_D = W_E\,(DE) = 500 \times 0.5 = 250$ N-m (Beam hogging, –ve)

$M_C = W_E\,(CE) + W_D\,(CD) = 500 \times 1 + 800 \times 0.5$

$= 900$ N-m (–ve)

$M_B = W_E\,(BE) + W_B\,(BD) + W_C\,(BC)$

$= 500 \times 1.5 + 800 \times 1 + 300 \times 0.5$

$= 1700$ N-m (–ve)

$M_D = 2700$ N-m (–ve)

Draw BMD using linear inclined lines.

Example 3.12: Draw SFD and BMD of a cantilever of length L carrying a uniformly distributed load of w per unit run over whole length.

Solution:

1. Draw free body diagram,

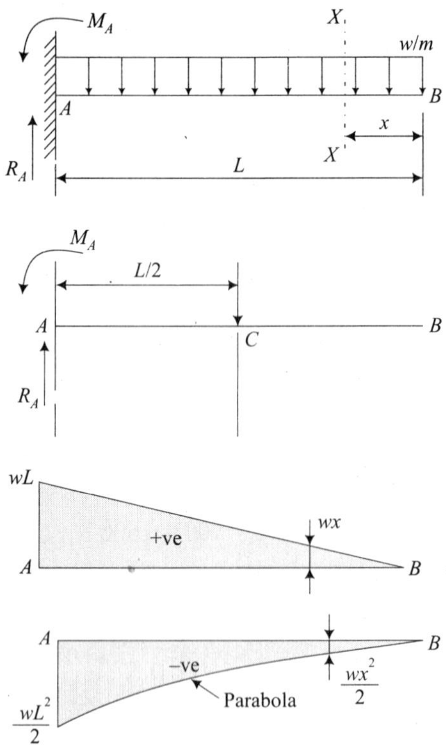

LD

SFD

BMD

2. Draw equivalent load diagram,

$$W = wL \text{ (Area under the curve)}$$

Point of application = $L/2$

3. *Support Reactions*

Apply equilibrium conditions

(i) $\Sigma F_y = 0$

$\therefore$ $R_A = W$ (Left up positive)

$= WL$

(ii) $\Sigma M_A = 0$

$\therefore$ $$M_A - W \times \frac{L}{2} = WL\left(\frac{L}{2}\right) = \frac{WL^2}{2} \text{ (Hogging beam ve)}$$

4. *Shear Force*

Shear force at section X-X which is x distance from B

$$F_X = wx$$

At $x = 0$, $F = 0$

At $x = L$, $F = wL$

The SFD is an inclined linear diagram under UDL.

5. *Bending Moment*

B.M. at section *X-X*,

$$M_X = wx\left(\frac{x}{2}\right) = \frac{Wx^2}{2} \text{ (–ve)}$$

At $x = 0$, $M = 0$

At $x = L$, $M = \frac{WL^2}{2}$

The BMD is a parabola under UDL.

Example 3.13: A cantilever of length L is carrying uniformly distributed load of w per unit run over a distance 'a' from free end. Draw SFD and BMD.

Solution:

1. Draw load diagram.

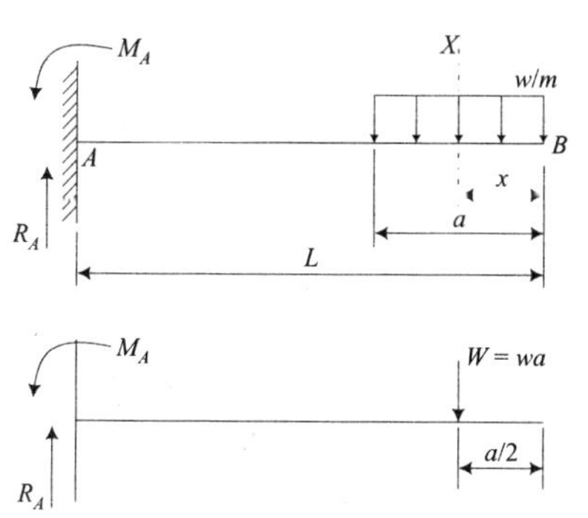

LD

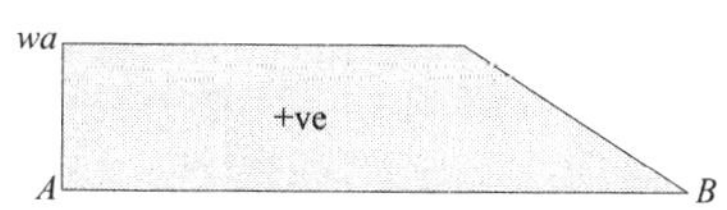

SFD

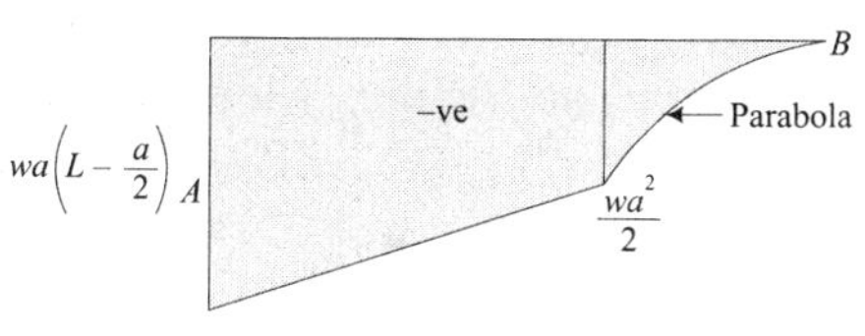

BMD

$$W = wa \text{ (Area under load curve)}$$

Point of application $= \dfrac{a}{2}$ from free end.

2. *Support Reactions*

   Apply equilibrium conditions

   (i) $\Sigma F_y = 0$

   $$R_a = W = wa$$

   (ii) $\Sigma M_A = 0$

   $$M_A = W\left(L - \frac{a}{2}\right)$$

   $$= wa\left(L - \frac{a}{2}\right) \text{ (Hogging –ve)}$$

3. *Shear Force*

   Shear force at section *X-X*

   $$F_X = wx$$

   At $x = 0,$ $F = 0$

   At $x = a,$ $F = wa$

   At $x = L,$ $F = wa$

   SFD is an inclined line under UDL and horizontal line under no load.

4. *Bending Moment*

   Bending moment at section *X-X*,

   $$M_X = wx\left(\frac{x}{2}\right) = \frac{wx^2}{2} \text{ (–ve, beam hogging)}$$

   At $x = 0,$ $M = 0$

   At $x = a$ $M = \dfrac{wa^2}{2}$

   At $x = L$ $M = M_A = wa\left(L - \dfrac{a}{2}\right)$

   From $x = a$ to $x = L$, BMD is an inclined straight line under no load. BMD is a parabola under UDL from $x = 0$ to $x = a$.

Example 3.14: Draw the Shear Force and Bending Moment diagrams for the cantilever beam shown in figure. [U.P.T.U. II Sem., 2005–06]

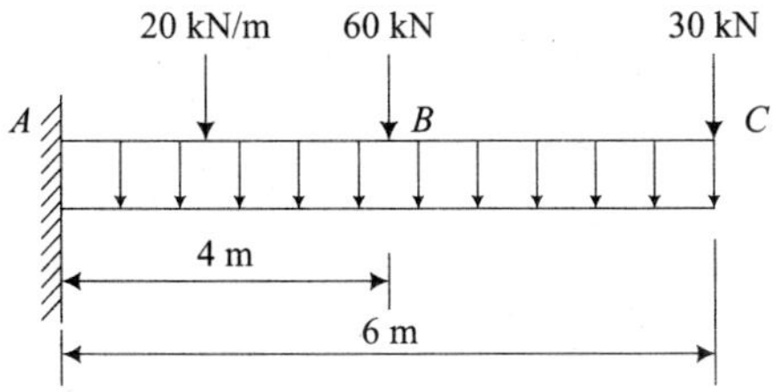

Solution:

1. Draw equivalent loading diagram.

   A load of 120 kN (area under the load curve equal to 20 × 6 kN will be acting at 3 m from *A* i.e., c.g. of load curve).

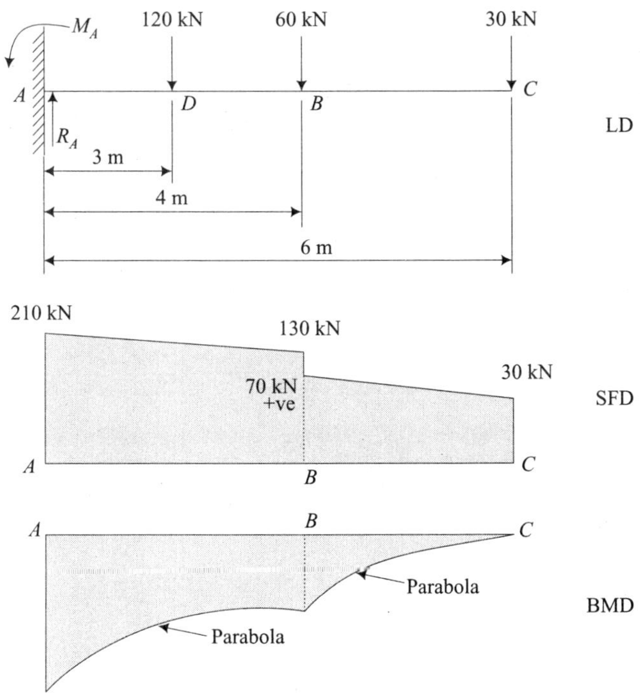

2. *Support Reactions*

   Apply equilibrium conditions.

   (i) $\Sigma F_y - 0$

$$\therefore \quad R_A = \Sigma W = 120 + 60 + 30 = 210 \text{ kN}$$

   (ii) $\Sigma M_A = 0$

$$\therefore \quad M_A = 120\,(3) + 60\,(4) + 30\,(6)$$

$$= 360 + 240 + 180 = 780 \text{ kN-m}$$

3. *Shear Force*

$$F_C = 30 \text{ kN}$$
$$F_B = 30 + 60 + (20 \times 2) = 130 \text{ kN}$$
$$F_A = \Sigma W = 120 + 60 + 30 = 210 \text{ kN}$$

SFD will be an inclined straight line under UDL.

4. *Bending Moment*

$$M_C = 0 \text{ (Free end)}$$
$$M_B = 30 \times 2 + 20 \times 2 \times 1 = 60 + 40 = 100 \text{ kN-m}$$
$$M_A = 780 \text{ kN-m}$$

BMD will be parabolas under UDL.

Example. 3.15: A cantilever of length L carries a load whose intensity varies uniformly from zero at the free end to w per unit length at the fixed end. Draw SFD and BMD.

Solution:

1. Draw load diagram.

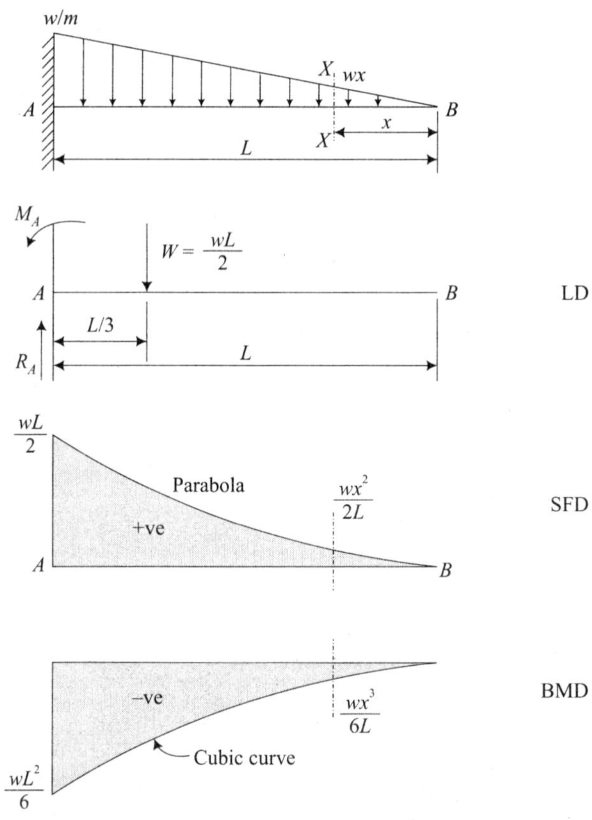

2. Draw equivalent load diagram.

   W is area under load curve

$$W = \int_{o}^{L} w_x \, dx = \frac{WL}{2}$$

   where, $w_X = \dfrac{wx}{L}$

   Location W will be acting at a distance $L/3$ from A, i.e., c.g. of load curve.

3. *Support Reactions*

   Apply equilibrium conditions,

   (i) $\Sigma F_y = 0$

$$\therefore \qquad R_A = W = \frac{wL}{2}$$

   (ii) $\Sigma M_A = 0$

$$M_A = W \times \frac{L}{3} = \frac{wL}{2} \times \frac{L}{3} = \frac{wL^2}{6}$$

4. *Shear Force*

   Shear force at section X–X,

$$F_X = w_X \cdot \frac{x}{2} \text{ (Area under load curve on the spare } X\text{-}B\text{)}$$

$$= \frac{wL}{2} \cdot \frac{x}{2} = \frac{wx^2}{2L}$$

   At $x = 0$ $\qquad F = 0$

   At $x = L$ $\qquad F = \dfrac{wL}{2}$

   SFD is a parabola under uniformly varying load.

5. *Bending Moment*

   Bending moment at section X-X is equal to area under the load curve X distance of centroid,

$$M_X = \frac{wx^2}{2L} \cdot \frac{x}{3} = \frac{wx^3}{6L}$$

   At $x = 0$ $\qquad M = 0$

   At $x = L$ $\qquad M = \dfrac{wL^2}{6}$ [Hogging beam –ve]

   BMD is a cubic under a uniformly varying load.

Example 3.15(a): Calculate the values of shear force and bending moments for the cantilever shown. Also draw the shear force and bending moment diagram.

[U.P.T.U. 2009-10]

Solution:

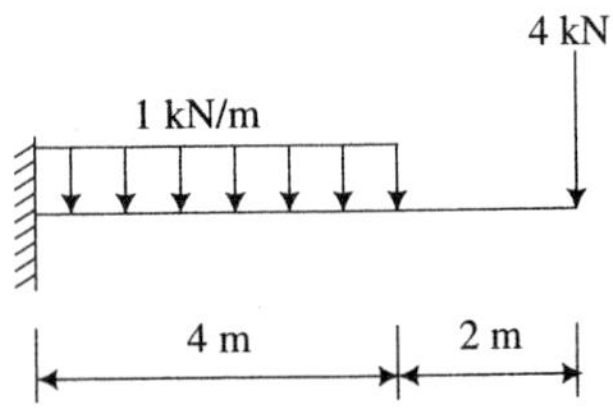

Draw FBD of the cantilever.

1. *Support Reactions*

$$\Sigma F_y = 0$$

$$\therefore \quad R_A = 4 + 4 = 8 \text{ kN}$$

$$\Sigma M_A = 0$$

$$\therefore \quad M_A = 4 \times 6 + 4 \times 2 = 32 \text{ kNm}$$

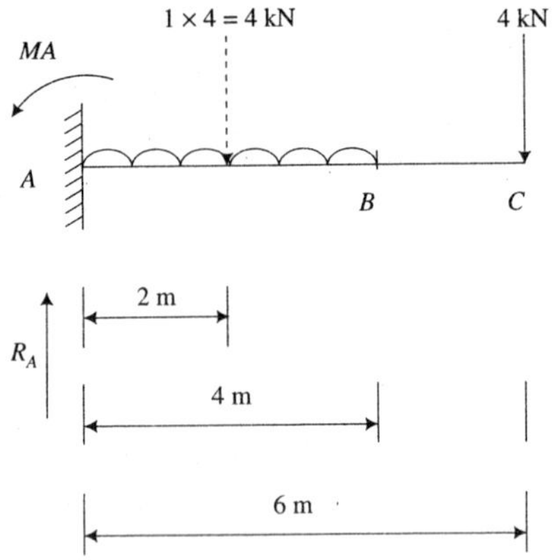

2. *SFD*

$$F_A = R_A = +\,8 \text{ kN}$$

$$F_B = F_A - W_{AB} = 8 - 4 = +4 \text{ kN}$$

$$F_C = P_C = +\,4 \text{ kN}$$

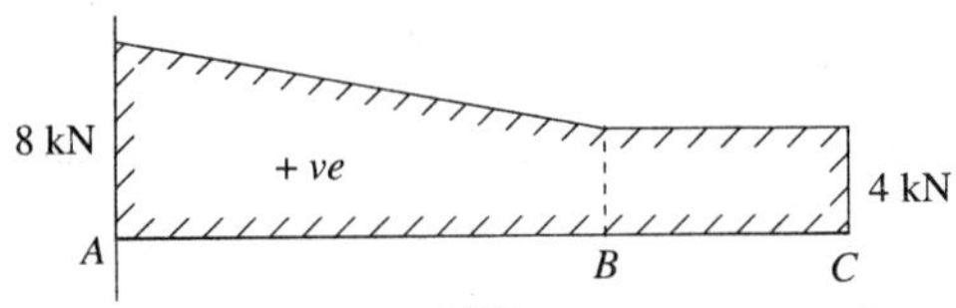

5 FD

Shear force varies linearly under UDL between A and B and is constant between B and C.

3. *BMD*

$$M_C = 0 \text{ (Free end)}$$

$$M_B = P_C \times BC = 4 \times 2 = -\,8 \text{ kNm (Hogging)}$$

$$M_A = -\,32 \text{ kNm.}$$

Bending moment varies linearly between B and C (point load) and parabolically under UDL between B and A.

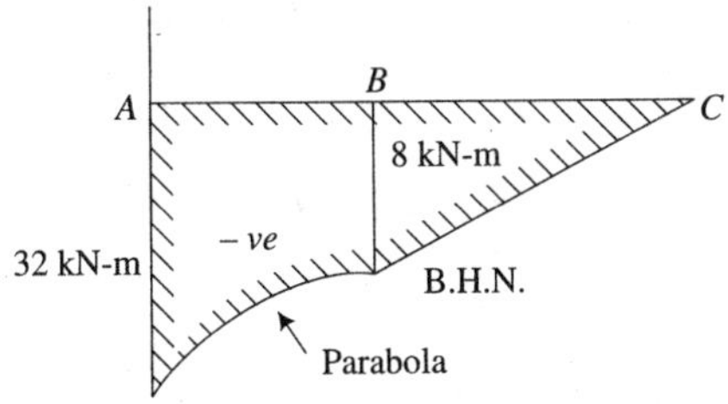

Example 3.15(b): Calculate the values of shear force and bending moments for the cantilever beam shown. Also draw the shear force and bending moment diagram. [U.P.T.U. 2008-09]

Solution:

Draw FBD of the cantilever.

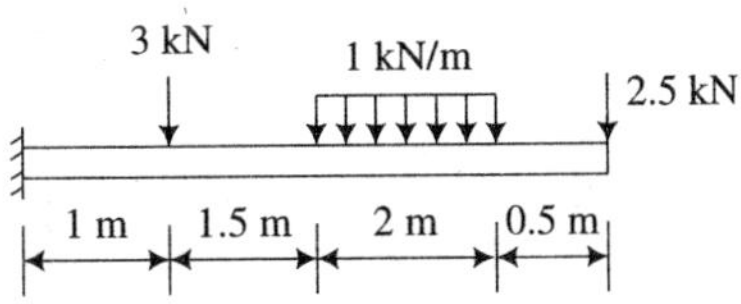

1. *Support Reactions*

$$\Sigma F_y = 0$$

$$R_A = 3 + (1 \times 2) + 2.5 = 7.5 \text{ kN.}$$

$$\Sigma MA = 0$$

$$MA = 3 \times 1 + 2 \times 3.5 + 2.5 \times 5$$

$$= 22.5 \text{ kNm}$$

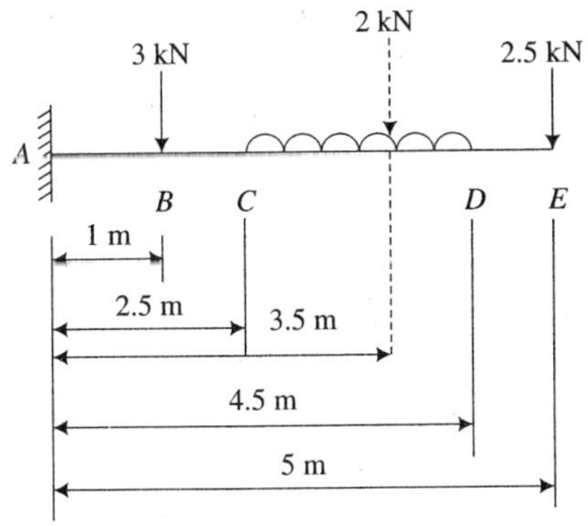

2. *SFD*

$$F_A = R_A = +\,7.5 \text{ kN}$$

$$F_B = F_A - P_B = 7.5 - 3 = +\,4.5 \text{ kN}$$

$$F_C = F_B = +\,4.5 \text{ kN.}$$

$$F_D = F_C - W_{CD} = 4.5 - 2 = +\,2.5 \text{ kN.}$$

$$F_E = P_E = +\,2.5 \text{ kN.}$$

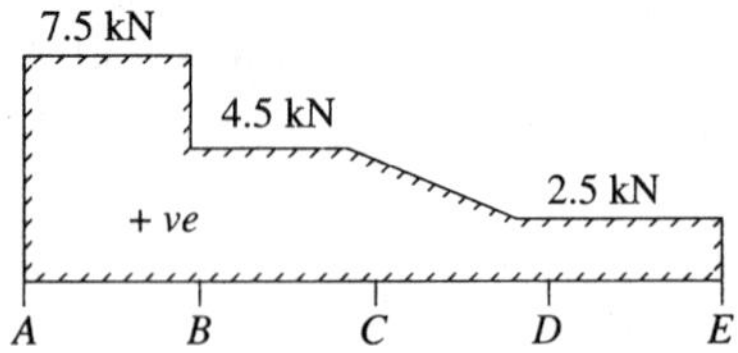

Shear force is constant between $A-B$, $B-C$ and $D-E$ and varies linearly between C and D under UDL.

3. *BMD*

$$M_E = 0 \text{ (Free end)}$$

$$M_D = P_E \times DE = 2.5 \times 0.5$$

$$= -1.25 \text{ kNm (Hogging)}$$

$$M_C = P_E \times EC + W_{CD} \times \frac{CD}{2}$$

$$= 2.5 \times 2.5 + 2 \times 1 = -8.25 \text{ kNm.}$$

$$M_A = -22.5 \text{ kNm}$$

$$M_B = M_A - R_A \times AB$$

$$= -22.5 + 7.5 \times 1 = -15 \text{ kNm}$$

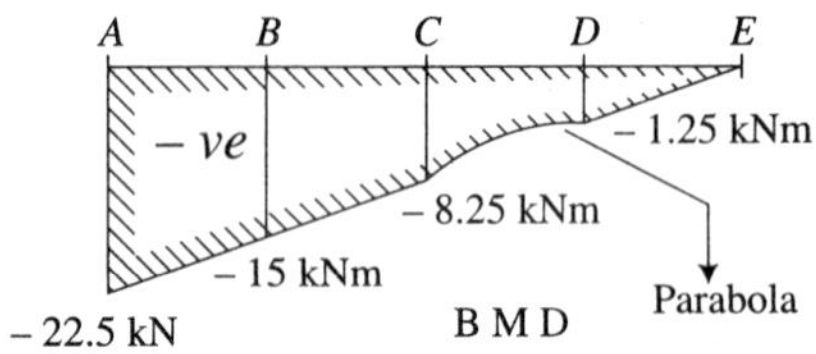

Bending moment varies linearly between $A-B$, $B-C$ and DE and varies parabolically under UDL between C and D.

Example 3.15(c): Find the shear force and bending moment for the cantilever shown. Also sketch the shear force and bending moment diagram.

[U.P.T.U. 2008-09]

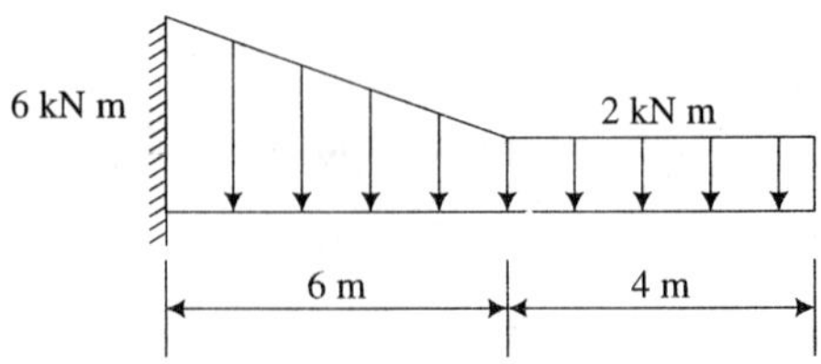

Solution:

Draw FBD of the cantilever.

1. *Support Reaction*

$$\Sigma F_y = 0 \quad R_A = 12 + 20 = 32 \text{ kN}$$

$$\Sigma M_A = 0 \quad M_A = 12 \times 2 + 20 \times 5 = 124 \text{ kNm.}$$

The load can be divided into two parts.

1. Uniformly distributed load of 2 kNm over a span of 10 m.
2. Uniformly varying load from zero to 4 kNm over a span of 6 m.

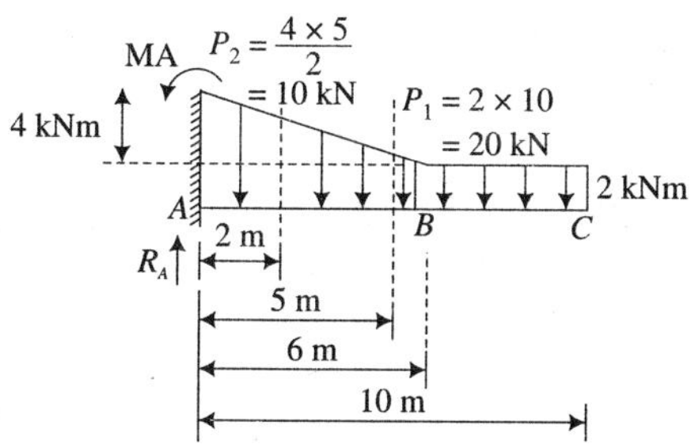

2. *SFD*

$$F_A = R_A = +\,32 \text{ kN}$$

$$F_C = 0$$

$$F_B = W_{BC} = 2 \times 4 = +\,8 \text{ kN}$$

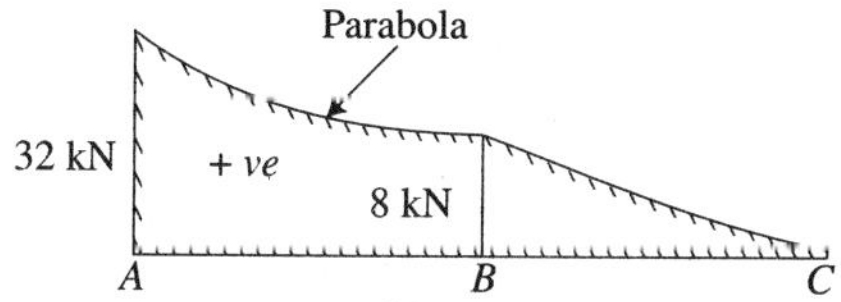

SFD

SF varies linearly from *B* – *C* under UDL and parabolically from *B* – *A* under uniformly varying load.

3. *BMD*

$$M_C = 0 \text{ (free end)}$$

$$M_B = W_{BC} \times \frac{BC}{2} = 2 \times 4 \times \frac{4}{2} = -16 \text{ kNm (hogging)}$$

$$M_A = -124 \text{ kNm.}$$

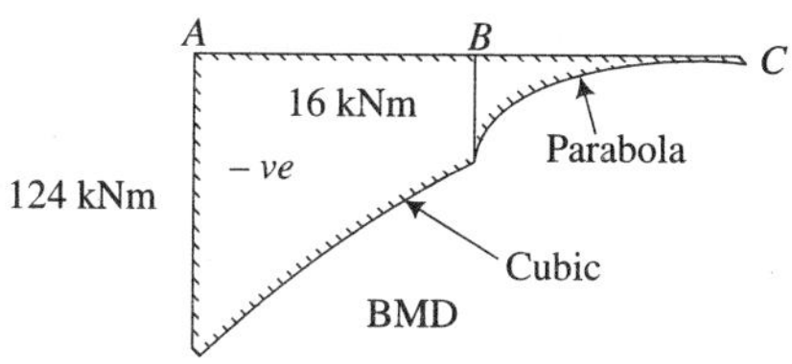

BMD

Example 3.16: A simply supported beam of span L carries a concentrated load W at the mid-span. Draw SFD and BMD.

Solution:

1. Draw load diagram.

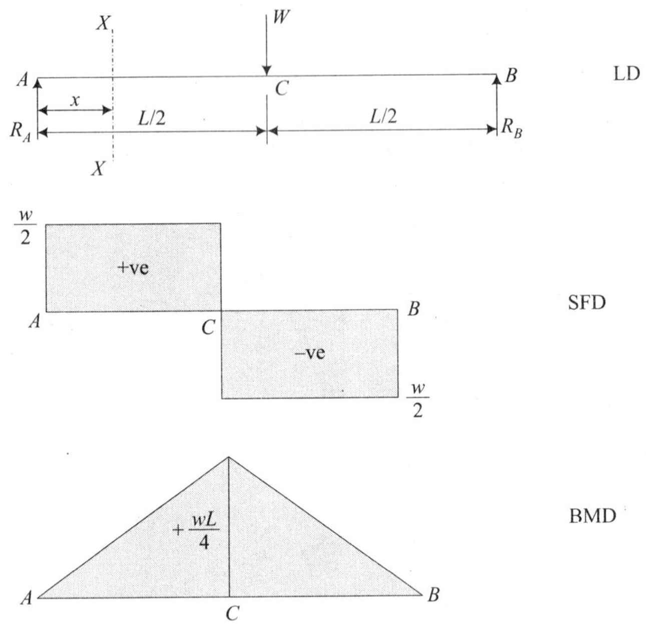

2. *Support Reactions*

   The loading is symmetrical

$$\therefore \qquad R_A = R_B = \frac{W}{2}$$

$$M_A = M_B = 0 \text{ at free supports.}$$

3. *Shear Force*

$$F_A = R_A = +\frac{W}{2}$$

$$F_C = F_A - W_C = \frac{W}{2} - W = -\frac{W}{2}$$

$$F_B = R_B = -\frac{W}{2}$$

   SFD is horizontal straight lines under concentrated loads.

4. *Bending Moment*

   The beam is sagging, bending moment is +ve.

   Bending moment at section *X-X*,

$$M_X = R_A x = \frac{W}{2} x$$

$$\left.\begin{array}{lll} \text{At} & x = 0 & M = 0 \\ \text{At} & x = L & M = 0 \end{array}\right\} \text{Free supports}$$

$$\text{At} \quad x = \frac{L}{2} \quad M = \frac{WL}{4}$$

BMD is inclined straight line under concentrated load.

Example 3.17: The loading of a simply supported beam AB of span 8 m is shown. Draw SFD and BMD.

Solution:

1. Draw load diagram.

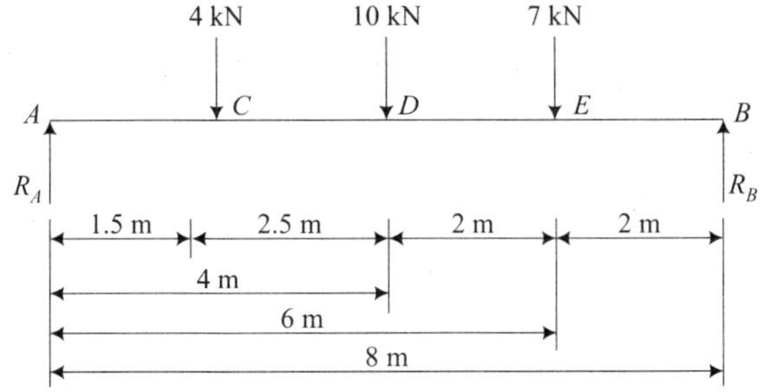

LD

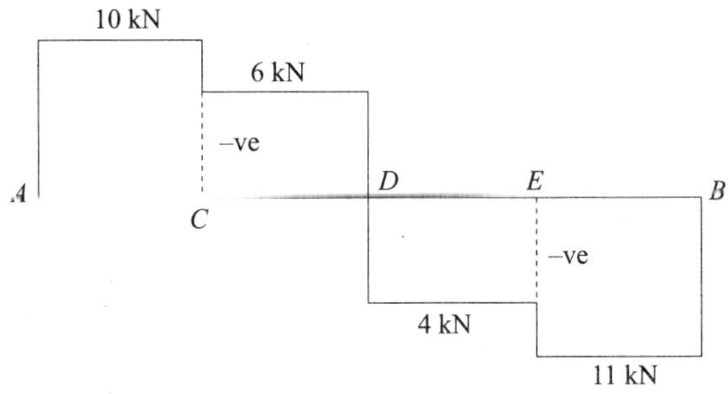

SFD

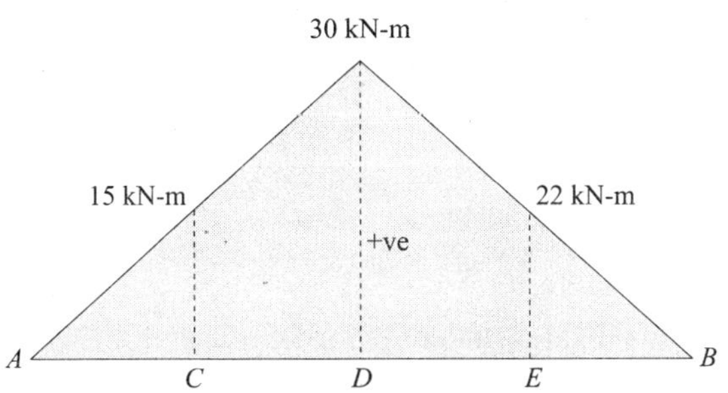

BMD

2. *Support Reactions*

   Apply equilibrium conditions,

   (i) $\Sigma M_A = 0$

$$R_B \times 8 = 4 \times 1.5 + 10 \times 4 + 7 \times 6$$

$$R_B = 11 \text{ kN}$$

   (ii) $\Sigma F_y = 0$

$$R_A + R_B = 4 + 10 + 7 = 21 \text{ kN}$$

$$\therefore \quad R_A = 21 - R_B = 10 \text{ kN}$$

3. *Shear Force*

$$F_A = R_A = 10 \text{ kN}$$

$$F_C = F_A - W_C = 11 - 4 = 6 \text{ kN}$$

$$F_D = F_C - W_D = 6 - 10 = -4 \text{ kN}$$

$$F_E = F_D - F_E = -6 - 7 = -11 \text{ kN}$$

$$F_B = R_B = -11 \text{ kN}$$

   SFD is horizontal straight lines under concentrated loads.

4. *Bending Moment*

   Bending moment is +ve as beam is sagging.

$$M_A = M_B = 0 \text{ (Free supports)}$$

$$M_C = R_A \times AC = 10 \times 1.5 = 15 \text{ kN-m}$$

$$M_D = R_A \times AD - W_C \times CD = 10 \times 4 - 4 \times 2.5 = 30 \text{ kN-m}$$

$$M_E = R_B \times BE = 11 \times 2 = 22 \text{ kN-m}$$

   BMD is straight inclined lines between concentrated loads.

Example 3.18: A simply supported beam is shown. Draw SFD and BMD.

[U.P.T.U. I Sem., 2004–05]

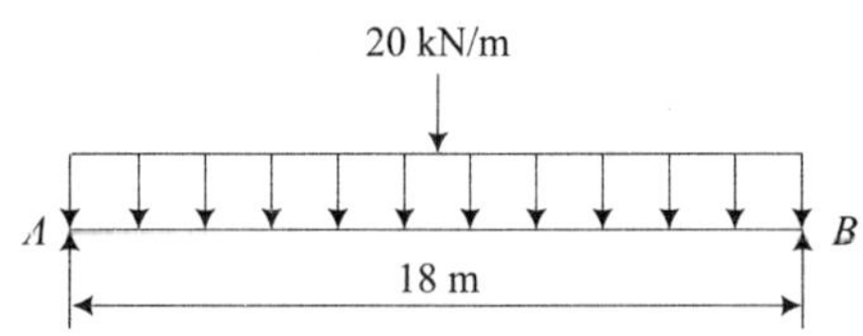

Solution:

1. Draw equivalent load diagram.

Equivalent load W is equal to area under the load curve and acts at c.g. of load curve.

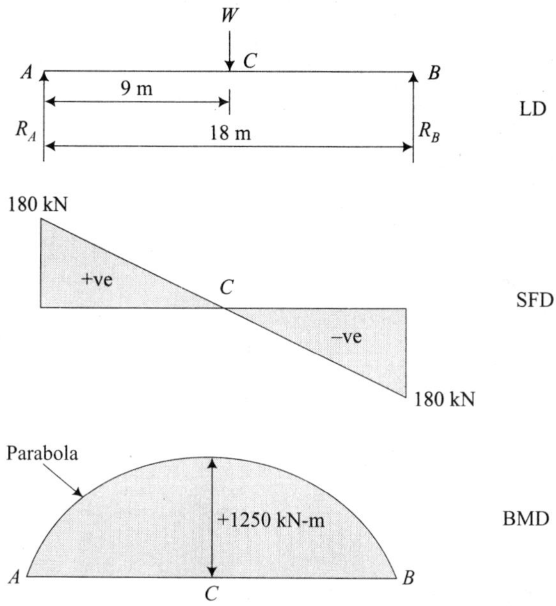

$$\therefore \qquad W = 20 \times 18 = 360 \text{ kN}$$

and acts at mid-span.

2. *Support Reactions*

The beam is symmetrically loaded,

$$\therefore \qquad R_A = R_B = \frac{W}{2} = 180 \text{ kN}$$

3. *Shear Force*

$$F_A = R_A = 180 \text{ kN.}$$

$$F_B = R_B = -180 \text{ kN}$$

$$F_C = F_A - \frac{wL}{2} = 0$$

SFD is a straight inclined line under UDL.

4. *Bending Moment*

$$M_A = M_B = 0 \text{ (Free supports)}$$

$$M_C = R_A \times \frac{L}{2} - \frac{WL}{2} \times \frac{L}{4}$$

$$= 180 \times 9 - 180 \times \frac{18}{4} = 1620 - 810$$

$$= 810 \text{ kN-m}$$

Check: $$M_C = \frac{WL^2}{8} = \frac{20 \times 18^2}{8} = +810 \text{ kN-m}$$

The beam is sagging. Therefore, bending moment is +ve. BMD is a parabola under UDL.

Example 3.19: A simply supported beam is subjected to various loadings as shown. Sketch the shear force and bending moment diagrams showing their values at significant locations. [U.P.T.U.I Sem., 2005–06]

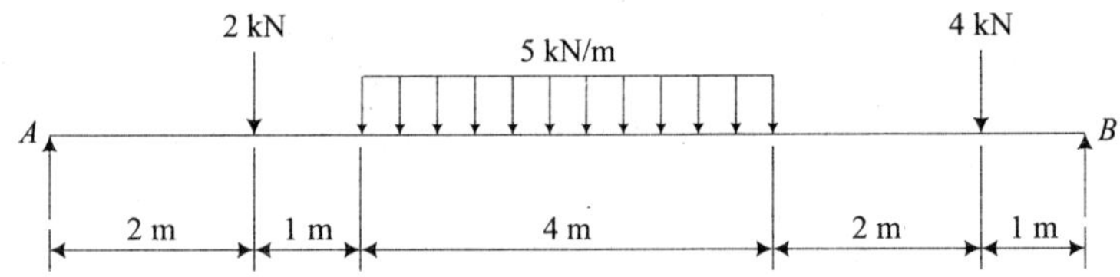

Solution:

1. Draw equivalent load diagram.

   The equivalent load of UDL will be $5 \times 4 = 20$ kN. i.e., area under UDL curve and will act at c.g of load curve.

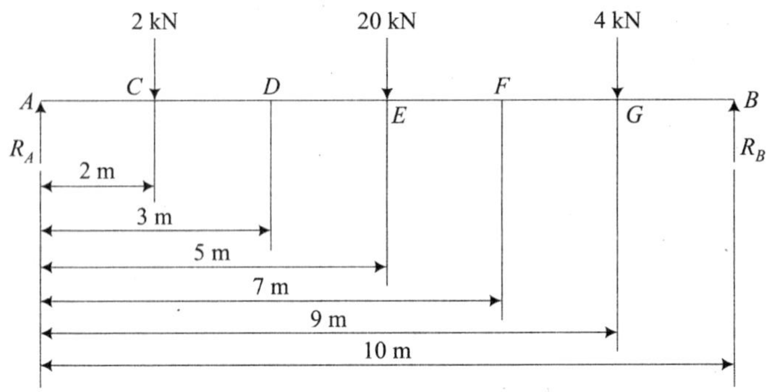

LD

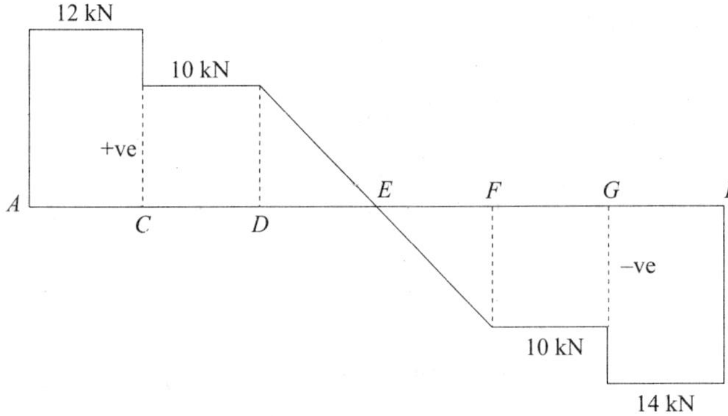

SFD

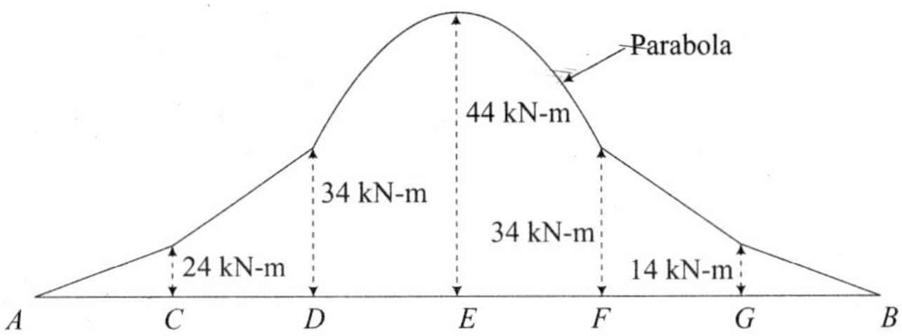

2. *Support Reactions*

   Apply equilibrium condition

(i) $\Sigma F_Y = 0$

$$R_A + R_B = 2 + 20 + 4 = 26 \text{ kN}$$

(ii) $\Sigma M_A = 0$

$$R_B \times 10 = 2 \times 2 + 20 \times 5 + 4 \times 9 = 4 + 100 + 36 = 140 \text{ kN-m}$$

$$R_B = 14 \text{ kN}$$

$\therefore$ $$R_A = 26 - 14 = 12 \text{ kN}$$

3. *Shear Force*

$$F_A = R_A = 12 \text{ kN}$$

$$F_C = F_A - W_C = 12 - 2 = 10 \text{ kN}$$

$$F_D = F_C = 10 \text{ kN}$$

$$F_B = R_B = -14 \text{ kN}$$

$$F_G = F_B - W_G = -14 + 4 = -10 \text{ kN}$$

$$F_F = F_G = -10 \text{ kN}$$

$$F_E = F_D - W_{DE} = 10 - 5 \times 2 = 0$$

SFD is horizontal straight line between *AD* and *FB* and straight inclined line under UDL between *D* and *F*.

4. *Bending Moment*

   Bending moment is +ve for sagging beam.

$$M_A = M_B = 0 \text{ (Free supports)}$$

$$M_C = R_A \times AC = 12 \times 2 = 24 \text{ kN-m}$$

$$M_D = R_A \times AD - W_C \times CD = 12 \times 3 - 2 \times 1 = 34 \text{ kN-m}$$

$$M_G = R_B \times GB = 14 \times 1 = 14 \text{ kN-m}$$

$$M_F = R_B \times BF - WG \times GF = 14 \times 3 - 4 \times 2 = 34 \text{ kN-m}$$

$$M_E = R_A \times AE - W_C \times CE - W_{DE} \times \frac{DE}{2}$$

$$= 12 \times 5 - 2 \times 3 - 10 \times 1$$

$$= 44 \text{ kN-m}$$

BMD is straight oblique line between AD and FB under concentrated loads and a parabola between D and F under UDL.

Example 3.20: Draw the shear force and bending moment diagrams for the beam shown.

[U.P.T.U. II Sem., 2007–08]

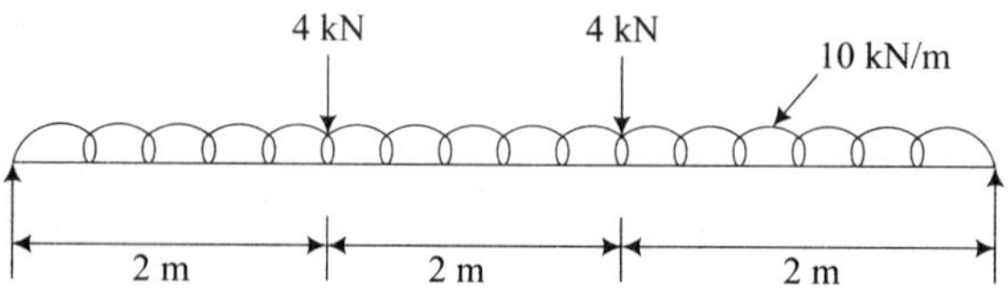

Solution: (a) *Draw equivalent load diagram*

$$W_B = 4 \text{ kN}$$

$$W_C = 4 \text{ kN}$$

$$W_{AB} = W_E = 10 \times 2 = 20 \text{ kN}$$

$$W_F = W_{BC} = 2 \times 10 = 20 \text{ kN}$$

$$W_G = W_{CD} = 2 \times 10 = 20 \text{ kN}$$

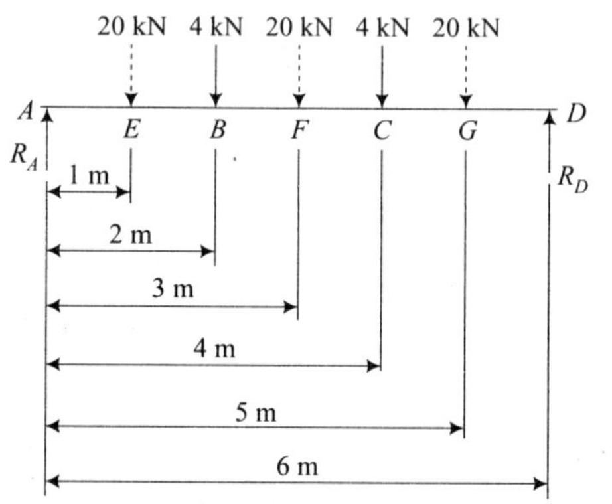

ELD

(b) *Reactions*

$$R_A = R_D \text{ due to symmetry}$$

$\therefore$
$$R_A + R_D = 20 + 4 + 20 + 4 + 20$$

$$= 68 \text{ kN}$$

$$R_A = R_D = 34 \text{ kN}$$

(c) *Shear Force Diagram*

$$F_A = +R_A = +34 \text{ kN}$$
$$F_{BL} = F_A - W_{AB} = +34 - 20 = +14 \text{ kN}$$
$$F_{BR} = F_{BL} - W_B = +14 - 4 = +10 \text{kN}$$
$$F_D = -R_D = -34 \text{ kN}$$
$$F_{CR} = F_D - W_{CD} = -34 + 20 = -14 \text{ kN}$$
$$F_{CL} = F_{CR} - W_C = -14 + 4 = -10 \text{ kN}$$

SFD is straight inclined line under *UDL*

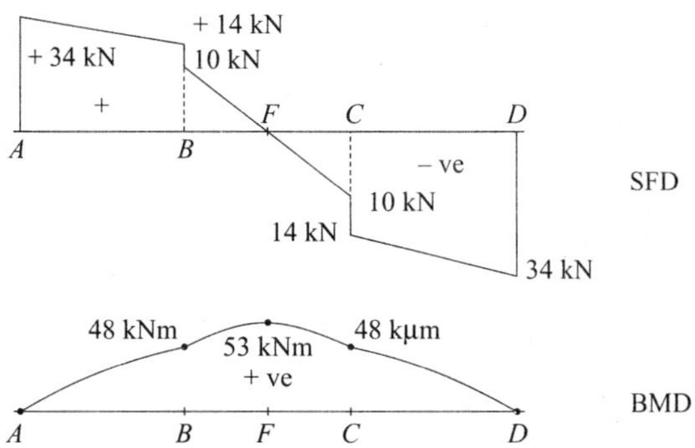

(d) *Bending Moment Diagram*

$$M_A = M_D = 0 \text{ (Free supports)}$$

$$M_B = R_A(AB) - \frac{w(AB)^2}{2} = 34 \times 2 - \frac{10(2)^2}{2} = +48 \text{ kNm}$$

(Sagging)

$$M_F = R_A(AF) - 4(BF) - \frac{w(AF)^2}{2}$$

$$= 34 \times 3 - 4 \times 1 - \frac{10(3)^2}{2} = +53 \text{ kN-m}$$

$$M_C = M_B = +48 \text{ kN-m (Symmetry)}$$

BMD is parabola under *UDL*.

Example 3.21: Draw SFD and BMD of the overhanging beam *AE* as shown and find out point of contraflexure.

Solution:

1. Draw the load diagram.
2. *Support Reactions*

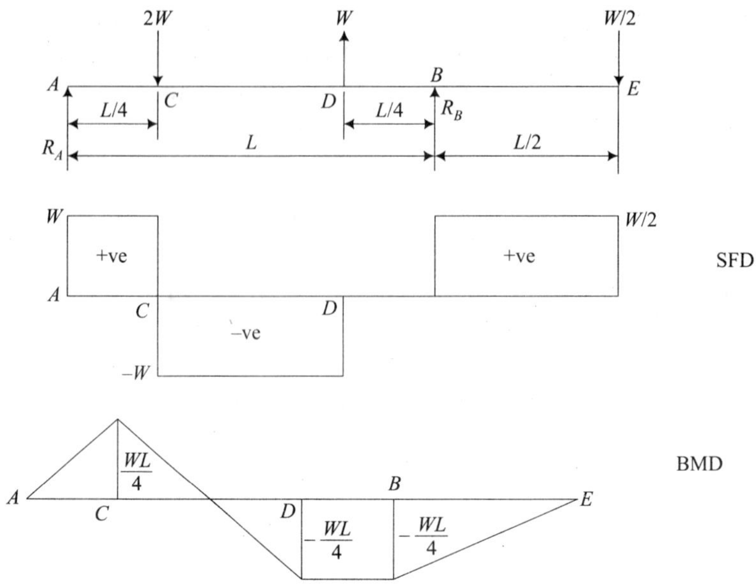

Apply equilibrium conditions

(i) $\Sigma F_Y = 0$

$$R_A + R_B = 2w - w + \frac{W}{2} = \frac{3W}{2}$$

(ii) $\Sigma M_A = 0$

$$R_B \times L + W \times \frac{3L}{4} - 2W \times \frac{L}{4} - \frac{W}{2} \times \frac{3L}{2} = 0$$

$$R_B = \frac{W}{2}$$

$$R_A = \frac{3W}{2} - \frac{W}{2} = W$$

3. *Shear Force*

$$F_A = R_A = +W$$

$$F_C = R_A - W_C = W - 2W = -W$$

$$F_D = F_C - W_D = -W + W = 0$$

$$F_B = R_B = \frac{W}{2}$$

$$F_E = W_E = \frac{W}{2}$$

4. *Bending Moment*

$$M_A = 0 \qquad \text{Free support}$$

$$M_E = 0 \qquad \text{Free end}$$

$$M_C = R_A \times A_C = W \times \frac{L}{2} = + \frac{WL}{4} \text{ (sagging)}$$

$$M_D = R_A \times A_D - W_C\,(CD)$$

$$= W \times \frac{3L}{4} - 2W \times \frac{L}{2} = - \frac{WL}{4} \text{ (hogging)}$$

$$M_B = W_E\,(BE) = \frac{W}{2} \times \frac{L}{2} = - \frac{WL}{4} \text{ (hogging)}$$

The point of contraflexure is between C and D.

Example 3.21(a): A beam is loaded as shown. Draw its shear force and bending moment diagrams. [U.P.T.U. 2009-10]

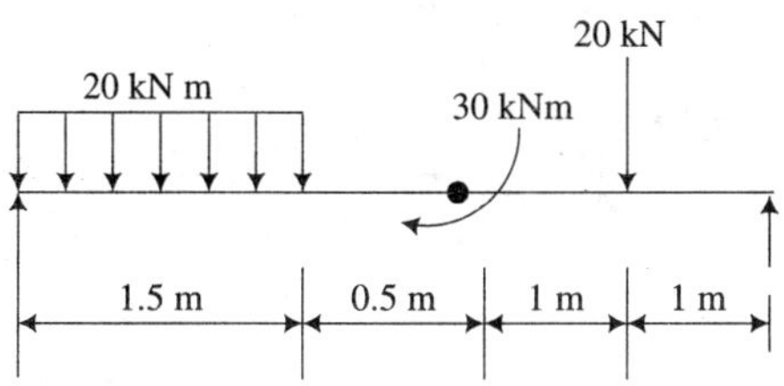

Solution:

Draw FBD diagram.

1. *Supply Reactions*

$$\Sigma F_y = 0 \quad R_A + R_E = 30 + 20 = 50 \text{ kN}$$

$$\Sigma M_A = 0$$

$$R_E \times 4 = 30 \times 0.75 + 30 + 20 \times 3$$

$$= 22.5 + 30 + 60 = 112.5 \text{ kNm}$$

$$R_E = \frac{112.5}{4} = 28.125 \text{ kN.}$$

$$R_A = 50 - 28.125 = 21.875 \text{ kN.}$$

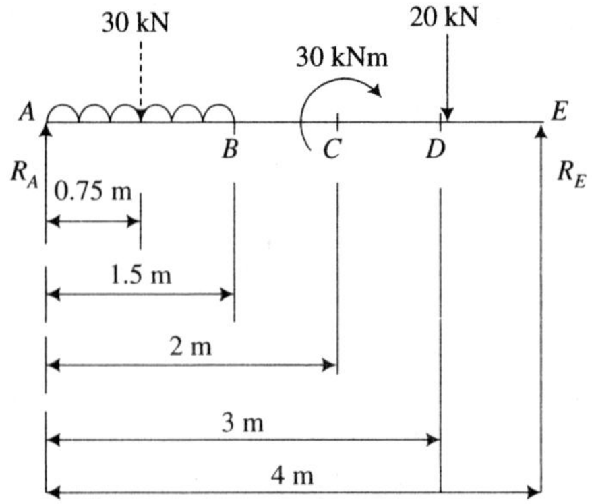

2. *SFD*

$$F_A = R_A = +\,21.875 \text{ kN}$$

$$F_B = F_A - W_{AB} = 21.875 - 30 = -\,8.125 \text{ kN}$$

$$F_C = F_B = -\,8.125 \text{ kN}$$

$$F_D = F_C - P_D = -8.125 - 20 = -\,28.125 \text{ kN}$$

$$F_E = R_E = -\,28.125 \text{ kN}$$

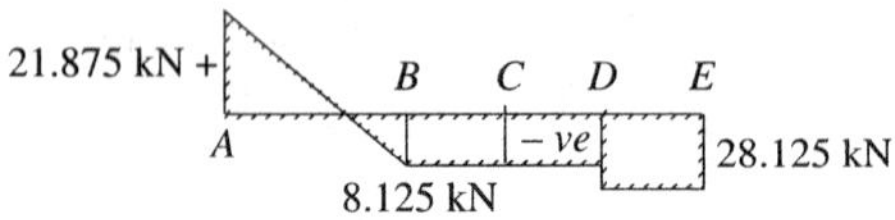

SFD

Shear force varies linearly under UDL over span *AB* and is constant over span *BD* and suddenly changes under concentrated load P_D.

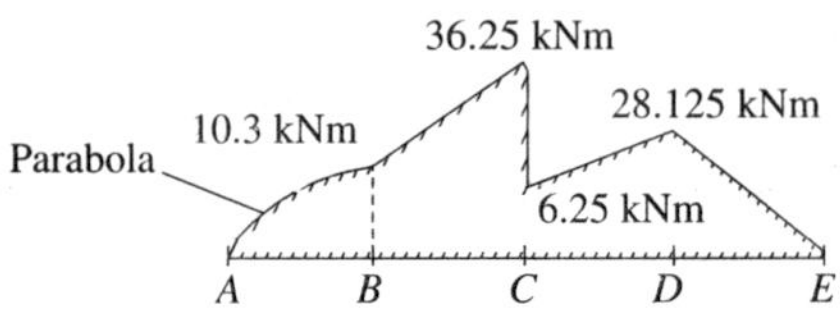

BMD

3. *BMD*

$$M_A = 0 \text{ (Freely supported end)}$$

$$M_E = 0 \text{ (Freely supported end)}$$

$$M_B = R_A \times AB - W_{AB} \times \frac{AB}{2}$$

$$= 21.875 \times 1.5 - 30 \times 0.75$$

$$= 32.8125 - 22.5 = +\,10.3 \text{ kNm}$$

$$M_D = R_E \times ED = 28.125 \times 1 = +\,28.125 \text{ kNm}$$

$$M_C = R_E \times CE - 20 \times CD.$$

$$= 28.125 \times 2 - 20 \times 1 = 36.25 \text{ kNm}$$

Bending moment changes parabolically under UDL over the span *AB* and varies linearly over the spans *BC*, *CD* and *DE*. There is a sudden drop in bending moment at point *C*.

Example 3.21(b): Draw the shear force and bending moment diagram for the beam loaded as shown. [U.P.T.U. 2008-09]

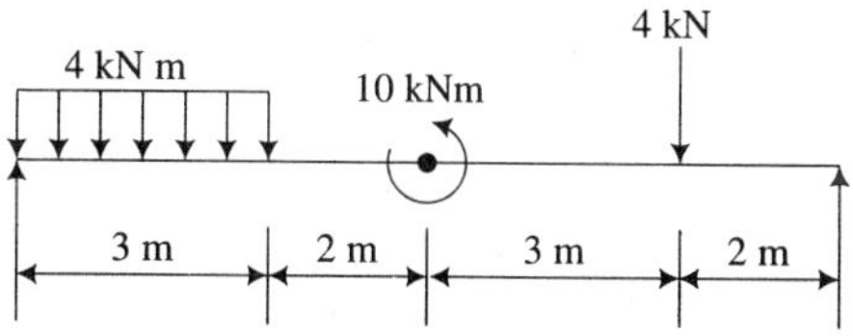

Solution:

Draw FBD of the beam.

1. *Support Reaction*

$$\Sigma F_y = 0$$

$$R_A + R_E = 12 + 4 = 16 \text{ kN}.$$

$$\Sigma M_A = 0$$

$$R_E \times 10 + 10 = 4 \times 8 + 12 \times 1.5$$

$$R_E = \frac{32 + 18 - 10}{10} = 4 \text{ kN}$$

$$R_A = 16 - 4 = 12 \text{ kN}.$$

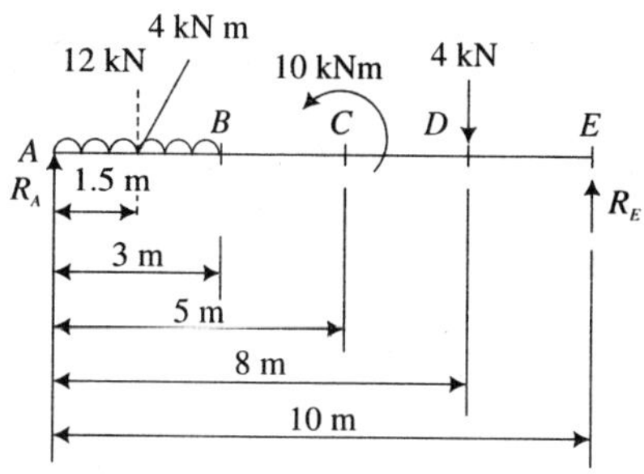

2. *SFD*

$$F_A = R_A = +\ 12 \text{ kN}$$

$$F_B = F_A - W_{AB} = 12 - 12 = 0$$

$$F_C = F_B = 0$$

$$F_D = F_C = 0$$

$$F_E = R_E = -\ 4 \text{ kN}$$

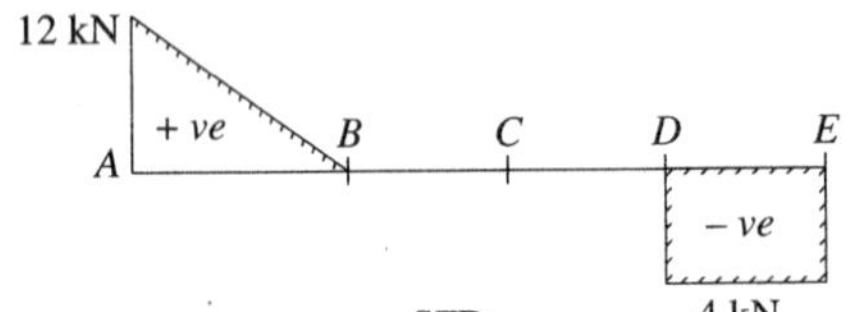

SFD

Shear force varies linearly under UDL over the span *AB* and is constant over the spaces *BC*, *CD*, and *DE*.

3. *BMD*

$$M_A = 0 \text{ (Free support end)}$$

$$M_E = 0 \text{ (Free support end)}$$

$$M_B = R_A \times AB - W_{AB} \times \frac{AB}{2}$$

$$= 12 \times 3 - 12 \times 1.5 = +\ 18 \text{ kNm}$$

$$M_C = M_B - 10 = +\ 8 \text{ kNm}$$

$$M_D = R_E \times DE = 4 \times 2 = +\ 8 \text{ kNm}$$

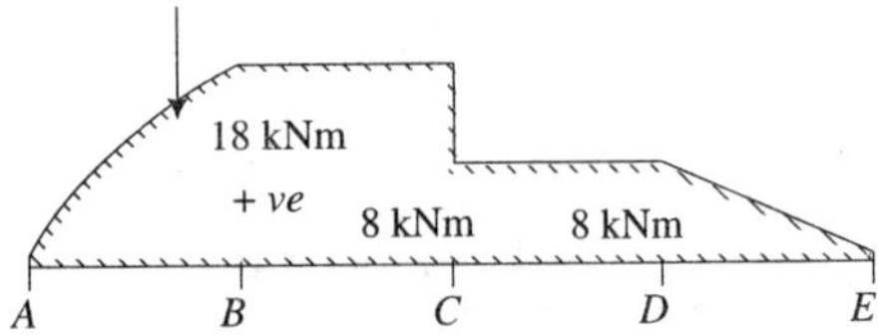

BMD

Bending moment varies parabolically under UDL over the span *AB* and is constant over the spans *BC* and *CD* as shear force is zero. It varies linearly over the span *DE*.

Example 3.21(c): Draw the shear force and bending moment diagrams for the beam shown in figure. [U.P.T.U. 2008-09]

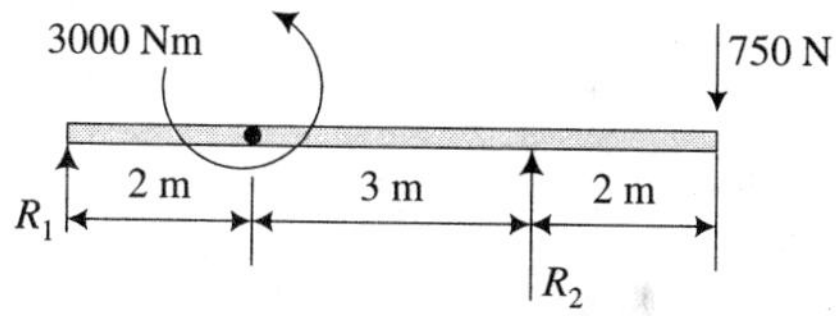

Solution:

Draw FBD of the beam.

1. *Support Reaction*

$$\Sigma Fy = 0$$

$$R_1 + R_2 = 750 \text{ N.}$$

$$\Sigma M_A = 0$$

$$R_2 \times 5 + 3000 - 750 \times 7 = 0$$

$$R_2 = \frac{750 \times 7 - 300}{5} = 450 \text{ N.}$$

$$R_1 = 750 - 450 = 300 \text{ N}$$

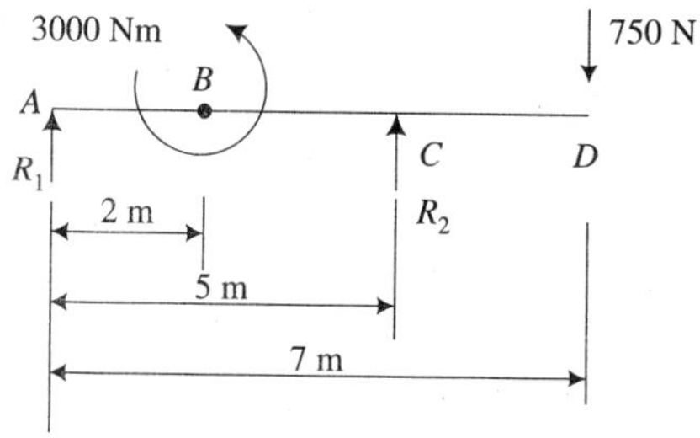

2. *SFD*

$$F_A = R_A = +\,300 \text{ N.}$$

$$F_B = F_A = +\,300 \text{ N.}$$

$$F_D = 750 \text{ N} = F_C$$

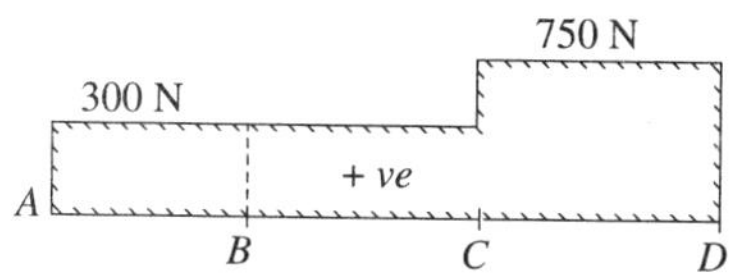

3. *BMD*

$$M_A = 0 \text{ (Free support end)}$$

$$M_D = 0 \text{ (Free end)}$$

$$M_B = R_1 \times AB = 300 \times 2 = +\,600 \text{ Nm.}$$

$$M_C = 750 \times 2 = -\,1500 \text{ Nm}$$

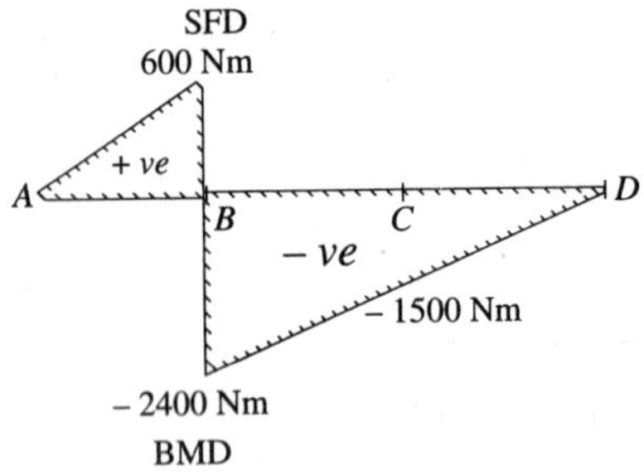

Bending movement varies linearly over the spans AB, BC and CD. There is a sudden decrease of 300 Nm at point B.

Example 3.22: The bending moment diagram (BMD) of a simply supported

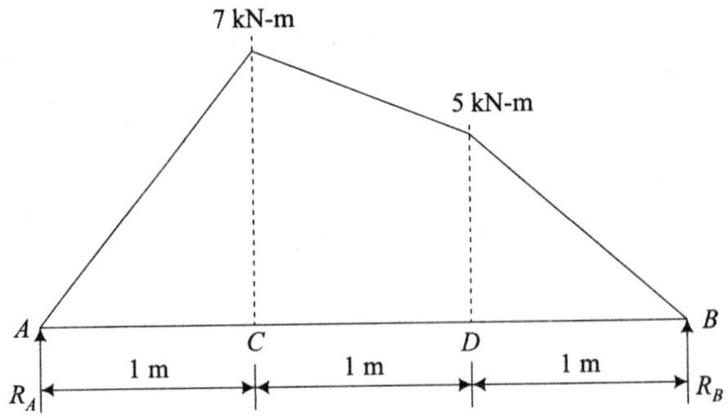

beam is shown. Calculate the support reactions of the beam.

[U.P.T.U. I Sem., 2000–01]

Solution: The BMD is drawn.

$$M_C = R_A \times AC$$

$$\therefore \qquad R_A = \frac{M_C}{AC} = \frac{7}{1} = 7 \text{ kN}$$

$$M_D = R_B \times DB$$

$$\therefore \qquad R_B = \frac{M_D}{DB} = \frac{5}{1} = 5 \text{ kN}$$

Example 3.23: The shear force diagram (SFD) of a simply supported beam is shown. Calculate the support reactions of the beam and also draw BMD.

Solution:

1. Draw load diagram.

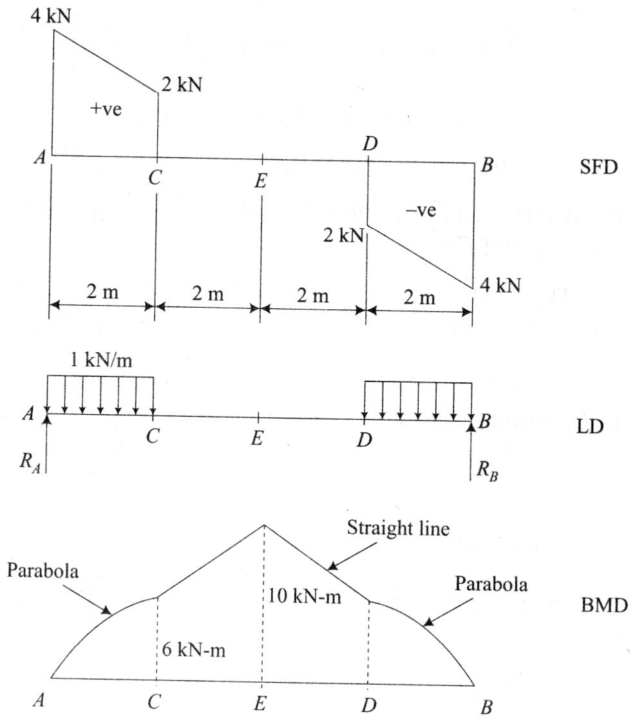

2. *Support Reactions* $F_A = R_A = 4$ kN

$$F_C = F_A - W_C$$

$$\therefore \quad W_C = F_A - F_C = 4 - 2 = 2 \text{ kN}$$

The SFD is straight inclined line. Therefore, there is UDL between *AC*.

$$W = wL$$

$$\therefore \quad w = \frac{W}{L} = \frac{2}{2} = 1 \text{ kN/m}$$

$$F_B = R_B = 4 \text{ kN}$$

Similarly, there is UDL of 1 kN/m between. *B* and *D*.

3. *Bending Moment*

$$M_A = M_B = 0 \text{ (Free support)}$$

$$M_C = w\,(AC).\left(\frac{AC}{2}\right) - R_A\,(AC)$$

$$= -1 \times 2 \times 1 + 4 \times 2$$

$$= 6 \text{ kN-m (sagging)}$$

$$M_E = R_A (AE) = w(AC) \times \left(CE + \frac{AC}{2} \right)$$

$$= 4 \times 4 - 1 \times 2 \times 3 = 10 \text{ kN-m}$$

$$M_D = M_C = 6 \text{ kN-m.}$$

BMD is a parabola in the portion AC and BD and straight inclined lines in the portion CE and ED.

Example 3.24: The SFD of a simply supported beam is shown. Calculate the support reactions of the beam and draw BMD. [U.P.T.U. I Sem., 2001–02]

Solution:

1. *Loading Diagram*

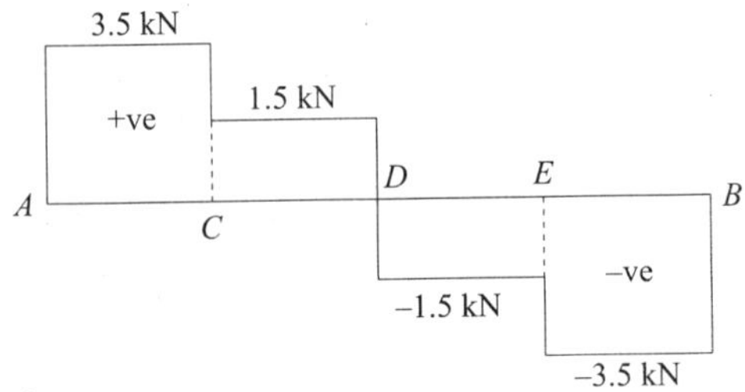

SFD

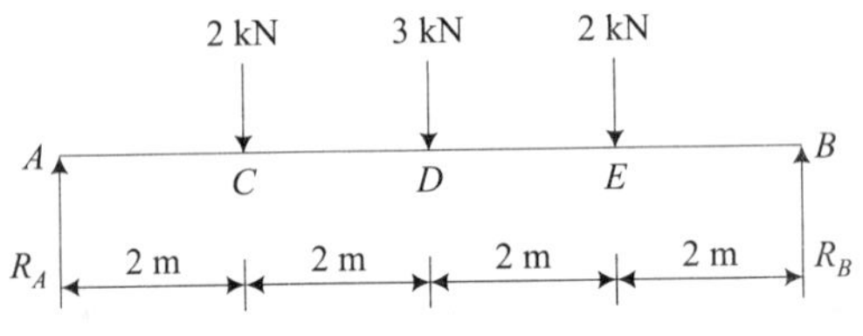

LD

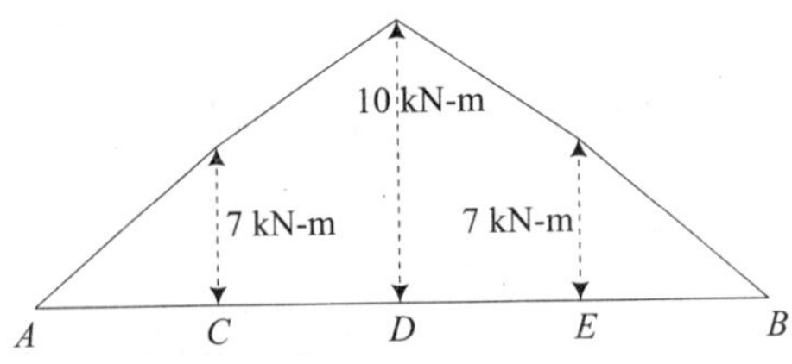

BMD

$$F_A = R_A = 3.5 \text{ kN}$$

$$F_C = F_A - W_C$$

$\therefore$ $$W_C = F_A - F_C$$

$$= 3.5 - 1.5 = 2 \text{ kN}$$

$$F_D = F_C - W_D$$

$$\therefore \quad W_D = F_D - F_C = 1.5 - (-1.5)$$

$$= 3.0 \text{ kN}$$

$$F_B = R_B = -3.5 \text{ kN}$$

$$F_E = F_D - W_E = -1.5 - (-3.5)$$

$$= 2 \text{ kN}$$

2. *Bending Moment*

$$M_A = M_B = 0 \text{ (Free supports)}$$

$$M_C = R_A (AC) = 3.5 \times 2 = 7 \text{ kN-m (+ve sagging)}$$

$$M_D = R_A (AD) - W_C (CD)$$

$$= 3.5 \times 4 - 2 \times 2 = 10 \text{ kN-m (+ve sagging)}$$

$$M_E = M_C = 7 \text{ kN-m}$$

BMD is straight inclined lines between concentrated loads.

Example 3.25: The shear force diagram of a simply supported beam at A and B is shown. Calculate the support reactions and draw the bending moment diagram.

[U.P.T.U. I Sem., 2002–03]

Solution:

1. The load diagram is drawn.

$$F_C = W_C = 10 \text{ kN (hogging)}$$

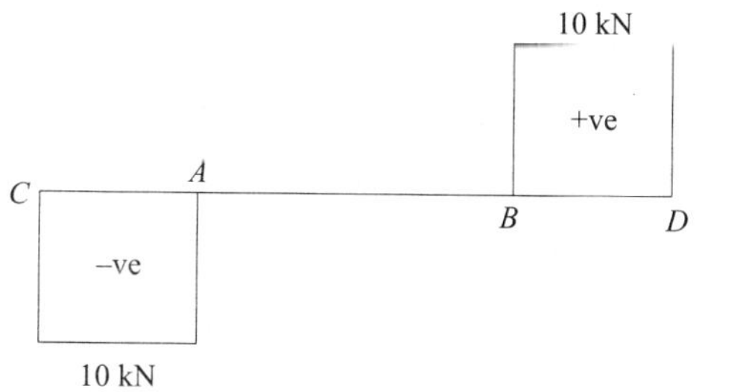

SFD

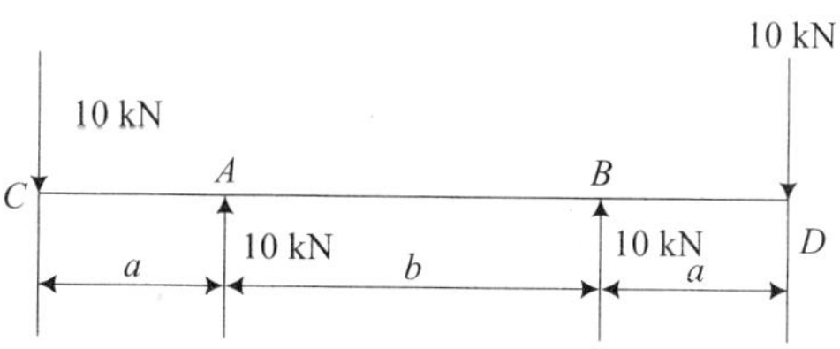

LD

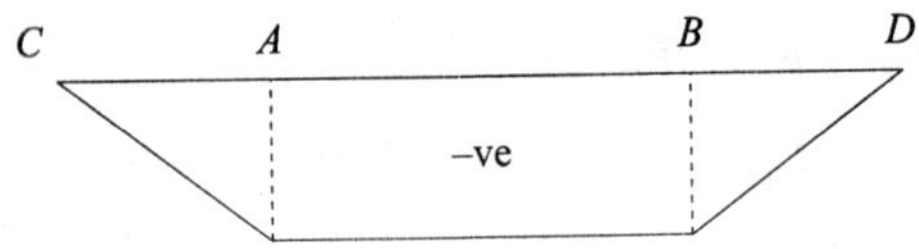

BMD

$$F_A = F_C - R_A = 0$$

$$\therefore \quad R_A = F_C = 10 \text{ kN}$$

By symmetry

$$W_D = 10 \text{ kN}$$

$$R_B = 10 \text{ kN}$$

2. *Bending Moment*

$$M_C = M_D = 0 \text{ (Free ends)}$$

$$M_A = R_A \times (a) = -\ 10a \text{ kN-m (Hogging)}$$

$$M_B = M_A = -\ 10a \text{ kN-m}$$

Example 3.26: Draw the load diagram for the SF diagram shown for a simply supported beam. Calculate the maximum bending moment and its location.

[U.P.T.U. I Sem., 2001–02]

Solution:

1. *Load Diagram*

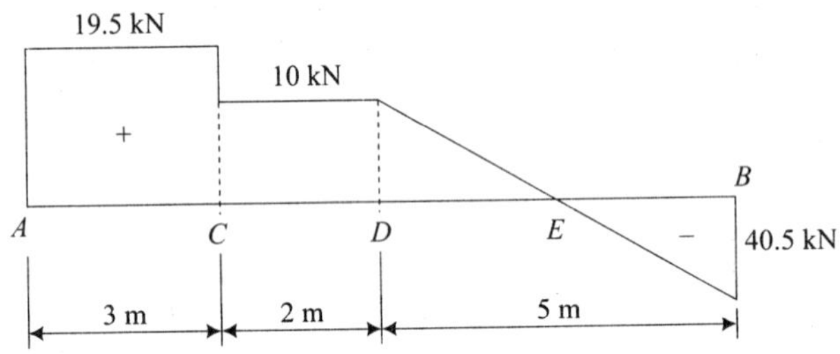

SFD

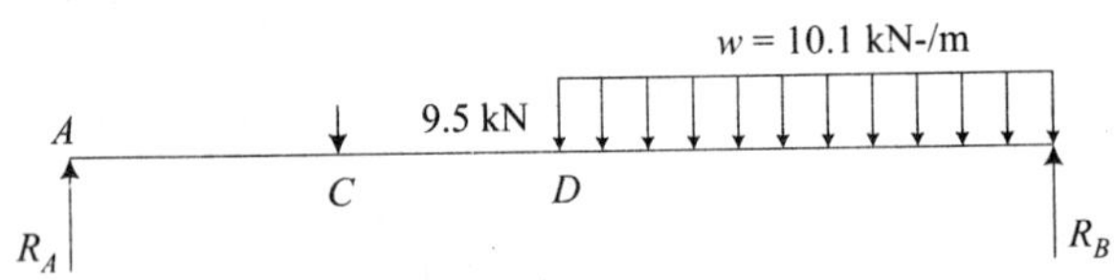

LD

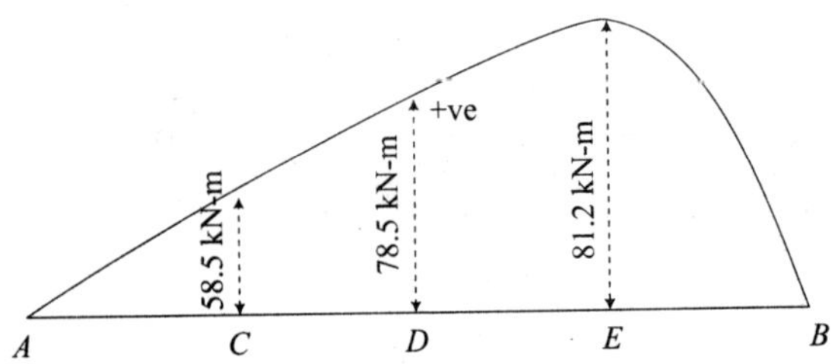

BMD

$$F_A = R_A = 19.5 \text{ kN}$$

$$F_B = R_B = 40.5 \text{ kN}$$

$$F_C = F_A - W_C$$

$$W_C = F_A - F_C = 19.5 - 10 = 9.5 \text{ kN}$$

$$R_A + R_B = W_C + w\,(DB)$$

$$\therefore \quad w = \frac{R_A + R_B - W_C}{DB}$$

$$= \frac{19.5 + 40.5 - 9.5}{5} = 10.1 \text{ kN}$$

2. *Bending Moment*

$$M_A = M_B = 0 \text{ (Free supports)}$$

$$M_C = R_A\,(CA)$$

$$= 19.5 \times 3 = +\,58.5 \text{ kN-m (sagging)}$$

$$M_D = R_A\,(AD) - W_C\,(CD)$$

$$= 19.5 \times 5 - 9.5 \times 2 = 78.5 \text{ kN-m (sagging)}$$

Maximum Bending Moment will be at E where shear force is zero.

$$\frac{DE}{BE} = \frac{10}{40.5}$$

$$DE = \frac{10}{40.5}\,(5 - DE)$$

$$\therefore \quad DE + 0.2469\,DE = 1.2345$$

$$\therefore \quad DE = 0.99 \text{ m or } 4.01\text{m from } B$$

$$M_{max} = M_E = R_B\,(BE) - w\,(BE) \times \frac{BE}{2}$$

$$= 40.5\,(4.01) - 10.1\,(4.01) \times \frac{4.01}{2}$$

$$= \mathbf{81.2 \text{ kN-m}} \quad \textbf{Ans.}$$

QUESTION BANK NO. 3

1. Define a beam and classify different types of beams on the basis of:
   (i) Support conditions
   (ii) Loadings.
2. What do you understand by 'shear force' and 'bending moment' and what is their importance in beam design?
3. What do you understand by statically determinate beam?
4. How shearing force and bending moment diagrams are drawn for a beam?
5. What do you understand by the term 'point of contraflexure'?
6. Explain rules for shear force and bending moment diagram.
7. What is the relationship between shear force and bending moment?
8. State true or false:
   (i) The shear force changes suddenly at a section where, there is a vertical point load.
   (ii) The shear force between any two vertical loads remains constant.
   (iii) BM is maximum at a section where SF is zero.
   (iv) Where a beam is subjected to a couple at a section, BM changes suddenly at the section but SF remains unaltered.
   (v) At point of contraflexure, BM is zero.

QUICK REVIEW INFORMATION NO. 3

1. The values of maximum bending moment are as follows:

Beam	*Loading*	*Maximum Bending Moment*	*Position*
Cantilever	Concentrated load at free end (W)	WL	Fixed end
Cantilever	Uniformly distributed load (w)	$wL^2/2$	Fixed end
Simply supported beam	Concentrated load at midspan	$WL/4$	Midspan
Simply supported beam	Uniformly distributed load	$wL^2/8$	Midspan

2. Variation of shear force and bending moment with type of load.

Type of loading	*Shear Force*	*Bending Moment*
No load	Constant	Varies linearly
Concentrated load	Varies suddenly	Varies linearly

Table contd....

Table contd....

Uniformiy distributed load	Varies linearly	Varies parabolically
Uniformly varying load	Varies parabolically	Varies cubically

3. In a cantilever beam carrying a concentrated load at free end, the bending moment will be zero at the free end and maximum (WL) at the fixed end.
4. A cantilever beam of length 2 m carries 10 kN force at a distance of 1 m from the support, the bending moment will be zero at a distance of 1.5 m from the support.
5. A moment of 10 Nm is applied in the middle of a simply supported beam, the magnitude of bending moment in the beam at the middle will be 5 Nm.

CHAPTER 4

Trusses

4.1 INTRODUCTION

A truss is a system of members that are fastened together at their ends to support stationary and moving loads. These are used in bridges, buildings, roofs of industrial sheds, railway platforms, godown, transmission towers, etc.

The main characteristics of a truss are as follows:

1. Each member of a truss is usually of uniform cross-section along its length.
2. Members of a truss have different cross-sections and cross-sectional areas as they must transmit different forces.
3. The common cross-sections of truss members are angles (L), I-section (I); channels ([ ), hollow tubes made from rolled steel.
4. The members of a truss are always connected at their ends. A truss is different than a frame. A frame may have some members connected along their length in addition to end connections.
5. In order to maximize the load-carrying capacity of a truss, the external loads must be applied at the joints, because the members of a truss are long and slender, the compression members may buckle under transverse loads.
6. The members of the trusses are fastened together by welding, riveting or bolting, through gusset plates in plane trusses.

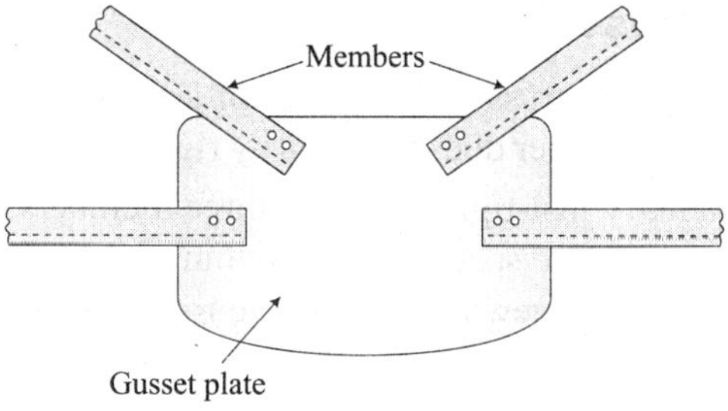

Fig. 4.1 Truss joint

7. For analysis of forces, the joint is taken as pin connection for coplanar concurrent members and a simple ball and socket connection for a space truss.

Fig. 4.2 Different joints

8. The weight of the member of a truss is small as compared to external loads applied on a truss. The weight of the member is usually neglected. In case the weight of the member has to be considered, half load is considered acting at each joint.
9. Each member of a truss is two-force member, i.e., a tensile member or compression member. A compression member pushes the joint whereas a tensile member pulls the joint where it is connected.

Fig. 4.3 Loading of truss members

4.2 TYPES OF TRUSSES

There are two categories of trusses according to *geometry*:

1. *Plane truss* consists of coplanar system of members.

   Examples: Roof truss, sides of a bridge.
2. *Space truss* consists of a three-dimensional system of members.

   Example: Electric power transmission tower.

4.2.1 A Simple Truss

A simple truss is just-rigid truss and removal of any of its members destroys its rigidity. If removing a member does not destroy rigidity, the truss is over-rigid.

The most elementary just-rigid truss has three members connected to form a triangle (ABC). The plane and space trusses are built-up from this triangle by adding for each new joint three new members. The trusses so built are called simple trusses.

A simple relationship exists between the number of joints, j and the number of members, m in a simple truss.

Simple Space Truss, $m = 3j - 6$

Simple Plane Truss, $m = 2j - 3$

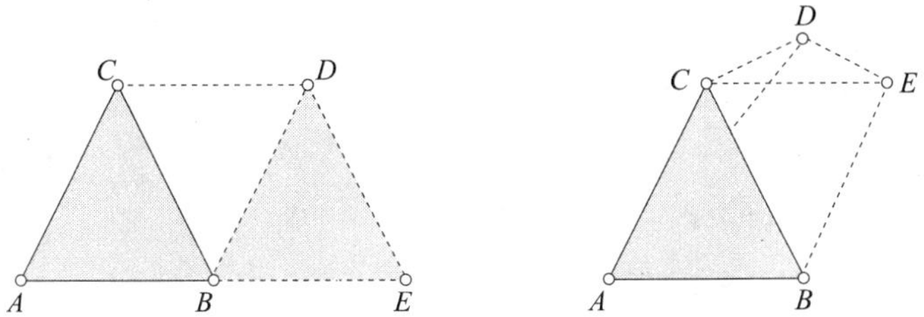

Plane truss Space truss

Fig. 4.4 The simple truss

4.2.2 Classification of Simple Truss

A simple truss can be classified as follows:

(a) Perfect truss (stable truss)

In a perfect truss the above relationship holds good. The truss does not collapse under the loading.

Example: In a triangular truss *ABC*,

No. of joints, $j = 3$

No. of members, $m = 3$

For plane truss,

$$m = 2\,j - 3$$
$$= 2 \times 3 - 3 = 3$$

A triangular truss satisfies the relationship and is a perfect truss.

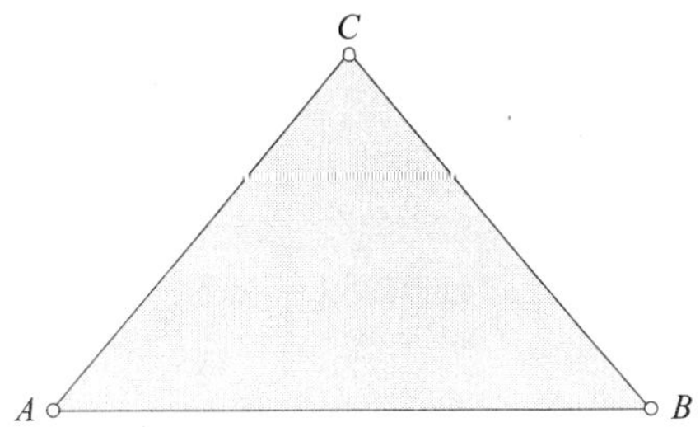

Fig. 4.5 Perfect truss

(b) Imperfect truss (unstable truss)

A truss which does not satisfy the above relationship is called imperfect truss. A truss which collapses when loaded is called unstable truss.

(c) Redundant truss (over-rigid truss)

In a redundant truss,

$$m > 2j - 3$$

Example: In Fig. 4.6 a redundant truss *ABCD* is shown.

No. of joints, $j = 4$

No. of members, $m = 6$

$$2j - 3 = 2 \times 4 - 3 = 5$$

Therefore, $m > 2j - 3$

One member is surplus and can be removed.

Fig. 4.6 Redundant truss

(d) Deficient truss

A truss in which $m < 2j - 3$ is called deficient truss and will collapse under loading.

Example: In Fig. 4.7, *ABCD* is a truss which has

Members, $m = 4$

Joints, $j = 4$

Now, $2j - 3 = 2 \times 4 - 3 = 5$

Therefore, one member is short.

To make the truss just-rigid or perfect, add a member *AC*.

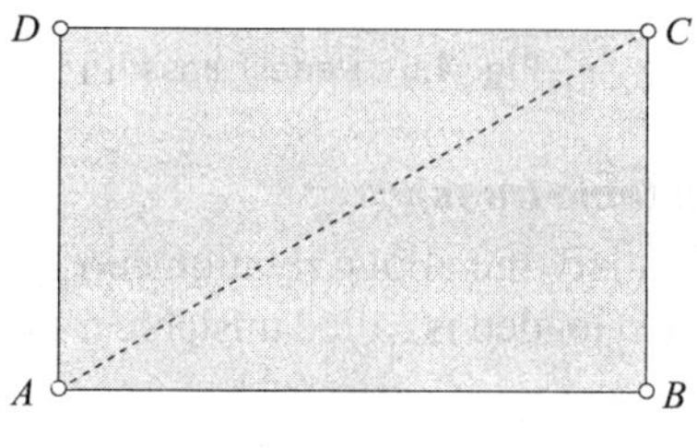

Fig. 4.7 Deficient truss

4.3 ANALYSIS OF SIMPLE PLANE TRUSS

When a truss is loaded, the loads are transferred to the supports and members are subjected to longitudinal forces (either tensile or compressive). The *solution* of simple truss consists of :

1. Computation of supporting reactions that must exist to keep the truss in equilibrium.
2. Computation of forces (in magnitude and direction) in each member of the truss so that every joint should be in equilibrium and the forces acting at every joint should form a system in equilibrium.

4.3.1 Assumptions

The following assumptions are made for the analysis of simple plane truss.

1. The ends of the members are pin-connected (hinged) and frictionless and cannot resist moments.
2. The loads act only at the joints.
3. Self-weights of the members are neglected.
4. The members of the truss are straight and their cross-sections are uniform and are two-force members.
5. The truss is a perfect truss.
6. The truss can be considered coplanar force system.
7. The truss is statically *determinate* and the equations of external loads, member forces and support reactions can be completely solved using equations of equilibrium only.

4.3.2 Methods of Analysis

Broadly there are two methods of analysis, namely, graphical as analytical. The analytical methods are:

(a) Method of Joints or Method of Resolution.

(b) Method of Sections.

(a) Method of Joints

This method is very widely used to find forces in all the members or most of the members of a truss. A plane truss can be a simply supported truss or a cantilever truss.

Following steps are used in the analysis of a truss.

(i) Check the stability of the truss by using the following equation.

For simply supported truss,

$$m = 2j - 3$$

For a cantilever truss,

$$m = 2j - 4$$

(ii) Draw free-body diagram of entire truss using the basic principles.

(iii) Use three equations of equilibrium to find out the reactions at the supports. Determination of support reactions may not be necessary in case of cantilever truss.

Example 4.1: Find out the support reactions of the truss shown.

Solution:

(a) Draw free-body diagram of the entire truss as shown.

(b) Apply equilibrium conditions.

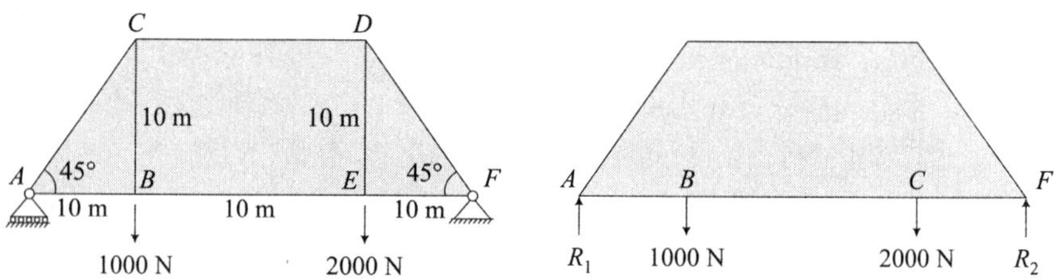

Fig. 4.8 Calculation of support reactions

A simply supported truss has one hinged support (F) and one roller support (A) to take care of thermal expansion and contraction.

$$\Sigma F_X = 0$$

$$\Sigma F_y = 0$$

∴ $$R_1 + R_2 = 1000 + 2000 = 3000$$

Taking moments about point A,

$$\Sigma M_A = 0$$

∴ $$R_1 \times 0 + R_2 \times 30 - 1000 \times 10 - 2000 \times 20 = 0$$

∴ $$R_1 = \frac{10,000 + 40,000}{30} = 1666 \text{ N}$$

and $$R_2 = 3000 - 1666 = 1334 \text{ N}$$

Example 4.2: Find the reactions of the cantilever truss shown.

Solution:

(a) Draw free-body diagram

(b) Apply equilibrium conditions

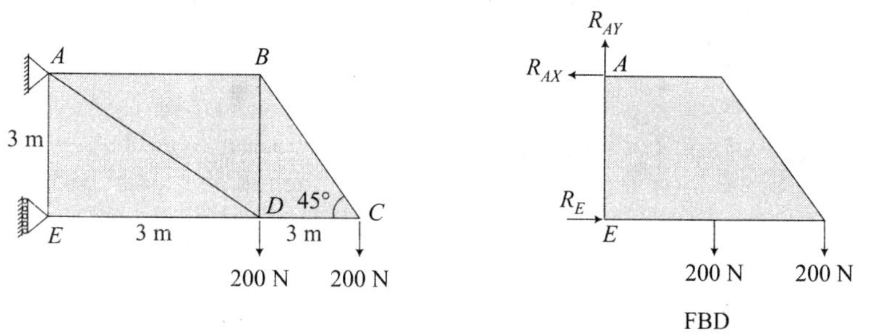

Fig. 4.9 Calculation of support reaction

$$\Sigma F_X = 0$$
$$\Sigma F_Y = 0$$
$$R_{AY} = 200 + 200 = 400 \text{ N}$$
$$\Sigma M_A = 0$$
$$R_E \times 3 = 200 \times 3 + 200 \times 6$$
$$\therefore \quad R_E = 600 \text{ N}$$
$$\therefore \quad R_{AX} = -R_E = -600 \text{ N}$$

(iv) Select a joint such that maximum two unknown forces exist.

(v) Draw FBD of selected joint by assuming forces in the members to be tensile.

(vi) The forces at the joint form a planar concurrent system in equilibrium. Apply equilibrium conditions:

$$\Sigma F_x = 0$$
$$\Sigma F_y = 0$$

(vii) Find out some forces by inspection.

(viii) Repeat the procedure by selecting different joints and find out forces in all members of the truss.

(ix) Tabulate the results in a Force Table using positive as tensile and negative as compressive.

Force Table

Sl. No.	*Member*	*Magnitude*	*Nature*

(x) Draw final FBD of truss showing magnitude and nature of all external and internal forces.

Important Rules

1. When two non-collinear members of a truss meet at a point and no external force or reaction act, then the force in both the members will be zero.

$$F_1 = F_2 = 0$$

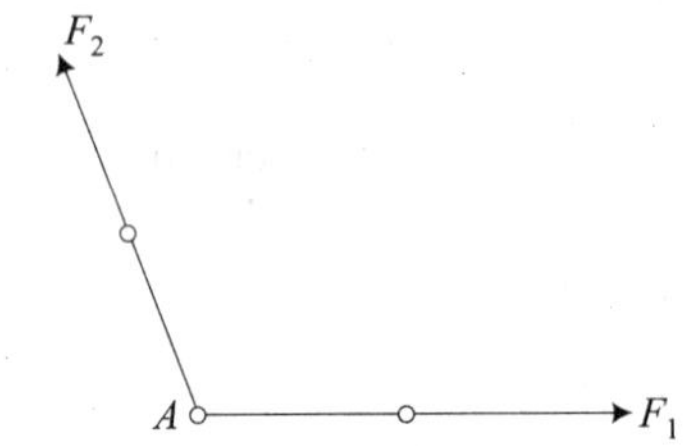

Fig. 4.10 Two-force joint

2. When two collinear members of a truss meet at a point, and a third member is connected at this point, then the force in the third member will be zero.

$$F_1 = F_2 \quad \text{and} \quad F_3 = 0$$

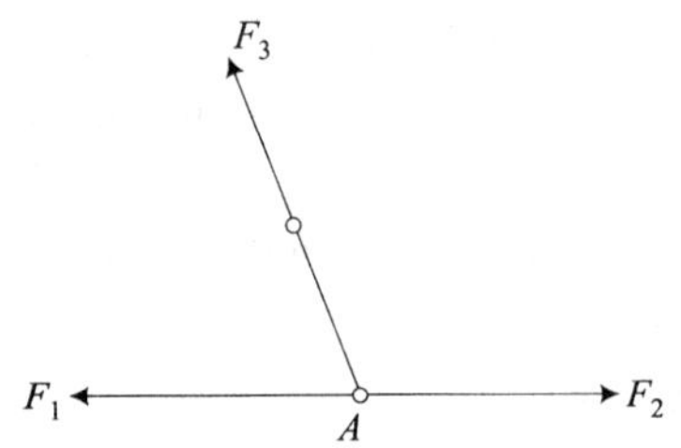

Fig. 4.11 Three-force joint

3. If a joint carries four forces with two pairs of collinear forces (cross joint), the collinear force must be equal and opposite.

$$F_1 = F_2 \quad \text{and} \quad F_3 = F_4$$

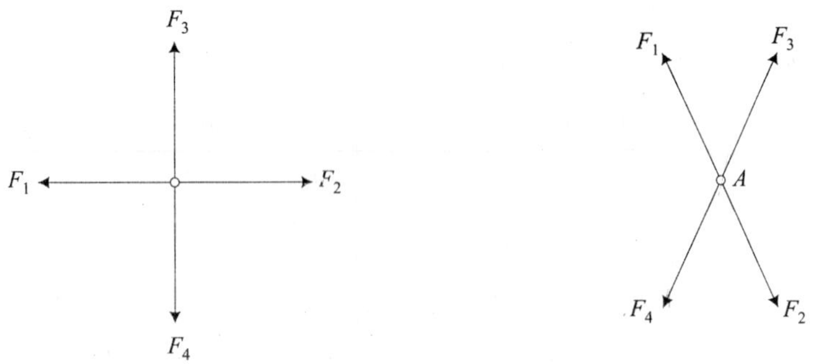

Fig. 4.12 Four-force joint

Example 4.3: Find whether the truss shown is a simple truss.

Solution:

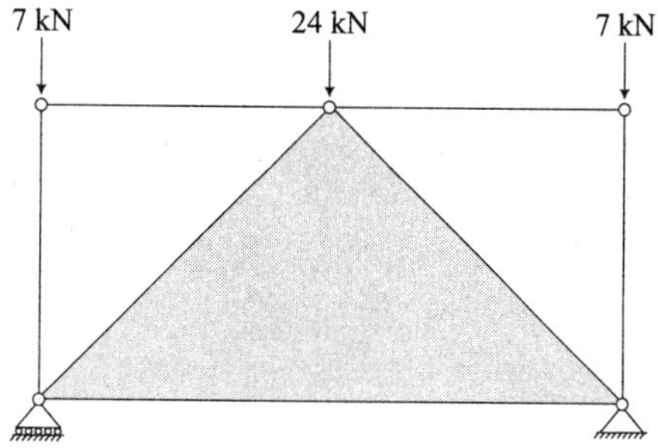

No. of joints, $j = 5$

No. of members, $m = 7$

For a simple truss,

$$m = 2j - 3$$
$$= 2 \times 2 - 3$$
$$= 7$$

Therefore, it is a simple truss.

Example 4.4: Identify the zero load members in the plane truss shown.

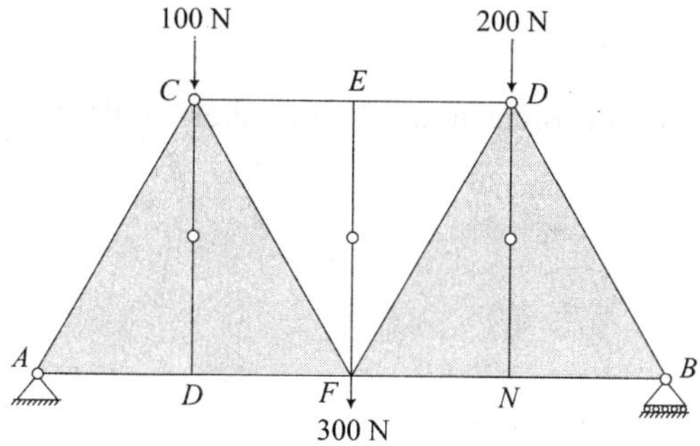

Solution:

Joint D

Three forces acting at joint D are F_{AD}, F_{FD} and F_{CD}.

Forces F_{AD} and F_{FD} are collinear and no external force or reaction is acting at joint D.

Therefore, $F_{AD} = F_{FD}$ and $F_{CD} = 0$

Similarly, forces F_{EF} and F_{GH} are zero.

Therefore, member, CD, EF and HG are zero force members.

Example 4.5: Identify the zero load members in the cantilever truss.

Solution:

Joint D

Forces, F_{AD} and F_{ED} are collinear and the joint D is unloaded.

$$\therefore \quad F_{AD} = F_{ED} \text{ and } F_{CD} = 0$$

Member CD is unloaded and redundant.

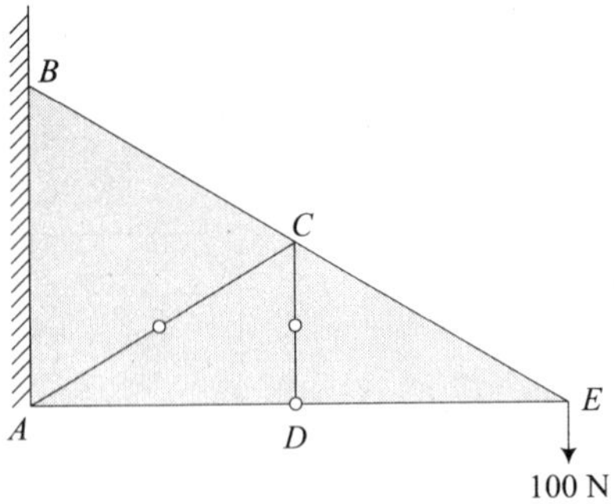

Joint C

Forces F_{BC} and F_{EC} are collinear and joint is unloaded ($F_{DC} = 0$)

$\therefore \qquad F_{AC} = 0$

Member AC is unloaded.

Example 4.6: Find out forces in all the members of the truss ABC shown.

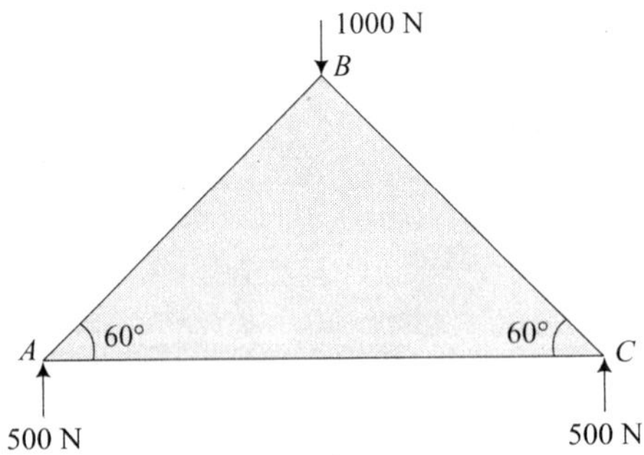

Solution:

(i) Check the stability of truss

No. of joints, $j = 3$

No. of members, $m = 3$

For a stable truss,

$$m = 2j - 3$$

$$= 2 \times 3 - 3 = 3$$

$\therefore$ The truss is stable.

(ii) *Reactions*. The truss is symmetrical,

$$\therefore \qquad R_A = R_B = \frac{1000}{2} = 500 \text{ N}$$

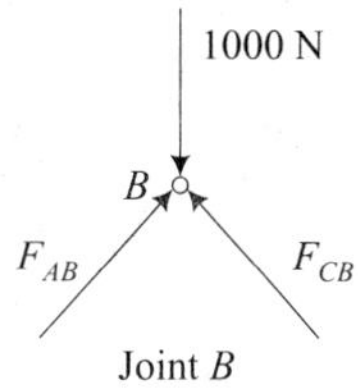

(iii) *FBD of joint B*

$$\Sigma F_X = 0$$

$$\Sigma F_Y = 0$$

The forces are pushing against joint B, therefore, are compressive.

$F_{AB} \sin 60° + F_{CB} \sin 60° = 1000$

But, $$F_{AB} = F_{CB} = \frac{1000}{2 \sin 60°}$$

$$= 577 \text{ N}$$

(iv) *FBD of joint A*

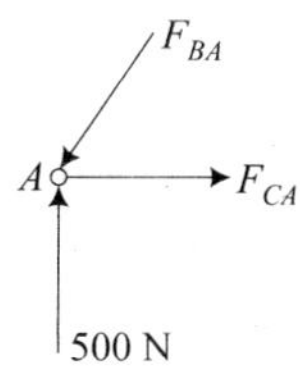

$$\Sigma F_X = 0$$

$$\Sigma F_Y = 0$$

$$F_{CA} = F_{BA} \cos 60°$$

$$= 577 \cos 60°$$

$$= 288.5 \text{ N}$$

(v) *Force Table*

Member	*Force*	*Nature*
AB	577 N	*C*
BC	577 N	*C*
AC	288.5 N	*T*

(vi) *Force diagram*

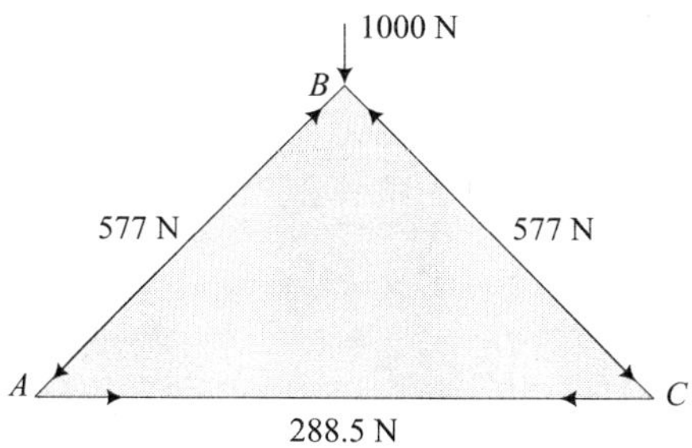

(b) Method of Section

This method is useful for calculating forces in some members of the truss and avoid laborious process of proceeding joint by joint reaching a joint on which the desired unknown force acts.

Following steps are used in the analysis of a truss.

(i) Check the stability of a truss.

(ii) Draw FBD and calculate reactions by applying three conditions of equilibrium.

(iii) Select a section, cutting maximum three members (maximum three unknown forces) including the member in which force is to be determined.

(iv) Draw FBD of any cut portion of truss and show forces in the cut members as towards the section (away from joints).

The FBD should include only external forces acting on that part and internal tensile forces in the cut members.

(v) Apply equations of equilibrium for a coplanar forces system.

Important Rules

1. A section can cut more than three members but the unknown forces should not exceed three.
2. In case of cantilever truss, draw FBD of part not containing the wall.

Example 4.7: Calculate the force in the member *CE* only.

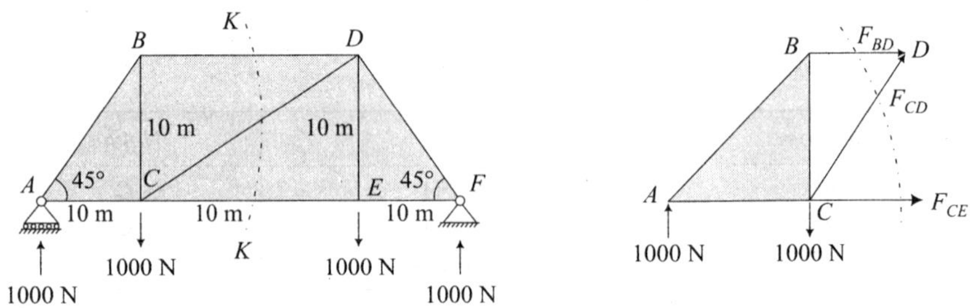

Solution:

1. Draw FBD of entire truss.
2. Reaction $R_A = R_B = 1000$ N by symmetry.
3. Cut section *K-K*.
4. Cut FBD of position *ABCD* showing the internal tensile forces F_{BD}, F_{CD} and F_{CE}.
5. Apply the condition,

$$\Sigma M_D = 0$$

Forces F_{BD} and F_{CD} pass through joint D.

$\therefore$ $$F_{CE} \times DE + 1000 \times CE - 1000 \times AE = 0$$

$$F_{CE} \times 10 + 1000 \times 10 - 1000 \times 20 = 0$$

$\therefore$ $$F_{CE} = 1000 \text{ N } (T).$$

Example 4.8: Find the forces in the members AB, AC and BC of the truss shown. [U.P.T.U. II Sem., 2002–03]

Solution:

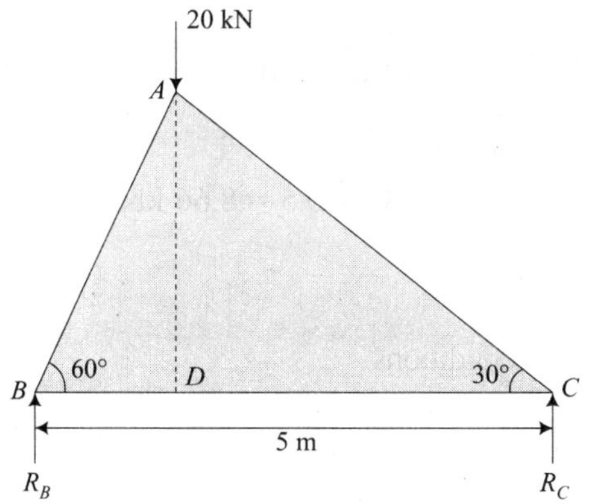

1. *Dimensions*

$$BC = 5 \text{ m}$$

$$AB = BC \cos 60°$$

$$= 5 \times 0.5 = 2.5 \text{ m}$$

$$BD = AB \cos 60°$$

$$= 2.5 \times 0.5 = 1.25 \text{ m}$$

2. *Reactions*

$$\Sigma F_y = 0$$

$$R_A + R_C = 20 \text{ kN}$$

$$\Sigma M_B = 0$$

$\therefore$ $$R_C \times 5 = 20 \times 1.25$$

$$R_C = \frac{20 \times 1.25}{5} = 5 \text{ kN}$$

$\therefore$ $$R_B = 20 - 5 = 15 \text{ kN}$$

3. *Consider Joint B*

(i) Draw FBD

(ii) Apply equilibrium conditions

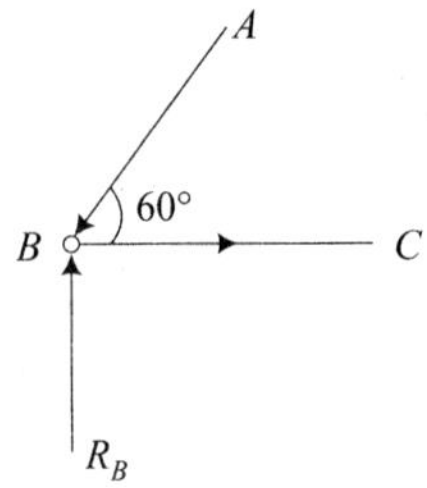

FBD of joint *B*

$$\Sigma F_X = 0$$

$$\therefore \quad F_{CB} - F_{AB} \cos 60° = 0$$

$$\Sigma F_Y = 0$$

$$\therefore \quad F_{AB} \sin 60° - R_B = 0$$

$$\therefore \quad F_{AB} = \frac{R_B}{\sin 60°} = \frac{15}{0.866} = 17.32 \text{ kN } (C)$$

$$F_{CB} = F_{AB} \sin 60°$$

$$= 17.32 \times 0.5 = 8.66 \text{ kN } (T)$$

4. *Consider Joint C*

(i) Draw FBD.

(ii) Apply equilibrium conditions

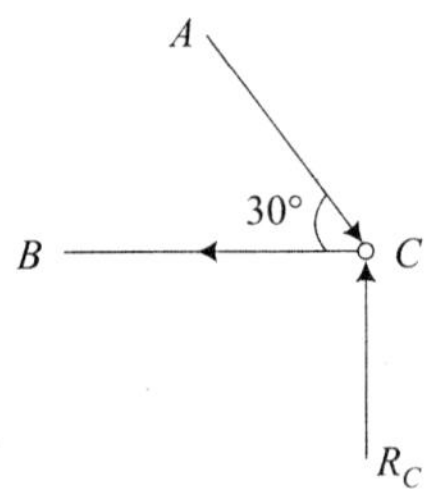

FBD of joint *C*

$$\Sigma F_X = 0$$

$$F_{AC} \cos 30° - F_{BC} = 0$$

$$\therefore \quad F_{AC} = \frac{F_{BC}}{\cos 30°} = \frac{8.66}{0.866} = 10 \text{ kN } (C)$$

Force Table

Member	Force
AB	17.32 kN (C)
AC	10.00 kN (C)
BC	8.66 kN (*T*)

Force Diagram

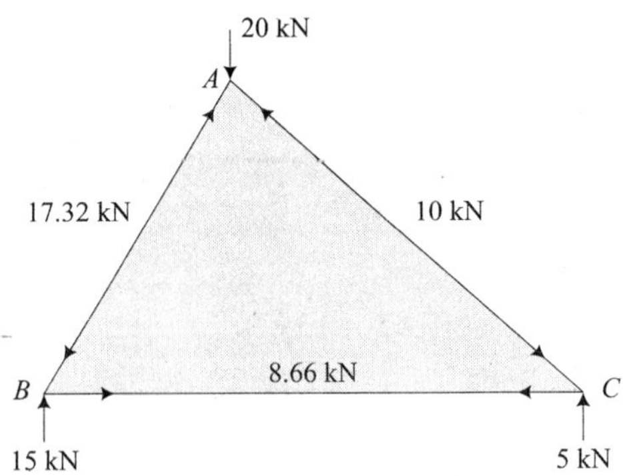

Example 4.9: A truss having a span of 6 m, carries a load of 30 kN and is shown below. Find the forces in members, *AB*, *AC*, *BC* and *AD*.

[U.P.T.U. II Sem., 2005–06]

Solution:

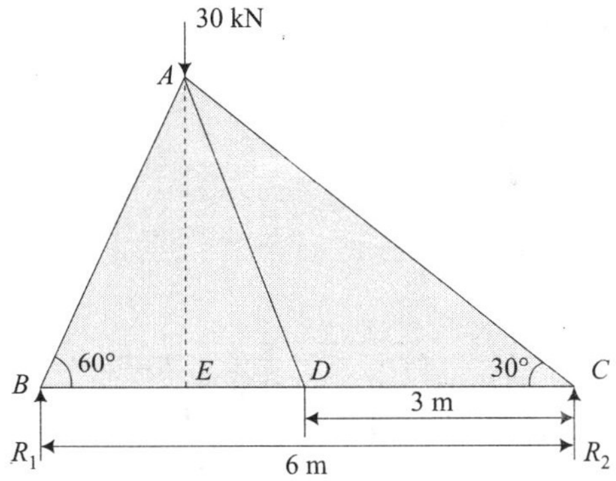

(i) *Check the stability of the truss,*

No. of members, $m = 5$

No. of joints, $j = 4$

For a stable truss,

$$m = 2j - 3$$
$$= 2 \times 4 - 3$$
$$= 5$$

Therefore, the truss is stable and just rigid.

(ii) *Dimensions*

Draw a vertical *AE*,

$$BD = 3 \text{ m and } BC = 6 \text{ m}$$
$$AB = BC \cos 60° = 6 \times 0.5 = 3 \text{ m}$$
$$BE = AB \cos 60° = 3 \times 0.5 = 1.5 \text{ m.}$$

(iii) *Reactions*

$$\Sigma F_Y = 0$$

$\therefore$ $$R_1 + R_2 = 30$$

$$\Sigma M_B = 0$$

$\therefore$ $$R_2 \times 6 = 30 \times 1.5$$

$$\therefore \quad R_2 = \frac{3.0 \times 1.5}{6} = 7.5 \text{ kN}$$

$$R_1 = 30 - 7.5 = 22.5 \text{ kN}$$

(iv) *Joint B*

Consider equilibrium of joint B. Draw its FBD,

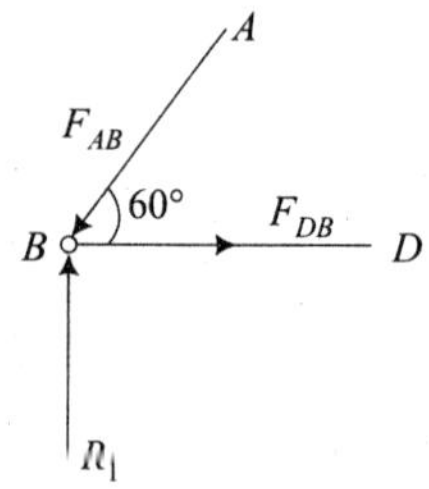

FBD of joint B

$\Sigma F_X = 0$

$$\therefore \quad F_{DB} - F_{AB} \cos 60° = 0$$

$\Sigma F_Y = 0$

$$\therefore \quad F_{AB} \sin 60° = R_1 = 22.5$$

$$\therefore \quad F_{AB} = \frac{22.5}{\sin 60°} = \frac{22.5}{0.866} = 26 \text{ kN}$$

$$\therefore \quad F_{DB} = 26 \times 0.5 = 13 \text{ kN}$$

(v) *Joint C*

Consider equilibrium of joint C. Draw its FBD,

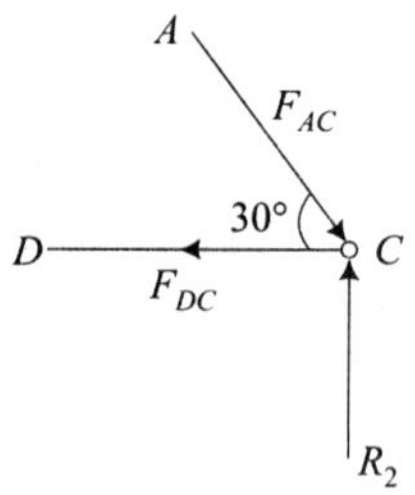

FBD of joint C

$\Sigma F_X = 0$

$$F_{DC} - F_{AC} \cos 30° = 0$$

$\Sigma F_Y = 0$

$$F_{AC} \sin 30° = R_2 = 7.5$$

$$\therefore \quad F_{AC} = \frac{7.5}{0.5} = 15 \text{ kN}$$

$$\therefore \quad F_{DC} = 15 \times 0.866 = 13 \text{ kN}$$

(vi) *Joint A*

Consider equilibrium of joint A and B draw its FBD,

From geometry,

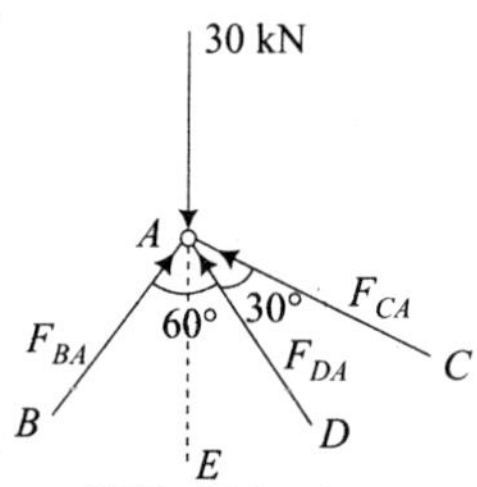

FBD of joint A

$$\angle BAD = 60°$$

$$\angle CAD = 30°$$

$$\therefore \quad \angle BAE = 30°$$

$$\angle CAE = 60°$$

$$\Sigma F_X = 0$$

$$F_{BA} \cos 30° + F_{DA} \cos 30° + F_{CA} \cos 60° = 30$$

$$26 \times 0.866 + F_{DA} \times 0.866 + 15 \times 0.5 = 30$$

$$\therefore \quad F_{DA} = 30 - 22.5 - 7.50 = 0$$

(vii) The force table and force diagram are shown,

Member	Force
AB	26 kN (C)
AC	7.5 kN (C)
BC	13 kN (T)
AD	0

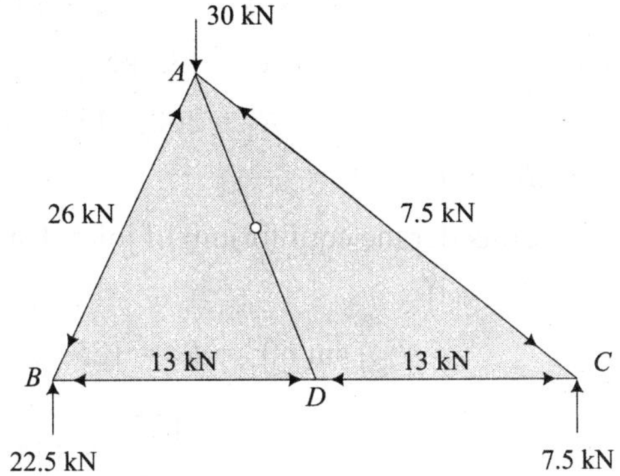

Example 4.10: Each member of the following truss is 2 m long. The truss is simply supported at the ends. Determine forces in all members clearly showing whether they are in tension or compression. [U.P.T.U. I Sem., 2005–06]

Solution:

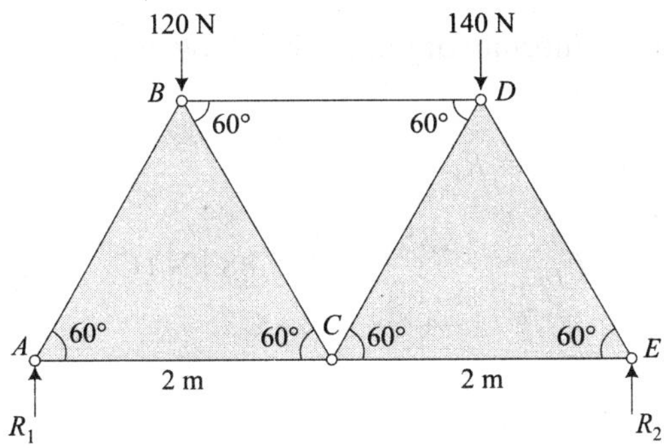

(i) *Check the stability of the truss,*

No. of members, $m = 7$

No. of joints, $j = 5$

For a stable truss,

$$m = 2j - 3$$
$$= 2 \times 5 - 3 = 7$$

∴ The truss is just-rigid and stable.

(ii) *Reactoins*

$\Sigma F_Y = 0$

∴ $$R_1 + R_2 = 120 + 140 = 260 \text{ kN}.$$

$\Sigma M_A = 0$

$$R_2 \times 4 = 120 \times 1 + 140 \times 3 = 540$$

$$\therefore \quad R_2 = 135 \text{ kN}$$

$$R_1 = 260 - 135 = 125 \text{ kN}$$

(iii) *Joint A*

Consider the equilibrium of joint A and draw its FBD.

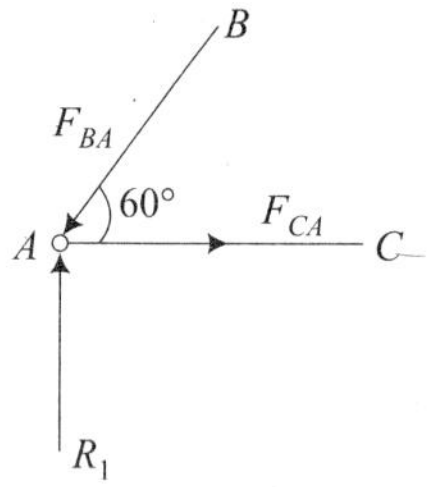

FBD of joint A

$\Sigma F_Y = 0$

$$F_{BA} \sin 60° = R_1 = 125$$

$$\therefore \quad F_{BA} = \frac{125}{0.866} = 144.3 \text{ kN } (C)$$

$\Sigma F_X = 0$

$$F_{CA} = F_{BA} \cos 60°$$

$$= 144.3 \times 0.5 = 72.15 \text{ kN } (T)$$

(iv) *Joint E*

Consider the equilibrium of joint E and draw its FBD.

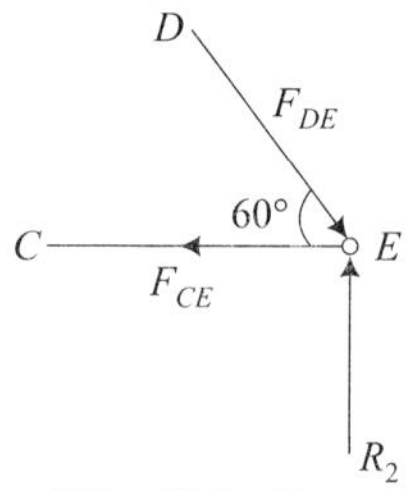

FBD of joint E

$\Sigma F_Y = 0$

$$F_{DE} \sin 60° = R_1 = 135$$

$$\therefore \quad F_{DE} = \frac{135}{\sin 60°} = 155.88 \text{ kN } (C)$$

$\Sigma F_X = 0$

$$F_{CE} = F_{DE} \cos 60°$$

$$= 155.88 \times 0.5 = 77.94 \text{ kN } (T)$$

(v) *Joint C*

Consider the equilibrium of joint C and draw its FBD.

$$F_{AC} = 72.15 \text{ kN } (C)$$

$$F_{EC} = 77.94 \text{ kN } (T)$$

$\Sigma F_Y = 0$

$$F_{BC} \cos 30° + F_{DE} \cos 30° = 0$$

$$F_{BC} = -F_{DC}$$

$\Sigma F_X = 0$

$$-F_{AC} + F_{EC} = F_{BC} \cos 60° - F_{DC} \cos 60°$$

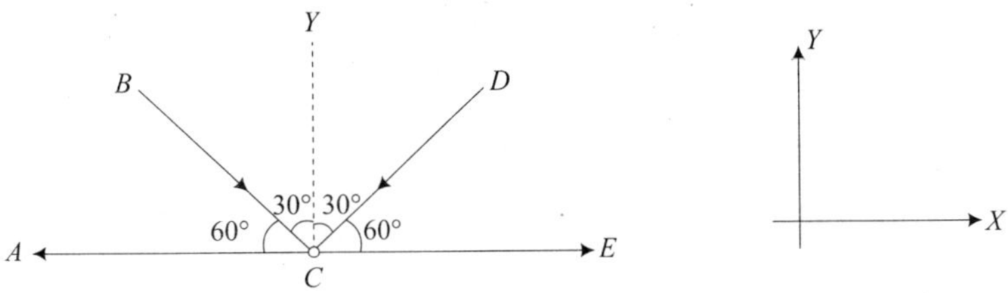

FBD of joint C

$$F_{BC} = \frac{F_{EC} - F_{AC}}{2\cos 60°} = \frac{77.94 - 72.15}{2 \times 0.5} = 5.79 \text{ kN } (C)$$

$\therefore$ $$F_{DC} = 5.79 \text{ kN } (T)$$

(vi) *Joint D*

Consider the equilibrium of joint D and draw its FBD,

$$F_{CD} = 5.79 \text{ kN } (T)$$
$$F_{ED} = 155.88 \text{ kN } (C)$$
$$\Sigma F_X = 0$$
$$F_{BD} = -F_{CD} \cos 60° - F_{ED} \cos 60°$$
$$= -5.79 \times 0.5 + 155.88 \times 0.5$$
$$= 75 \text{ kN } (C)$$

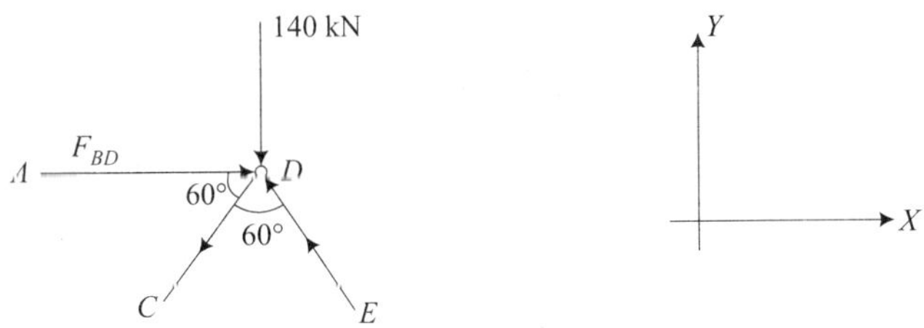

FBD of joint D

(vii) *Force Table and Diagram*

Member	*Force*
AB	144.3 kN (*C*)
BC	5.79 kN (*C*)
CD	5.79 kN (*T*)
DE	155.88 kN (*C*)
CE	77.94 kN (*T*)
AC	72.15 kN (*T*)

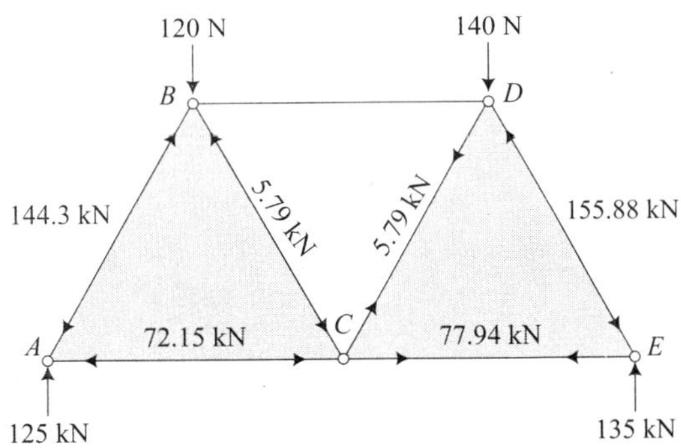

Example 4.11: Determine the forces in all the members of the cantilever truss shown.

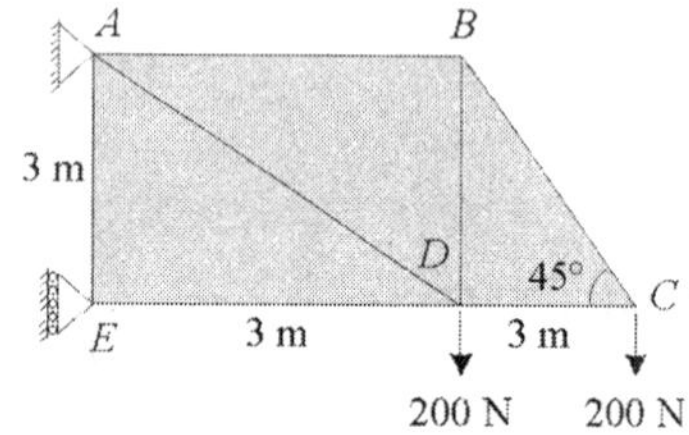

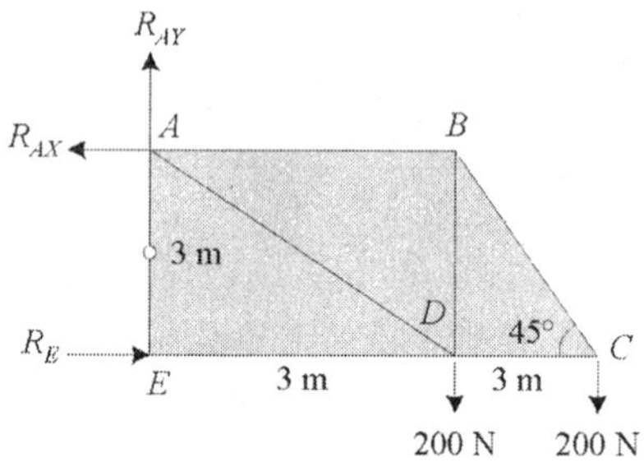

FBD of truss

Solution:

(i) Draw FBD of the truss.

(ii) Check the stability of the truss,

No. of members, $m = 6$

No. of joints, $j = 5$

For a stable truss,

$$m = 2j - 4 = 2 \times 5 - 4 = 6$$

The truss is stable.

(iii) *Reactions*

$\Sigma F_Y = 0$

$$\therefore \quad R_{AY} = 200 + 200 = 400 \text{ N}$$

$\Sigma M_A = 0$

$$R_E \times 3 = 200 \times 3 + 200 \times 6$$

$$R_E = 600 \text{ N}$$

$$R_{AX} = -R_E = -600 \text{ N}$$

(iv) *Joint E*

Consider the equilibrium of joint E and draw its FBD.

$\Sigma F_X = 0$

$$\therefore \quad F_{DE} = R_E = 600 \text{ N } (C)$$

$$F_{AE} = 0$$

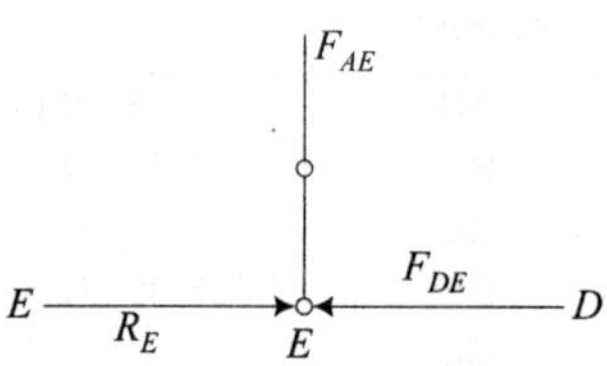

FBD of joint *E*

(v) *Joint C*

Consider the equilibrium of joint C and draw its FBD.

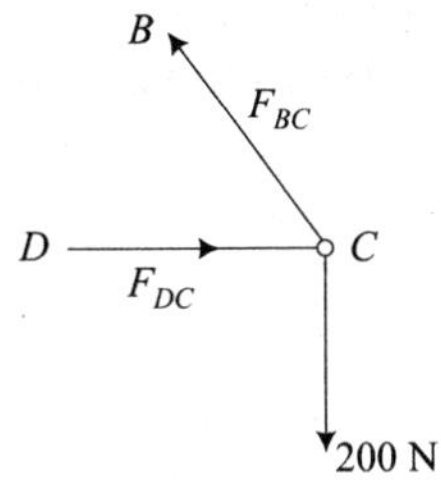

FBD of joint C

$$\Sigma F_Y = 0$$

$$F_{BC} \sin 45° = 200$$

$$F_{BC} = \frac{200}{\sin 45°} = 282.88 \text{ N } (T)$$

$$\Sigma F_X = 0$$

$$F_{DC} = F_{BC} \cos 45°$$

$$= 282.88 \times 0.707$$

$$= 200 \text{ N } (C)$$

(vi) *Joint B*

Consider the equilibrium of joint B and draw its FBD.

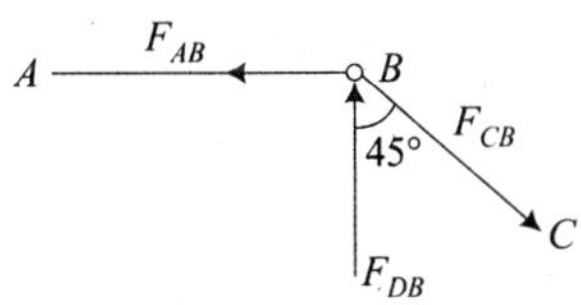

FBD of joint B

$$\Sigma F_Y = 0$$

$$F_{DB} = F_{CB} \cos 45°$$

$$= 282.88 \times 0.707$$

$$= 200 \text{ N } (C)$$

$$\Sigma F_X = 0$$

$$F_{AB} = F_{CB} \cos 45° = 200 \text{ N } (T)$$

(vii) *Joint D*

Consider the equilibrium of joint D and draw its FBD.

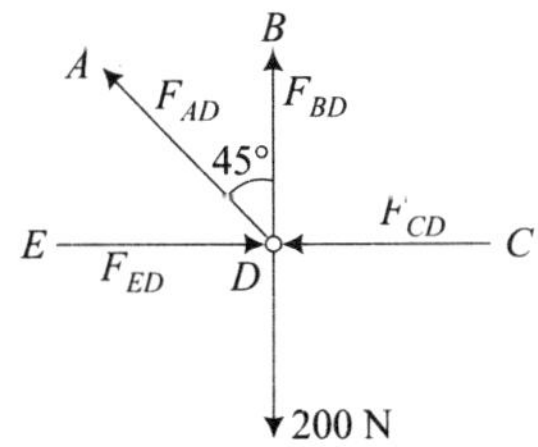

FBD of joint D

$$\Sigma F_X = 0$$

$$F_{AC} \sin 45° + F_{ED} - F_{CD} = 0$$

$$F_{AD} \sin 45° + 600 - 200 = 0$$

$$\therefore \quad F_{AC} = \frac{400}{\sin 45°} = 565.77 \text{ } (T)$$

(vii) *Draw Force Table and Diagram*

Member	*Force*
AB	200 N (*T*)
BC	28.88 N (*T*)
CD	200 N (*C*)
DE	600 N (*C*)
EA	0
DA	565.77 N (*T*)
DB	200 N (*C*)

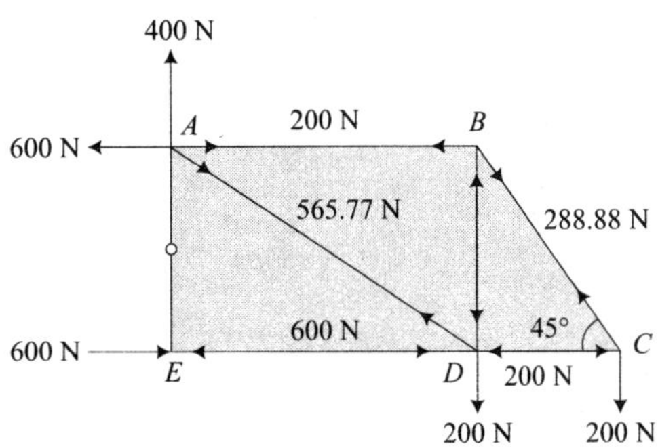

Example 4.12: Determine the support reactions and nature and magnitude of forces shown in the truss. [U.P.T.U I Sem., 2001–02]

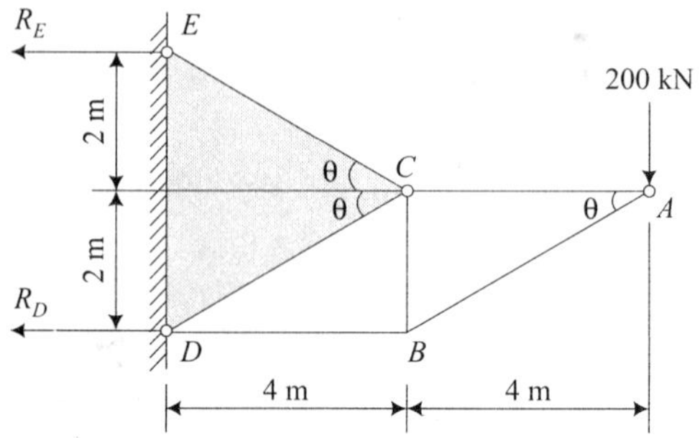

Solution:

(i) Check the stability of cantilever truss,

No. of members (excluding wall), $m = 6$

No. of joints = 5

For stability of cantilever truss,

$$m = 2j - 4 = 2 \times 5 - 4 = 6$$

∴ The truss is stable.

(ii) *Geometrical configuration*

$$AB = \sqrt{AC^2 + BC^2} = \sqrt{4^2 + 2^2} = 4.472 \text{ m}$$

$$\sin\theta = \frac{BC}{AB} = \frac{2}{4.472} = 0.447$$

$$\cos\theta = \frac{AC}{AB} = \frac{2}{4.472} = 0.892$$

(iii) *Pin A*

Consider the equilibrium of pin A and draw its FBD,

$$\Sigma F_X = 0$$

$$F_{BA}\cos\theta = F_{CA}$$

$$\Sigma F_y = 0$$

$$F_{BA}\sin\theta = 200$$

$$F_{BA} = \frac{200}{0.447} = 447.43 \text{ kN } (C)$$

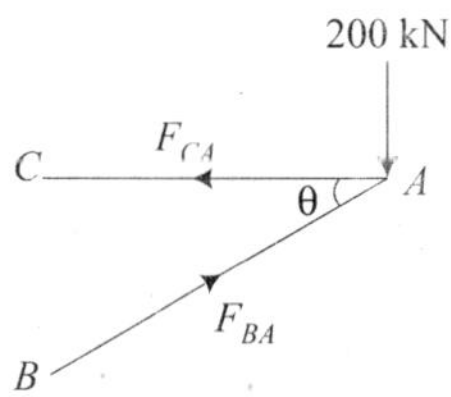

FBD of pin A

$$F_{CA} = F_{BA}\cos\theta = 447.43 \times 0.84 = 400 \text{ kN } (T)$$

(iv) *Pin B*

Consider the equilibrium of pin B and draw its FBD,

$$\Sigma F_X = 0$$

$$F_{DB} - F_{AB}\cos\theta = 0$$

$$F_{DB} = 447.43 \times 0.894$$

$$= 400 \text{ kN } (C)$$

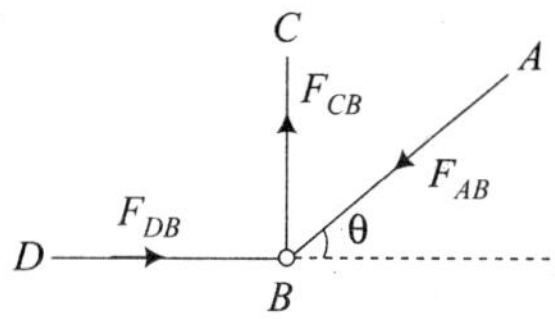

FBD of pin B

$$\Sigma F_Y = 0$$

$$F_{CB} - F_{AB}\sin\theta = 0$$

$$F_{CB} = 447.43 \times 0.047$$

$$= 200 \text{ kN } (C)$$

(v) *Pin C*

Consider the equilibrium of pin C and draw its FBD,

$$\Sigma F_X = 0$$

$$F_{AC} - F_{DC}\cos\theta - F_{EC}\cos\theta = 0$$

$$F_{DC} + F_{EC} = \frac{400}{0.894} = 447.427 \text{ kN} \quad \ldots(1)$$

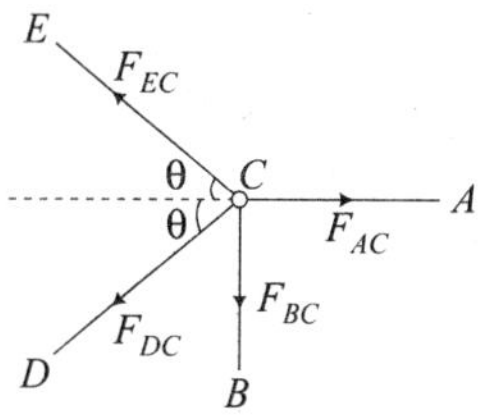

FBD of pin C

$$\Sigma F_Y = 0$$

$$F_{EC}\sin\theta - F_{DC}\sin\theta - F_{BC} = 0$$

$$F_{EC} - F_{DC} = \frac{F_{BC}}{\sin\theta} = \frac{200}{0.447} = 447.427 \text{ kN} \quad \ldots(2)$$

From equations (1) and (2)

$$F_{EC} = 447.427 \text{ kN } (T)$$

$$F_{DC} = 0$$

(vi) *Joint D*

Consider the equilibrium of joint D and draw its FBD,

$$\Sigma F_X = 0$$

$$\therefore \quad R_D = F_{BD} = 400 \text{ kN}$$

$$\therefore \quad R_E = -R_D = -400 \text{ kN}$$

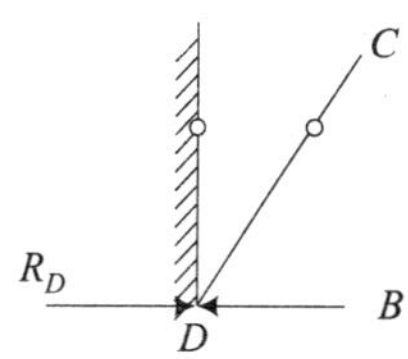

FBD of joint D

(vii) *Draw Force Table and Diagram.*

Member	Force
AC	400 kN (*T*)
AB	447.43 kN (*C*)
BC	200 kN (*T*)
BD	400 kN (*C*)
CD	0
CE	447.43 kN (*T*)

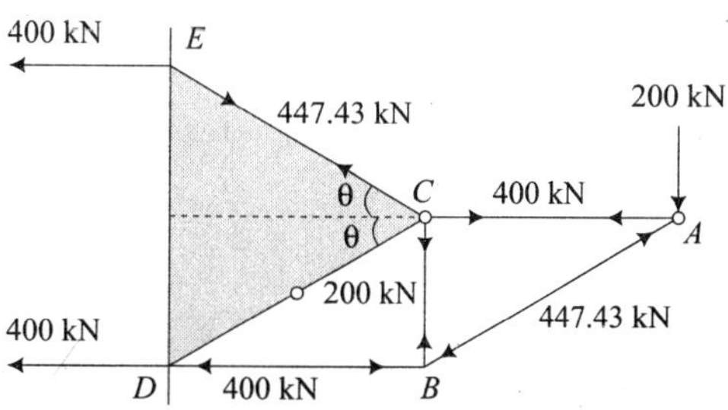

Example 4.13: A cantilever truss is loaded and supported as shown. Find the value of load P which would produce an axial force of magnitude 3 kN in the member AC using methods of section. [U.P.T.U. II Sem., 2002–03]

Solution: Let section K-K cut the members AC, AD and BD. Consider right hand part of cut.

Take moments about point C through which unknown forces F_{AD} and F_{BD} are passing.

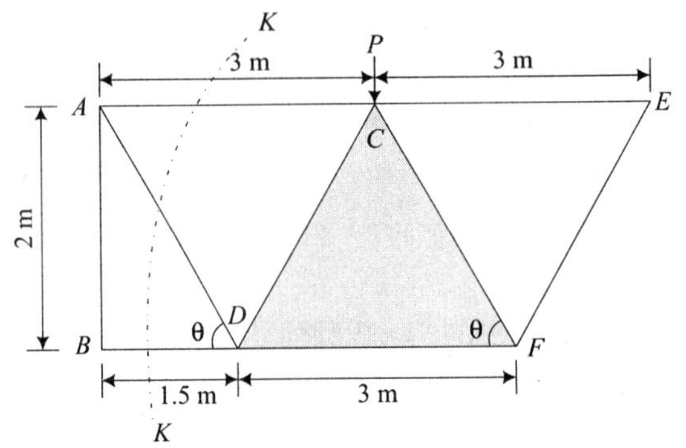

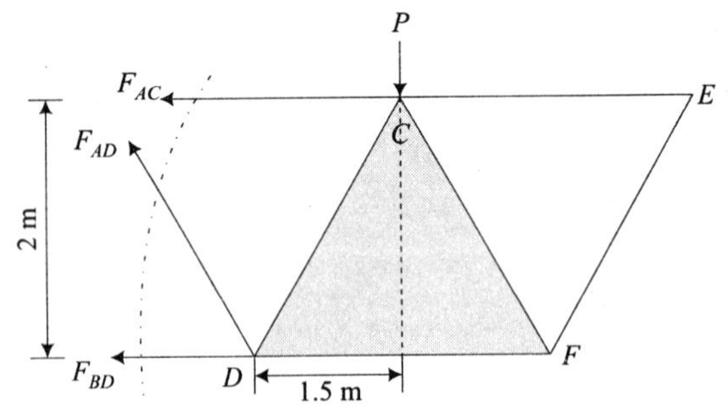

$$\Sigma M_D = 0$$

$$P \times 1.5 + F_{BD} \times 0 + F_{AD} \times 0 = F_{AC} \times 2$$

$$1.5\ P = 2\ F_{AC}$$

$$F_{AC} = 3 \text{ kN (given)}$$

$$\therefore \quad P = \frac{3 \times 2}{1.5} = 4 \text{ kN}$$

Example 4.14: Find the forces in the members DE, DC and AC respectively in the roof truss shown.

Solution:

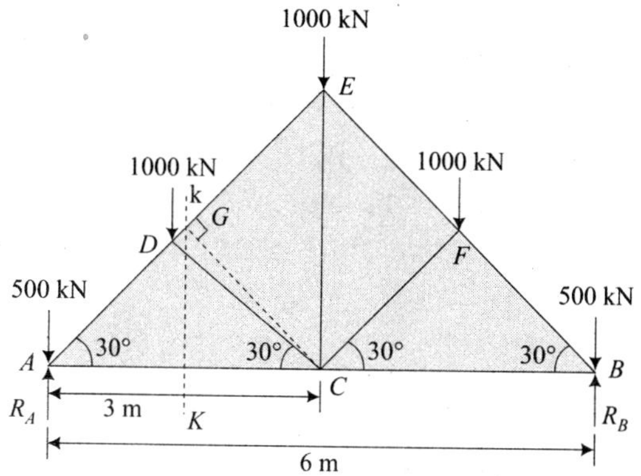

(i) *Reactions*

By symmetry, $R_A = R_B = \dfrac{(500 + 1000)\,2 + 1000}{2} = 200$ kN.

(ii) Cut the truss by section *K-K* through *DE*, *DC* and *AC*.

(iii) Draw the free body diagrams of left part of cut truss.

(iv) Draw *CG* perpendicular to *AE*.

$$CG = AC \sin 30° = 3 \times 0.5 = 1.5 \text{ m}$$

(v) In order to find F_{DE} take members about point *C*.

$$\Sigma M_C = 0$$

$\therefore$ $$F_{DE} \times CG + 1000 \times 1.5 + 500 \times 3 = 2000 \times 3$$

$$F_{DE} \times 1.5 + 1000 \times 1.5 + 500 \times 3 = 2000 \times 3$$

$\therefore$ $$F_{DE} = -2000 \text{ kN (C)}$$

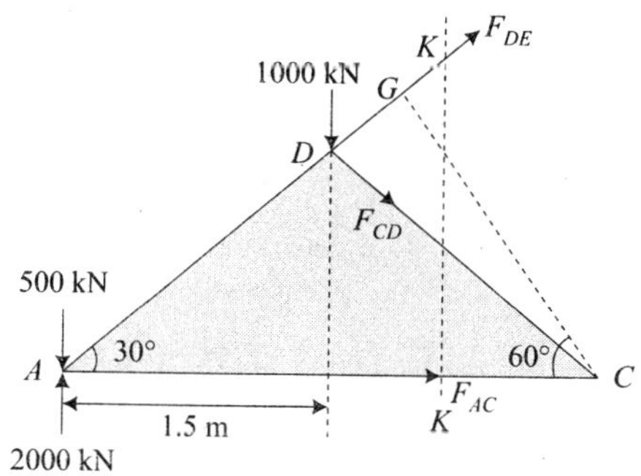

FBD of cut section *K-K*

(vi) In order to find F_{CD}, take moments about point *A*.

$$F_{CD} \times AG + 1000 \times 1.5 = 0$$

$$F_{CD} \times 3 \cos 60° + 1000 \times 1.5 = 0$$

$\therefore$ $$F_{CD} = -1000 \text{ kN } (C)$$

(vii) In order to find F_{AC} take moments about point D,

$$F_{AC} \times 1.5 \tan 30° + 500 \times 1.5 = 2000 \times 1.5$$

$$\therefore \quad F_{AC} = 2598 \text{ kN } (T)$$

Example 4.15: It is desired to know the force in the member AB of the plane truss shown. The supporting forces have been determined and shown in the diagram.

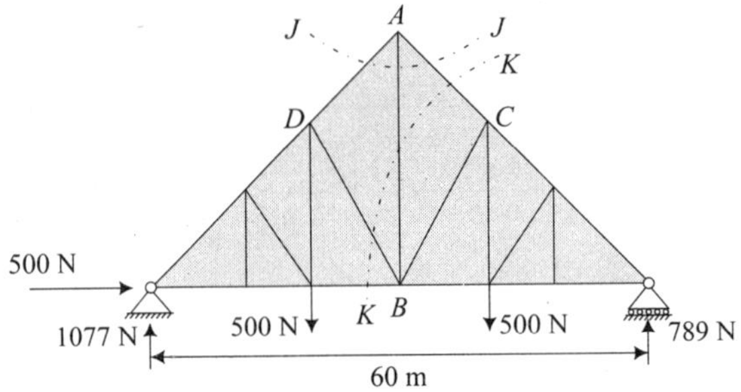

Solution:

(i) Draw FBD of section J-J.

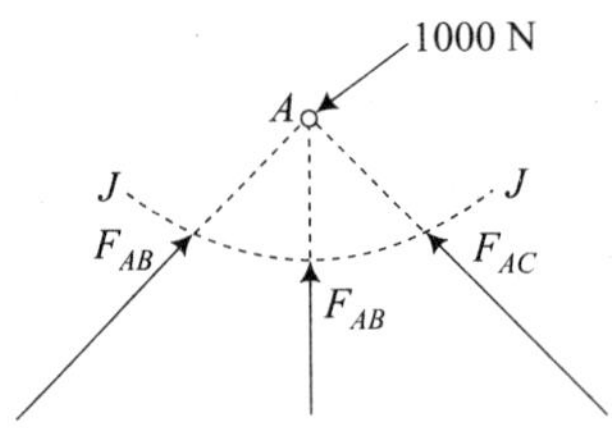

FBD of cut section J-J

There are three unknown forces F_{DA}, F_{AB} and F_{AC}, but only two equilibrium equations are available.

$$\Sigma F_X = 0 \text{ and } \Sigma F_Y = 0$$

Therefore, cut section K-K.

(ii) Draw FBD of section K-K. Take moments about point B.

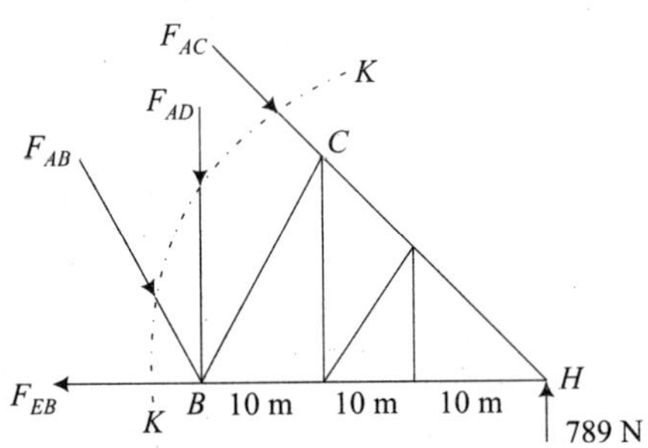

FBD of section K-K

$\Sigma M_B = 0$

$(-500)(10) + (789)(30) - F_{AC}(\sin 30°)(30) = 0$

The force F_{AC} has been transmitted to joint H.

$\therefore \quad F_{AC} = 1245$ N.

(iii) *Section (J-J)*

Consider the equilibrium of section J-J again.

$\Sigma F_X = 0$

$\therefore \quad F_{DA} \cos 30° - F_{AC} \cos 30° - 1000 \sin 30° = 0$

$F_{DA} = 1822$ N

$\Sigma F_Y = 0$

$F_{DA} \sin 30° + F_{AC} \sin 30° + F_{AB} - 1000 \cos 30° = 0$

$\therefore \quad F_{AD} = -667$ N (Tension member)

Example 4.16: Determine the forces and their nature in each member of the truss loaded as shown in Fig. [U.P.T.U. II Sem. 2007–08]

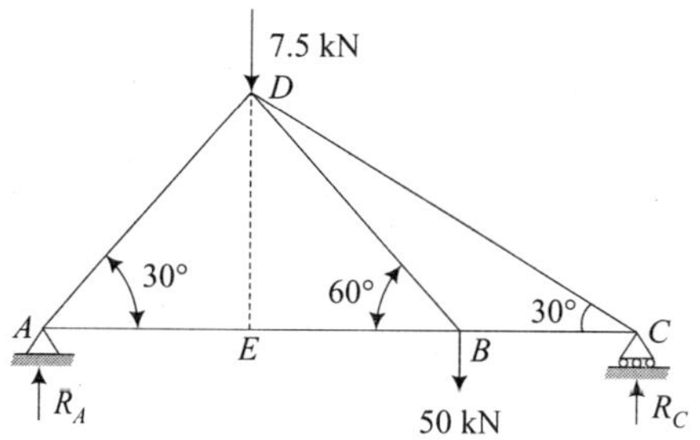

Solution: Let $AC - 1$

$\therefore \quad AE = EC = 0.5$

$$\tan 30° = \frac{DE}{AE}$$

$\therefore \quad DE = AE \tan 30° = 0.2887$

$$\frac{DE}{BE} = \tan 60°$$

$$\therefore \quad BE = \frac{DE}{\tan 60°} = 0.16668$$

$BC = 0.5 - 0.16668 = 0.3333$

(a) Reactions

$$R_A + R_C = 7.5 + 5.0 = 57.5$$

$$\Sigma M_A = 0$$

$$\therefore \quad 7.5 \times 0.5 + 50\ (AB) = R_C\ (AC)$$

$$\therefore \quad R_C = \frac{7.5 \times 0.5 + 50(0.5 + 0.16668)}{1} = 37 \text{ kN}$$

$$\therefore \quad R_A = 57.5 \times 37 = 20.5 \text{ kN}$$

(b) Joint A

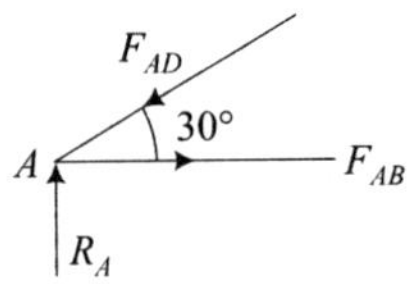

Joint *A*.

$$\Sigma F_y = 0$$

$$F_{AD} \sin 30° = R_A$$

$$F_{AD} = \frac{20.5}{\sin 30°} = 41 \text{ kN (C)}$$

$$\Sigma F_x = 0$$

$$F_{AB} = F_{AD} \text{ Cos } 30° = 35.5 \text{ kN (T)}$$

(c) Joint C

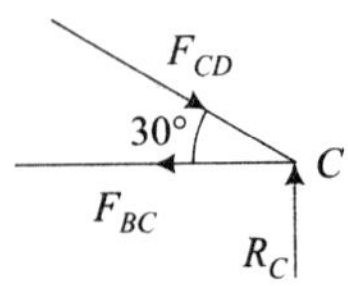

Joint *C*.

$$\Sigma F_y = 0$$

$$F_{CD} \sin 30° = R_C$$

$$\therefore \quad F_{CD} = \frac{Rc}{\sin 30°} = \frac{37}{0.5} = 74 \text{ kN (C)}$$

$$\Sigma F_x = 0$$

$$F_{BC} = F_{CD} \cos 30° = 64 \text{ kN (T)}$$

(d) Joint B

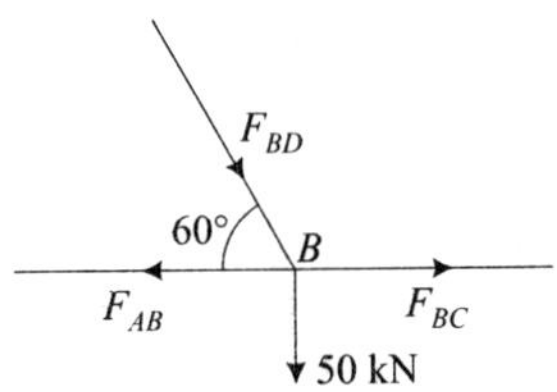

Joint *B*.

$$\Sigma F_y = 0$$

$$F_{BD} \sin 30° = 50$$

$$\therefore \quad F_{BD} = 57.7 \text{ kN (T)}$$

(e) The force table

Member	Force	Nature
AB	35.5 kN	Tensile
AD	41 kN	Compressive
CD	74 kN	Compressive
BC	64 kN	Tensile
BD	57.7	kN Tensile

Example 4.17: Find the forces in all members of a truss as shown below which carries a horizontal load of 12 kN at point *D* and a vertical load of 18 kN at point *C*.
[U.P.T.U. 2009-10]

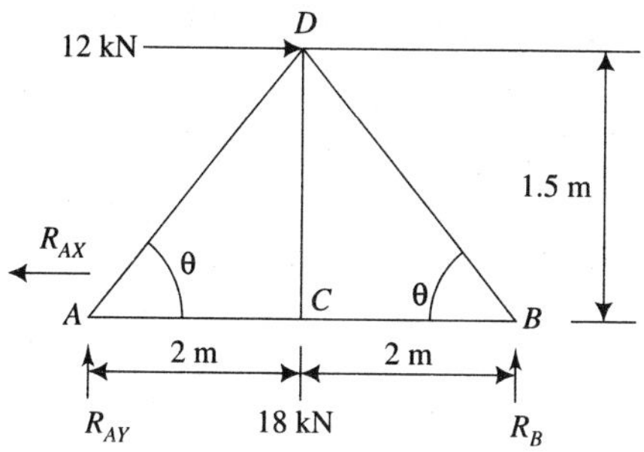

Solution:

(i) *Support Reactions*

$$\tan\theta = \frac{1.5}{2} = 0.75$$

$$\therefore \quad \theta = 36.87°$$

$$\sin\theta = 0.60$$

$$\cos\theta = 0.80.$$

$$\Sigma F_x = 0 \quad \therefore \quad R_{AX} = 12 \text{ kN}$$

$$\Sigma F_y = 0 \quad \therefore \quad R_{AY} + R_B = 18 \text{ kN}.$$

$$\Sigma M_A = 0 \quad R_B \times 4 = 18 \times 2 + 12 \times 1.5$$

$$\therefore \quad R_B = \frac{36+18}{4} = 13.5 \text{ kN}.$$

$$R_{AY} = 18 - 13.5 = 4.5 \text{ kN}$$

(ii) *Joint B*.

$$\Sigma F_y = 0 \quad F_{BD} \sin\theta = R_B.$$

$$\therefore \quad F_{BD} = \frac{R_B}{\sin\theta} = \frac{13.5}{0.60} = 22.5 \text{ kN } (C)$$

$$\Sigma F_y = 0 \quad F_{BC} = F_{BD} \cos\theta$$

$$= 22.5 \times 0.80 = 18 \text{ kN}(T)$$

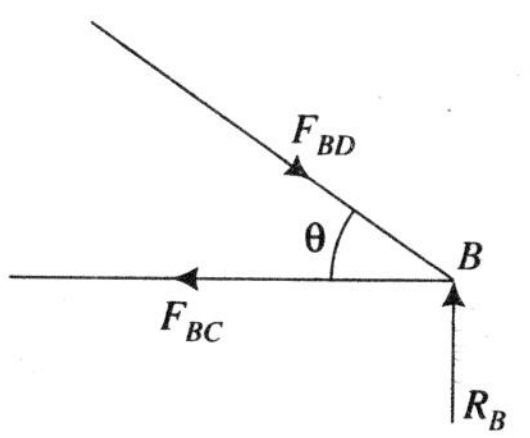

(iii) *Joint C*

$$\Sigma F_y = 0, F_{CD} = 18 \text{ kN}(T)$$

$$\Sigma F_x = 0, F_{CA} = F_{BC} = 18 \text{ kN}(T)$$

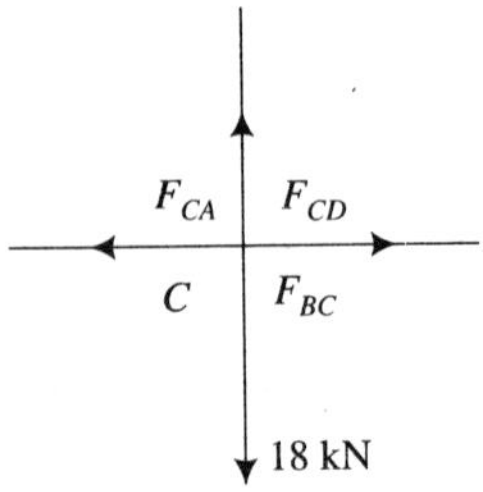

(iv) *Joint A*

$$\Sigma F_y = 0$$

$$F_{AD} \sin\theta = R_{AY}$$

$$F_{AD} = \frac{R_{AY}}{\sin\theta} = \frac{4.5}{0.6} = 7.5 \text{ kN}(C)$$

$$\Sigma F_x = 0 \quad F_{AC} = R_{AX} + F_{AD} \cos\theta$$

$$= 12 + 7.5 \times 0.8 = 18 \text{ kN}(T)$$

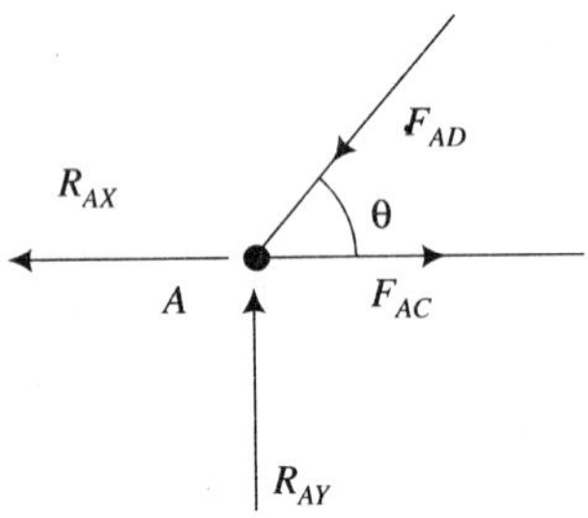

Member	Force	Nature
AD	7.5 kN	C
AC	18 kN	T
CD	18 kN	T
BC	18 kN	T
BD	22.5 kN	C

Example 4.18: Find the axial forces in all members of a truss shown below.
[U.P.T.U. 2008-09]

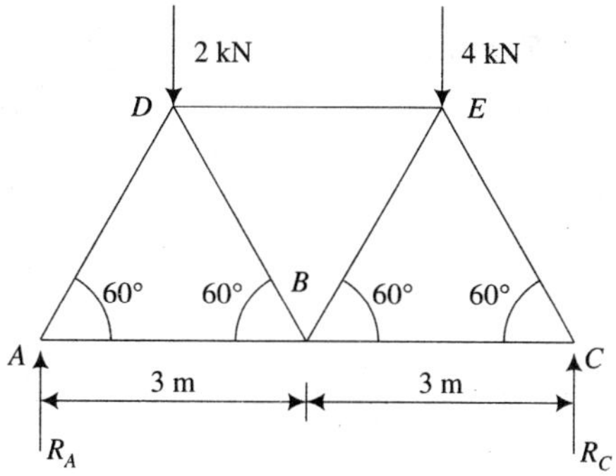

Solution:

(i) *Support Reaction*

$$\sin 60° = 0.866$$

$$\cos 60° = 0.5$$

$$\Sigma F_y = 0 \quad R_A + R_C = 2 + 4 = 6 \text{ kN.}$$

$$\Sigma M_A = 0 \quad R_C \times 6 = 2 \times 1.5 + 4 \times 4.5 = 3 + 18 = 21$$

$$\therefore \quad R_C = \frac{21}{6} = 3.5 \text{ kN}$$

$$\therefore \quad R_A = 6 - 3.5 = 2.5 \text{ kN}$$

(ii) *Joint A*

$$\Sigma F_y = 0 \quad F_{AD} \sin \theta = R_A$$

$$F_{AD} = \frac{R_A}{\sin\theta} = \frac{2.5}{0.866} = 2.887 \text{ kN}(C)$$

$$\Sigma F_x = 0 \quad F_{AB} = F_{AD} \cos \theta = 2.887 \times 0.5$$

$$= 1.44 \text{ kN}(T)$$

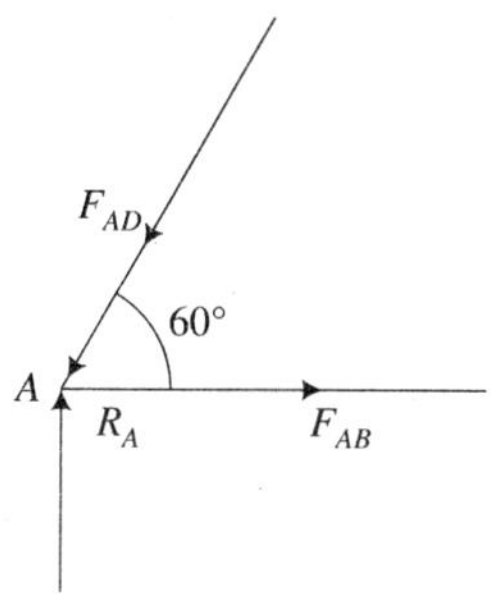

(iii) *Joint C*

$$\Sigma F_y = 0 \quad F_{CE} \sin \theta = R_C$$

$$F_{CE} = \frac{R_C}{\sin\theta} = \frac{3.5}{0.866} = 4 \text{ kN}(C)$$

$$\Sigma F_x = 0 \quad F_{CB} = F_{CE} \cos \theta = 4 \times 0.5 = 2 \text{ kN}(T).$$

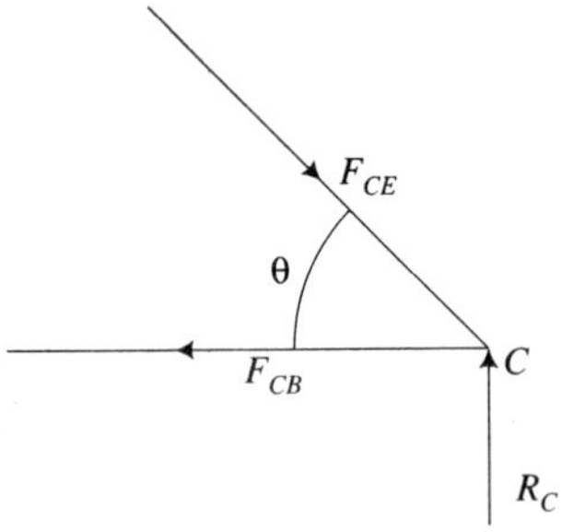

(iv) *Joint B*

$$F_{BA} = 1.44 \text{ kN}(T)$$

$$F_{BC} = 2 \text{ kN } T$$

$$\Sigma F_y = 0$$

$$F_{BD} \sin 60° + F_{BE} \sin 60° = 0$$

$$\therefore \quad F_{BD} = -F_{BE}$$

$$\Sigma F_x = 0$$

$$-F_{BA} + F_{BC} = F_{BD} \cos 60° - F_{BE} \cos 60°$$

$$F_{BD} = \frac{F_{BC} - F_{BA}}{2 \cos 60°} = \frac{2 - 1.44}{2 \times 0.5} = 0.56 \text{ kN.}(C)$$

$$F_{BE} = 0.56 \text{ kN}(T)$$

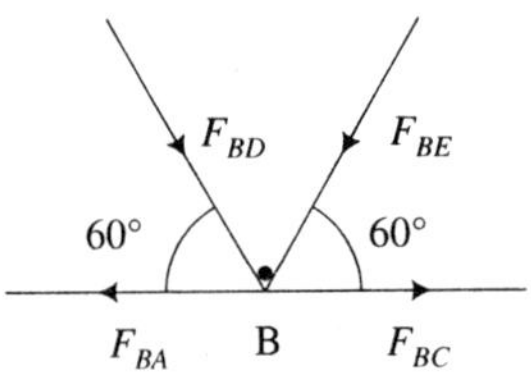

(v) *Joint E*

$$F_{BE} = 0.56 \text{ kN}(T)$$

$$F_{EC} = 4 \text{ kN}(C)$$

$$\Sigma F_x = 0$$

$$F_{ED} = -F_{BE} \cos 60° + F_{EC} \cos 60°$$

$$= -0.56 \times 0.5 + 4 \times 0.5$$

$$= 1.72 \text{ kN}(C)$$

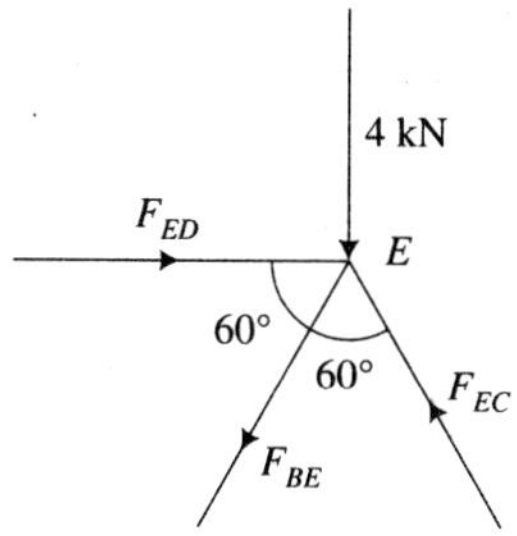

Member	*Force (kN)*	*Nature*
AB	1.44	*T*
BC	2	*T*
AD	2.887	*C*
CE	4	*C*
BD	0.56	*C*
BE	0.56	*T*
DE	1.72	*C*

Example 4.19: Find the forces in the members of truss shown.

[U.P.T.U. 2008-09]

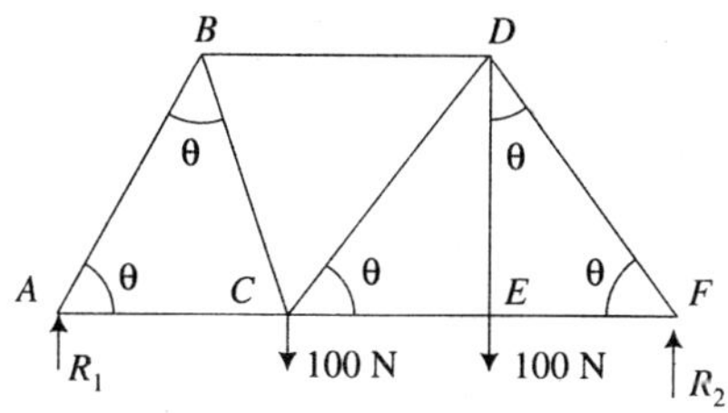

Solution:

By inspection

$$\theta = 45°$$

$$R_1 = R_2 = 100 \text{ N}$$

$$\sin 45° = \cos 45° = 0.7$$

Joint A

$$\Sigma F_y = 0 \quad F_{AB} \sin 45° = R_1$$

$$F_{AB} = \frac{100}{0.7} = 142.857 \text{ N}(C)$$

$$\Sigma F_x = 0 \quad F_{AC} = F_{AB} \cos 45°$$

$$= 142.857 \times 0.7 = 100 \text{ N}(T)$$

The truss is symmetrical

$$\therefore \quad F_{FE} = F_{AC} = 100 \text{ N}(T)$$

$$F_{AD} = F_{AB} = 142.857 \text{ N}(C)$$

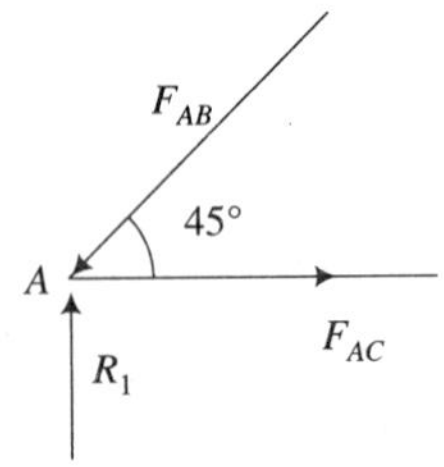

Joint E

$$\Sigma F_y = 0 \quad \therefore \quad F_{ED} = 100 \text{ N}(T)$$

$$\Sigma F_x = 0 \quad F_{EF} = F_{EC} = 100 \text{ N}(T)$$

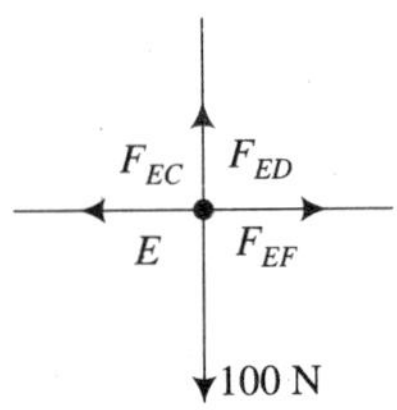

Joint C

$$F_{CA} = 100 \text{ N}(T)$$

$$F_{CE} = 100 \text{ N}(T)$$

$$\therefore \quad F_{CD} = 0$$

$$F_{CB} = 100 \text{ N}(T)$$

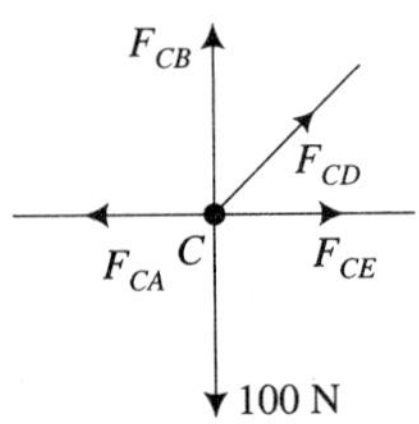

Joint B

$$\Sigma F_x = 0 \quad F_{BD} = F_{BA} \cos 45°$$

$$= 142.857 \times 0.7$$

$$= 100 \text{ N}(C)$$

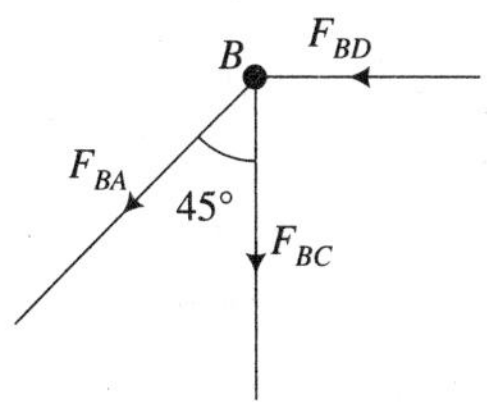

Member	Force (N)	Nature
AB	142.857	C
AC	100	T
BC	100	T
BD	100	C
CE	100	T
CD	0	
EF	100	T
ED	100	T

QUESTION BANK NO. 4

1. Define a truss. What is the difference between a frame and truss? Where do you find trusses in use?
2. How the trusses are classified? Define the following trusses:
   (i) Perfect truss
   (ii) Imperfect truss
   (iii) Deficient truss
   (iv) Redundant truss
3. What is a simple truss? What assumptions are made in the analysis of a simple truss?
4. What are various methods of analysis of trusses? What is basically found when analysis of truss is done? What is the advantage of method of section over the method of joint?
5. What is the difference between a simply supported truss and a cantilever truss? Discuss the method of finding out reactions in both the cases. Is it essential to find out the reactions in a cantilever truss before analysing it?
6. State true or false:
   (i) All the upper members of a simply supported truss carrying only vertical loads are in compression.
   (ii) When two collinear members of a truss meet at a point, and a third member is connected at this point, then the force in the third member will be zero.

(iii) When two collinear members of a truss meet at a point and no external force or reaction act at this point, then the force in both the members will be zero.

(iv) Equation of equilibrium can be used to determine forces in the members of statically determinate truss.

QUICK REVIEW INFORMATION NO. 4

1. A perfect truss has minimum three joints and three members.
2. A truss is said to be rigid in nature when there is no collapsing on application of any external load.
3. In method of section, the section must pass through not more than three members and equations of equilibrium are given as for coplanar force system.
4. In truss analysis:

   (i) The weight of the members is assumed to be zero.

   (ii) The stress induced on application of force is longitudinal.

   (iii) The members are connected at their ends by frictionless hinges.

   (iv) All forces acting on truss are applied at the joints only and also lie in the plane of the truss.

Unit III: Moment of Inertia

CHAPTER 5

Centroid and Moment of Inertia

5.1 INTRODUCTION

Centroid and moment of inertia are important properties of a section for the analysis of many engineering problems. Determination of these properties are needed for the study of strength of materials, structural design and machine design.

1. *Centroid* is a point in a plane area or arc or volume where the total area or length or volume of a geometrical quantity is assumed to be concentrated. Centroid is a property of the area or volume itself and is independent of reference axes. The moment of an area about any axis through the centroid is zero.

   Centroidal axes are the axes passing through the centroid of an area.

   Centroidal coordinates $(\overline{X}, \overline{Y})$ are the coordinate of centroid from some reference axes.

2. *Axis of Symmetry* Centroid of an area lies on the axis of symmetry of a section if it exists. This definition is used to locate the centroid of an area.

   Example, centroid of a circle is its centre and centroid of a rectangle is the intersection point of its diagonals.

3. *First Moment of an Area* is the moment of the area of a section about reference axes.

   First moment of area A about x-axis is:

   $$M_x = \int_A y\, dA \qquad \text{and}$$

   First moment of area A about y-axis is:

   $$M_y = \int_A x\, dA$$

   From the definition of centroid,

$$M_x = A\bar{y} = \int_A y\, dA$$

$$M_y = A\bar{x} = \int_A x\, dA,$$

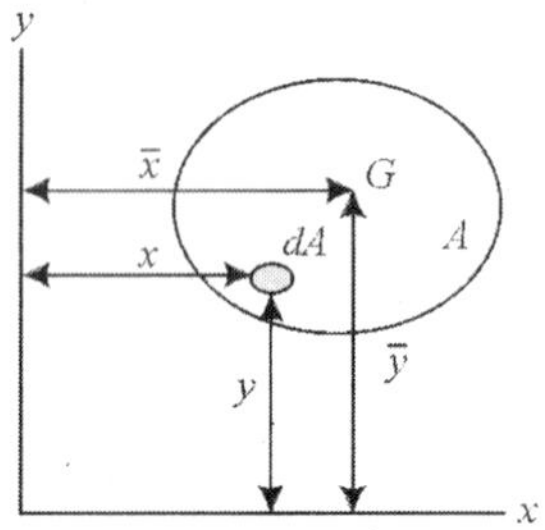

Fig. 5.1 Centroid

Therefore, the centroidal coordinates can be found out as:

$$\bar{x} = \frac{\int_A x\, dA}{A} \quad \text{and}$$

$$\bar{y} = \frac{\int_A y\, dA}{A}$$

The first moment of an area about any of is centroidal axes must be zero.

5.2 CENTROID OF COMMON SECTIONS

The definition of centroidal coordinates by the use of principle of first moment of an area can be used to find the centroid of simple sections.

5.2.1 Centroid of a Triangle

A triangle *ABC* has a base width *b* and height *h*. An elemental strip of width b_1 and thickness *dy* is located at a distance of *y* from base.

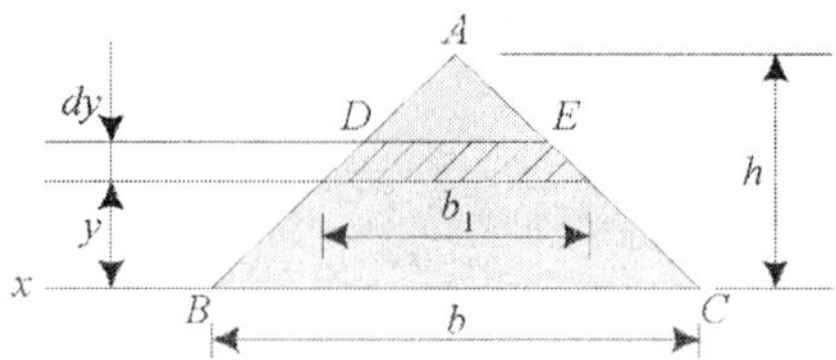

Fig. 5.2 Centroid of a triangle

From similar Δs ABC and ADE

$$\frac{b_1}{b} = \frac{h-y}{h}$$

$$\therefore \qquad b_1 = \left(\frac{h-y}{h}\right)b = \left(1-\frac{y}{h}\right)b.$$

Area of elemental strip,

$$dA = b_1 dy = \left(1-\frac{y}{h}\right)b\,dy.$$

Moment of area about base,

$$M_x = \int y\,dA = \int_0^h y\left(1-\frac{y}{h}\right)b\,dy = \int_0^h \left(y-\frac{y^2}{h}\right)b\,dy$$

$$= b\left|\frac{y^2}{2}-\frac{y^3}{3h}\right|_o^h = b\frac{h^2}{6}$$

$$M_x = A\bar{y} = \frac{1}{2}bh\,\bar{y}$$

$$\therefore \qquad \frac{1}{2}bh\,\bar{y} = \frac{bh^2}{6}$$

$$\therefore \qquad \bar{y} = \frac{h}{3}$$

5.2.2 Centroid of Semicircle

A semicircle of radius R is shown in Fig. 5.3. The centroid lies on y-axis due to symmetry. Consider an element of radius r and thickness dr.

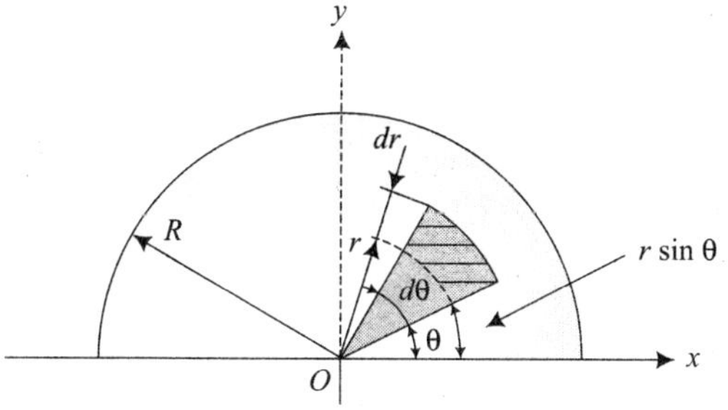

Fig. 5.3 Centroid of semicircle

Area of element,

$$dA = r\,d\theta\,dr.$$

First moment of element about x-axis,

$$dM_x = rd\theta dr \times r\sin\theta = r^2 \sin\theta\, dr\, d\theta$$

Total first moment of area,

$$M_x = \int_o^{\pi}\int_o^{R} r^2 \sin\theta\, dr\, d\theta = \int_o^{\pi} \left|\frac{r^3}{3}\right|_o^R \sin\theta\, d\theta$$

$$= \frac{R^3}{3}\left|-\cos\theta\right|_o^{\pi} = \frac{R^3}{3}[1+1] = \frac{2R^3}{3}$$

Area of semicircle, $A = \dfrac{\pi}{2}R^2.$

$$\therefore \qquad \bar{y} = \frac{\int y\,dA}{A} = \frac{2R^3/3}{\frac{\pi}{2}R^2} = \frac{4R}{3\pi} \qquad \textbf{Ans.}$$

5.2.3 Centroid of Sector of a Circle

A sector of circle of radius R and included angle 2α is shown. By symmetry, centroid lies on x-axis.

Consider an element of radius r and thickness dr. Area of element,

$$dA = r\,d\theta\,dr.$$

Moment of dA about y-axis

$$dM_y = dA\; r\cos\theta$$
$$= r^2 \cos\theta\, dr\, d\theta$$

$$M_y = \int_{-\alpha}^{\alpha}\int_o^{R} r^2 \cos\theta\, dr\, d\theta$$

$$= \left|\frac{r^3}{3}\right|_o^R \left|\sin\theta\right|_{-\alpha}^{\alpha}$$

$$= \frac{R^3}{3}\cdot 2\sin\alpha.$$

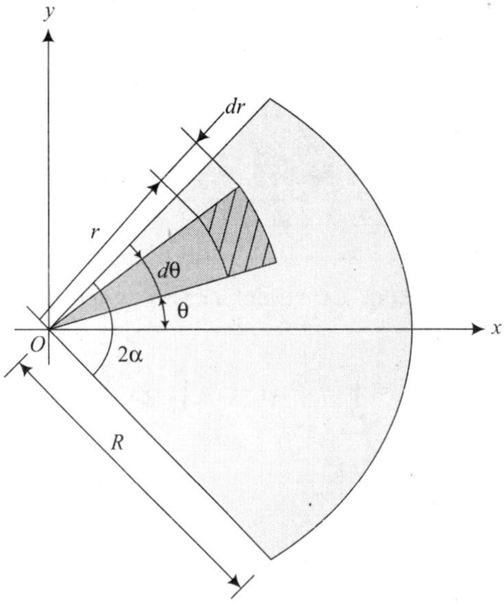

Fig. 5.4 Centroid of sector of a circle

Total area of the sector,

$$A = \int_{-\alpha}^{\alpha}\int_{0}^{R} r\,dr\,d\theta = \int_{-\alpha}^{\alpha}\left|\frac{r^2}{2}\right|_0^R d\theta = \frac{R^2}{2}|\theta|_{-\alpha}^{\alpha}$$

$$= R^2\alpha$$

$$\text{The distance of centroid from } O = \frac{My}{A} = \frac{\frac{2R^3}{3}\sin\alpha}{R^2\alpha}$$

$$= \frac{2R}{3\alpha}\sin\alpha \quad \textbf{Ans.}$$

5.2.4 Centroid of Parabolic Spendrel

A parabolic spendrel enclosed by a parabola $y = kx^2$, x-axis and a line parallel to y-axis at a distance of b is shown

Area of element,

$$dA = y\,dx = kx^2\,dx$$

Total area,

$$A = \int_a^b kx^2\,dx = \left|\frac{kx^3}{3}\right|_0^b$$

$$= \frac{kb^3}{3}.$$

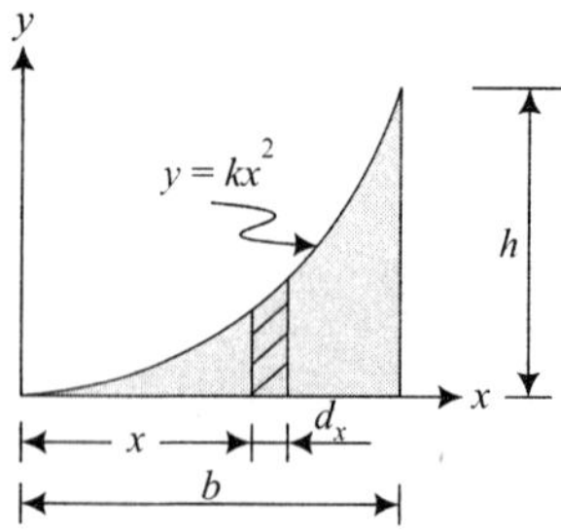

Fig. 5.5(a) Centroid of a parabolic spendrel

$$M_y = \int_0^b kx^2\, dx,\ x = \int_0^b kx^3\, dx$$

$$= \frac{kb^4}{4}.$$

$$y = kx^2$$

$\therefore$
$$x = \sqrt{\frac{y}{k}}$$

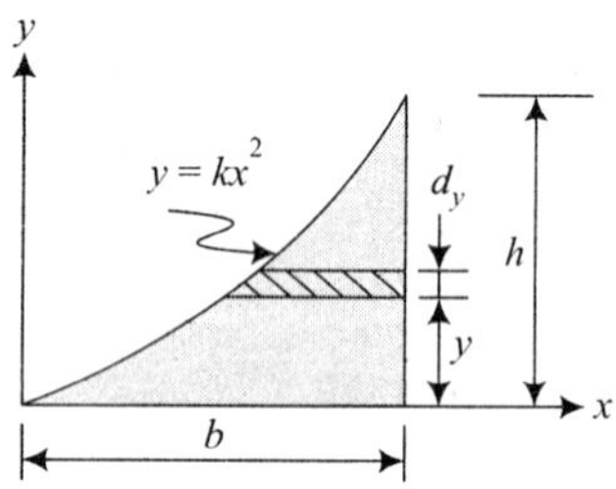

Fig. 5.5(b) Centroid of parabolic spendrel

Moment of area about x-axis,

$$M_x = \int_0^h dyx \cdot \text{y}$$

$$= \int_0^h \sqrt{\frac{k}{k}}\ y\, dy = \int_0^h \frac{y^{\frac{3}{2}}}{k^{\frac{1}{2}}}\, dy = \frac{1}{\sqrt{k}} \left| \frac{y^{\frac{5}{2}}}{\frac{5}{2}} \right|_0^h$$

$$= \frac{2}{5\sqrt{k}} \cdot h^{\frac{5}{2}}$$

At $x = b, y = h$

$$\therefore \quad h = kb^2$$

$$\text{or} \quad k = \frac{h}{b^2}.$$

$$\therefore \quad M_y = \frac{h}{b^2}\cdot\frac{b^4}{4} = \frac{hb^2}{4}$$

$$M_x = \frac{2}{5}\frac{h^{5/2}}{\sqrt{h/b^2}} = \frac{2bh^2}{5}$$

$$A = \frac{kb^3}{3} = \frac{h}{b^2}\cdot\frac{b^3}{3} = \frac{bh}{3}$$

$$\overline{X} = \frac{M_y}{A} = \frac{hb^2}{4}\times\frac{3}{bh} = \frac{\mathbf{3}}{\mathbf{4}}\cdot\boldsymbol{b} \quad \textbf{Ans.}$$

$$\overline{Y} = \frac{M_x}{A} = \frac{2h^2b}{5}\times\frac{3}{bh} = \frac{\mathbf{6}}{\mathbf{5}}\cdot\boldsymbol{h} \quad \textbf{Ans.}$$

Example 5.1: A plane area is bounded by x-axis, the arc $y^2 = 25x$ and a vertical line parallel to y-axis at a distance of 10 m. Find out the first moment of inertia of area about x-axis and y-axis and also centroid coordinates.

Solution: 1. Draw the plane area as shown in Fig. 5.6.

2. When $x = 10$ m, $y^2 = 25x$

$$y = 5\sqrt{x} = \sqrt{25\times 10} = \sqrt{250}\text{ m}$$

3. Use a vertical strip of area, $dA = 5\sqrt{x}\ dx$.

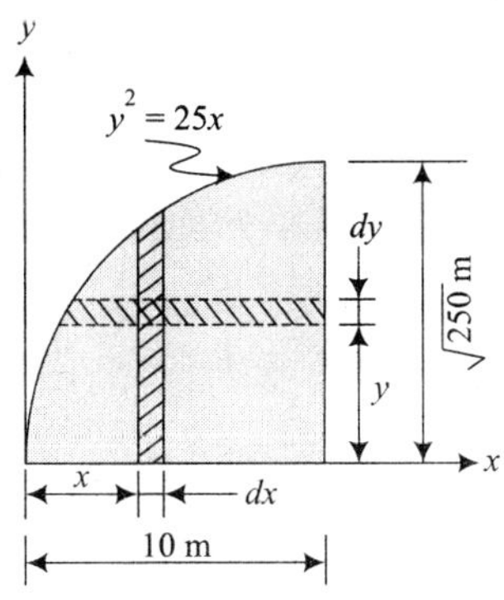

Fig. 5.6 (Example 5.1)

4. Moment of area about y-axis,

$$M_y = \int_0^{10} x(y\,dx)$$

$$= \int_0^{10} x(5\sqrt{x}\,)dx$$

$$= \int_0^{10} 5\,x^{\frac{3}{2}}\,dx$$

$$= \left|\frac{5x^{\frac{5}{2}}}{5/2}\right|_0^{10} = 632\text{ m}^3$$

5. Use horizontal strip of area

$$dA = (10 - x)dy$$

$$M_x = \int_0^{\sqrt{250}} y(10 - x)dy$$

$$= \int_0^{\sqrt{250}} y\left[10 - \frac{y^2}{25}\right]dy \qquad \left[\because x = \frac{y^2}{25}\right]$$

$$= \int_0^{\sqrt{250}} \left[10y - \frac{y^3}{25}\right]dy$$

$$= \left|\frac{10y^2}{2} - \frac{y^4}{25\times 4}\right|_0^{\sqrt{250}} = 625\text{ m}^3$$

6. Area of the enclosed surface

$$A = \int_0^{10} y\,dx = \int_0^{10} 5\sqrt{x}\,dx = \left|\frac{5x^{\frac{3}{2}}}{\frac{3}{2}}\right|_0^{10}$$

$$= 105.4\text{ m}^2$$

7. The centroidal coordinates are:

$$\left.\begin{aligned}\bar{x} &= \frac{M_y}{A} = \frac{632}{105.4} = \mathbf{6\ m} \\ \bar{y} &= \frac{M_x}{A} = \frac{625}{105.4} = \mathbf{5.93\ m}\end{aligned}\right\} \textbf{Ans}$$

5.3 CENTROID OF A CIRCULAR CURVE

A circular curve of 90° angle and radius R is shown in Fig. 5.7.

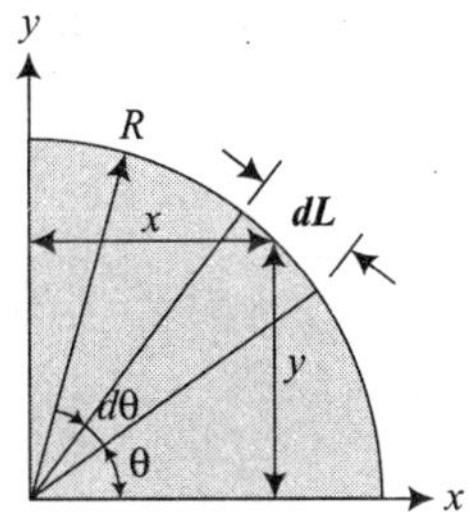

Fig. 5.7 Centroid of a curve

Consider a segment of arc of length dL.

$$dL = R\,d\theta$$

$$x = R\cos\theta$$

$$y = R\sin\theta$$

$$\bar{y} = \frac{\int y\,dL}{L} = \frac{\int y\,dL}{\int dL}$$

$$= \frac{\int_0^{\frac{\pi}{2}} R\sin\theta.(R d\theta)}{\int_0^{\frac{\pi}{2}} R d\theta}$$

$$\bar{y} = \frac{R^2\left|-\cos\theta\right|_0^{\frac{\pi}{2}}}{R\left|\theta\right|_0^{\frac{\pi}{2}}} = \frac{-R\left(\cos\frac{\pi}{2} - \cos 0°\right)}{\left(\frac{\pi}{2} - 0\right)} = \frac{2R}{\pi}$$

$$\bar{x} = \frac{\int x\,dL}{\int dL} = \frac{\int_0^{\pi/2} (R\cos\theta)\,(R d\theta)}{\int_0^{\pi/2} R d\theta} = \frac{R^2\left|\sin\theta\right|_0^{\pi/2}}{R\left|\theta\right|_0^{\pi/2}}$$

$$\bar{x} = \frac{R\left(\sin\frac{\pi}{2} - \sin 0°\right)}{\left(\frac{\pi}{2} - 0\right)} = \frac{2R}{\pi} \quad \textbf{Ans.}$$

5.4 CENTROID OF VOLUMES

5.4.1 Centroid of a Cone

A cone of base radius R and height h is shown in Fig. 5.8.

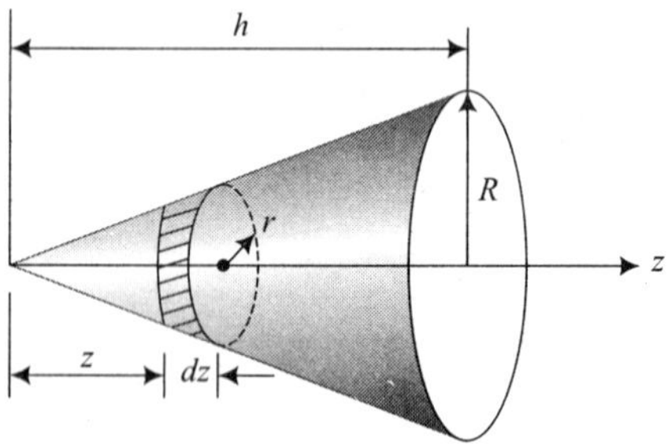

Fig. 5.8 Centroid of a cone

Consider an elementary strip of radius r and thickness dZ

The volume of strip,

$$dv = \pi r^2\, dz$$

From similar triangles,

$$\frac{R}{r} = \frac{h}{z}$$

$$r = \frac{RZ}{h}$$

$\therefore$
$$dv = \pi\left(\frac{RZ}{h}\right)^2 dz$$

Now
$$\bar{z} = \frac{\int z dv}{\int dv} = \frac{\int_0^h z\pi\left(\frac{RZ}{h}\right)^2 dz}{\int_0^h \pi\left(\frac{RZ}{h}\right)^2 dz}$$

$$\bar{z} = \frac{\frac{\pi R^2}{h^2}\int_0^h z^3 dz}{\frac{\pi R^2}{h^2}\int_0^h z^2 dz}$$

$$= \frac{\left|\frac{z^4}{4}\right|_0^h}{\left|\frac{z^3}{3}\right|_0^h} = \frac{3h}{4} \quad \textbf{Ans.}$$

5.4.2 Centroid of a Hemisphere

A hemisphere of radius R is shown in Fig. 5.9. Consider a small strip of radius r and thickness dz. The volume of strip, $dv = \pi r^2\, dz$

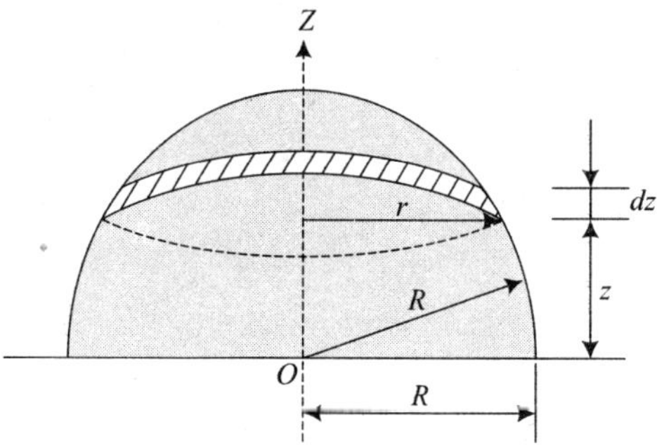

Fig. 5.9 Centroid of hemisphere

From geometry,

$$R^2 = r^2 + z^2$$
$$r^2 = R^2 - z^2$$

The volume of strip,

$$dv = \pi r^2\, dz$$
$$= \pi(R^2 - Z^2)dz$$

$$\bar{z} = \frac{\int z dv}{\int dv}$$

$$\bar{z} = \frac{\int_0^R z\pi(R^2 - z^2)dz}{\int_0^R \pi(R^2 - z^2)dz}$$

$$= \frac{\pi\left|\frac{R^2z^2}{2} - \frac{z^4}{4}\right|_0^R}{\pi\left|R^2z - \frac{z^3}{3}\right|_0^R} = \frac{\frac{R^4}{2} - \frac{R^4}{4}}{R^3 - \frac{R^3}{3}}$$

$$= \frac{\frac{R^4}{4}}{\frac{2}{3}R^3} = \frac{3}{8}R \quad \textbf{Ans.}$$

5.5 CENTROID OF COMMON FIGURES

The properties of common figures are tabulated in Table 5.1.

Table 5.1 Centroid of common figures

Shape	*Figure*	*Area*	*Centroid Coordinates*	
			$\bar{x}$	$\bar{y}$
Triangle	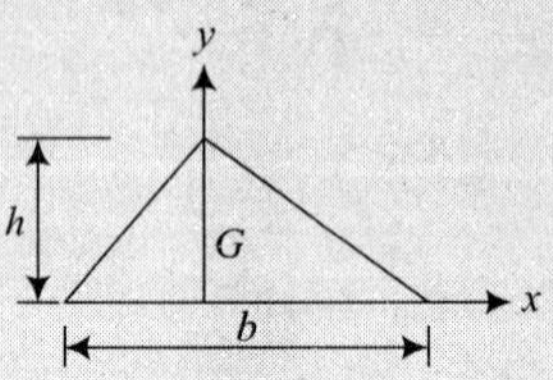	$\frac{bh}{2}$	–	$\frac{h}{3}$
Semicircle	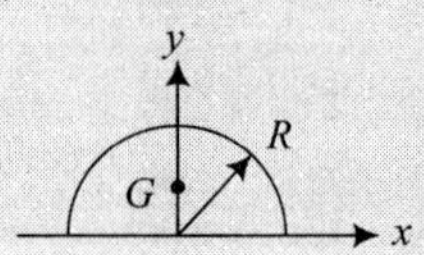	$\frac{\pi R^2}{2}$	0	$\frac{4R}{3\pi}$
Quarter circle	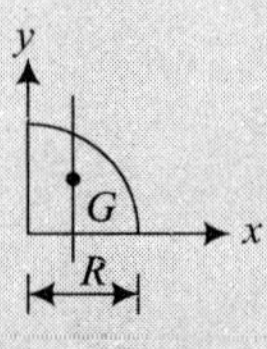	$\frac{\pi R^2}{4}$	$\frac{4R}{3\pi}$	$\frac{4R}{3\pi}$
Sector of a circle	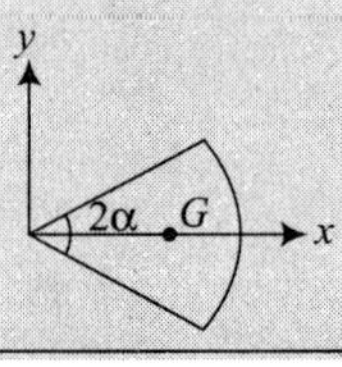	αR^2	$\frac{2R}{3\alpha}\sin\alpha$	0

5.6 CENTROID OF COMPOSITE BODIES

In engineering, composite bodies are made up of simple figures such as rectangle, triangle, circle, semicircle, etc. as solid or hollow section. The centroid of composite bodies are formed out as following.

1. Draw the section of composite body.
2. Split the area into suitable simple figures.
3. Calculate the area and locate the centroidal coordinates of simple figures.
4. Calculate first moment of area of simple figures by multiplying the area with the distance between the centroid of simple figure and reference axis.
5. Sum up the total moment as well as total area of simple figures.
6. Calculate the centroid of composite body as follows.

$$\bar{x} = \frac{\Sigma M_y}{\Sigma A}$$

$$\bar{y} = \frac{\Sigma M_x}{\Sigma A}$$

Note: Use of table for calculation can make solution easy.

Example 5.2: Determine the centroid of T-section shown.

Solution: The centroid lies on y-axis due to symmetry. Divide into reactangles 1 and 2.

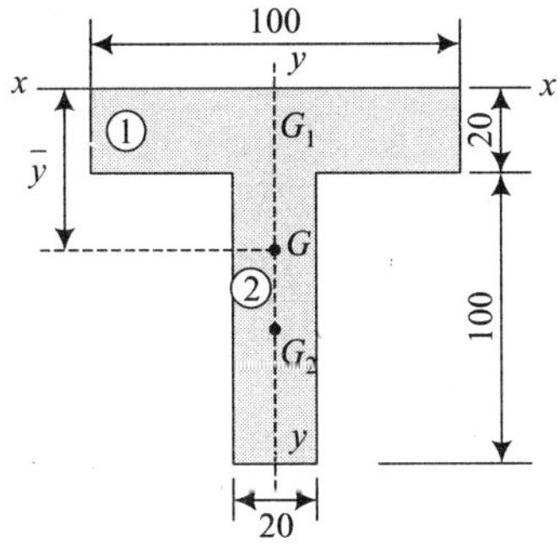

Fig. 5.10 (Example 5.2)

Area	A (mm²)	Y (mm)	M_x
Rectangle 1	100 × 20 = 2000	10	20000
Rectangle 2	100 × 20 = 2000	70	140000
Σ	4000		160000

$$\bar{y} = \frac{\Sigma M_x}{\Sigma A} = \frac{160000}{4000} = \mathbf{40\ mm} \quad \textbf{Ans.}$$

Example 5.3: Determine the centroid of unequal angle shown.

Solution: Divide the area into rectangles 1 and 2.

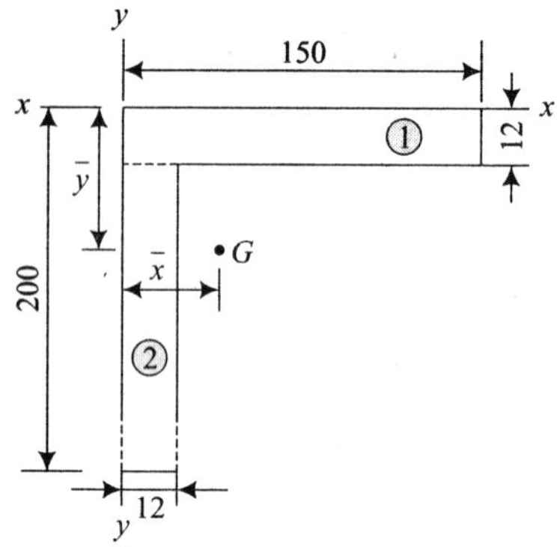

Fig. 5.11 (Example 5.3)

Area	*A* *(mm²)*	*X* *(mm)*	*Y* *(mm)*	$M_x = AY$	$M_y = AX$
Rectangle 1	150 × 12 = 1800	75	6	10800	135000
Rectangle 2	(200 – 12)12 = 2256	6	$\left(\frac{200-12}{2}\right) + 12$ = 106	239136	13536
Σ	4056			249936	148536

$$\bar{x} = \frac{\Sigma M_y}{\Sigma A} = \frac{148536}{4056} = \mathbf{36.62\ mm}$$

$$\bar{y} = \frac{\Sigma M_x}{\Sigma A} = \frac{249936}{4056} = \mathbf{61.62\ mm} \qquad \textbf{Ans.}$$

Example 5.4: Determine the centroid of I-section shown.

Solution: Divide the section into three rectangles. This section is symmetrical, the centroid lies on *y*-axis.

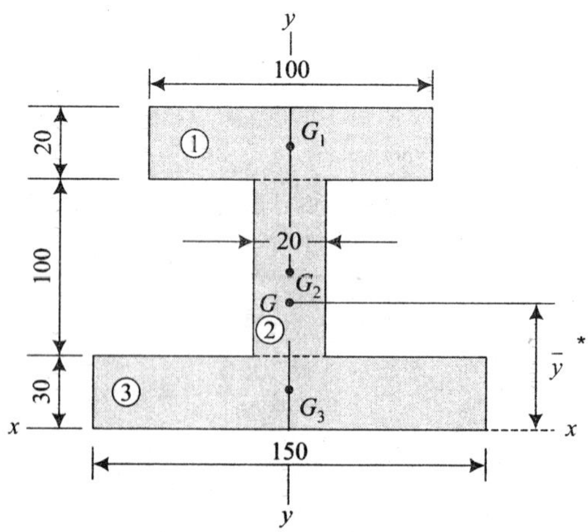

Fig. 5.12 (Example 5.4)

Area	*A* *(mm^2)*	*Y* *(mm)*	M_x *(mm^3)*
Rectangle 1	100 × 20 = 2000	10 + 100 + 30 = 140	280000
Rectangle 2	100 × 20 = 2000	50 + 30 = 80	160000
Rectangle 3	150 × 30 = 4500	$\frac{30}{2}$ = 15	67500
Σ	8500	–	507500

$$\bar{y} = \frac{\Sigma M_x}{\Sigma A} = \frac{507500}{8500}$$

$$= \textbf{59.71 mm} \quad \textbf{Ans.}$$

Example 5.5: Determine the centroid of the section with respect to axes shown.

Solution: Divide the section into a triangle, rectangle and a semicircle.

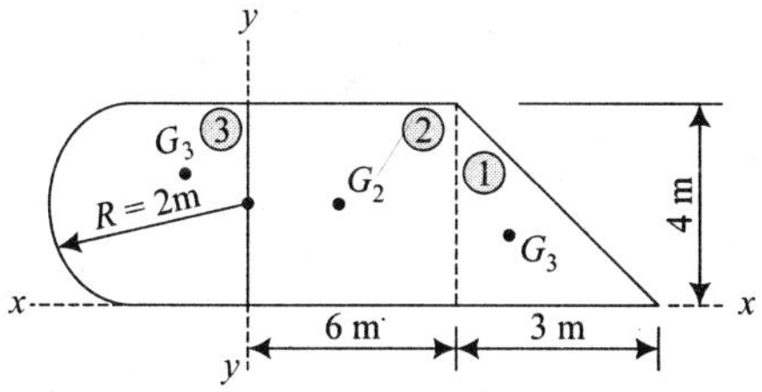

Fig. 5.13 (Example 5.5)

Area	*A* *(m^2)*	*X* *(m)*	*Y* *(m)*	M_x *(m^3)*	M_y *(m^3)*
Triangle 1	$\frac{4 \times 3}{2} = 6$	$6 + \frac{3}{3} = 7$	$\frac{4}{3}$	8	42
Rectangle 2	6 × 4 = 24	3	2	48	72
Semicircle 3	$\frac{\pi \times 2^2}{2}$ = 6.2832	$\frac{-4 \times 2}{3\pi}$ = –0.8488	2 m	12.5664	–5.83
Σ	36.2832	–	–	68.5664	108.17

$$\left.\begin{aligned} \bar{x} &= \frac{\Sigma M_y}{\Sigma A} = \frac{108.17}{36.2832} = \textbf{2.995 m} \\ \bar{y} &= \frac{\Sigma M_x}{\Sigma A} = \frac{68.5664}{36.2832} = \textbf{1.89 m} \end{aligned}\right\} \textbf{Ans.}$$

Example 5.6: A semicircle area is removed from a trapezium as shown. Determine the centroid of the remaining area.

Solution: Divide the area into a rectangle, a triangle and semicircle.

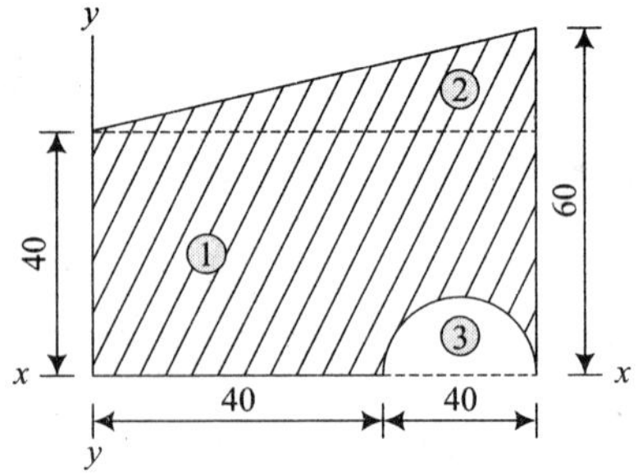

Fig. 5.14 (Example 5.6)

Area	*A* *(mm²)*	*X* *(mm)*	*Y* *(mm)*	M_x *(mm³)*	M_y *(mm³)*
Rectangle 1	80 × 40 = 3200	40	20	64000	128000
Triangle 2	$\frac{80 \times 20}{2} = 800$	$\frac{2}{3} \times 80$ = 53.33	$40 + \frac{20}{3}$ = 46.67	37336	42664
Semicircle 3	$\frac{\pi}{2} \times 20^2$ = 628.3	40 + 20 = 60	$\frac{4 \times 20}{3\pi}$ = 8.49	–5134.27	–13698
Σ	3371.7	–	–	96201.63	156866

$$\bar{x} = \frac{\Sigma M_y}{\Sigma A} = \frac{156866}{3371.7} = \mathbf{39.44\ mm} \quad \textbf{Ans.}$$

$$\bar{y} = \frac{\Sigma M_x}{\Sigma A} = \frac{96201.63}{33717.7} = \mathbf{28.47\ mm} \quad \textbf{Ans.}$$

Example 5.6(a): Locate the centroid of channel section shown.

[U.P.T.U. 2009-10]

Solution:

Divide the area into rectangles 1, 2 and 3.

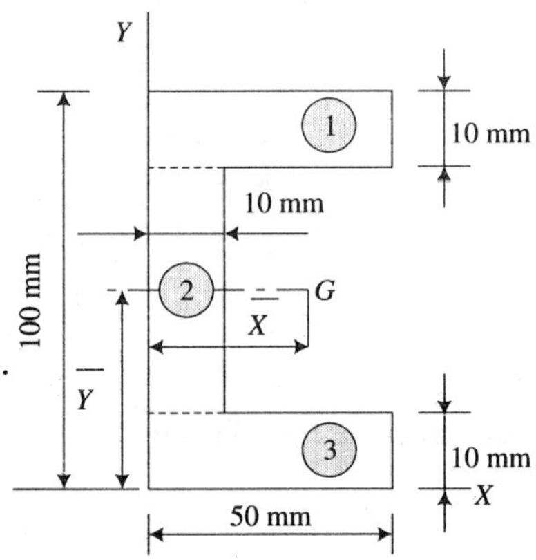

Area	*A(mm²)*	*X(mm)*	*Y(mm)*	$M_x = Ay$	$M_y = Ax$
Rectangle 1	50 × 10 = 500	50	95	47,500	25000
Rectangle 2	80 × 10 = 800	5	50	40,000	4,000
Rectangle 3	50 × 10 = 500	50	5	2,500	25,000
Σ	1800			90,000	54,000

$$\left.\begin{aligned}\bar{x} &= \frac{\Sigma M_y}{\Sigma A} = \frac{54{,}000}{1800} = 30 \text{ mm} \\ \bar{y} &= \frac{\Sigma M_x}{\Sigma A} = \frac{90{,}000}{1800} = \mathbf{50\ mm}\end{aligned}\right\} \textbf{Ans.}$$

Example 5.6(b): Find the centroid of the following figure. [U.P.T.U. 2008-09]

Solution:

Divide the figure into a triangle, a semicircle and a rectangle.

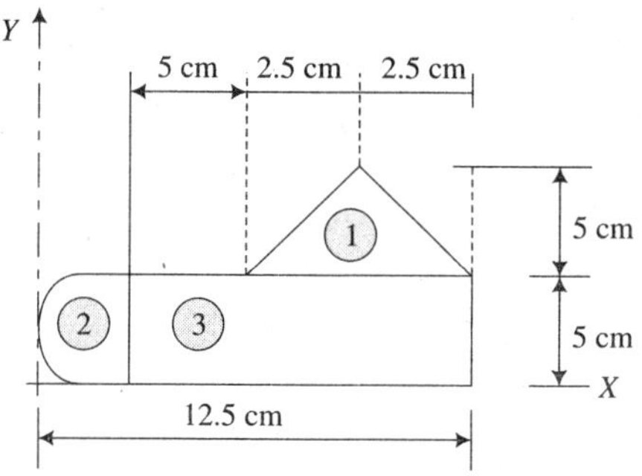

Area	*A(cm²)*	*X(cm)*	*Y(cm)*	$M_x = Ay(cm^3)$	$M_y = Ax\ (cm^3)$
Triangle 1	$\frac{5\times 5}{2} = 12.5$	10	$5 + \frac{5}{3} = 6.667$	83.33	125
Semicircle 2	$\frac{\pi(2.5)^2}{2} = 9.8$	$\frac{4\times 2.5}{3\pi} = 1.438967$	2.5	24.5	14.10
Rectangle 3	12.5 × 5 = 62.5	5	2.5	156.25	312.5
Σ	84.8			264.08	451.6

$$\left.\begin{aligned}\bar{x} &= \frac{\Sigma M_y}{\Sigma A} = \frac{451.6}{84.8} = 5.325 \text{ cm} \\ \bar{y} &= \frac{\Sigma M_x}{\Sigma A} = \frac{264.08}{84.8} = \mathbf{3.1\ cm}\end{aligned}\right\} \textbf{Ans.}$$

Example 5.6(c): A semicircle is removed from the trapezoid shown. Determine the centroid of they remaining area. [U.P.T.U. 2009-10]

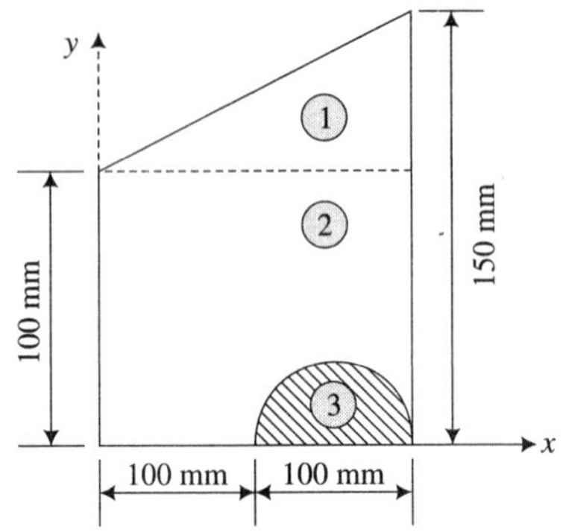

Solution:

Divide the area into a triangle, a rectangle and a semicircle.

Area	$A(mm^2)$	$X(mm)$	$Y(mm)$	$M_x = Ay(mm^3)$	$M_y = Ax(mm^3)$
Triangle 1	$\frac{200 \times 50}{2}$ = 5000	$\frac{2}{3} \times 200 = \frac{400}{3}$	$100 + \frac{50}{3} = \frac{350}{3}$	5,83,333.3	6,66,666.67
Rectangle 2	200 × 100 = 20,000	100	50	10,00,000	20,00,000
Semicircle 3	(–) $\frac{\pi \times 50^2}{2}$ = 3927	100 + 50 = 150	$\frac{4 \times 50}{3\pi}$ = 21.22	(–) 83,330	(–) 589050
Σ	21073			1500003.3	2077616.7

$$\bar{x} = \frac{\Sigma M_y}{\Sigma A} = \frac{2077616.7}{21073} = 98.6 \text{ mm}$$

$$\bar{y} = \frac{\Sigma M_x}{\Sigma A} = \frac{1500003.3}{21073} = \mathbf{71.18 \text{ mm}}$$

Ans.

Example 5.6(d): A triangle is removed from a semicircle as shown. Locate the centroid of the remaining object. [U.P.T.U. 2008-09]

Solution:

Divide the object into a semicircle 1, triangle 2, and triangle 3. The height of the triangle,

$$BC = \sqrt{AB^2 - AC^2}$$

$$= \sqrt{(4.5)^2 - (1.5)^2} = 4.243 \text{ cm}$$

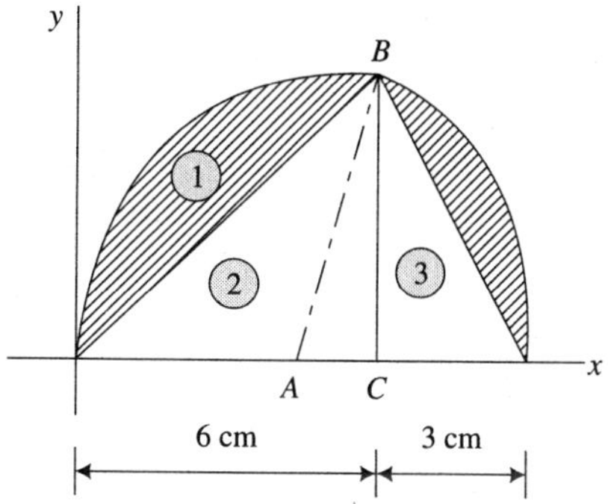

Area	$A(cm^2)$	$X(cm)$	$Y(cm)$	$M_x = Ay\ (cm^3)$	$M_y = Ax\ (cm^3)$
Semicircle 1	$\frac{\pi}{2} \times (4.5)^2 = 31.80$	4.5	$\frac{2 \times 4.5}{3\pi} = 0.955$	30.369	143.1
Triangle 2	$(-)\frac{6 \times 4.243}{2} = 12.73$	$\frac{2}{3} \times 6 = 4$	$\frac{4.243}{3} = 1.41$	(–) 17.95	(–) 50.92
Triangle 3	$(-)\frac{3 \times 4.243}{2} = 6.36$	$6 + \frac{3}{3} = 7$	$\frac{4.243}{3} = 1.41$	– 8.967	(–) 44.52
Σ	12.71			3.452	47.66

$$\bar{x} = \frac{\Sigma M_y}{\Sigma A} = \frac{47.66}{12.71}$$

$$- \mathbf{3.75\ cm} \quad \textbf{Ans.}$$

$$\bar{y} = \frac{\Sigma M_x}{\Sigma A} = \frac{3.452}{12.71}$$

$$= \mathbf{0.27\ cm} \quad \textbf{Ans.}$$

Example 5.6(e): A wire is bent into a closed loop $A - B - C - D - E - A$ as shown in which AB is a circular arc. Determine the centroid of the wire.

[U.P.T.U. 2010-11]

Solution:

Divide the wire into the following lengths: AB, BC, CD, DE and EA.

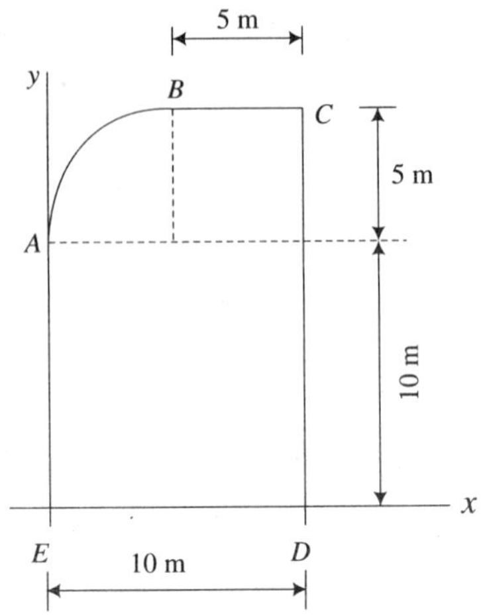

Wire	Length(m)	X(m)	Y(m)	LX(m^2)	LY(m^2)
AB	$\frac{\pi \times 10}{4} = 7.85$	$5 - \frac{2 \times 5}{\pi} = 1.81$	1.81 + 10 = 11.81	14.2	92.7
BC	5	5 + 2.5 = 7.5	15	37.5	75
CD	15	10	7.5	150	112.5
DE	10	5	0	50	0
EA	10	0	5	0	50
Σ	47.85			251.7	330.2

$$\bar{x} = \frac{\Sigma Lx}{\Sigma L} = \frac{251.7}{47.85}$$

$$= \mathbf{5.26\ m} \quad \textbf{Ans.}$$

$$\bar{y} = \frac{\Sigma Ly}{\Sigma L} = \frac{330.2}{47.85}$$

$$= \mathbf{6.90\ m} \quad \textbf{Ans.}$$

5.7 MOMENT OF INERTIA

Moment of inertia is the moment of first moment of an area or second moment of an area about an axis.

It is a property of the section of a structural member and is a measure of resistance of the member against bending, i.e., deflection or deformation due to bending.

A plane area A with centroid $G(\bar{x}, \bar{y})$ is shown in the figure. Consider a small area dA at distances y and x from x-axis and y-axis respectively.

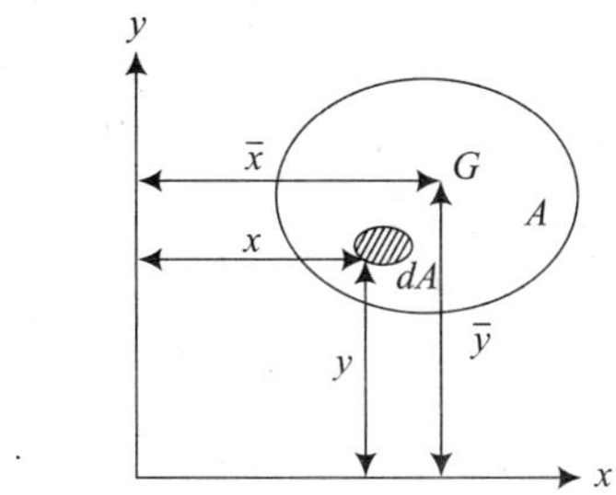

Fig. 5.15 Moment of inertia

Moment of inertia about x-axis,

$$I_{xx} = A(\bar{y})^2 = \int_A dA.y^2$$

Similarly, moment of inertia about y-axis,

$$I_{yy} = A(\bar{x})^2 = \int_A dA.x^2$$

The moment of inertia for standard simple sections is given in Table 5.2.

Table 5.2 Moment of Inertia of simple sections

S.No.	*Shape*	*Figure*	I_{xx}	I_{yy}
1.	Rectangle	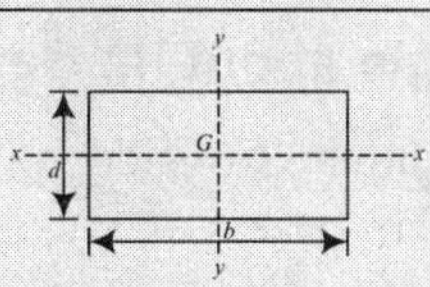	$\frac{bd^3}{12}$	$\frac{db^3}{12}$
2.	Triangle	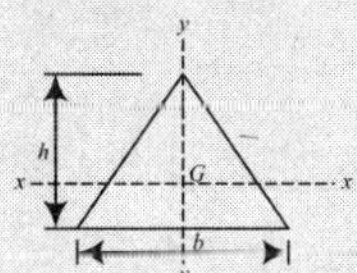	$\frac{bh^3}{36}$	–
3.	Circle	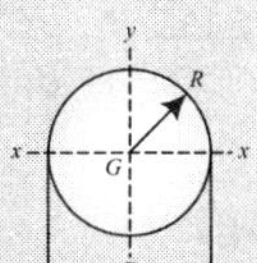	$\frac{\pi R^4}{4}\left(\frac{\pi D^4}{64}\right)$	$\frac{\pi R^4}{64}\left(\frac{\pi D^4}{64}\right)$
4.	Semicircle	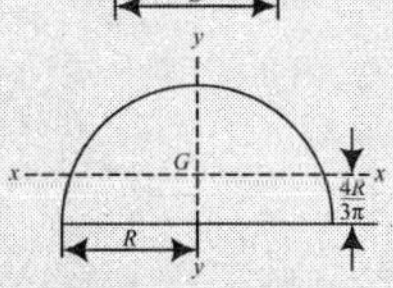	$0.11\ R^4$	$\frac{\pi}{8} R^4$
5.	Quarter circle	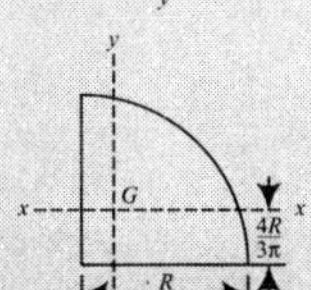	$0.055\ R^4$	$0.055\ R^4$

5.8 MOMENT OF INERTIA OF SIMPLE PLANE AREAS

5.8.1 Moment of Inertia of Rectangle about Centroidal Axis

A rectangle of base width b and height d is shown. An elemental strip of width dy is taken at a distance y from x-axis passing through the centroid.

$$I_{xx} = \int_{-d/2}^{d/2} y^2\, dA = \int_{-d/2}^{d/2} y^2 \cdot b\, dy$$

$$\therefore \quad I_{xx} = b\left|\frac{y^3}{3}\right|_{-d/2}^{d/2}$$

$$= b\left[\frac{d^3}{24} + \frac{d^3}{24}\right]$$

$$\therefore \quad I_{xx} = \frac{bd^3}{12}$$

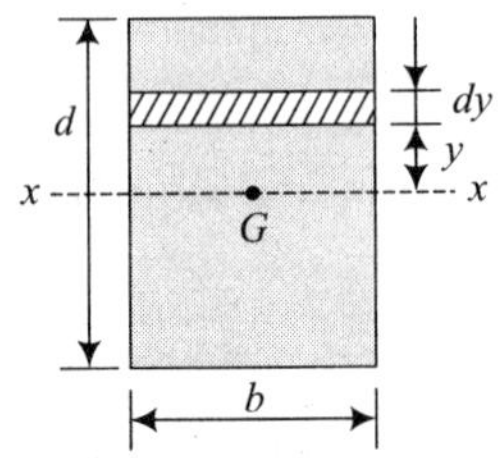

Fig. 5.16 MOI of rectangle

5.8.2 Moment of Inertia of a Triangle about its Base

A triangle of base width b and height h is shown. Consider an elemental strip of width b_1 and thickness dy at a distance y from base, i.e., x-axis

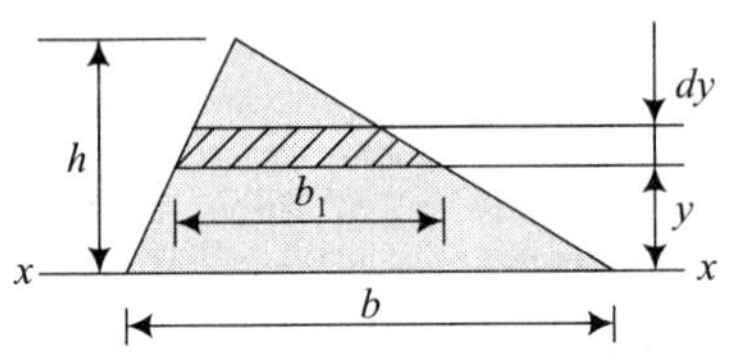

Fig. 5.17 MOI of triangle

$$\frac{b_1}{b} = \frac{(h-y)}{h}$$

$$\therefore \quad b_1 = \frac{(h-y)}{h}\, b$$

$$I_{xx} = \int_0^h y^2 dA$$

$$= \int_0^h y^2 \frac{(h-y)}{h} b\, dy$$

$$= b\left|\frac{y^3}{3} - \frac{y^4}{4h}\right|_0^h = b\left[\frac{h^3}{3} - \frac{h^4}{4h}\right]$$

$$= \frac{bh^3}{12}$$

5.8.3 Moment of Inertia of a Circle about Diametral Axis

A circle of radius R and diameter D is shown. Consider an elemental strip of radius r and thickness dr.

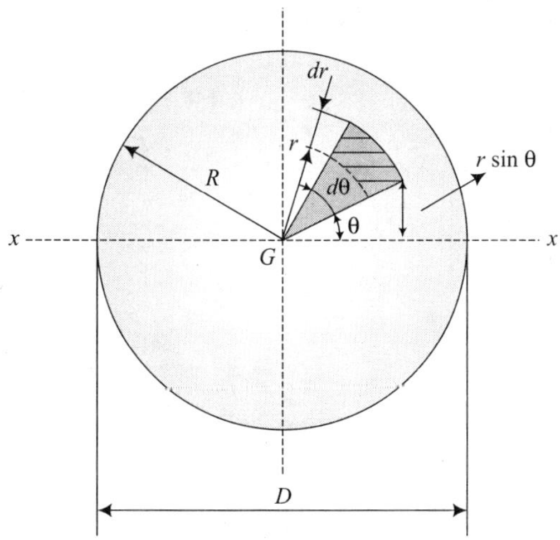

Fig. 5.18 MOI of a circle

$$dA = r d\theta\, dr$$

$$y = (r \sin \theta)$$

$$I_{xx} = \int_0^R \int_0^{2\pi} y^2\, dA$$

$$= \int_0^R \int_0^{2\pi} (r \sin \theta)^2\, r\, d\theta\, dr$$

$$I_{xx} = \int_0^R \int_0^{2\pi} r^3 \left(\frac{1-\cos 2\theta}{2}\right) d\theta\, dr$$

$$= \int_0^R \left|\theta - \frac{\sin 2\theta}{2}\right|_0^{2\pi} dr$$

$$= \left|\frac{r^4}{8}\right|_0^R (2\pi - 0 + 0 - 0)$$

$$= \frac{2\pi}{8} R^4 = \frac{\pi}{4} R^4 = \frac{\pi}{64} D^4.$$

5.9 PARALLEL AXES THEOREM

Moment of inertia about any axis in the plane of area is equal to the sum of moment of inertia about a parallel centroidal axis and the product of area and square of the distance between the two parallel axes.

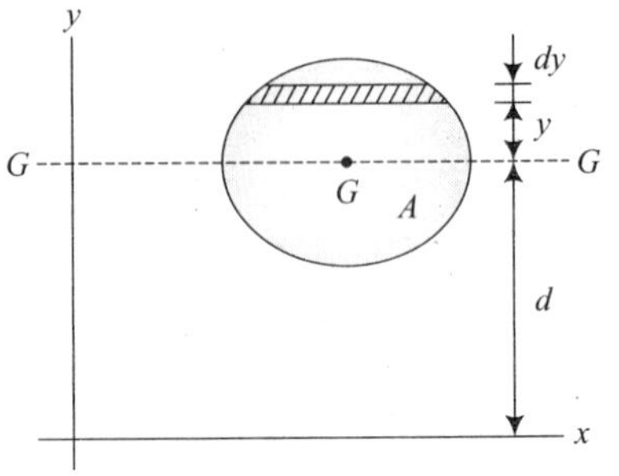

Fig. 5.19 Parallel axes theorem

$$I_{xx} = I_{GG} + Ad^2$$

where I_{xx} = Moment of inertia about axis X-X

I_{GG} = Moment of inertia about centriodal axis GG parallel to axis X-X

A = The area of plane figure

d = Distance between the axis X-X and centroidal axis G-G.

Consider an elemental parallel strip of area dA at a distance y from centroidal axis

$$I_{xx} = \int (y + d)^2\, dA$$

$$I_{xx} = \int (y^2 + 2yd + d^2)dA$$

$$= \int y^2\, dA + 2\int y\, d\, dA + \int d^2\, dA$$

$$\int y^2\, dA = I_{GG} = \text{Moment of inertia about axis } GG$$

$$2\int yd\, dA = 2d\int ydA = 2dA\int \frac{ydA}{A}$$

Here $2d.A = \text{constant}$

$$\frac{\int ydA}{A} = \text{Distance of centroid from axis } GG$$

$$\int d^2\, dA = d^2 \int dA = d^2 A.$$

∴ $I_{xx} = I_{GG} + Ad^2$

Example 5.7: Determine the moment of inertia of a triangle about centroidal axis using parallel axes theorem. The moment of inertia about the base $I_{xx} = \dfrac{bh^3}{12}$.

Solution:

$$I_{xx} = I_{GG} + A\left(\frac{h}{3}\right)^2$$

∴
$$I_{GG} = I_{xx} - A\left(\frac{h}{3}\right)^2$$

$$= \frac{bh^3}{12} - \left(\frac{bh}{2}\right)\left(\frac{h}{3}\right)^2$$

$$= \frac{bh^3}{12} - \frac{bh^3}{18}$$

$$= \frac{bh^3}{36} \quad \textbf{Ans.}$$

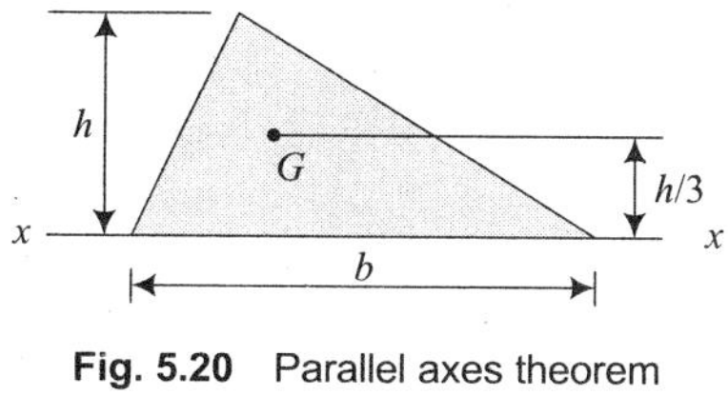

Fig. 5.20 Parallel axes theorem

Example 5.8: Find the moment of inertia of a quarter circle about the base using parallel axes theorem. The moment of inertia of the quarter circle about centroidal axis, $I_{GG} = 0.055\ R^4$.

Solution:

$$I_{xx} = I_{GG} + Ad^2$$

$$A = \frac{\pi}{4} R^2$$

$$d = \frac{4R}{3\pi}$$

$$\therefore \quad I_{xx} = 0.055\ R^4 + \frac{\pi}{4} R^2 \times \left(\frac{4R}{3\pi}\right)^2$$

$$= 0.055\ R^4 + \frac{4}{9\pi} R^4$$

$$= (0.055 + 0.141)R^4 = \mathbf{0.196\ R^4} \qquad \textbf{Ans.}$$

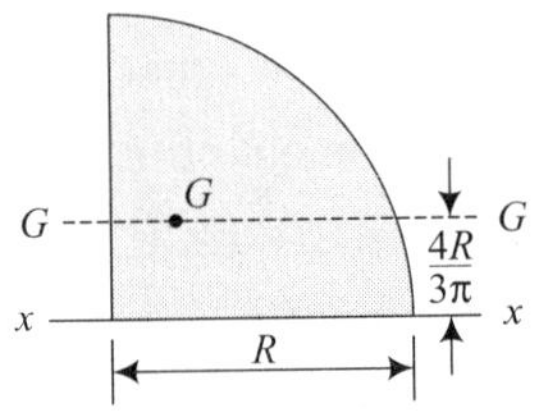

Fig. 5.21 (Example 5.8)

5.10 PERPENDICULAR AXES THEOREM

The moment of inertia of an area about an axis perpendicular to its plane at any point 0 is equal to the sum of moments of inertia about any two mutually perpendicular axis through the same point in the plane of the area. This moment of inertia is called *polar moment of inertia* and is designated as J or I_{zz}.

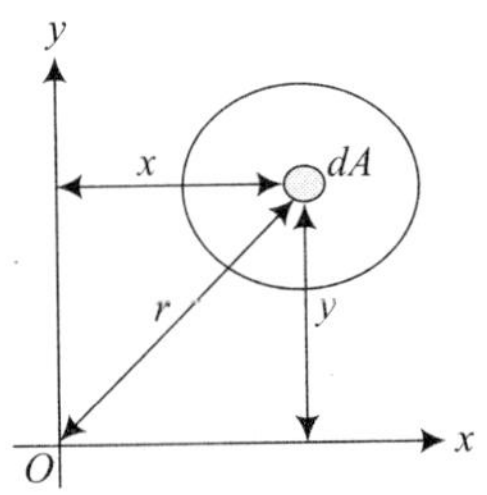

Fig. 5.22 Perpendicular axes theorem

$$I_{zz} = I_{xx} + I_{yy}$$

Let the coordinates of an elemental area dA be x and y. The distance of dA from 0 is r.

$$I_{zz} = \int r^2 \, dA$$

$$= \int (x^2 + y^2) dA$$

$$= \int x^2 \, dA + \int y^2 \, dA$$

$$I_{zz} = I_{xx} + I_{yy}$$

The polar moment of inertia represents the rigidity of the section of a shaft transmitting power or motion.

5.11 PRINCIPAL MOMENT OF INERTIA

If an area dA is at a distance of x and y respectively, from y-axis and x-axis, the *product moment of inertia* is defined as:

$$I_{xy} = \int_A xy \, dA$$

If the two axes about which the product moment of inertia becomes zero, the two axes are called the *principal axes*. The moment of inertia about a principal axis is called the principal moment of inertia. The direction of principal axes can be located as:

$$\tan \theta = \frac{2I_{xy}}{I_{yy} - I_{xx}} \qquad \text{...(1)}$$

where, θ = Angle between principal axis and x-axis or y-axis

I_{xy} = Product moment of inertia

I_{xx} = Moment of inertia about x-axis

I_{yy} = Moment of inertia about y-axis

The above equation gives two principal axes by giving the values of θ differing by 90°. There will be two values of principal moment of inertia, i.e., minimum principal moment of inertia and maximum principal moment of inertia.

The maximum principal moment of inertia,

$$I_{uu} = \frac{(I_{yy} + I_{xx})}{2} + \sqrt{\left(\frac{I_{yy} - I_{xx}}{2}\right)^2 + (I_{xy})^2} \qquad \text{...(2)}$$

The minimum principal moment of inertia

$$I_{vv} = \frac{(I_{yy} + I_{xx})}{2} - \sqrt{\left(\frac{I_{yy} - I_{xx}}{2}\right)^2 + (I_{xy})^2} \qquad \text{...(3)}$$

$$I_{uu} + I_{vv} = I_{xx} + I_{yy}.$$

Sum of moments of inertia about any two perpendicular axes remains constant.

The principal moments of inertia and direction of principal axes can be found out from equations (2), (3) and (1). These can also be found out from Mohr's circle for principal moments of inertia.

5.12 MOMENT OF INERTIA OF COMPOSITE BODIES

The following procedure can be followed to determine the moment of inertia of composite body.

1. Draw the composite area.
2. Divide the area into simple plane areas.
3. The moment of inertia of simple area about centroid axes are known from standard tables.
4. Find out the centroid of composite area.
5. Find distances between centroid of composite area and centroids of individual simple areas.
6. Apply parallel axes theorem to find moment of inertia of simple areas about centroidal axis of composite area.
7. The sum of moments of inertia of simple areas will be the moment of inertia of composite body.

Example 5.9: Determine the moment of inertia of I-section shown about centroidal axes. Also find the polar moment of inertia.

Solution: Divide the composite area into three rectangles and tabulate the data.

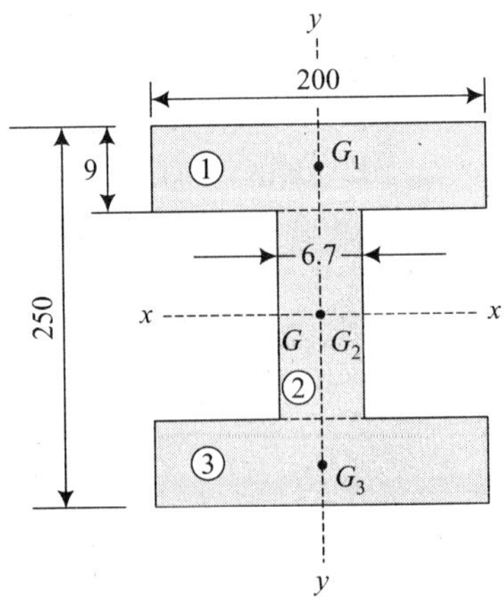

Fig. 5.23 (Example 5.9)

Area	A (mm^2)	I_{GG}= (mm^4)	d = ($G - G_i$)	d^2	Ad^2	I_{xx} (mm^4)	I_{yy} (mm^4)
Rectangle 1	200 × 9 = 1800	$\frac{200\times9^3}{12}$ = 12150	125 – 4.5 = 120.5	14520.25	26136450	26148600	$\frac{9\times200^3}{12}$ = 6000000
Rectangle 2	(250 – 2×9) × 6.7 = 1554.4	$\frac{6.7\times(250-18)^3}{12}$ = 7180121.6	0	0	0	7183121.6	$\frac{232\times6.7^3}{12}$ = 5814.75
Rectangle 3	200 × 9 = 1800	$\frac{200\times9^3}{12}$ = 12150	120.5	14520.25	26136450	26148600	6000000
Σ	5154.4					59480322	12005815

$$= 59480322 + 12005815$$

$$= \mathbf{71486137\ mm^4} \quad \textbf{Ans.}$$

Example 5.9(a): Determine the moment of inertia of T-section shown about the horizontal and vertical axes, passing through the c.g. of the section.

[U.P.T.U. 2009-10]

Solution:

Divide the composite area into two rectangles and tabulate the data. The c.g. lies on the y-axis, i.e., axis of symmetry

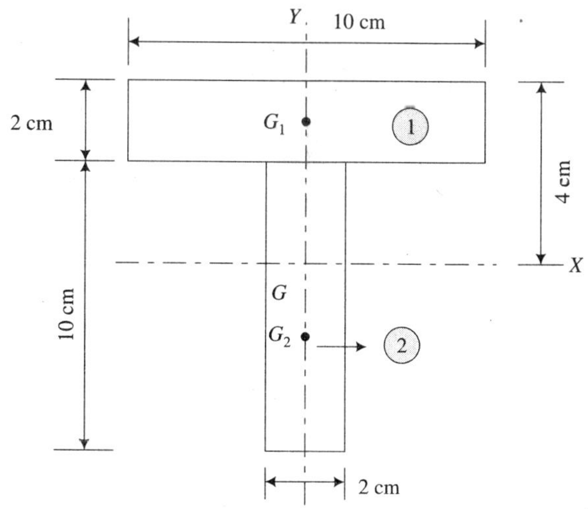

Area	*A (cm²)*	*Y cm*	*$M_x = Ay$ cm³*	*I_{GG} (cm⁴)*	*d = (G – G₁)*	*d²*	*Ad²*	*$I_{XX} = I_{GG} + Ad^2$*	*I_{yy} (cm⁴)*
Rectangle 1	$10 \times 2 = 20$	1	20	$\frac{10 \times 10^3}{12} = 6.667$	$4 - 2 = 2$	4	80	86.667	$\frac{2 \times 10^3}{12} = 166.667$
Rectangle 2	$2 \times 10 = 20$	7	140	$\frac{2 \times 10^3}{12} = 166.667$	$7 - 4 = 3$	9	180	340.667	$\frac{10 \times 2^3}{12} = 6.667$
Σ	40		160					427.33	173.334

$$\bar{x} = \frac{\Sigma M_x}{\Sigma A} = \frac{160}{40} = 4 \text{ cm.}$$

$$\therefore \quad \left.\begin{aligned} I_{xx} &= 427.33 \text{ cm}^4 \\ I_{yy} &= 173.33 \text{ cm}^4 \end{aligned}\right\} \quad \textbf{Ans.}$$

Example 5.10: Find the moment of inertia of the shaded area about *x*-axis.

Solution: Divide the composite area into three parts and tabulate the data.

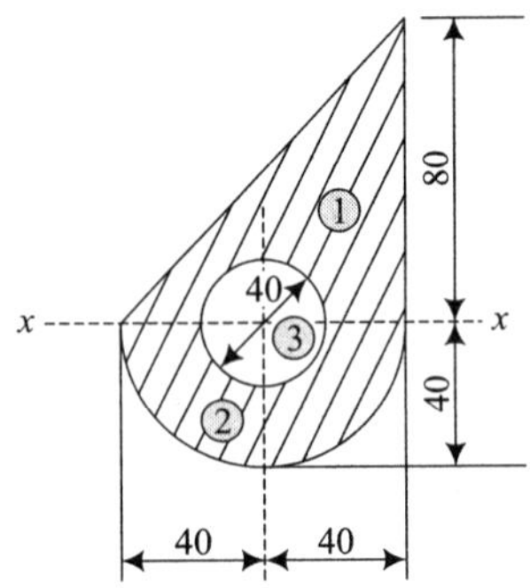

Fig. 5.24 (Example 5.10)

Area	*I_{xx}(mm⁴)*	*Formula*
Triangle (1)	$\frac{80 \times 80^3}{12} = 3413333.3$	MOI about base $\frac{bd^3}{12}$
Semicircle (2)	$\frac{\pi}{128} \times 80^4 = 1005309.6$	MOI about base $\frac{\pi}{128} D^4$
Circle (3)	$-\frac{\pi}{64} \times 40^4 = -125663.7$	MOI about centroidal axes $\frac{\pi}{64} D^4$
Σ	4292979.2	

$$\therefore \quad I_{xx} = \mathbf{4292979.2 \ mm^4} \quad \textbf{Ans.}$$

Example 5.11: Find the moment of inertia of the area shown shaded about the edge *AB*. [U.P.T.U. 2008-09]

Solution:

The area may be divided into a rectangle and a semicircle.

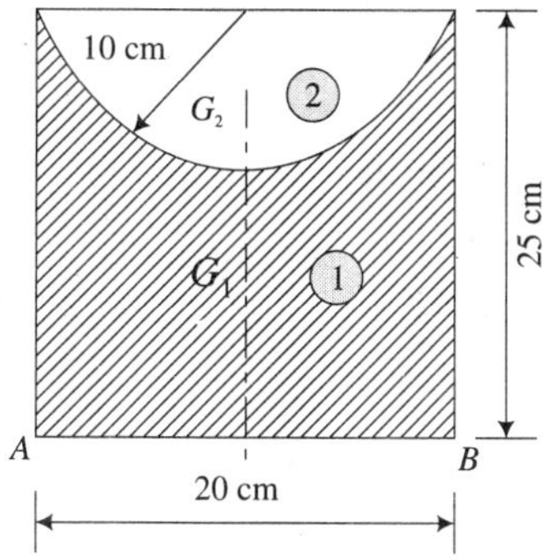

Area	*A* (cm^2)	G_i (cm)	*d* ($AB - G_i$)	d^2	Ad^2	I_{GG}	$I_{AB} = I_{GG} + Ad^2$
Rectangle 1	25×20 = 500	12.5	12.5	156.25	78125	$\frac{20 \times 25^3}{12}$ = 26047.667	104172.67
Semicircle 2	$\frac{\pi}{2}(10)^2$ = 157	$\frac{4 \times 10}{3\pi}$ = 4.244	20.756	430.8	67635.6	$0.11(10)^4$ = 1100	(–) 68735.6
Σ							35437(cm^4)

Example 5.12: For the Z-section shown, the moment of inertia with respect to *x* and *y* axes are given as $I_x = 1548$ cm^4 and $I_y = 2668$ cm^4. Determine the principal axes of the section about *O* (centroid of vertical section and point *O* coincides) and values of principal moments of inertia. [U.P.T.U. 2009-10]

Solution:

The product moment of inertia,

$$I_{xy} = \int Axy$$

Divide the area of *Z*-section into three rectangles and tabulate the data.

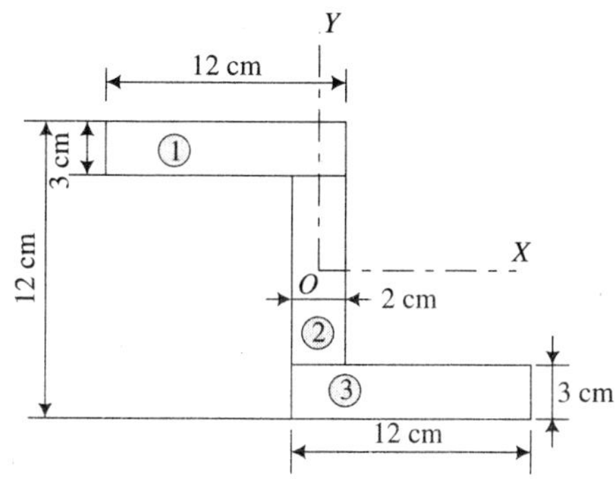

Area	$A(cm^2)$	X(cm)	Y(cm)	$Axy(cm^4)$
Rectangle 1	12 × 3 = 36	–5.0	+ 4.5	–810
Rectangle 2	2 × 6 = 12	0.0	0.0	0
Rectangle 3	12 × 3 = 36	+ 5.0	– 4.5	–810
Σ				–1620

The principal moment of inertia

$$I_{uu} = \frac{I_x + I_y}{2} + \sqrt{\left(\frac{I_x - I_y}{2}\right)^2 + I_{xy}^2}$$

$$= \frac{1548 + 2668}{2} + \sqrt{\left(\frac{1548 - 2668}{2}\right)^2 + (-1620)^2}$$

$$= 2108 + 1714 = 3822 \text{ cm}^4$$

$$I_{vv} = \frac{I_x + I_y}{2} - \sqrt{\left(\frac{I_x - I_y}{2}\right)^2 + I_{xy}^2}$$

$$= 2108 - 1714 = 394 \text{ cm}^4$$

$$\tan 2\theta_1 = \frac{2\,I_{xy}}{I_y - I_x} = \frac{-2 \times 1620}{2668 - 1548} = -\,2.892857$$

$$2\theta_1 = -70.93$$

$$\therefore \quad \theta_1 = -\,35.46° \text{ and } \theta_2 = (90 - \theta_1) = 54.54°$$

Example 5.13: Find the principal moment of inertia about the origin of the area shown. All dimensions are in mm. [U.P.T.U. 2010-11]

Solution:

Divide the section into two rectangles.

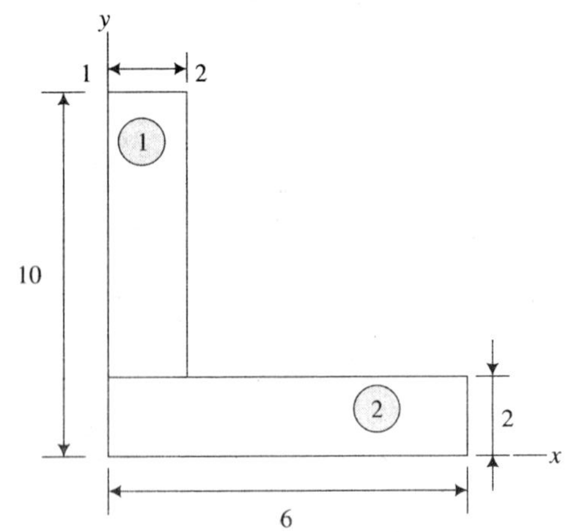

Area	*A* *(mm^2)*	I_{GG_x} *(mm^4)*	I_{GG_y}	$I_{xx} = I_{GG_x} + Ad^2$	$I_{yy} = I_{GG_y} + Ad^2$	$I_{xy} = A_{xy}$ *(mm^4)*
Rectangle 1	8×2 = 16	$\frac{2 \times 8^3}{12}$ = 85.33	$\frac{8 \times 8^3}{12}$ = 5.33	$85.33 + 16 \times 6^2$ = 661.33	$5.33 + 16 \times 1^2$ = 21.33	$16 \times 6 \times 1$ = 96
Rectangle 2	6×2 = 12	$\frac{6 \times 2^3}{12}$ = 4	$\frac{2 \times 6^3}{12}$ = 36	$4 + 12 \times 1^2$ = 16	$36 + 4 \times 3^2$ = 72	$12 \times 1 \times 3$ = 36
Σ	28	89.33	41.33	677.33	93.33	13.2

$$I_x = 677.33 \text{ mm}^4$$

$$I_y = 93.33 \text{ mm}^4$$

$$I_{xy} = 132 \text{ mm}^4$$

$$I_{uu} = \frac{I_x + I_y}{2} + \sqrt{\left(\frac{I_x - I_y}{2}\right)^2 + (I_{xy})^2}$$

$$= \frac{677.33 + 93.33}{2} + \sqrt{\left(\frac{677.33 - 93.33}{2}\right)^2 + (132)^2}$$

$$= 385.33 + 320.45$$

$$= 705.78 \text{ mm}^4$$

$$I_{vv} = 385.53 - 320.45$$

$$- 64.88 \text{ mm}^4$$

$$\tan 2\theta_1 = \frac{2I_{xy}}{I_y - I_x} = \frac{2 \times 132}{93.33 - 677.33}$$

$$= -0.4520$$

$$2\theta_1 = -24.32°$$

$$\therefore \quad \theta_1 = -12°$$

$$\theta_2 = (90 - \theta_1)$$

$$= 78°$$

QUESTION BANK NO. 5

1. Define centroid, centroidal axes, first moment of inertia centroidal coordinates.
2. Explain the following:
   (i) Moment of Inertia
   (ii) Polar Moment of Inertia
   (iii) Product Moment of Inertia
   (iv) Principal Moment of Inertia
3. State and prove the following theorems.
   (i) Parallel axes theorem
   (ii) Perpendicular axes theorems

QUICK REVIEW INFORMATION NO. 5

1. First moment of an area about an axis is zero if the axis is symmetrical and passes through centroid (c.g.).
2. If a concentric circular plate of radius R/2 is removed from a solid circular plate of radius R, the c.g. will remain unchanged.
3. A wire of length L is bent, the e.g. will change as per shape.
4. The centre of gravity for various shapes along vertical axis from the base is as follows.

Shape	*Centroid (c.g.)*
Triangle	$h/3$
Semicircle	$4R/3\pi$
Quarter circle	$4R/3\pi$
Hemisphere	$3R/5$
Cone	$3h/4$
Circular curve	$3R/\pi$

5. The moment of inertia about centroidal axis.

Shape (Area)	*M.O.I.*
Rectangle	$bh^3/12$
Triangle	$bh^3/36$
Circle	$\pi D^4/64$
Semicircle	$\pi R^4/8$ $(0.11R^4)$
Quarter circle	$\pi R^4/16$ $(0.055R^4)$

CHAPTER 6

Mass Moment of Inertia

6.1 INTRODUCTION

Mass moment of inertia, centre of gravity, radius of gyration and axis of symmetry of solid bodies are important properties required in the study of motion of rigid bodies.

1. *Mass moment of inertia* of a body about an axis is the sum total of product of its masses and square of their distances from the axis.

$$I_{xx} = \int r^2 \, dm$$

where, dm = element of mass

r = distance of element of mass from the axis

The unit of mass moment of inertia is N-m sec^2.

2. *Centre of gravity* of a solid body is the point through which the resultant weight of the body passes in whichever position the body is kept. For a body made of uniform material, centre of gravity is same as centroid of the volume.

3. *Radius of gyration* is the distance at which the entire mass of a solid body is assumed to be concentrated. The mass moment of inertia,

$$I = MK^2$$

$$K = \sqrt{\frac{I}{M}}$$

where, M = mass of the body

K = radius of gyration (m).

4. *Axis of symmetry* Centroid always lies on the plane of symmetry. If there are two planes of symmetry in a solid body, the centre of gravity lies on the line of intersection of the two symmetrical planes. If there are three planes of symmetry, the centre of gravity lies on the point of intersection of planes of symmetry.

The axis of rotation of solid body normally coincides with the axis of symmetry.

5. *Determination of mass moment of inertia* Following procedure may be followed for determining the mass moment of inertia of simple bodies from the first principle. The simple bodies may be a disc, ring, cylinder, sphere and cone.
    1. Draw the figure of the solid body.
    2. Write down the expression for mass of element, dm, and its distance, r, from the axis.
    3. Integrate $\int r^2\, dm$ between suitable limits to cover the total mass of the solid body.

6.2 MASS MOMENT OF INERTIA OF A SOLID DISC

A uniform circular disc (lamina) of radius R and thickness t is shown in Fig. 6.1. It is required to find mass moment of inertia about centroidal axis.

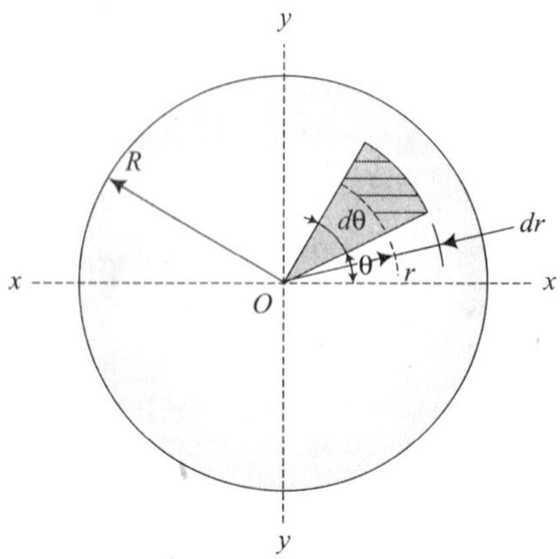

Fig. 6.1 Circular disc

Consider an elemental area $r d\theta\, dr$ and thickness t.

Mass of element

$$dm = \rho r\, d\theta\, dr\, t$$
$$= \rho t\, r\, d\theta\, dr$$

Distance from $0 = r$

$$I_{zz} = \int_M r^2\, dm$$

$$I_{zz} = \int_0^R \int_0^{2\pi} r^2 \rho t \, r dr \, d\theta$$

$$= \rho t \int_0^R r^3 |\theta|_0^{2\pi} dr$$

$$= \rho t \int_0^R 2\pi r^3 \, dr$$

$$= \rho t \, 2\pi \left| \frac{r^4}{4} \right|_0^R$$

$$= \rho t \, 2\pi \frac{R^4}{4} = \rho t \frac{\pi R^4}{2}$$

Total mass of disc,

$$M = \rho t \pi R^2$$

$\therefore$

$$I_{zz} = \frac{MR^2}{2}$$

6.3 MASS MOMENT OF INERTIA OF A CIRCULAR RING

A uniform circular ring of mean radius R is shown in Fig. 6.2. The mass of ring per unit length is m.

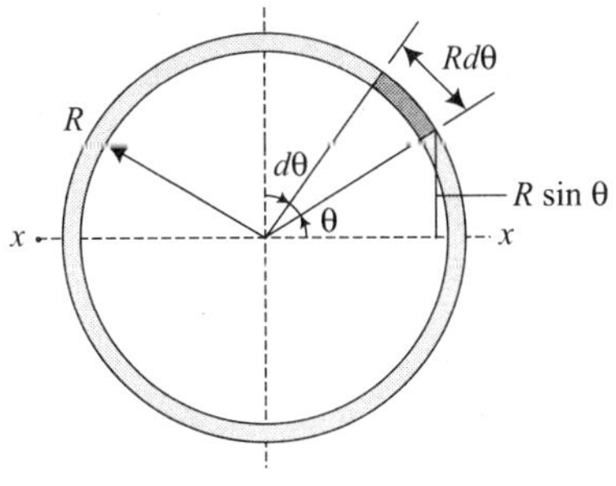

Fig. 6.2 Circular Ring

Total mass of the ring,

$$M = 2\pi Rm.$$

Consider an elemental length $dS = R\,d\theta$ at an angle θ from x-axis. The distance of the element from x-axis is $R \sin \theta$.

Mass of element $= mRd\theta$

The mass moment of inertia, about x-axis

$$I_{xx} = \int_0^{2\pi} (R \sin \theta)^2 \, mRd\theta$$

$$= mR^3 \int_0^{2\pi} \sin^2\theta \, d\theta$$

$$= mR^3 \int_0^{2\pi} \left(\frac{1 - \cos 2\theta}{2}\right) d\theta$$

$$= \frac{mR^3}{2} \left| \theta - \frac{\sin 2\theta}{2} \right|_0^{\pi} = mR^3\pi$$

$$M = 2\pi Rm$$

$$\therefore \quad I_{xx} = \frac{MR^3}{2}$$

The mass moment of inertia about z-axis

$$I_{zz} = \int_0^{2\pi} R^2 \, mRd\theta$$

$$= mR^3 \int_0^{2\pi} d\theta = mR^3 \, |\theta|_0^{2\pi}$$

$$= 2\pi mR^3 = MR^2$$

6.4 MASS MOMENT OF INERTIA OF A CYLINDER

A uniform solid cylinder of radius R and height l is shown in Fig. 6.3.

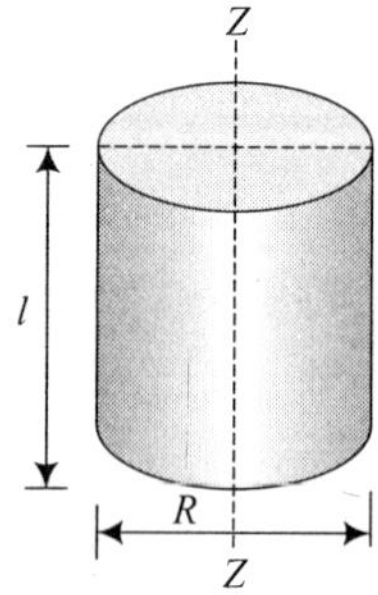

Fig. 6.3 Solid cylinder

Mass of cylinder $\quad M = \pi R^2 l\rho$

Solid cylinder can be treated as a solid disc of radius R and thickness l.

$$\therefore \qquad I_{zz} = \rho l \frac{\pi R^4}{2}$$

$$= \frac{MR^2}{2}$$

6.5 MASS MOMENT OF INERTIA OF A SOLID SPHERE

A solid sphere of radius R is shown in Fig. 6.4.

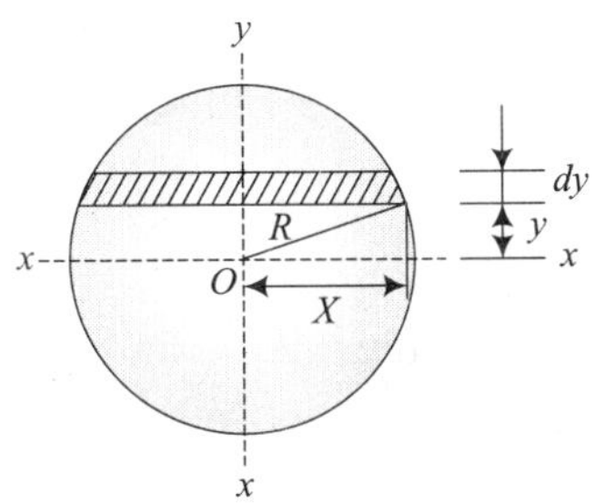

Fig. 6.4 Solid sphere

Consider an elemental plate of thickness dy at a distance y from diametral axis x-x.

Radius of elemental circular plate

$$x^2 = R^2 - y^2$$

Mass of elemental plate

$$dm = \rho\pi x^2\, dy$$

$$= \rho\pi(R^2 - y^2)\, dy$$

Momcnt of incrtia of this circular platc about y-axis

$$I_{yy} = \frac{MR^2}{2}$$

$$= \frac{1}{2}\rho\pi x^2\, dy \times x^2$$

$$= \rho\frac{\pi}{2}x^4\, dy$$

$$= \rho\frac{\pi}{2}(R^2 - y^2)^2 dy$$

$$= \frac{\pi}{2}\rho(R^4 + y^4 - 2R^2y^2)dy$$

Moment of inertia of circular plate about y-axis

$$I_{yy} = 2\int_0^R \rho \frac{\pi}{2}(R^4 - 2R^2y^2 + y^4)dy$$

$$= \pi\rho \left| R^4 y - \frac{2R^2y^3}{3} + \frac{y^5}{5} \right|_0^R$$

$$= \pi\rho R^5 \left| 1 - \frac{2}{3} + \frac{1}{5} \right|$$

$$= \frac{8}{15}\rho\pi R^5$$

Mass of sphere

$$M = 2\int_0^R dm = 2\int_0^R \rho\pi x^2\, dy$$

$$= 2\int_0^R \rho\pi(R^2 - y^2)dy$$

$$= 2\rho\pi \left| R^2 y - \frac{y^3}{3} \right|_0^R$$

$$= 2\pi\rho \left[R^3 - \frac{R^3}{3} \right]$$

$$= \frac{4\pi R^3}{3}$$

$$\therefore \qquad I_{yy} = \frac{2}{5} MR^2$$

6.6 MASS MOMENT OF INERTIA OF A SOLID CONE

A solid cone of base radius R and height h is shown in Fig. 6.5. Consider an elemental plate of thickness dx at a distance x from vertex O. Let radius of plate be r.

Mass of elementary plate,

$$dm = \rho\pi r^2 dx$$

The mass moment of inertia of a plate about polar axis

$$I_{zz} = \frac{MR^2}{2}$$

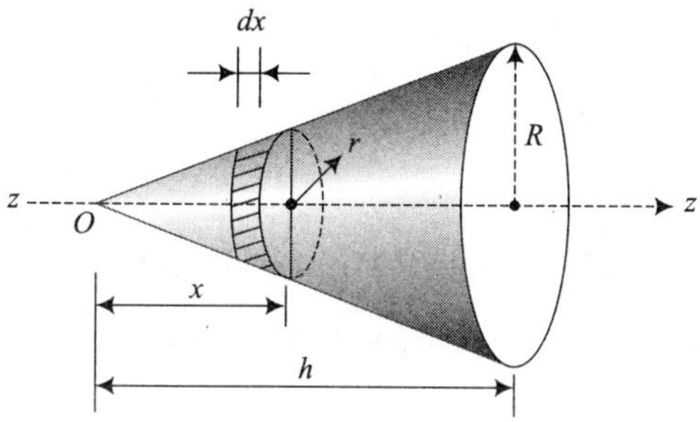

Fig. 6.5 Solid cone

$$I_{zz} = \frac{1}{2}\rho\pi r^2 dx . r^2$$

$$= \frac{\pi\rho}{2} r^4 \, dx$$

But $$\frac{r}{R} = \frac{x}{h}$$

∴ $$r = \left(\frac{x}{h}\right) R.$$

$$I_{zz} = \int_0^R \frac{\rho\pi}{2} R^4 \frac{x^4}{h^4} dx$$

$$= \frac{\rho\pi}{2} \frac{R^4}{h^4} \int_0^h x^4 dx$$

$$= \frac{\pi\rho}{2} \frac{R^4}{h^4} \left| \frac{x^5}{5} \right|_0^h$$

$$= \frac{\rho\pi R^4 h}{10} \qquad \ldots(1)$$

Total mass of the cone,

$$M = \int_0^h \rho\pi r^2 dx$$

$$= \int_0^h \pi R^2 \frac{x^2}{h^2} dx = \frac{\pi R^2}{h^2} \left| \frac{x^3}{3} \right|_0^h$$

$$= \frac{\pi R^2 h}{3} \qquad \ldots(2)$$

From equations (1) and (2)

$$\therefore \qquad I_{zz} = \frac{3}{10}MR^2$$

6.7 MASS MOMENT OF INERTIA OF COMPOSITE BODIES

The following procedure is followed for determination of mass moment of inertia for composite bodies.

1. Divide the composite body into simple bodies.
2. The centre of gravity and mass moment of inertia of simple bodies is known from standard table about centroidal axis.
3. Use parallel axis theorem to find the mass moment of inertia of simple bodies about the required axis.

$$I_A = I_G + Md^2$$

where, I_A = Mass moment of inertia about the required axis.
I_G = Mass moment of inertia about the centroidal axis.
M = Mass of the body
d = Distance of required axis from the centroidal axis

4. Sum up the mass moment of inertia of each simple body about the required axis to get mass moment of inertia of composite body.

Example 6.1: Determine the mass moment of inertia of a uniform rod of length L about its:

(a) centroidal axis normal to rod, and
(b) axis at the end of the rod and normal to it.

Solution:

(a) *Mass moment of inertia about the centroidal axis*

The mass of small element

$$dm = m\,dx$$

where m is the mass of the rod per unit length.

Mass moment of inertia

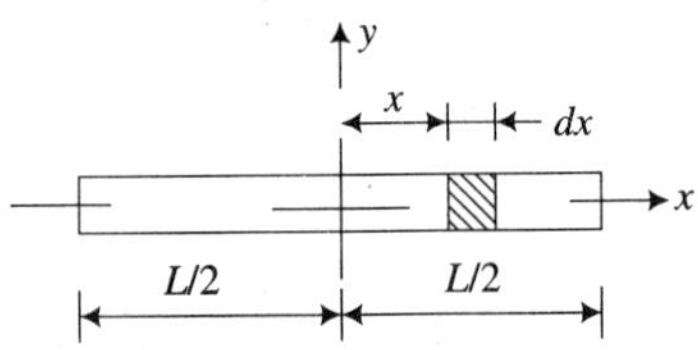

$$I_{yy} = \int_{-L/2}^{L/2} x^2\,dm = \int_{-L/2}^{L/2} x^2\,m\,dx$$

$$= m\left|\frac{x^3}{3}\right|_{-L/2}^{L/2} = \frac{mL^3}{12}$$

or
$$I_{yy} = \frac{ML^2}{12}$$

where $M = mL$ = Total mass of rod

(b) *Mass moment of inertia about axis at the end of the rod*

Mass moment of inertia,

$$I_{yy} = \int_0^L x^2 \, m \, dx$$

$$= m \left| \frac{x^3}{3} \right|_0^L = \frac{mL^3}{3}$$

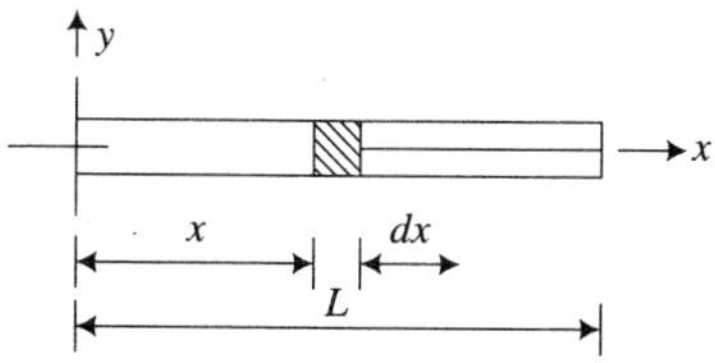

or
$$I_{yy} = \frac{ML^2}{3}$$

Example 6.2: Determine the mass moment of inertia of a rectangular plate of size $a \times b$ and thickness t about its centroidal axis. [U.P.T.U. 2010-11]

Solution:

(i) *Mass moment of inertia about X-axis*

The mass of elemental strip,

$$dm = \rho \, bt \, dy$$

where ρ = mass density of material

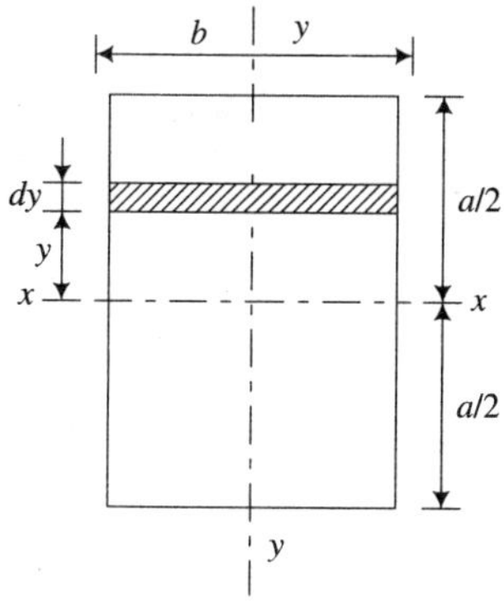

$$\therefore \quad I_{xx} = \int_{-a/2}^{a/2} y^2 \, dm$$

$$= \int_{-a/2}^{a/2} y^2 \rho bt \, dy$$

$$\therefore \quad I_{xx} = \rho bt \left| \frac{y^3}{3} \right|_{-a/2}^{a/2}$$

$$= \frac{\rho bt a^3}{12}$$

Total mass of the plate,

$$M = \rho \, bta$$

$$\therefore \quad I_{xx} = \frac{Ma^2}{12}$$

(ii) *Mass moment of inertia about y-axis*

$$I_{yy} = \int_{-b/2}^{b/2} x^2 \, dm$$

$$= \int_{-b/2}^{b/2} x^2 \rho at \, dx$$

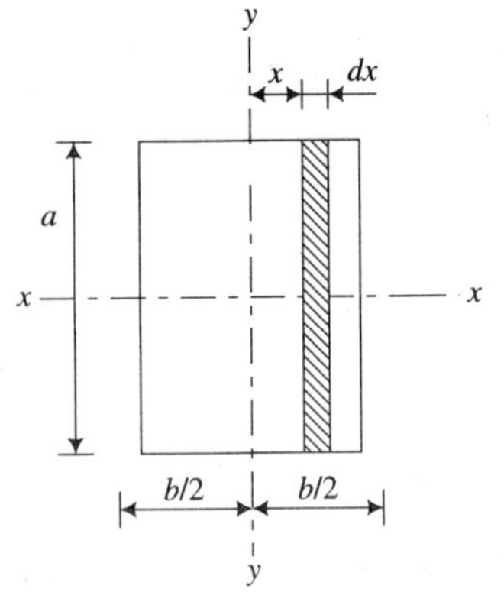

$$= \rho at \left| \frac{x^3}{3} \right|_{-b/2}^{b/2}$$

$$= \frac{\rho atb^3}{12}$$

$$= \frac{Mb^2}{12}$$

where M = Total mass of the plate

(iii) *Mass moment of inertia about z-axis*

$$r^2 = x^2 + y^2$$

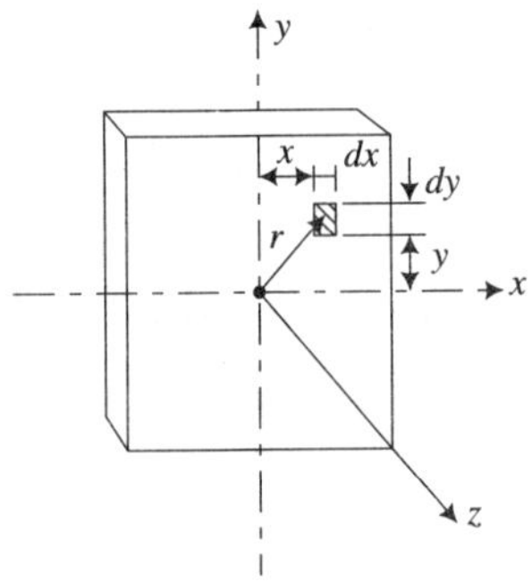

$$I_{zz} = \int r^2 \, dm$$

$$= \int (x^2 + y^2) \, dm$$

$$= \int x^2 \, dm + \int y^2 \, dm$$

$$= I_{xx} + I_{yy}$$

$$= \frac{Ma^2}{12} + \frac{Mb^2}{12}$$

$$= \frac{M}{12} (a^2 + b^2)$$

Example 6.3: Determine the mass moment of inertia of a triangular plate of base width b and height h about the base. The plate is of uniform thickness t and mass density ρ .

Solution:

Mass of plate

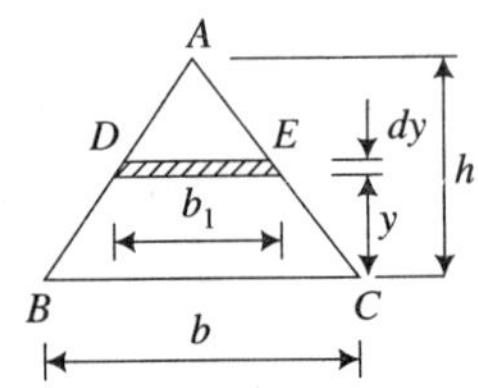

$$M = \rho \frac{bht}{2}$$

The mass of elemental strip

$$dm = \rho b_1 t \, dy$$

Mass moment of inertia about base BC

$$I_{base} = \int dm \, y^2$$

From Δs ABC and ADE

$$\frac{b_1}{b} = \frac{h-y}{h}$$

$$\therefore \quad b_1 = b\left(\frac{h-y}{h}\right)$$

$$\therefore \quad I_{base} = \int_0^h \rho b\left(\frac{h-y}{h}\right) t\, y^2\, dy$$

$$= \frac{\rho bt}{h}\int_0^h (h-y)\, y^2\, dy$$

$$= \frac{\rho bt}{h}\left|\frac{hy^3}{3} - \frac{y^4}{4}\right|_0^h$$

$$= \frac{\rho bt}{h}\left(\frac{h^4}{3} - \frac{h^4}{4}\right)$$

$$= \frac{\rho bth^3}{12}$$

$$= \frac{\rho bht}{2} \times \frac{h^2}{6}$$

$$= \frac{1}{6} Mh^2$$

QUESTION BANK NO. 6

1. Define and explain the following:
   (a) Mass moment of inertia
   (b) Centre of gravity
   (c) Axis of symmetry
   (d) Radius of gyration
2. Derive the expression for mass moment of inertia for a solid circular disc from the first principle. How can this expression be used for a solid cylinder?
3. Explain step by step how to find mass moment of inertia for composite bodies.
4. Prove that mass moment of inertia about polar axis is double of that about diametral axis.

QUICK REVIEW INFORMATION NO. 6

The mass moment of inertia about polar axis:

Solid Body	*Mass Moment of inertia*
1. Thin disc (circular lamina)	$MR^2/2$
2. Circular ring	MR^2
3. Cylinder	$MR^2/2$
4. Sphere	$2/5\ MR^2$
5. Cone	$3/10\ MR^2$

Unit IV: Dynamics of Rigid Body

CHAPTER 7

Kinematics of Rigid Body

7.1 INTRODUCTION

Kinematics is the branch of dynamics dealing with the problems of motion of bodies without the consideration of forces causing the motion of the body.

A body may move in any direction in space. Plane motion is the motion of a body in single plane. Plane motion may the classified as follows:

1. *Translation:* If a straight line drawn on the moving body remains parallel to its original position at any time, the motion is called translation. If the path traced by any point on the body is a straight line, the motion is called *rectilinear translation*. If the path traced by the point is a curve, the motion is called *curvilinear translation*. Rectilinear translation is also known as linear motion and curvilinear translation as curvilinear motion.
2. *Rotation:* If all particles of a rigid body move in a concentric circle, the motion is called rotation.
3. *General Plane Motion:* Combination of both translation and rotation is called general plane motion.

Example 1: Wheel rolling on straight line.

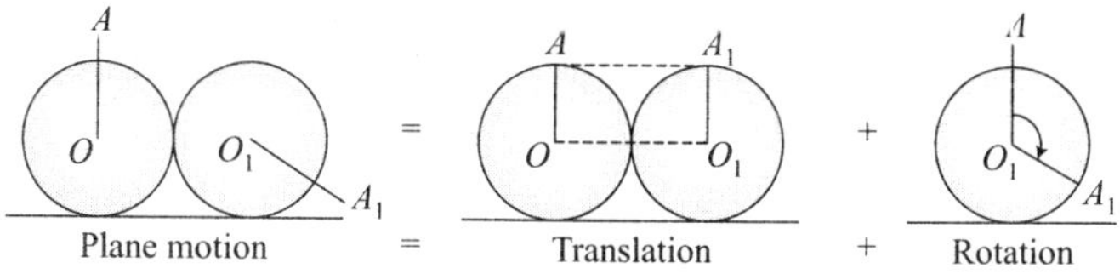

Fig. 7.1 Plane motion

Example 2: Ladder sliding against the wall.

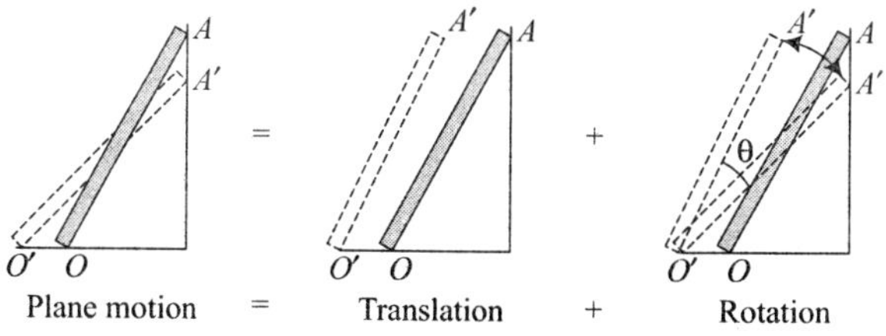

Fig. 7.2 Plane motion

7.2 LINEAR MOTION

A body is said to be in linear motion, if the path traced by it is a straight line. Different types of linear motion problems may be

1. Motion with uniform velocity
2. Motion with uniform acceleration
3. Motion under acceleration due to gravity
4. Motion with varying acceleration

Many kinematic problems can be solved in linear motion by using the definition of velocity and acceleration.

A body in motion changes its position relative to the surrounding. *Velocity* is the rate of change of displacement (s) with respect to its surroundings in a particular direction. Velocity is a vector quantity.

$$v = \frac{ds}{dt}$$

Acceleration is the rate of change of velocity. The rate of change of velocity may be uniform and the motion is said to be with uniform acceleration. The body may move with variable acceleration if the rate of change of velocity is not uniform.

$$a = \frac{dv}{dt}$$

If the motion is with uniform velocity

$$\frac{ds}{dt} = v$$

$$\therefore \quad S = \int v\, dt = Vt \qquad (\because V = \text{constant})$$

7.3 MOTION WITH UNIFORM ACCELERATION

The acceleration

$$a = \frac{dv}{dt}$$

$$dv = a\, dt$$

$$\therefore \quad v = at + C_1 \qquad (\because a = \text{constant})$$

C_1 = Constant of integration

When $t = 0$, $v = u$ = initial velocity

$$\therefore \quad u = 0 + C_1$$

$$\therefore \quad v = u + at \qquad \text{...(1)}$$

$$v = \frac{ds}{dt} = u + at$$

$\therefore \quad ds = (u + at)\, at$

$\therefore \quad S = ut + \frac{1}{2} at^2 + C_2$

When $t = 0, \quad S = 0 \quad \therefore C_2 = 0$

$$S = ut + \frac{1}{2} at^2 \qquad ...(2)$$

Now, $\quad a = \frac{dv}{dt} = \frac{dv}{ds} \cdot \frac{ds}{dt} = \frac{dv}{ds} \cdot v \qquad [\because \frac{ds}{dt} = v]$

$\therefore \quad ads = v\, dv$

Integrating,

$$a \int ds = \int_u^v v\, dv$$

$$as = \left| \frac{v^2}{2} \right|_u^v = \frac{v^2}{2} - \frac{u^2}{2}$$

$\therefore \quad v^2 - u^2 = 2as \qquad ...(3)$

Equations (1), (2) and (3) are called equations of motion.

7.4 ACCELERATION DUE TO GRAVITY

The acceleration due to gravity is constant near earth's surface.

$a = -g = -9.81$ m/s^2 for vertically upward motion

$a = g = 9.81$ m/s^2 for vertically downward motion

Example 7.1: A steel ball is thrown vertically upward with an initial velocity of 4 m/s. Find

(i) the time taken to reach the maximum height

(ii) the maximum height reached

(iii) time required for descending

(iv) velocity when it strikes the ground

Solution: (i) When maximum height is reached,

Final velocity, $v = 0$

$g = -9.81$ m/s^2

Initial velocity $= u$

$$v = u - gt$$

$$\therefore \quad t = \frac{u}{g}$$

(ii) Let maximum height reached = h

$$v^2 - u^2 = -2gh \qquad v = 0$$

$$\therefore \quad h = \frac{u^2}{2g}$$

(iii) For downward motion of ball

$$u = 0$$

$$g = 9.81 \text{ m/s}^2$$

Now,
$$v^2 - u^2 = 2gh$$

$$v^2 = 2gh$$

$$v = \sqrt{2gh}, \text{ (velocity when it strikes the ground)}$$

(iv)
$$v = u + gt$$

$$u = 0$$

$$\therefore \quad t = \frac{v}{g} = \frac{u}{g}$$

Time taken for upward motion is same as the time taken for the downward motion.

Example 7.2: A ball is dropped from the top of a tower having a height of 30 m. At the same moment, another ball is thrown upward from the ground with initial velocity of 15 m/s. Find the relative velocity when the two balls cross. Where and when do the balls cross?

Solution:

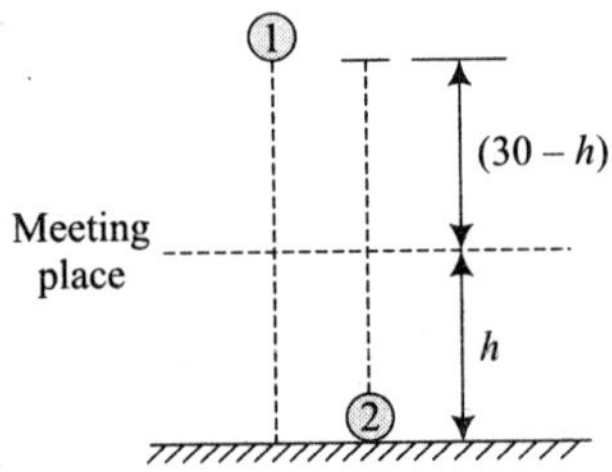

Fig. 7.3 (Example 7.2)

(1) For the first ball,

$$u = 0$$
$$S = (h - 30)$$
$$g = 9.81 \text{ m/s}^2$$
$$S = ut + \frac{1}{2}gt^2$$
$$(30 - h) = 0 + \frac{9.81}{2}t^2 \qquad ...(1)$$

(2) For the second ball,

$$u = 15 \text{ m/s}$$
$$S = h$$
$$g = -9.81 \text{ m/s}^2$$

$$\therefore \qquad h = 15t - \frac{9.81}{2}t^2 \qquad ...(2)$$

From equations (1) and (2),

$$t = 2 \text{ sec. } h = 10.38 \text{ m.}$$

(3) At $t = 2$ sec.

Downward velocity of first ball,

$$v_1 = u + gt = 0 + 9.81 \times 2 = 19.62 \text{ m/s}$$

Upward velocity of second ball,

$$v_2 = u - gt$$
$$= 15 - 9.81 \times 2 = -4.62 \text{ m/s}$$
$$= 4.62 \text{ m/s downward}$$

Relative velocity = 19.6 – 4.62 = **15 m/s** **Ans.**

7.5 MOTION WITH VARYING ACCELERATION

A vehicle normally moves with varying acceleration. The following equations can be used to solve the problem for motion with varying acceleration

$$v = \frac{ds}{dt}$$
$$a = \frac{dv}{dt} = \frac{d^2s}{dt^2} = v\frac{dv}{ds}$$

The variation of displacement or velocity or acceleration with time is given.

Example 7.3: A car was moving with a velocity of 72 kmph. A child was stopped on the road and brakes were applied. The vehicle stopped in a distance of 15 m. Assume that retardation produced is proportional to the distance from the point where brakes were applied, find the equation of retardation.

Solution: Assume, $a = -kS$

where, $k = \text{constant}$

$S = \text{distance travelled}$

$$\therefore \quad a = v \cdot \frac{dv}{ds} = -kS$$

$$v \cdot dv = -kS\, dS$$

Integration gives,

$$\frac{v^2}{2} = -\frac{kS^2}{2} + C_1$$

When brakes are applied, $S = 0$ and $v = 72$ km/s

$$= \frac{72 \times 1000}{36 \times 36} = 20 \text{ m/s}$$

$$\therefore \quad \frac{20^2}{2} = -0 + C_1$$

$$\therefore \quad C_1 = 200$$

$$\frac{v^2}{2} = -\frac{kS^2}{2} + 200$$

When vehicle stops, $v = 0$, $S = 15$ m

$$\therefore \quad 0 = -k\frac{15^2}{2} + 200$$

$$\therefore \quad k = 1.778.$$

The expression for retardation is

$$\boldsymbol{a = -1.7785} \quad \textbf{Ans.}$$

Example 7.4: A body, starting from rest, moves in a straight line, whose equation of motion is given by $S = t^3 - 2t^2 + 3$. Find the velocity and acceleration after 5 seconds.

Solution: $S = t^3 - 2t^2 + 3$

$$v = \frac{dS}{dt} = 3t^2 - 4t$$

When $t = 5$ sec.

$$v = 3(5)^2 - 4(5) = \mathbf{55\ m/s} \quad \textbf{Ans.}$$

$$a = \frac{dv}{dt} = \frac{d^2S}{dt^2} = 6t - 4$$

When $t = 5$ sec.

$$a = 6(5) - 4 = \mathbf{26\ m/s^2} \quad \textbf{Ans.}$$

Example 7.5: A body, starting from rest, moves in straight line, whose acceleration is given by the equation,

$$a = 10 - 0.006\, S^2$$

where a is in m/s^2 and S is in m.

Determine:

(i) the velocity of the body, when it has travelled 50 m.

(ii) the distance travelled by the body, when it comes to rest.

Solution:

$$a = 10 - 0.006 S^2$$

or

$$v\frac{dv}{dS} = 10 - 0.006 S^2$$

$\therefore$

$$vdv = (10 - 0.006 S^2)\, dS.$$

(i) $S = 50$ m

Integrate,

$$\frac{v^2}{2} = 10S - \frac{0.006 S^3}{3} + C$$

$$v^2 = 20S - 0.004 S^3 + C$$

When $s = 0,\quad v = 0 \qquad \therefore\ C = 0$

$\therefore$

$$v^2 = 20\,S - 0.004 S^3$$

$$v^2 = 20(50) - 0.004(50)^3 = 500$$

$$v = \sqrt{500} = \mathbf{22.36\ m/s} \quad \textbf{Ans.}$$

(ii) Distance travelled before coming to rest

$$v = 0$$

$$v^2 = 20S - 0.004S^3$$

$$0 = 20S - 0.004S^3$$

$$0.004\, S^2 - 20 = 0$$

$\therefore$

$$S = \sqrt{\frac{20}{0.004}} = \sqrt{5000}$$

$$= \mathbf{70.7\ m} \quad \textbf{Ans.}$$

Example 7.5(a): The motion of a particle is given by $a = t^3 - 3t^2 + 5$ where a is the acceleration in m/s^2 and t is the time in seconds. The velocity of the particle at $t = 1$ sec. is 6.25 m/s and the displacement is 8.30 m. Calculate the displacement and velocity at $t = 2$ s. [U.P.T.U. 2009-10]

Solution:

$$\alpha = t^3 - 3t^2 + 5.$$

$$= \frac{dv}{dt}$$

$\therefore$ $$dv = (t^3 - 3t^2 + 5)\ dt$$

By integration,

$$v = \frac{t^4}{4} - \frac{3t^3}{3} + 5t + c_1$$

When $t = 1\text{s}.\ V = 6.25$ m/s

$\therefore$ $$6.25 = \frac{1}{4} - 1 + 5 + c_1$$

$\therefore$ $$c_1 = 6.25 - 4.25 = 2.$$

$\therefore$ $$v = \frac{t^4}{4} - t^3 + 5t + 2 \qquad \text{...(1)}$$

$$= \frac{ds}{dt}$$

$\therefore$ $$ds = \left(\frac{t^4}{4} - t^3 + 5t\right) dt$$

By integration,

$$s = \frac{t^5}{20} - \frac{t^4}{4} + \frac{5t^2}{2} + c_2$$

When $t = 1$ s, $s = 8.30$ m.

$\therefore$ $$8.30 = \frac{1}{20} - \frac{1}{4} + \frac{5}{2} + c_2$$

$\therefore$ $$c_2 = 4$$

$\therefore$ $$s = \frac{t^5}{20} - \frac{t^4}{4} + \frac{5t^2}{2} + 4 \qquad \text{...(2)}$$

When $t = 2$ s

From equation (1)

$$v = \frac{16}{4} - 8 + 10 + 2 = 8 \text{ m/s}$$

From equation (2)

$$s = \frac{32}{20} - \frac{16}{4} + \frac{20}{2} + 4 + 4$$

$$= \mathbf{15.6\ m} \quad \textbf{Ans.}$$

Example 7.5(b): The motion of a particle is defined by the relation.

$$s = 6t^4 - 2t^3 - 12t^2 - 3t + 3$$

Determine the time, position, velocity and distance travelled when acceleration is zero. [U.P.T.U. 2010-11]

Solution:

$$s = 6t^4 - 2t^3 - 12t^2 - 3t + 3$$

$$v = \frac{ds}{dt} = 24t^3 - 6t^2 - 12t - 3$$

$$a = \frac{dv}{dt} = 72t^2 - 12t - 12.$$

When $a = 0$

$$72\,t^2 - 12t - 12 = 0$$

or $6t^2 - t - 1 = 0$

$$\therefore \qquad t = \frac{1 \pm \sqrt{1+24}}{12} = \frac{1+5}{12}$$

$$= \mathbf{0.55} \quad \textbf{Ans.}$$

$$s = 6\,(0.5)^4 - 2\,(0.5)^3 - 12\,(0.5)^2 - 3\,(0.5) + 3$$

$$= -1.375\ \text{m}$$

$$v = 24 \times (0.5)^3 - 6\,(0.5)^2 - 12 \times 0.5 - 3$$

$$= -7.5\ \text{m/s}$$

Therefore, when acceleration is zero, time is 0.5s, the particle is 1.375 m to the left and velocity is 7.5 m/s.

7.6 ANGULAR MOTION

When a rigid body rotates about a fixed axis, every particle of the rigid body moves along concentric circles. When a particle on the rigid body moves from position A to B, the angular displacement θ is measured in radians.

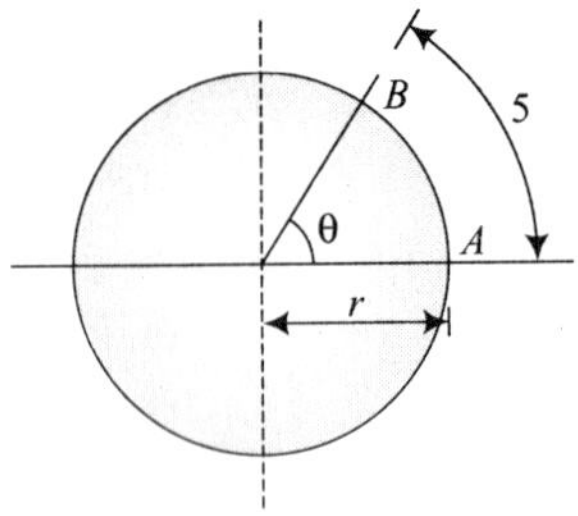

Fig. 7.4 Rotation of a body

The rate of change of displacement with time is called *angular velocity* (ω).

$$\omega = \frac{d\theta}{dt}$$

The rate of change of angular velocity with time is called *angular acceleration* (α).

$$\alpha = \frac{d\omega}{dt} = \frac{d^2\theta}{dt^2}$$

Also,
$$\alpha = \frac{d\omega}{dt} = \frac{d\omega}{d\theta} \cdot \frac{d\theta}{dt} = \omega \cdot \frac{d\omega}{d\theta}$$

7.6.1 Relationship between Angular Motion and Linear Motion

From Fig. 7.4,

$$S = r\theta$$

The tangential velocity of particle is called *linear velocity* (v).

$$v = \frac{ds}{dt} = r\frac{d\theta}{dt}$$

The tangential acceleration of the particle is called *linear acceleration* (α_t).

$$\alpha_t = \frac{dv}{dt} = r\frac{d^2\theta}{dt^2}$$

The radical acceleration,

$$\alpha_n = \frac{v^2}{r} = r\,\omega^2.$$

7.6.2 Uniform Angular Velocity

The angular distance moved in t sec with uniform angular velocity (ω),

$$\theta = \omega t \text{ radian}$$

Angular acceleration

$$\alpha = 0$$

Angular velocity

$$\omega = \frac{2\pi N}{60} \text{ rad/s}$$

where, N = No. of revolutions per minute

$= \text{rpm}$

The time taken for one revolution (2π)

$$2\pi = \omega T$$

or
$$T = \frac{2\pi}{\omega}$$

7.7 ROTATION WITH UNIFORM ACCELERATION

Angular acceleration,

$$\alpha = \frac{d\omega}{dt}$$

$\therefore$
$$d\omega = \alpha\, dt$$

$\therefore$
$$\omega = \alpha t + C_1 \qquad (\because \alpha \text{ is constant})$$

When $t = 0$, $\omega = \omega_0$ = initial velocity

$\therefore$
$$C_1 = \omega_0$$

$\therefore$
$$\omega = \omega_0 + \alpha t \qquad \text{...(1)}$$

The angular velocity

$$\omega = \frac{d\theta}{dt} = (\omega_0 + \alpha t)$$

$$d\theta = (\omega_0 + \alpha t)\, dt$$

$$\theta = \omega_0 t + \frac{1}{2}\alpha t^2 + C_2.$$

When $t = 0$, $\theta = 0$, $C_2 = 0$

$\therefore$
$$\theta = \omega_0 t + \frac{1}{2}\alpha t^2 \qquad \text{...(2)}$$

Angular acceleration,

$$\alpha = \omega\frac{d\omega}{d\theta}$$

$$\therefore \qquad \alpha d\theta = \omega d\omega$$

Integrating,

$$\alpha\theta = \frac{\omega^2}{2} + C_3$$

When $\theta = 0$, $\omega = \omega_0$.

$$\alpha \times 0 = \frac{{\omega_0}^2}{2} + C_3$$

$$\therefore \qquad C_3 = -\frac{\omega_0^2}{2}$$

$$\therefore \qquad \alpha\theta = \frac{\omega^2}{2} - \frac{{\omega_0}^2}{2}$$

or $\qquad \omega^2 - {\omega_0}^2 = 2\alpha\theta \qquad$ ....(3)

For rotation with uniform acceleration

$$\omega = \omega_0 + \alpha t$$

$$\theta = \omega_0 t + \frac{1}{2}\alpha t^2$$

$$\omega^2 - \omega_0^2 = 2\alpha\theta$$

Example 7.6: The rotation of a flywheel is governed by the equation $\omega = 3t^2 - 2t + 2$ where ω is in rad/s and time t in sec. The displacement was 4 radians after one second after start. Determine the angular displacement, angular velocity and angular acceleration of flywheel after 3 sec.

Solution: $\qquad \omega = 3t^2 - 2t + 2$

$$\omega = \frac{d\theta}{dt} = 3t^2 - 2t + 2$$

$$d\theta = (3t^2 - 2t + 2)\, dt$$

$$\theta = \frac{3t^3}{3} - \frac{2t^2}{2} + 2t + C$$

$$\therefore \qquad \theta = t^3 - t^2 + 2t + C$$

When $t = 1$ sec, $\theta = 4$

$$4 = 1 - 1 + 2 + C$$

$$\therefore \qquad C = 2$$

$$\therefore \qquad \theta = t^3 - t^2 + 2t + 2$$

When $t = 3$ sec

$$\theta = 3^3 - 3^2 + 2 \times 3 + 2 = \textbf{26 rad} \quad \textbf{Ans.}$$

$$\omega = 3(3)^2 - 2(3) + 2 = \textbf{23 rad/sec} \quad \textbf{Ans.}$$

Angular acceleration

$$\alpha = \frac{d\omega}{dt} = 6t - 2$$

$$= 6 \times 3 - 2 = \textbf{16 rad/sec}^2 \quad \textbf{Ans.}$$

Example 7.7: A wheel was rotating at 900 rpm when power supply was cutoff It made 360 revolutions before coming to rest. Determine

(i) angular retardation

(ii) time taken to come to rest after supply was cut off.

Solution: $w_0 = 900 \text{ rpm} = \dfrac{900 \times 2\pi}{60} = 30\pi \text{ rad/sec}$

Final velocity, $w = 0$

$$\theta = 360 \text{ revolutions} = 2\pi \times 360 = 720\pi \text{ rad.}$$

Now equation of motion,

$$w^2 - w_0^{\,2} = 2\alpha\theta$$

$$0 - (30\pi)^2 = 2\alpha \times 720\,\pi$$

$$\alpha = \textbf{–1.9635 rad/sec}^2 \quad \textbf{Ans.}$$

From equation of motion,

$$w = w_0 + \alpha t$$

$$0 = 30\pi - 1.9635t$$

$$\therefore \quad t = \frac{30\pi}{1.9635} = \textbf{48 sec} \quad \textbf{Ans.}$$

Example 7.7(a): A wheel rotates for 5 seconds with constant angular acceleration and describes 100 radians during this time. It then rotates with a constant angular velocity and during the next 5 seconds describes 80 radians. Find the initial angular velocity and angular acceleration. [U.P.T.U. 2008-09]

Solution:

$$\omega_2 = \omega_1 + \alpha_1 t_1 \quad \text{...(1)}$$

and

$$\omega_2^2 = \omega_1^2 + 2\alpha_1\,\theta_1 \quad \text{...(2)}$$

$$\omega_2 = \frac{\theta_2}{t_2} = \frac{80}{5} = 16 \text{ rad/s}$$

∴ From equation (1)

$$16 = \omega_1 + 5\alpha_1$$

$$\therefore \qquad \alpha_1 = \left(\frac{16 - w_1}{5}\right)$$

From equation (2),

$$(16)^2 = \omega_1^2 + 2\alpha_1 \times 100$$

$$256 = \omega_1^2 + \frac{2}{5}(16 - \omega_1)' \times 100$$

or

$$\omega_1^2 - 40\,\omega_1 + 384 = 0$$

$$\therefore \qquad \omega_1 = \frac{40 \pm \sqrt{1600 - 4 \times 384}}{2} = 20 \pm \frac{\sqrt{64}}{2} = 20 \pm 4$$

$$= 24 \text{ and } 16 \text{ rad/s}$$

When $\omega_1 = 24$ rad/s

$$\alpha_1 = \frac{16 - 24}{5} = -\,1.6 \text{ rad/s}$$

When $\omega_1 = 16$ rad/s

$$\alpha_1 = \frac{16 - 16}{5} = 0. \text{ This is not possible.}$$

$$\therefore \qquad \omega_1 = 24 \text{ rad/s}$$

and $\alpha_1 = \mathbf{-1.6}$ **rad/s** **Ans.**

Example 7.7(b): A horizontal bar 1.5 m long and of small cross-section rotates about vertical axis through one end. It accelerates uniformly from 1200 rpm to 1500 rpm in an interval of 5 seconds. What is the linear velocity at the beginning and end of the interval? What are the normal and tangential components of acceleration of the midpoint of the bar after 5 seconds after the acceleration begins? [U.P.T.U. 2009-10]

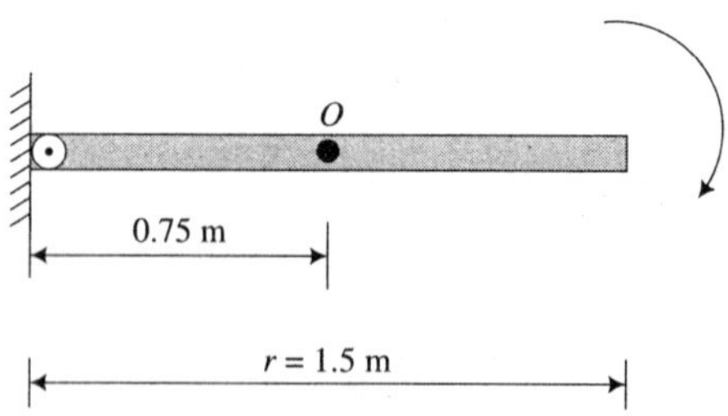

Solution:

The angular velocity with uniform acceleration,

$$\omega = \omega_0 + \alpha t$$

$$\omega_o = \frac{2\pi No}{60} = \frac{2\pi \times 1200}{60}$$

$$= 125.66 \text{ rad/s}$$

$$\omega = \frac{2\pi N}{60} = \frac{2\pi \times 1500}{60}$$

$$= 157 \text{ rad/s}$$

$$\alpha = \frac{w - w_0}{t} = \frac{157 - 125.66}{5}$$

$$= 6.28 \text{ rad/s}^2$$

Linear velocity, $v = \omega r$

Linear velocity of midpoint of bar at the beginning of interval,

$$v_o = \omega_o \times \frac{r}{2} = 125.66 \times 0.75 = 94.245 \text{ m/s}$$

Linear velocity of midpoint of bar at the end of interval,

$$v = \omega_o \times \frac{r}{2} = 157 \times 0.75 = 117.8 \text{ m/s}$$

Tangential acceleration,

$$a_t = r\,\alpha = 0.75 \times 6.28 = 4.69 \text{ m/s}^2$$

Normal acceleration,

$$a_n = \omega^2 r = (157)^2 \times 0.75 = \mathbf{18486.75 \text{ m/s}^2} \quad \textbf{Ans.}$$

7.8 LINEAR VELOCITY AND ACCELERATION OF ROTATION

A body in rotation is shown.

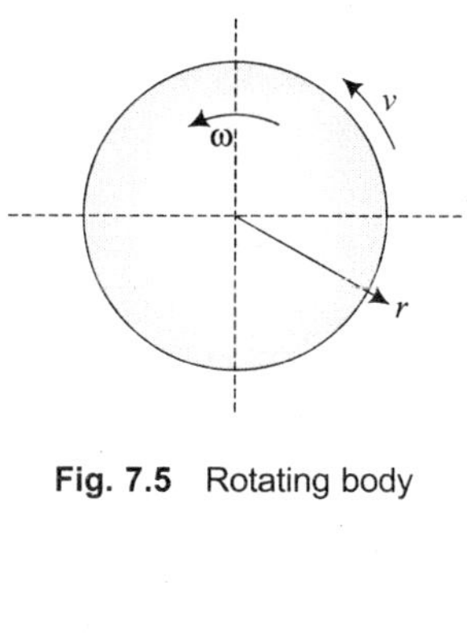

Fig. 7.5 Rotating body

Linear velocity of a particle on the periphery of the wheel,

$$v = wr \text{ [m/sec]}$$

where ω = Angular velocity of the body [rad/sec]

r = radius of circular path $|m|$

Linear acceleration,

$$a = \frac{dv}{dt}$$

$$= \frac{d}{dt}(\omega r)$$

$$= r\frac{d\omega}{dt}$$

$$= r\alpha$$

or $$\alpha = \frac{a}{r}$$

Example 7.8: A car with wheel diameter 75 cm is running at 72 kmph. Calculate the angular velocity of wheel about its axis of rotation.

The car is brought to rest with uniform retardation and it covers 20 metres. Find the angular retardation of the wheel.

Solution: (i) *Angular velocity of wheel*

$$v = 72 \text{ kmph} = \frac{72 \times 1000}{60 \times 60} = 20 \text{ m/sec}$$

$$r = \frac{d}{2} = \frac{75}{2 \times 100} = 0.375 \text{ m}$$

$$\omega = \frac{v}{r} = \frac{20}{0.375}$$

$$= \mathbf{53.3 \text{ rad/sec}} \quad \textbf{Ans.}$$

(ii) *Angular retardation of wheel*

$$v^2 = u^2 + 2aS$$

$$0 = (20)^2 + 2 \times a \times 20$$

$$= 400 + 40a$$

$$\therefore \qquad a = \frac{-400}{40}$$

$$= -10 \text{ m/sec}^2$$

$$\alpha = \frac{a}{r} = -\frac{10}{0.375}$$

$$= \mathbf{-26.7 \text{ rad/sec}^2} \quad \textbf{Ans.}$$

Example 7.8(a): A cord is wrapped around a wheel of radius 0.2 m which is initially at rest. If a force is applied to the cord and gives it an acceleration, $a = (4t)$ m/s^2, where t is in seconds, determine the angular velocity of the wheel and angular position of line *OP*, both as a function of time. [U.P.T.U. 2009-10]

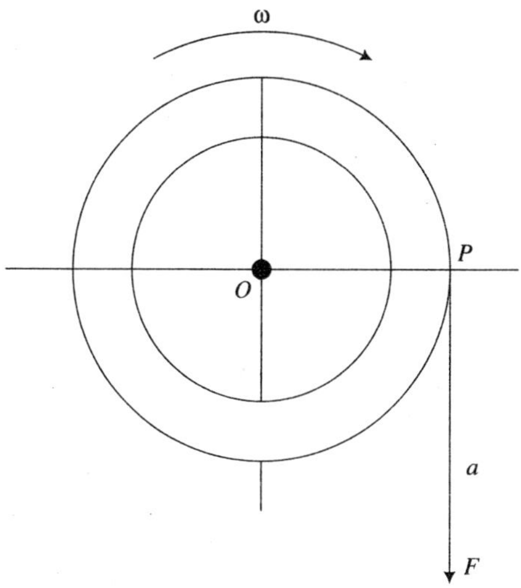

Solution:

The initial velocity of cord, $V = 0$ and initial acceleration, $a = 0$. After application of force,

$$a = (4\text{t}) \text{ m/s}^2$$

$$= \frac{dv}{dt}$$

$$\therefore \qquad dv = (4t)\, dt$$

By integration,

$$v = \frac{4t^2}{2} + c_1$$

When $\quad t = 0\,,\, v = 0$

$\therefore \quad c_1 = 0$

$\therefore \quad v = 2t^2 \qquad \text{....(1)}$

$$v = \frac{ds}{dt}$$

$\therefore \quad ds = (2t^2)\; dt$

By integration,

$$s = \frac{2t^3}{3} + c_2$$

$\therefore$ When $\quad t = 0,\, s = 0$

$\therefore \quad c_2 = 0$

$\therefore \quad s = \frac{2}{3}\; t^3 \qquad \text{...(2)}$

$$r = 0.2 \text{ m}$$

Angular velocity,

$$\omega = \frac{V}{r} = \frac{2t^2}{0.2}$$

$$= 10\; t^2 \text{ rad/s.}$$

$$s = r\,\theta$$

$\therefore \quad \theta = \frac{S}{r}$

$$= \frac{2t^3}{3 \times 0.2}$$

$$= \frac{10}{3}\; t^3 \text{ rad.}$$

7.9 ROTATION WITH VARIABLE ANGULAR ACCELERATION

All the relations of #7.5 for linear motion are also applicable to rotation with variable angular acceleration.

Example 7.9: The equation of motion of a particle moving on a circular path is given by:

$$\theta = 2t^3 + 0.5$$

where θ is in radian and t is in seconds. Find angular velocity, displacement and acceleration after 2 seconds.

Solution: (i) *Angular displacement after 2 seconds*

$$\theta = 2t^3 + 0.5$$

$$\theta = 2(2)^3 + 0.5$$

$$= \mathbf{16.5\ rad} \quad \textbf{Ans.}$$

(ii) *Angular velocity after 2 seconds*

$$\theta = 2t^3 + 0.5$$

$$\therefore \quad \frac{d\theta}{dt} = w = 6t^2$$

$$\therefore \quad w = 6(2)^2 = \mathbf{24\ rad/sec} \quad \textbf{Ans.}$$

(iii) *Angular acceleration after 2 seconds*

$$w = 6t^2$$

$$\alpha = \frac{dw}{dt} = 12t$$

$$= 12 \times 2 = \mathbf{24\ rad/sec^2} \quad \textbf{Ans.}$$

7.10 CIRCULAR MOTION

When a body moves with uniform velocity along a circular path, there is acceleration due to change in the direction of velocity. Such a motion is also called curvilinear motion.

The body is moving with uniform velocity. The tangential velocity

$$v_A = v_B = v$$

But there is change in the direction equal to $d\theta$. From velocity diagram, there is a change of velocity $= dv$ in radial inward direction

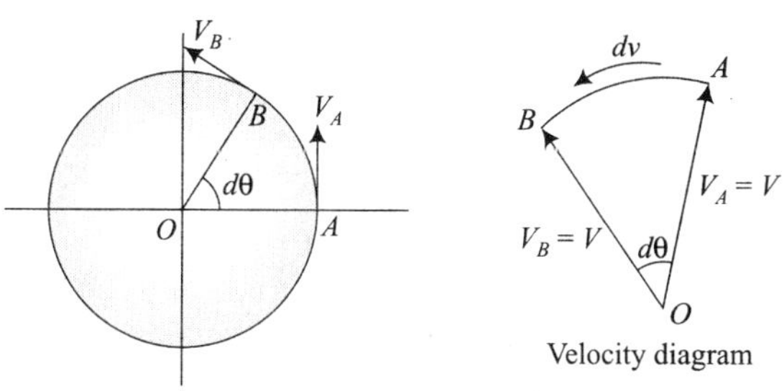

Fig. 7.6 Circular motion

$$dv = v\, d\theta$$

$$= v\,\frac{dS}{r} = \frac{v}{r}\, dS$$

Acceleration, $$a = \frac{dv}{dt} = \frac{v\, dS}{r\, dt} = \frac{v}{r}\cdot v = \frac{v^2}{r}.$$

The change of velocity directed towards centre of curvature of motion can be resolved into two components dv_x and dv_y.

$\therefore$ $$a = \frac{dv_x}{dt} + \frac{dv_y}{dt}$$

$$a_x = \frac{dv_x}{dt}$$

$$a_y = \frac{dv_y}{dt}$$

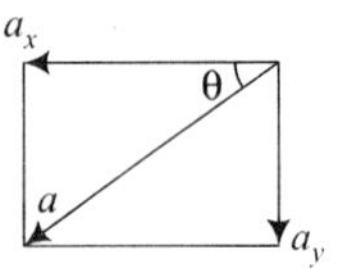

Fig. 7.7 Components of acceleration

or $$\vec{a} = \vec{a}_x + \vec{a}_y \text{ (vector equation)}$$

$$a = \sqrt{a_x^2 + a_y^2}$$

Direction $$\theta = \tan^{-1}\frac{a_y}{a_x}$$

The acceleration can also be resolved into component along normal and tangential direction

$$a = a_t + a_n \text{ (vector equation)}$$

where, a_t = tangential component of acceleration

$$= \frac{dv}{dt}$$

a_n = normal component of acceleration

$$= \frac{v^2}{r}$$

$$a = \sqrt{a_t^2 + a_n^2}$$

Example 7.10: A motorist is travelling on a curved road of radius 20 m at a speed of 72 kmph. Find the normal and tangential components of acceleration.

If he applies brakes to slow down his car uniformly to a speed of 36 kmph in 10 seconds, find the normal and tangential components of retardation just after brakes are applied.

Solution:

$$v_1 = 72 \text{ kmph}$$

$$= \frac{72 \times 1000}{60 \times 60} = 20 \text{ m/s}$$

$$v_2 = 36 \text{ kmph} = 10 \text{ m/s}$$

$$a_n = \frac{v^2}{r} = \frac{20 \times 20}{200}$$

$$= \mathbf{2\ m/s^2} \quad \textbf{Ans.}$$

$$a_t = \frac{dv}{dt} = 0$$

After the car is slowed down to 10 m/s

$$a_t = \frac{dv}{dt} = \frac{\Delta v}{\Delta t} = \frac{20 - 10}{10}$$

$$= \mathbf{1\ m/s^2}$$

$$a_n = \frac{v^2}{r} = \frac{(20)^2}{200}$$

$$= \mathbf{2\ m/s^2} \quad \textbf{Ans.}$$

Example 7.10(a): A train starts from rest and moves along a curved track of radius 600 m with uniform acceleration until it attains a velocity of 72 kmph at the end of third minute. Determine the tangential, normal and total acceleration of the train at the end of the second minute. [U.P.T.U. 2009-10]

Solution:

$$v_1 = 0$$

$$v_2 = 72 \text{ kmph} = \frac{72 \times 1000}{60 \times 60}$$

$$= 20 \text{ m/s.}$$

$$t = 3 \times 60 = 180 \text{ s.}$$

The tangential acceleration

$$a_t = \frac{dv}{dt} = \frac{\Delta v}{\Delta t} = \frac{20 - 0}{180}$$

$$= \frac{1}{9} \text{ m/s}^2$$

Velocity after 2 minutes,

$$v = v_1 + a_t{'}\times t = 0 + \frac{1}{9}{'}\times 120 = 13.3 \text{ m/s}$$

The normal acceleration

$$a_n = \frac{v^2}{r} = \frac{(13.3)^2}{600} = 0.295 \text{ m/s}^2$$

Total acceleration,

$$a = \sqrt{a_n^2 + a_t^2} = \sqrt{\left(\frac{1}{9}\right)^2 + (0.295)^2}$$

$$= \mathbf{0.315 \text{ m/s}^2} \quad \textbf{Ans.}$$

7.11 RELATIVE MOTION

If a body (*A*) is moving with a velocity V_A and body (*B*) is moving with a velocity v_B, then

v_A = Absolute velocity of *A* with respect to stationary earth

v_B = Absolute velocity of *B* with respect to stationary earth

The velocity of moving body *A* with respect to moving body *B* is called relative velocity of body *A* with respect to body *B*. It is the vector difference of absolute velocities v_A and v_B

$$\vec{v}_{A/B} = \vec{v}_A - \vec{v}_B$$

Example: 1. Bodies *A* and *B* are moving in the same parallel direction.

$$\vec{v}_{A/B} = \vec{v}_A - \vec{v}_B$$

If $v_A = 70$ m/s and

$v_B = 10$ m/s

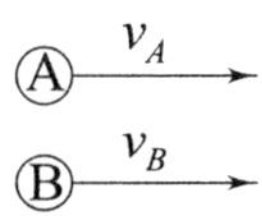

$$v_{A/B} = 70 - 10 = 60 \text{ m/s}$$

2. Bodies *A* and *B* are moving in the opposite parallel direction.

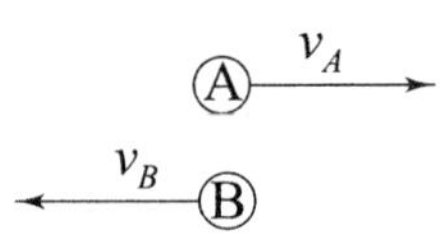

$$v_{A/B} = 70 - (-10) = 80 \text{ m/s}$$

7.11.1 Relative Velocity of Plane Motion

In order to find relative velocity in plane motion, find the components of velocities in x and y direction.

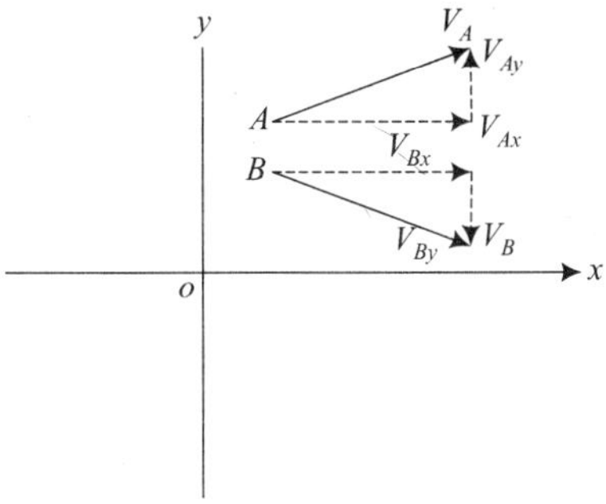

Fig. 7.8 Relative velocity

$$v_{A/B_x} = v_{A_x} - v_{B_x}$$

$$v_{A/B_y} = v_{A_y} - v_{B_y}$$

$$\therefore \quad v_{A/B} = \sqrt{(v_{A/B_x})^2 + (v_{A/B_y})^2}$$

$$\tan \alpha = \frac{v_{A/B_y}}{v_{A/B_x}}$$

where, α is the direction of relative velocity w.r.t. x-axis.

7.11.2 Relative Distance

The two bodies are moving with velocities v_A and v_B in a plane. The distance between the two bodies is AB. In order to find relative distance between A and B after time interval t, follow the procedure as given below:

1. Assume body B is stationary.
2. Move body A with relative velocity $v_{A/B}$.
3. Calculate distance travelled by A as $v_{A/B} \times t$ and move body A by the above distance to position A.
4. Join $A'B$. This is the relative distance between bodies A and B after time interval t.

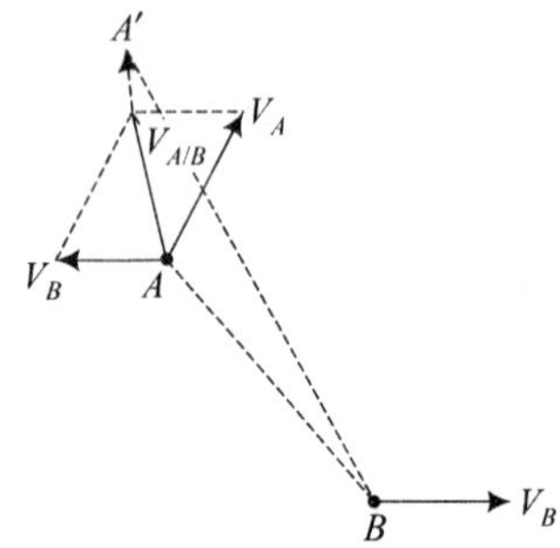

Fig. 7.9 Relative distance

7.11.3 Resultant Velocity

A body is moving with a velocity v_1. It is acted upon by an external force which imparts the body a velocity v_2 in its own direction. The combined effect of velocities v_1 and v_2 is the resultant velocity (v).

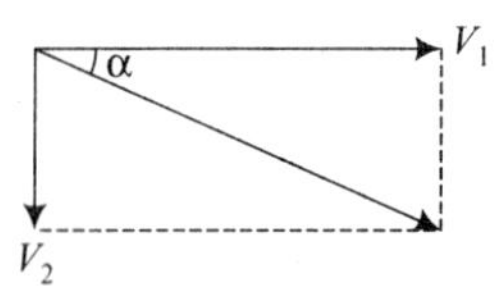

Fig. 7.10 Resultant velocity

Example 7.11: A passenger train 250 m long, moving with a velocity of 72 kmph, overtakes a goods train, moving on a parallel track in the same direction, completely in 45 seconds. If the length of goods train is 200 m, determine the speed of the goods train.

Solution:

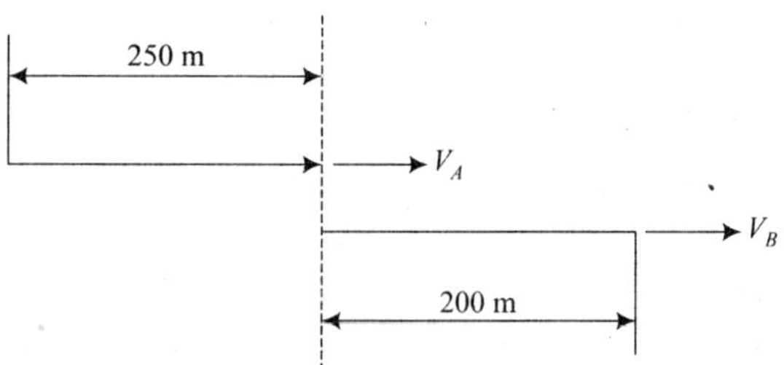

Fig. 7.11 (Example 7.11)

$$v_A = \frac{72 \times 1000}{60 \times 60} = 20 \text{ m/s}$$

The relative distance moved to overtake the goods train = 250 + 200 = 460 m

The relative velocity of A with respect to B

$$v_{A/B} = v_A - v_B = 20 - v_B$$

$$\therefore \quad v_{A/B} \times t = 450$$

$$(20 - v_B) \times 45 = 450$$

$$\therefore \quad v_B = 10 \text{ m/s} = \frac{10 \times 60 \times 60}{1000} = \mathbf{36 \text{ kmph}} \quad \textbf{Ans.}$$

Example 7.12: Train A is 240 m long and travelling in opposite parallel direction to train B of 300 m long. The velocity of A is twice that of B. Determine their velocities if they take 18 seconds to pass each other.

Solution: Relative distance = 240 + 300 = 540 m

Relative velocity, $v_{A/B} = v_A - (-v_B)$

$$= v - \left(-\frac{v}{2}\right) = 1.5\, v$$

Equation of motion,

$$1.5v \times 18 = 540$$

$$\therefore \quad v = 20 \text{ m/s} = 72 \text{ kmph}$$

$$\therefore \quad v_A = \mathbf{72 \text{ kmph}} \quad \textbf{Ans.}$$

$$v_B = \mathbf{36 \text{ kmph}} \quad \textbf{Ans.}$$

Example 7.13: A passenger is travelling in a train moving with a velocity of 36 kmph and observes raindrops at 30° to the vertical. The velocity of train is speeded up to 54 kmph and he observes raindrops at 45° to vertical. Determine the true velocity and direction of raindrops. Assume raindrops are in parallel vertical plane to that of train's vertical plane.

Solution:

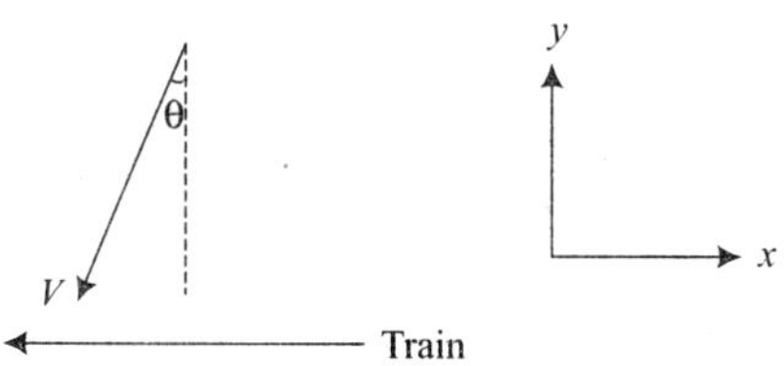

Fig. 7.12 (Example 7.13)

Assume true velocity of rain as v kmph and true angle θ with the vertical. The velocity components of rain,

$$v_{1x} = v \sin\theta$$

$$v_{1y} = v \cos\theta$$

(1) The velocity components of train when velocity of train is 36 kmph

$$v_{2x} = 36 \text{ kmph}$$

$$v_{2y} = 0$$

The relative velocity component of rain with respect to train are

$$v_{rx} = v \sin \theta - 36$$

$$v_{ry} = v \cos \theta$$

$$\therefore \qquad \tan\alpha = \frac{v\sin\theta - 36}{v\cos\theta}$$

But $\alpha = 30°$ (Given)

$$\therefore \qquad \tan 30° = \frac{v\sin\theta - 36}{v\cos\theta} = 0.577 \qquad ...(1)$$

(2) When velocity of train is 54 kmph, $\alpha = 45°$ (given)

$$\therefore \qquad \tan 45° = \frac{v\sin\theta - 45}{v\cos\theta} = 1$$

$$\therefore \qquad v\cos\theta = v\sin\theta - 54 \qquad ...(2)$$

From equations (1) and (2)

$$v \sin\theta = -11.402$$

$$v \cos\theta = -65.402$$

$$v = 66.388 \text{ kmph}$$

$$\sin\theta = -\frac{11.402}{66.388} = -9.89°$$

Example 7.14: Two ships leave a port at the same time: Ship A is moving in N30°W direction with a velocity of 30 kmph. Ship B is moving in south-west direction with a velocity of 40 kmph. Determine the relative velocity of A with respect to B and distance between them after 30 minutes.

Solution:

(1) The velocity components are:

$$v_{A_x} = 30 \sin 30° = 15 \text{ kmph}$$

$$v_{A_y} = 30 \cos 30° = 25.98 \text{ kmph}$$

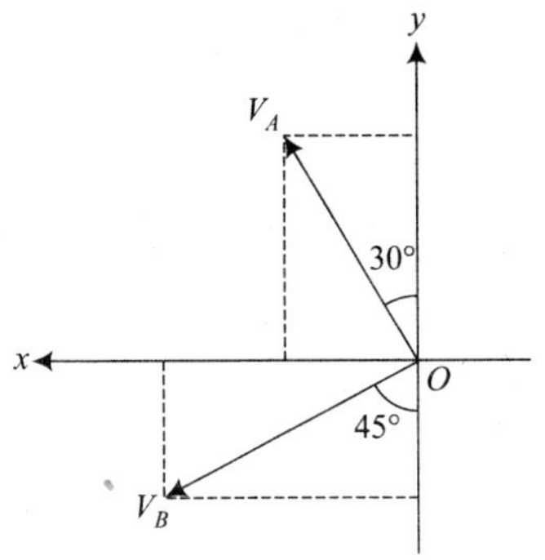

Fig. 7.13 (a) (Example 7.14)

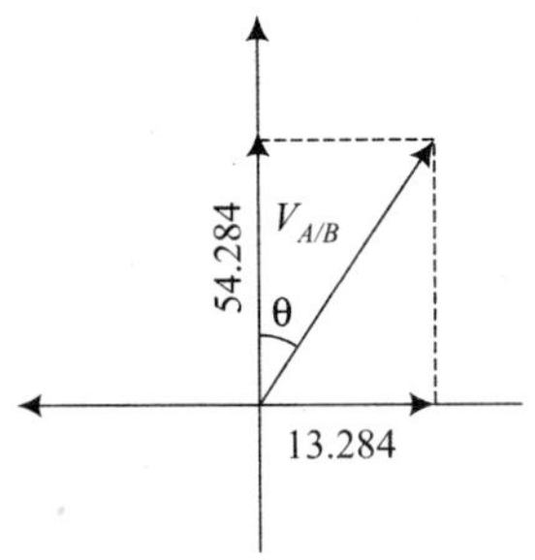

Fig. 7.13 (b) Velocity diagram

$$v_{B_x} = 40 \sin 45° = 28.284 \text{ kmph}$$

$$v_{B_y} = -40 \sin 45° = -28.284 \text{ kmph}$$

$\therefore$
$$v_{A/B_x} = 15 - 28.284 = -13.284 \text{ kmph}$$

$$v_{A/B_y} = 25.98 - (-28.284) = 54.264 \text{ kmph}$$

$$v_{A/B} = \sqrt{(13.284)^2 + (54.264)^2}$$

$$= 55.866 \text{ kmph}$$

$$\tan\theta = \frac{13.284}{54.264}$$

$\therefore$
$$\theta = 13.76°$$

Ship A appears to move with a velocity of 55.866 kmph in N 13.76°E direction with respect to ship B.

(2) Relative distance after 30 minutes

$$= v_{A/B} \times t = 55.866 \times 0.5 = \mathbf{27.933\ km} \quad \textbf{Ans.}$$

Example 7.15: Ship A is approaching a port in East direction with a velocity of 15 kmph. When ship A was 50 km away from the port, ship B sailed out of port in N 45°W direction with a velocity of 25 kmph. After how much time the two ships will be at minimum distance and how far each had travelled?

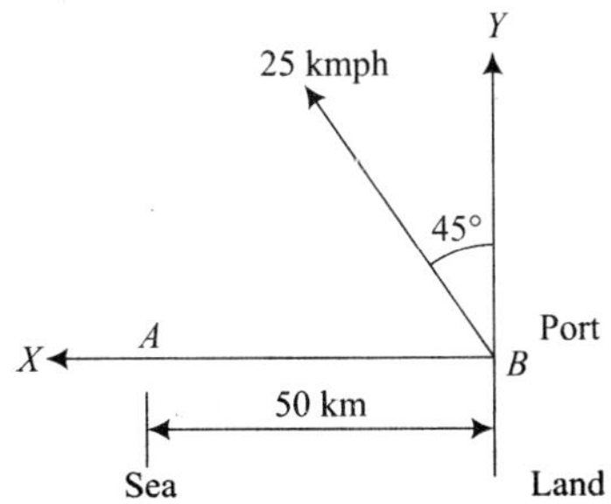

Fig. 7.14 (a) (Example 7.15)

Solution: Let West be x- and north be y-axis

The velocity components are

$$v_{B_x} = 25 \sin 45° = 17.678 \text{ kmph}$$

$$v_{B_y} = 25 \cos 45° = 17.678 \text{ kmph}$$

$$v_{A_y} = 0$$

$$v_{A_x} = -15 \text{ kmph}$$

$$v_{A/B_x} = 17.678 - (-15) = 32.678 \text{ kmph}$$

$$v_{A/B_y} = 17.678 \text{ kmph}$$

$$\therefore \quad v_{A/B} = \sqrt{(32.678)^2 + (17.678)^2} = 37.153 \text{ kmph}$$

$$\tan \alpha = \frac{v_{A/B_y}}{v_{A/B_x}} = \frac{17.678}{32.678}$$

$$\alpha = 28.41°.$$

The velocity diagram is shown in Fig. 7.14 (b).

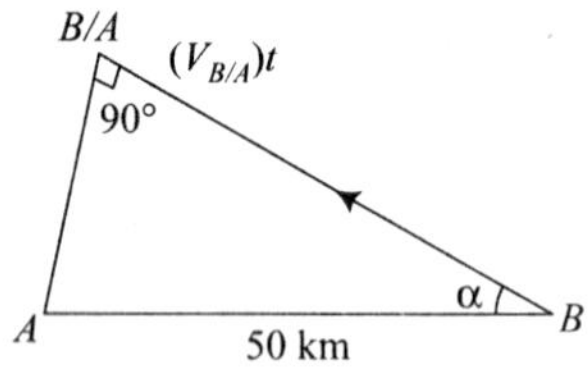

Fig. 7.14 (b) Velocity diagram

Holding A as stationary and let B move with relative velocity $v_{B/A}$ and two ships will be at minimum distance when angle at point B/A will be 90°.

$$\therefore \quad (v_{B/A})t = 50 \cos \alpha$$

$$37.153t = 50 \cos 28.41°$$

$$\therefore \quad t = \mathbf{1.1137 \text{ hous.}} \quad \textbf{Ans.}$$

The distance moved by ship A = 15 × 1.1837 = **17.750 km due East**

The distance moved by ship B = 25 × 1.1837 = **29.593 km in N 45°W direction. Ans.**

7.12 COMBINED MOTION OF ROTATION AND TRANSLATION

If a body possesses rotation and translation simultaneously, the body is said to have general plane motion. Common examples of such motion are a wheel rolling on straight line and a rod sliding against a wall at one end and the floor at the other.

For the analysis of plane motion, the motion should be splitted into translation and pure rotation. The two cases are analyzed separately and then combined to get the final motion.

7.12.1 Rolling of a Wheel without Slippage

A wheel of radius r is rolling with angular velocity ω and angular acceleration α.

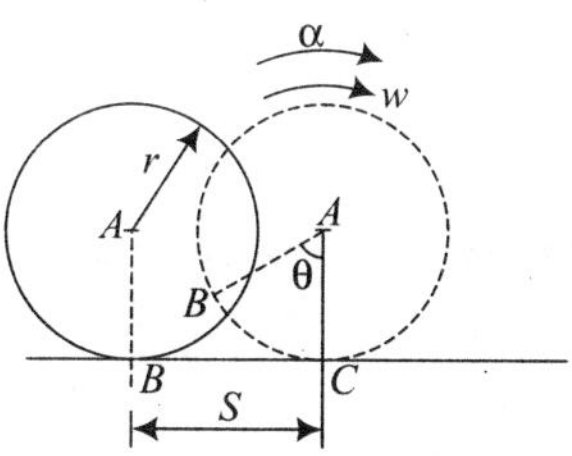

Fig. 7.15 General plane motion

It has rolled through angular distance θ in time t. There is no slippage.

∴ Linear distance, $S = r\theta$

$$\therefore \quad \frac{dS}{dt} = r\frac{d\theta}{dt}$$

$$\therefore \quad v_A = rw$$

$$\frac{d^2S}{dt^2} = r\frac{d^2\theta}{dt^2}$$

$$\therefore \quad a_A = r\alpha$$

When the wheel rotates without slip, the relationships between motion of geometric centre and its angular motion are

$$\left.\begin{aligned} S &= r\theta \\ v_A &= rw \\ a_A &= r\alpha \end{aligned}\right\}$$

7.12.2 Instantaneous Centre

The combined motion of rotation and translation can be assumed to be a motion of pure rotation about some centre. This is called *instantaneous center*. The position of this centre keeps on changing with the position of the rigid body. The locus of all instantaneous centres is called *centrode*. The position of instantaneous centre is located graphically. If instantaneous centre is located, the velocities of all other points on the rigid body at any instant can be determined by treating the plane motion as pure rotation about the instantaneous centre.

Example 7.16: A wheel of radius 1 m is rolling with angular velocity of 5 rad/sec. Determine the velocity of point *B* and *D* by instantaneous centre method.

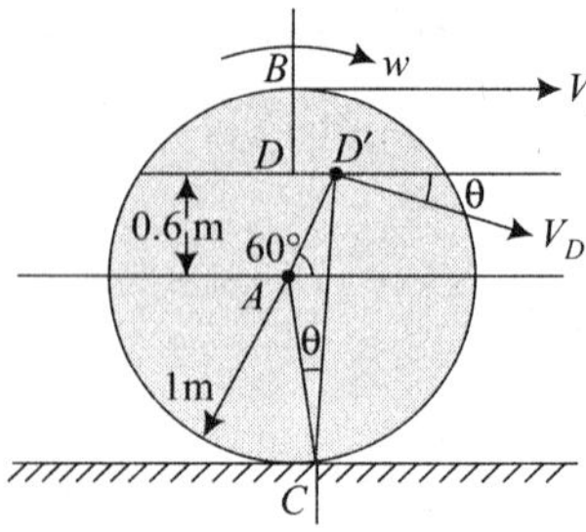

Fig. 7.16 (Example 7.16)

Solution: Velocity of geometric centre *A*,

$$v_A = wr = 5 \times 1 = 5 \text{ m/sec}$$

C is the instantaneous centre of rotation

$$v_A = wr$$

$$r = \frac{5}{5} = 1 \text{ m i.e., point } C$$

$\therefore$ $$v_B = CB \times w = 2 \times 5 = 10 \text{ m/sec}$$

$$v_D = CD \times w$$

$$CD = CA + AD$$

$$= 1 + 0.6 \sin 60° = 1.52 \text{ m}$$

$$DD' = 0.6 \cos 60° = 0.3 \text{ m}$$

$\therefore$ $$CD' = \sqrt{CD^2 + (DD')^2} = \sqrt{(1.52)^2 + (0.3)^2}$$

$$= 1.549 \text{ m}$$

$\therefore$ $$v_D = 1.549 \times 5 = \mathbf{7.745 \text{ m/sec}} \quad \textbf{Ans.}$$

Example 7.17: A rod 3 m long is standing against a wall making an angle 60° with the floor. Find the velocity of rod at wall end when end *A* at floor slides with a velocity of 2 m/sec. Solve the problem by the use of instantaneous centre method.

Solution: The velocity v_A is horizontal and v_B is vertical, drop perpendicular to v_A and v_B at points A and B, meeting at point C. Point C is the instantaneous centre.

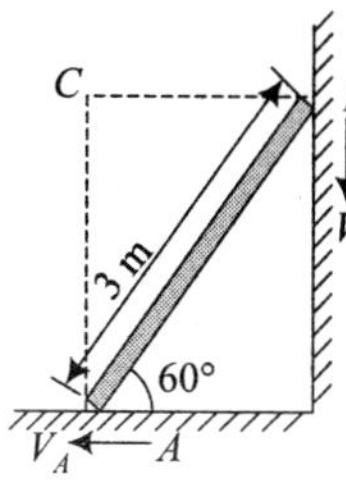

Fig. 7.17 (Example 7.17)

$$v_A = AC \times w$$

$$2 = 3 \sin 60 \times w$$

$$\therefore \quad w = \frac{2}{3\sin 60°}$$

$$= 0.77 \text{ rad/sec}$$

$$\therefore \quad v_B = BC \times w$$

$$= 3 \cos 60° \times 0.770$$

$$= \mathbf{1.555 \text{ m/sec}} \quad \textbf{Ans.}$$

Example 7.18: A straight rigid link AB of length 50 cm is shown. The end B of the link moves along x-axis with a velocity of 4 m/s and accelerates with an acceleration of 10 m/s^2. The end A is constrained to move along y-axis. Find the velocity and acceleration of the end A at the given instant. [U.P.T.U. 2010-11]

Solution:

The velocity V_B is horizontal and V_A is vertical. Drop perpendiculars to V_B and V_A to locate point C as instantaneous centre,

$$V_B = BC \times \omega$$

$$\therefore \quad BC = AB \sin 30°$$

$$= 0.5 \sin 30°$$

$$= 0.25 \text{ m}$$

$$\omega = \frac{V_B}{BC} = \frac{4}{0.25}$$

$$= 1.6 \text{ rad/s}$$

$$V_A = AC \times \omega$$
$$= 0.5 \cos 30° \times 16$$
$$= 6.93 \text{ m/s}$$
$$a_B = r\,\alpha$$
$$\therefore \quad \alpha = \frac{a_B}{r} = \frac{10}{BC} = \frac{10}{0.25}$$
$$= 40 \text{ rad/s}^2$$
$$a_A = AC \times \alpha$$
$$= 0.5 \cos 30° \times 40$$
$$= 17.32 \text{ m/s}^2$$

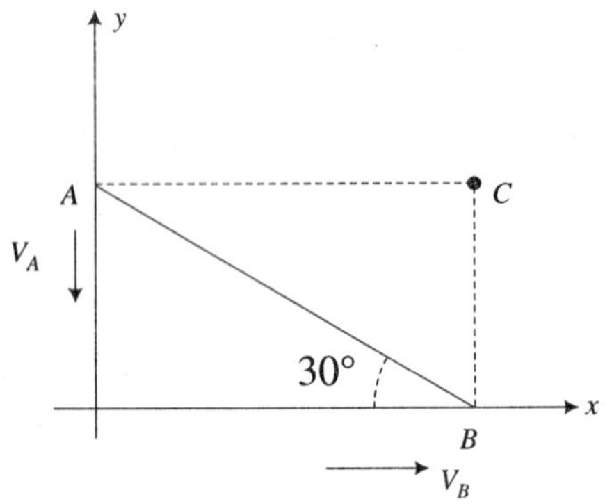

QUESTION BANK NO. 7

1. Explain different types of plane motion with suitable examples.
2. What is linear motion? Define velocity and acceleration for linear motion.
3. What are the equations of motion with uniform acceleration?
4. Establish relation between linear motion and angular motion with the help of equation of motion.
5. Describe circular motion and establish relationship between tangential and normal acceleration.
6. Explain the following:
   (a) Relative velocity
   (b) Relative distance
   (c) Resultant velocity
7. Define instantaneous centre and centroid. Explain with suitable example, the application of instantaneous centre for general plane motion.

QUICK REVIEW INFORMATION NO. 7

1. In rectangular motion, the particles remain parallel and straight.
2. In curvilinear motion, the particles remain parallel and in curve.
3. General plane motion is neither pure rotation nor pure translation.
4. Instantaneous motion is a pure rotary motion.
5. The angular velocity (rad/s) of a body rotating at N (r.p.m.) is $2\pi N/60$ and the linear velocity of a body rotating at ω (rad/s) along a circular path of radius r is $r\omega$.
6. The centripetal component of acceleration is measured perpendicular to the direction of velocity and acts along a line radial to the path of rotation towards the centre of curvature of path.
7. The radial acceleration (a_n) and linear acceleration (a_t) have the following values.

Type of motion	*Radial acceleration*	*Linear acceleration*
1. Uniform straight-line motion	$a_n = 0$	$a_t = 0$
2. Non-uniform straight-line motion	$a_n = 0$	$a_t \neq 0$
3. Uniform curvilinear motion	$a_n \neq 0$	$a_t = 0$
4. Non-uniform curvilinear motion	$a_n \neq 0$	$a_t \neq 0$

8. The components of acceleration for different types of motion are

Type of motion	*Acceleration component*	*Value*
1. Pure rotation	Tangential	$r\alpha$.
2. Pure rotation	Radial	$r\omega^2$
3. Curvilinear	Tangential	d^2s/dt^2
4. Curvilinear	Normal	$\frac{1}{r}\left(\frac{ds}{st}\right)^2$

CHAPTER 8

Kinetics of Rigid Body

8.1 INTRODUCTION

Kinetics is the branch of dynamics which deals with the motion of rigid bodies under the action of forces. The kinetic problems of a body with plane motion may be analysed by the following methods.

1. Work Energy Equation
2. Impulse Momentum Equation
3. D'Alemberts Principle of Dynamic Equilibrium

The work energy equation can be applied to problems involving force, velocity and displacement. Impulse momentum equation is useful for solving problems involving force, time and velocity. D'Alemberts Principle is useful for problems involving force and acceleration.

8.2 CHARACTERISTICS OF A FORCE

Newton's laws of motion forms the complete definition of force. From *Newton's first law* it is clear that if a body is subjected to a balanced system of forces, it will continue to be in the state of rest or of uniform motion. A body acted by a balanced system of forces has no acceleration.

Newton's Second Law of motion states: "The rate of change of momentum of a body is directly proportional to the impressed force and it takes place in the direction in which the force acts."

$$\text{Momentum} = \text{mass} \times \text{velocity}$$

$$\text{Rate of change of momentum} = \text{mass} \times \text{rate of change of velocity}$$

$$= \text{mass} \times \text{acceleration}$$

$$\text{Force} \propto \text{rate of change of momentum}$$

$$\therefore \quad \text{Force} \propto \text{mass} \times \text{acceleration}$$

$$F \propto m \times a$$

$$F = kma$$

where k = constant of proportionality

$$k = 1 \text{ in SI system}$$

$$\therefore \quad F = ma$$

Newton's second law states: "When an unbalanced system of forces act on a body, the body moves with an acceleration proportional to the resultant force and it takes in the direction of the resultant force."

Newton's third law states that for every action there is an equal and opposite reaction. This law will be used in treating kinetic problems.

A force is completely defined only when the following force characteristics are specified:

1. Magnitude in newton
2. Point of application
3. Line of action
4. Direction

Example: A person weighing 800 N is standing at point *C* of a ladder *AB* resting against a wall. Point *C* is 2 m from the floor.

The force applied by the person on the ladder has the following characteristics:

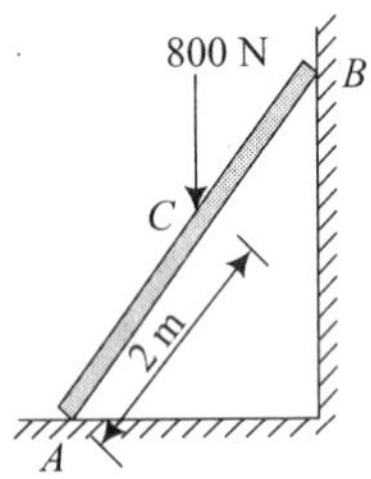

Fig. 8.1 Man and ladder

1. Magnitude of 800 N
2. Point of application *C*, 2 m from the floor along the ladder
3. The line of action is vertical (Line of arrow)
4. The direction is downward (Arrow head).

8.3 WORK ENERGY EQUATION

The *work* done by a force is the product of component of force in the direction of motion and the distance moved. The unit of work is joule (*J*). Joule is the amount of work done by a force of 1 newton (*N*) when the body moves a distance of 1 meter in the direction of force.

Energy is the capacity to do work. The unit of energy is same as that for work. The energy may be classified into potential energy and kinetic energy.

Potential energy is the capacity to do work due to mass and position of the body. A body of mass 'm' held at a height 'h' possesses an energy mgh.

Kinetic energy is the capacity to do work due to mass and velocity of the body. A body of mass 'm' and a velocity v as kinetic energy of $\frac{mv^2}{2}$.

Power is the time rate of doing work: Unit of power is watt (W) and defined as one joule of work done in one second.

The *work energy principle* states: "The work done by a system of forces acting on a body during a displacement is equal to the change in kinetic energy of the body during the same displacement."

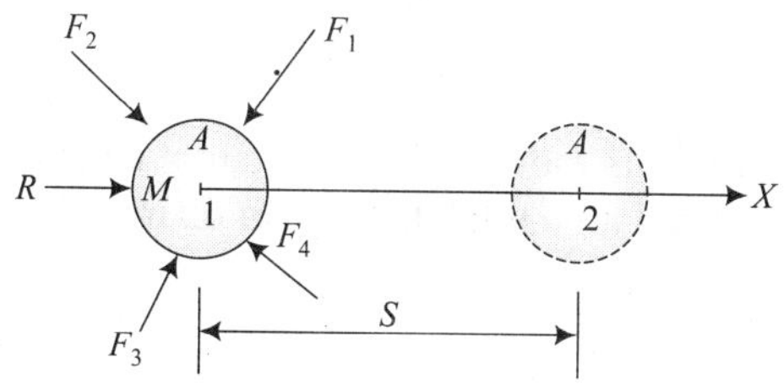

Fig. 8.2 Work energy equation

A body A of mass m is subjected to a system of four forces F_1, F_2, F_3 and F_4 which have resultant R acting in x-direction. The initial velocity of body at position 1 is u and final velocity at position 2 is v. The distance moved from position 1 to position 2 is S.

The resultant of forces in x-direction

$$R = \Sigma X_i$$

From Newton's second law of motion

$$R = ma = \frac{W}{g}a$$

Multiply both sides by elementary distance dS

$$\therefore \qquad RdS = \frac{W}{g}a\,dS$$

$$= \frac{W}{g}v\frac{dv}{dS}\,dS \qquad \left[\because a = v\frac{dv}{dS}\right]$$

$$= \frac{W}{g}\,v\,dv$$

Integrating for motion from position 1 to position 2,

$$\int_o^S R\,dS = \int_u^v \frac{W}{g} v\,dv$$

$$\therefore \qquad RS = \frac{W}{g}\left|\frac{v^2}{2}\right|_u^v = \frac{W}{2g}(v^2 - u^2)$$

Work done = Change of kinetic energy

Work energy equation can be used to solve the following kinetic problems.

1. Motion of body on horizontal plane
2. Motion of body on inclined plane
3. Motion of connected bodies
4. Work done by a spring

Example 8.1: A car was in motion along a straight level road. The brakes were applied and the car skidded for a total distance of 60 m before coming to halt. The coefficient of friction between tyres and road was 0.5. Calculate the speed of the car just before brakes were applied.

Solution: The free body diagram of the car is shown in Fig. 8.3.

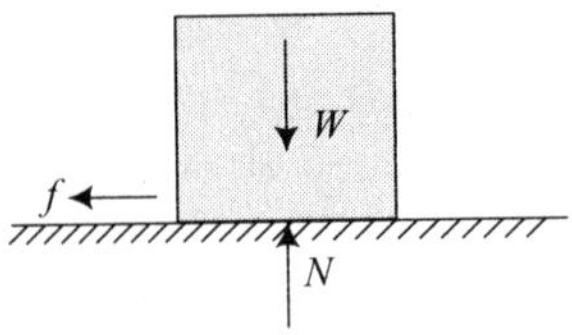

Fig. 8.3 Free body diagram of car

$$N = W$$

The force of friction

$$f = \mu N = \mu W = 0.5\ W$$

Final velocity of car,

$$v = 0$$

Displacement,

$$S = 60\ m$$

Applying work energy equation,

$$-f \times S = \frac{W}{2g}(v^2 - u^2)$$

$$-0.5\ W \times 60 = \frac{W}{2 \times 9.81}(0 - u^2)$$

$\therefore \quad u = 24.26$ m/s

$$= \frac{24.26 \times 60 \times 60}{1000} = \mathbf{87.3\ kmph} \quad \textbf{Ans.}$$

Example 8.2: A block of 300 N weight is pushed up on inclined plane of 30°. The force applied is 400 N. The initial velocity of block is 1.5 m/s and coefficient of friction is 0.2. Find the velocity of the block after it has been pushed by 6 m.

Solution: The free body diagram is shown in Fig. 8.4.

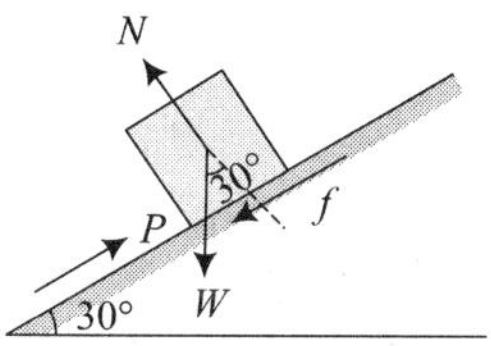

Fig. 8.4 Inclined plane

$$N = W \cos 30° = 300 \times \cos 30° = 259.81 \text{ N}$$

Force of friction,

$$f = \mu N = 0.2 \times 259.81 = 51.96 \text{ N}$$

Initial velocity, $u = 1.5$ m/s

Displacement, $S = 6$ m

Net force in the upward direction of motion is $(P - f - W \sin 30°)$

Applying work energy equation,

$$(P - f - W \sin 30°)S = \frac{W}{2g}(v^2 - u^2)$$

$$\left(400 - 51.96 - 300 \times \frac{1}{2}\right) 6 = \frac{300}{2 \times 9.81}(v^2 - 1.5^2)$$

$\therefore \quad v = \mathbf{8.94\ m/s}$ **Ans.**

Example 8.3: Two weights of 300 N and 450 N are hung to the two ends of a rope passing over a frictionless ideal pulley. What is the tension in the rope? Using work energy equation, calculate the distance moved by the blocks in increasing the velocity of the system from 2 m/s to 4 m/s.

Solution: 450 N weight moves down and 300 N weight moves up by the same velocity.

From work energy equation,

$$(450 - 300)S = \left[\frac{450}{2 \times 9.81}\right](v^2 - u^2) + \left[\frac{300}{2 \times 9.81}\right](v^2 - u^2)$$

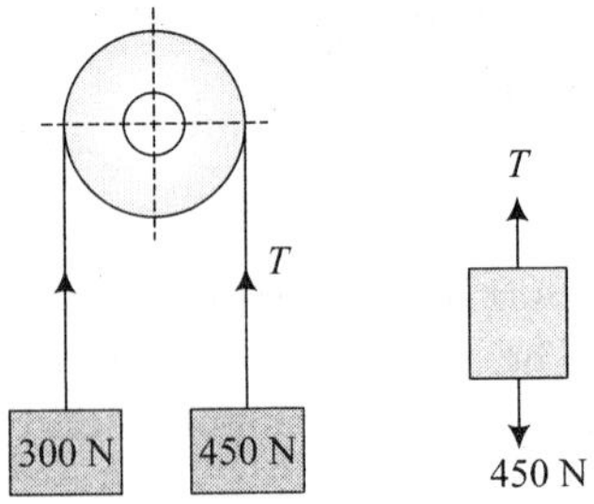

Fig. 8.5 Pulley Block

$$150\,S = \left(\frac{450}{2\times 9.81}\right)(4^2 - 2^2) + \left(\frac{300}{2\times 9.81}\right)(4^2 - 2^2)$$

$\therefore$ $S = \mathbf{3\ m}$ **Ans.**

Consider work energy equation for 450 N weight.

$$(450 - T)S = \left(\frac{450}{2\times 9.81}\right)(4^2 - 2^2)$$

$$(450 - T)\times 3 = \left[\frac{450}{2\times 9.81}\right](4^2 - 2^2)$$

$\therefore$ $T = \mathbf{360\ N}$ **Ans.**

Example 8.4: A wagon of 500 kN weight runs over 30 m distance along one percent inclined plane with a rolling resistance of 5 N/kN. The wagon starts from rest and strikes a spring bumper with a modulus of 15 kN/mm. Find the velocity of wagon at the time of striking the bumper and the compression of the spring.

Solution:

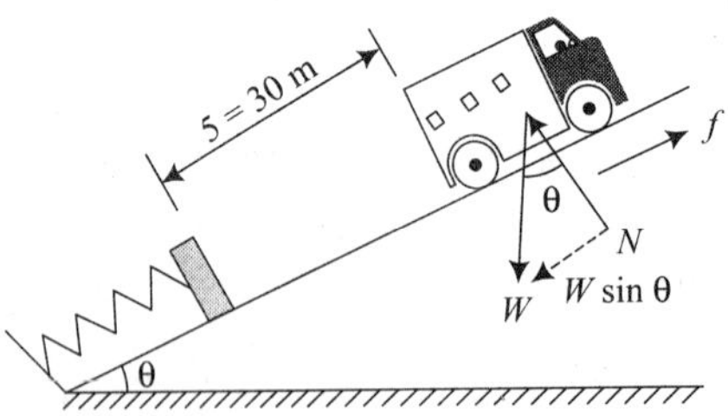

Fig. 8.6 (Example 8.4).

$$u = 0$$

$$S = 30\text{ m}$$

Apply work energy equation

$$(W \sin \theta - f)S = \frac{W}{2g}(v^2 - u^2)$$

Track resistance,

$$f = 5 \text{ N/kN} = 5 \times 500 = 2500 \text{ N} = 2.5 \text{ kN}$$

$$\left[500 \times \frac{1}{100} - 2.5\right] \times 30 = \frac{500}{2 \times 9.81}(v^2 - 0)$$

$$v = 1.7 \text{ m/s}$$

Apply work energy equation to bumper spring,

$$(W \sin \theta - f)(30 + x) - \frac{1}{2}kx^2 = \frac{W}{2g}(v^2 - u^2)$$

where x = spring compression assumed

k = spring constant

= 15 kN/mm = 15000 kN/m

$v = u = 0$

$$\therefore \quad (5 - 2.5)(30 + x) - \frac{1}{2} \times 1500x^2 = 0$$

$$3000x^2 - x - 30 = 0$$

$$x = \frac{-1 \pm \sqrt{1 + 4 \times 300 \times 30}}{2 \times 3000}$$

= **100 mm Ans.**

Example. 8.4(a): A sphere, a cylinder and a hoop, each having the same mass and radius are released from rest on an incline. Determine the velocity of each body after it has rolled through a distance corresponding to a change in elevation h.

[U.P.T.U. 2008-09]

Solution:

Loss of potential energy of all the articles = mgh

where m is the mass (same) and h is the fall (same)

$$\text{Gain in kinetic energy} = \frac{1}{2}mv^2 - \frac{1}{2}mu^2$$

$$= \frac{1}{2}mv^2$$

where initial velocity, $u = 0$ and v is the final velocity. From the principle of conservation of energy,

$$mgh = \frac{1}{2}mv^2$$

$$\therefore \qquad v = \sqrt{2gh}$$

Example 8.4(b): A vehicle weighing 50 kN moves with a velocity of 60 kmph along x-direction. Another vehicle weighing 25 kN moving along y-direction with a velocity of 90 kmph collides with it. If two vehicles get entangled after collision, determine their common velocity. [U.P.T.U. 2008-09]

Solution:

$$\text{Initial kinetic energy in } x\text{-direction} = \frac{1}{2}mv^2 = \frac{1}{2} \times \frac{50}{9.81}\left(\frac{60 \times 1000}{60 \times 60}\right)^2$$

$$= 707.9 \text{ kJ}$$

$$\text{Initial kinetic energy in } y\text{-direction} = \frac{1}{2} \times \frac{25}{9.81}\left(\frac{90 \times 1000}{60 \times 60}\right)^2$$

$$= 796.4 \text{ kJ}$$

$$\text{Resultant kinetic energy} = \sqrt{(707.9)^2 + (796.4)^2}$$

$$= 1065.54 \text{ kJ}$$

$$\text{Total weight after collision} = 50 + 25 = 75 \text{ kN}$$

$$\text{Final kinetic energy} = \frac{1}{2} \times \frac{75}{9.81} v^2$$

$$= 3.82\, v^2 \text{ kJ}$$

Applying the principle of conservation of energy,
Resultant initial kinetic energy = Final kinetic energy,

$$3.82\, v^2 = 1065.54$$

$$\therefore \qquad v = \sqrt{\frac{1065.54}{3.82}} = 16.70 \text{ m/s}$$

$$= \frac{16.70 \times 3600}{1000}$$

$$= \mathbf{60\ kmph} \quad \textbf{Ans.}$$

Example 8.4(c): A road roller has total mass of 12000 kg. The front roller has a mass of 2000 kg, a radius of gyration of 0.4 m and a diameter of 1.2 m. The rear axle, together with its wheels, has a mass of 2500 kg, a radius of gyration of 0.6 m and a diameter of 1.5 m. Calculate the kinetic energy of rotation of the wheels and axles at a speed of 9 kmph and total kinetic energy of road roller.

[U.P.T.U. 2009-10]

Solution:

$$\text{Kinetic energy of rotation} = \frac{1}{2} I\omega^2 = \frac{1}{2} mk^2 \omega^2.$$

(i) *Front roller*

Mass, $m = 2000$ kg.

Radius of gyration, $k = 0.4$ m

Radius $R = \frac{1.2}{2} = 0.6$ m

Velocity, $\omega = \frac{V}{R} = \frac{9 \times 1000}{60 \times 60 \times 0.6}$

$= 4.17$ rad/s

Kinetic energy $= \frac{1}{2} mk^2\omega^2$

$= \frac{1}{2} \times 2000 \times 0.4^2 \times (4.17)^2$

$=$ **2782.22 J Ans.**

(ii) *Rear wheel and axle*

Mass, $m = 2500$ kg

Radius of gyration $k = 0.6$

Radius, $R = \frac{1.5}{2} = 0.75$ m

∴ Velocity, $\omega = \frac{V}{R} = \frac{9 \times 1000}{60 \times 60 \times 0.75}$

$= 3.33$ rad/s

Kinetic energy $= \frac{1}{2} mk^2\omega^2 = \frac{1}{2} \times 2500 \times (0.6)^2 \times (3.33)^2$

$= \mathbf{4990.00\ J}$ **Ans.**

(iii) *Road Roller Body*

Mass, $m = 12000$ kg

Velocity, $v = \frac{9 \times 1000}{60 \times 60}$

$= 2.5$ m/s

Kinetic energy $= \frac{1}{2} mV^2 = \frac{1}{2} \times 12000 \times (2.5)^2$

$= \mathbf{37500\ J}$ **Ans.**

Total kinetic energy of road roller

$= 2782.22 + 4990 + 37500$

$= \mathbf{45272.22\ J}$ **Ans.**

Example 8.4(d): A horizontal sphere weighing 10 kg is attached to a slender rod of mass 2 kg . If the system is released from the horizontal position in rest condition, find the magnitude of angular acceleration. Also find the angular velocity of the system when it passes through the vertical position. [U.P.T.U. 2010-11]

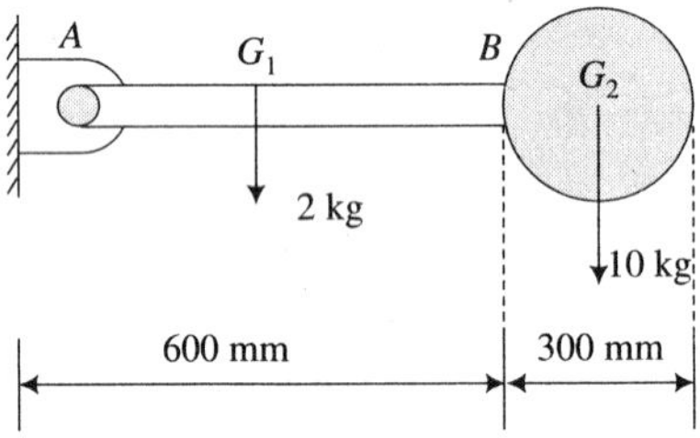

Solution:

Moment of inertia of rod about A,

$$I_1 = \frac{1}{3} m_1 l^2$$

$$= \frac{1}{3} \times 2 \times 0.6^2$$

$$= 0.24 \text{ kgm}^2$$

Moment of inertia of sphere about A,

$$I_2 = I_{G_2} + m_2 h^2$$

$$I_2 = \frac{2}{5} \times 10 \times (0.15)^2 + 10\ (0.6 + 0.15)^2$$

$$= 4.215 \text{ kgm}^2$$

Total moment of inertia about $A = 0.24 + 4.215 = 4.455$ kgm^2

Total moment about point A,

$$\Sigma M_A = 2 \times 9.81 \times 0.3 + 10 \times 9.81 \times 0.75$$

$$= 79.46 \text{ Nm}$$

For dynamic equilibrium,

$$\Sigma M_A = I\alpha$$

$\therefore$ Angular acceleration, $\alpha = \dfrac{\Sigma M_A}{I} = \dfrac{79.46}{4.455}$

$$= \mathbf{17.84 \text{ rad/s}} \quad \textbf{Ans.}$$

Now $\alpha = \dfrac{v^2}{r}$

$\therefore$ $v = \sqrt{\alpha r} = \sqrt{0.75 \times 17.84} = 3.66$ m/s.

Angular velocity, $\omega = \dfrac{v}{r} = \dfrac{3.66}{0.75}$

$$= \mathbf{4.88 \text{ rad/s}} \quad \textbf{Ans.}$$

8.4 IMPULSE MOMENTUM EQUATION

Impulse is the product of force and its duration of action.

$$\text{Impluse} = \int_o^t R\, dt$$

where R = Resultant force acting on a body

t = Time duration of action of force.

If the resultant force is in newtons and time in seconds, the unit of impulse is $N - s$.

If R is constant during time interval t, then

$$\text{Impulse} = R \times t$$

Momentum is the quantity of motion possessed by a body moving in a straight line and is the product of mass of the body and its velocity.

$$\text{Momentum} = mu$$

The unit of momentum is N – s.

8.4.1 Impulse Momentum Principle

The component of resultant linear impulse along any direction is equal to change in the component of momentum in that direction.

From Newton's second law of motion,

$$R = ma$$

where, $$a = \frac{dv}{dt}$$

$\therefore$ $$R = m\frac{dv}{dt}$$

or $$R dt = m\, dv$$

Integrating both sides

$$\int R\, dt = \int m\, dv$$

If u is the initial velocity and v is the final velocity after time interval t,

$$\int_o^t R\, dt = mv - mu$$

Impulse = Final momentum – Initial momentum

This is called impulse momentum equation and is very useful for solving kinetic problems in the following areas:

1. Horizontal blocks
2. Inclined plane
3. Connected bodies
4. Striking of balls

8.4.2 Conservation of Momentum

The principle of conservation of momentum may be stated as follows:

1. Momentum is conserved in a system if the resultant force is zero, or
2. If the resultant force in a system is zero, initial momentum will remain equal to final momentum.

Example: If m_1 is the mass of a person and m_2 that of the boat, v is the velocity of the person and boat before the person jumps out of the boat and v_1 and v_2 are the velocities of person and the boat after jumping, then

$$(m_1 + m_2)v = m_1v_1 + m_2v_2$$

The principle of conservation of momentum is useful to solve kinetic problem in the following area

1. Mass and boat
2. Collision of bodies

3. Gun and bullet
4. Pile and hammer

Example 8.5: A block of 1500 N weight is pushed with a force of 300 N on a horizontal plane, the coefficient of friction between block and plane being 0.1. How much time will be required before the block reaches a velocity of 16 m/s after starting from rest. For how long the block will continue after removal of pushing force? Use impulse momentum equation.

Solution: Draw free body diagram as shown in Fig. 8.7.

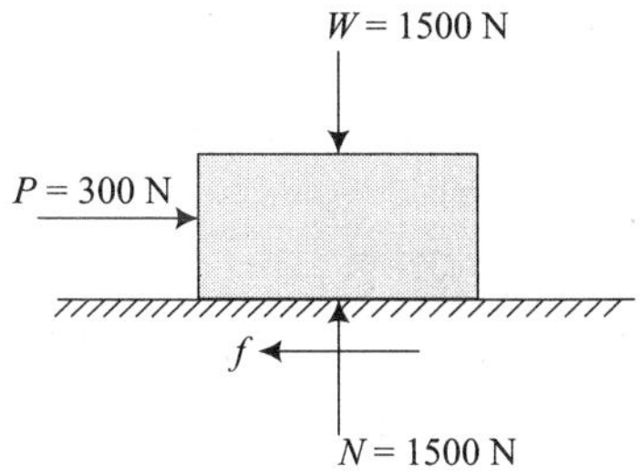

Fig. 8.7 (Example 8.5)

$$N = W = 1500 \text{ N}$$

Frictional force, $f = \mu N = 0.1 \times 1500$

$$= 150 \text{ N}$$

Applying impulse momentum equation in horizontal direction

$$(P - f)t = \left(\frac{W}{g}\right)(v - u)$$

$$(300 \quad 150)t - \frac{1500}{9.81}(16 \quad 0)$$

$\therefore$ $t = \mathbf{16.3\ sec}$ **Ans.**

When P is removed, $f = 150$ N is the only horizontal force. Applying impulse momentum equation,

$$-150t = \frac{1500}{9.81}(0 - 16)$$

$$t = 16.3 \text{ sec}$$

The block takes 16.3 sec before it comes to rest.

Example 8.6: A car is moving on a concrete road at a speed of 70 kmph. The total weight of the car with passengers is 20 kN. The coefficient of friction between wheels and car is 0.75. Determine the time to stop the car after application of brakes, if all the four wheels to skid.

Solution:

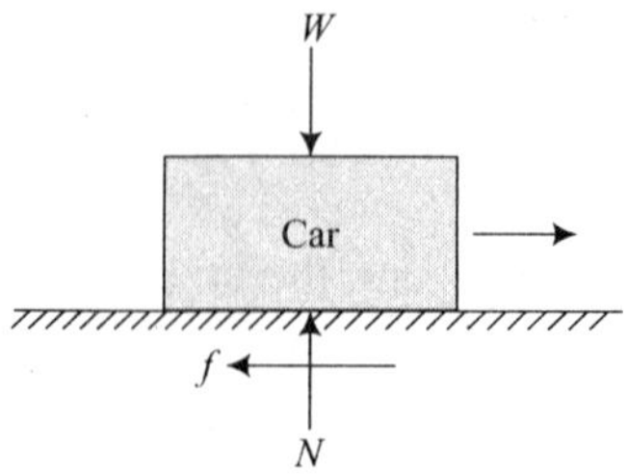

Fig. 8.8 FBD of car

Initial velocity of car, $u = 70$ kmph

$$= \frac{70 \times 1000}{60 \times 60} = 19.44 \text{ m/s}$$

Final velocity $\quad v = 0$

The force of friction,

$$f = \mu N$$

Applying impulse momentum equation,

$$-ft = \frac{W}{g}(v - u)$$

$$-20\ \mu t = \left(\frac{20}{9.81}\right)(19.44 - 0)$$

$\therefore \quad t =$ **2.64 seconds Ans.**

Example 8.7: A block of 130 N weight slides down an inclined plane with a slope 5 vertical to 12 horizontal. The coefficient of friction at contact surface is 0.3. Find out the velocity of block after 5 seconds if initial velocity is 2.4 m/s.

Solution:

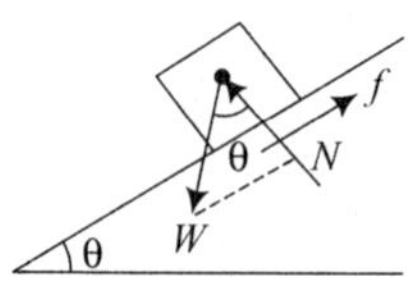

Fig. 8.9 Inclined plane

The angle of slope of inclined plane

$$\theta = \tan^{-1}\frac{5}{12} = 22.62°$$

$$N = W \cos \theta = 130 \cos 22.62°$$

$$= 120 \text{ N}$$

$$f = \mu N = 0.3 \times 120 = 36\text{N}$$

Total downward force

$$R = W \sin \theta - f = 130 \sin 22.62° - 36 = 14 \text{ N}$$

Applying impulse momentum equation,

$$Rt = \frac{W}{g}(v - u)$$

$$14 \times 5 = \frac{130}{9.81}(v - 2.4)$$

$\therefore$ $v =$ **7.68 m/s Ans.**

Example 8.8: A cricket ball is bowled horizontally with a velocity of 20 m/s. The ball was hit with a velocity of 48 m/s at an angle of 30° with the horizontal. Find the force exerted by the bat if the impact lasts for 0.02 sec and weight of ball is 1 N.

Solution: The free body diagram is shown in Fig. 8.10.

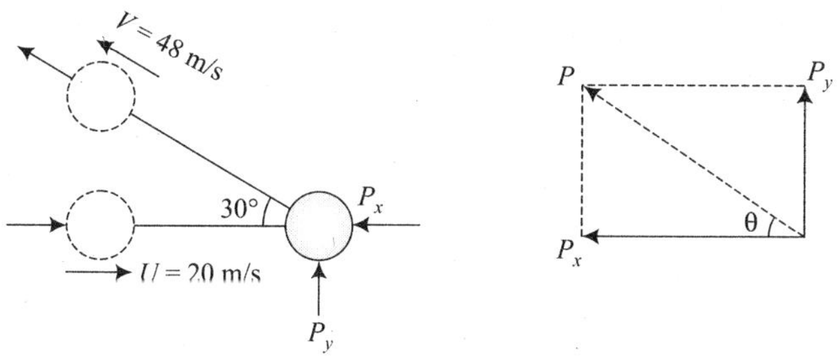

Fig. 8.10 Force diagram (Example 8.8)

Applying impulse momentum equation in horizontal direction,

$$P_x t = \frac{W}{g}(v - u)$$

$$P_x \times 0.02 = \frac{1}{9.81}[48 \cos 30° - (-20)]$$

$$P_x = 313.8 \text{ N}$$

Applying impulse momentum equation in vertical direction

$$P_y t = \frac{W}{g}(v - u)$$

$$P_y \times 0.02 = \frac{1}{9.81}[48 \sin 30° - 0]$$

$\therefore$
$$P_y = 122.3 \text{ N}$$

Resultant force

$$P = \sqrt{P_x^2 + P_y^2} = \sqrt{(313.8)^2 + (122.3)^2}$$

$$= 336.8 \text{ N}$$

Direction of resultant force

$$\theta = \tan^{-1}\frac{P_y}{P_x} = \tan^{-1}\left(\frac{122.3}{313.8}\right) = \mathbf{21.3°} \quad \textbf{Ans.}$$

Example 8.9: The block *A* is moving in the right direction with a velocity of 4 m/s due to block *B* pulling down. The coefficient of friction is 0.2 and pulley is ideal. A force *P* is applied to move the block *A* in left direction with a velocity of 6 m/s in a time interval of 20 sec. Calculate the value of force *P* applied.

Solution:

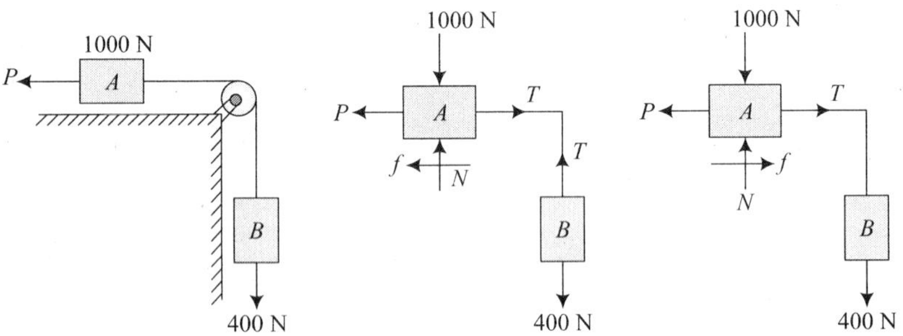

Fig. 8.11 System (Example 8.9)

Step I The block *A* is moving rightward and *f* is acting leftward. The force *P* is applied to bring the block *A* to stationary position.

Apply impulse momentum equation

$$N = W = 1000 \text{ N}$$

$$f = \mu N = 0.2 \times 1000 = 200 \text{ N}$$

$$T = 400 \text{ N}$$

$$(T - f - P)t_1 = \frac{W + T}{g}(v - u)$$

$$(400 - 200 - P)t_1 = \frac{1000 + 400}{9.81}(0 - 4)$$

$$(P - 200)t_1 = \frac{5600}{9.81} \qquad ...(1)$$

Step II The block *A* moves leftward with a velocity of 6 m/s after 20 seconds.

Apply impulse momentum equation

$$(P - f - T)(20 - t_1) = \frac{W + T}{g}(v - u)$$

$$(P - 200 - 400)(20 - t_1) = \frac{1000 + 400}{9.81}(6 - 0)$$

$$(P - 600)(20 - t_1) = \frac{1400}{9.81} \times 6 \qquad ...(2)$$

Solving equations (1) and (2) simultaneously

$$t_1 = 1.28 \text{ sec}$$

$P =$ **645.7 N Ans.**

Example 8.10: Three men each weighing 700 N are travelling in a car of 11 kN running at 10 m/s. The three men jump off the car simultaneously gaining a relative velocity of 5 m/s with the car. Find the speed of the car after the men have jumped off.

Solution: Applying principle of conservation of momentum,

$$(11000 + 700 \times 3)10 = 11000v + 3 \times 700\,(v - 5)$$

$$131000 = 13100v - 10500$$

$$v = \frac{131000 + 10500}{13100} = 10.8 \text{ s}$$ **Ans.**

Example 8.11: A gun fires a projectile with a velocity of 300 m/s. The weight of gun and projectile are 300 kN and 5 kN respectively. With what velocity the gun will recoil?

The recoil of the gun is stopped with a force of 600 kN. Calculate the distance travelled and time taken.

Solution: (a) Applying principle of conservation of momentum to the system of gun and projectile

Initial momentum = Final momentum

$$0 = 300 \times v + 5 \times 300$$

$$\therefore \quad v = -5 \text{ m/s}$$

The velocity of recoil of gun = 5 m/s in opposite direction.

(b) Using work energy equation

$$-600\,s = \frac{300}{2\times 9.81}\,(0-5^2)$$

$$s = 0.637 \text{ m}$$

Applying impulse momentum equation to gun only

$$-600t = \left(\frac{300}{9.81}\right)(0-5)$$

$$t = \mathbf{0.255\ sec} \quad \textbf{Ans.}$$

Example 8.12: A pile hammer of weight 20 kN drops on a pile of 10 kN from a height of 750 mm. The pile penetrates 100 mm/blow. Find the resistance of the ground against penetration if the resistance is constant.

Solution: Velocity of hammer at the time of strike

$$= \sqrt{2gh} + u$$

$$= \sqrt{2\times 9.81\times \frac{750}{1000}} + 0 = 3.84 \text{ m/s}$$

Applying principle of conservation of momentum to pile and hammer,

$$\frac{20}{9.81}\times 3.84 = \frac{20+10}{9.81}\,v$$

where, v = velocity of pile and hammer immediately after the impact

$\therefore$ $v = 2.56$ m/s

Applying work energy equation to the motion of hammer and pile

$$(20+10-R)\,S = \frac{20+10}{2g}\,(0-v^2)$$

where, R = Ground resistance against penetration

$$(30-R)\,0.1 = \frac{30}{2\times 9.81}\,(-2.56)^2$$

$$R = 130 \text{ kN}$$

8.5 D'ALEMBERT'S PRINCIPLE

If the sum of external forces, external couples and inertia forces and inertia couples is zero, the body is said to be in dynamic equilibrium. This concept is known as D'Alembert's Principle. Dynamic problem can be converted into static equilibrium problem by applying D'Alembert's Principle. This is done by introducing an additional force of inertia and applying equations of static equilibrium to solve the problem.

This principle is very useful in the solution of kinetic problems in the following areas.

1. Dynamic equilibrium of blocks moving horizontally.
2. Lifts and elevators
3. Motion of connected bodies on horizontal and inclined planes.
4. Pulleys and weights
5. Motion of vehicles

8.5.1 Equations of Dynamic Equilibrium

The equations of motion of a rigid body in plane motion are:

$$\Sigma \vec{F} = m\vec{a}$$

$$\Sigma \vec{M}_G = I_G \vec{\alpha}$$

For translatory motion,

$$F_x = ma_x$$

$$F_y = ma_y$$

For rotary motion,

$$M_G = I\alpha$$

where, F_x and F_y are external forces

M is the moment due to external forces

a_x and a_y are components of linear acceleration

α is the angular acceleration

I is the moment of inertia about axis of rotation

The equations of dynamics equilibrium are

$$F_x + (-ma_x) = 0$$

$$F_y + (-ma_y) = 0$$

$$M + (-I\alpha) = 0$$

where terms in the bracket are forces and couples of inertia.

The body is brought to dynamic equilibrium by applying inertia forces and inertia couples. Inertia forces are applied to the mass centre (or c.g.) of the body in a direction opposite to the direction of acceleration. Inertia couple can be applied anywhere in the plane of motion of the body but in a direction opposite to the direction of the angular acceleration of the body.

Example 8.13: A force of 250 N is acting on a body of mass 100 kg. Find the acceleration of the body.

Solution:

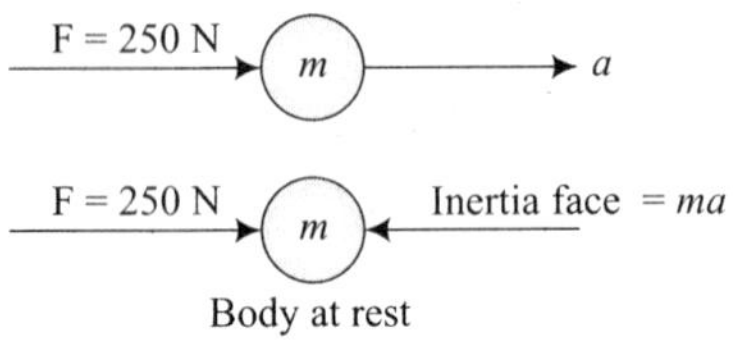

Fig. 8.12 (Example 8.13)

Applying a friction force equal to *ma* in the opposite direction of motion to bring it to rest.

Equation of dynamic equilibrium of the body,

$$\Sigma Fa - ma = 0$$

$$250 - 100a = 0$$

$\therefore$ $a = \mathbf{2.5\ m/s}$ **Ans.**

Example 8.14: Two bodies of mass 80 kg and 20 kg are connected by a thread and move along a rough horizontal plane under the action of a force of 400 N applied to first body of mass 80 kg. The coefficient of friction between the sliding surface of bodies and plane is 0.3. Determine the acceleration of the two bodies and tension in the thread, using D'Alembert's Principle.

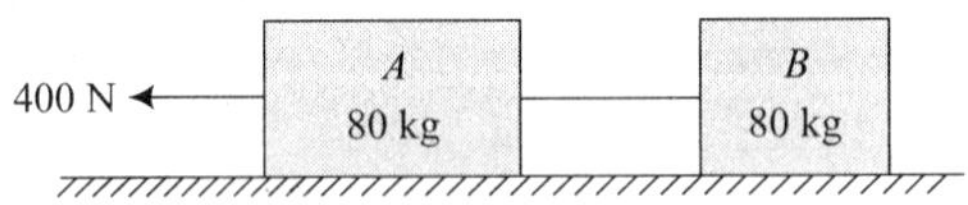

Solution: (1) Draw free body diagram of block *A*.

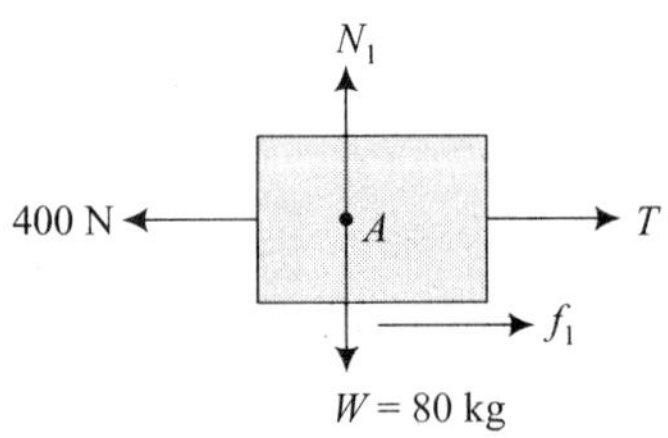

Fig. 8.13 FBD of block *A*

$$N_1 = W = 80 \times 9.81 = 784 \text{ N}$$

$$f_1 = \mu N_1 = 0.3 \times 784 = 235.2 \text{ N}$$

Resultant horizontal force

$$R_1 = -P + T + f_1$$

$$R_1 = -400 + T + 235.2$$
$$= T - 164.8$$

As per D'Alembert's Principle,

$$R_1 - m_1 a = 0$$
$$(T - 164.8) + 80a = 0$$
$$\therefore \quad T = 164.8 - 80a \qquad ...(1)$$

(2) Draw free body diagram of block B

$$N_2 = 20 \times 9.81$$
$$f_2 = \mu N_2 = 0.3 \times 20 \times 9.81$$
$$= 58.8 \text{ N}$$

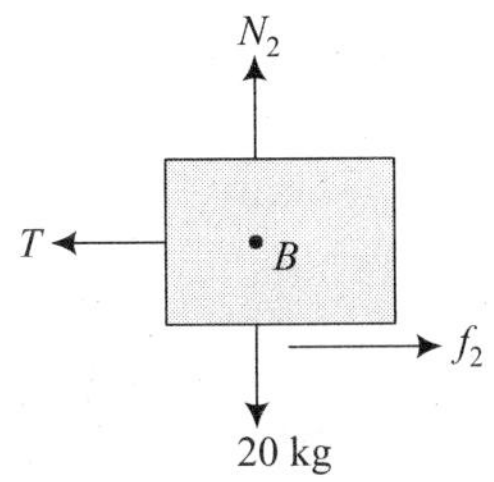

Fig. 8.14 FBD of block B

Resultant horizontal force,

$$R_2 = f_2 - T$$

As per D'Alembert's Principle,

$$R_2 - m_2 a = 0$$
$$(58.8 - T) - 20a = 0$$
$$\therefore \quad T = 58.8 + 20a \qquad ...(2)$$

From equations (1) and (2)

$$a = 1.06 \text{ m/s}^2$$
$$T = 58.8 + 20 \times 1.06 = \mathbf{80\ N} \quad \textbf{Ans.}$$

Example 8.15: Two weights of 300 N and 450 N are hung to the two ends of a rope passing over an ideal pulley as shown. With what acceleration the heavier weight comes down? What is the tension in the rope?

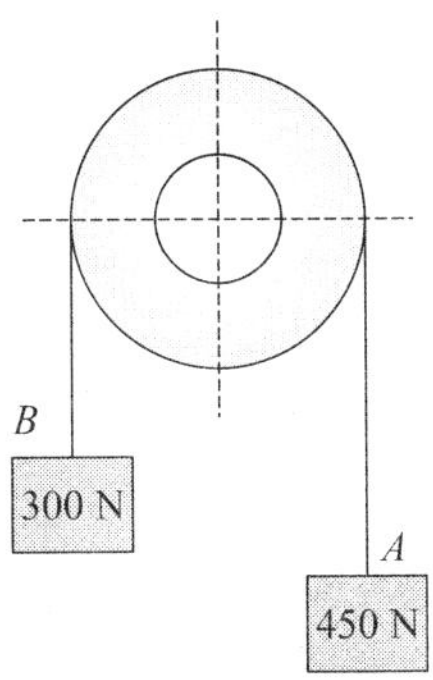

Solution: (1) Draw free body diagram of weight *A*.

Resultant vertical force

$$R_1 = 450 - T$$

As per D' Alembert's Principle,

$$R_1 - m_1 a = 0$$

$$(450 - T) - \frac{450}{9.81} a = 0$$

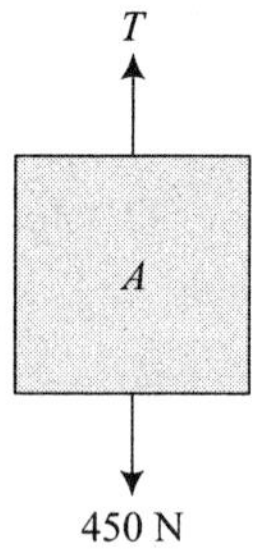

Fig. 8.15 FBD of *A*

$$T = 450 - \frac{450}{9.81} a \quad ...(1)$$

(2) Draw free body diagram of weight *B*.

Resultant vertical force

$$R_2 = T - 300$$

As per D' Alembert's Principle,

$$R_2 - m_2 a = 0$$

$$(T - 300) - \frac{300}{9.81} a = 0$$

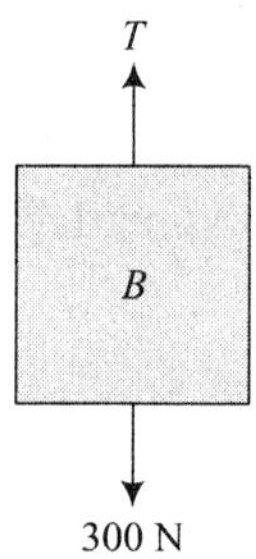

Fig. 8.16 FBD of *B*

From equation (1)

$$T = 450 - \frac{450}{9.81} a$$

$$\therefore \quad \left(450 - \frac{450}{9.81} a\right) - \frac{300}{9.81} a = 0$$

$$150 - \frac{750}{9.81} a = 0$$

$$\therefore \quad a = \frac{150 \times 9.81}{750} = 1.962 \text{ m/s}^2$$

$$T = 450 - \frac{450}{9.81} \times 1.962 = \mathbf{360\ N} \quad \textbf{Ans.}$$

Example 8.16: Determine tensions in strings and acceleration of blocks *P* and *Q* weighing 300 N and 60 N connected through string and passing over the frictionless, weightless pulley as shown.

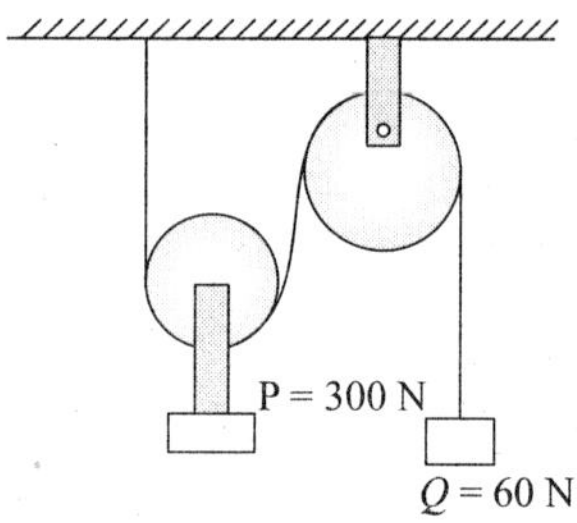

Solution: (1) Draw FBD of block Q

Resultant vertical force

$$R_1 = T - Q = T - 60$$

As per D' Alembert's Principle

$$R_1 - m_1 a = 0$$

$$(T - 60) - \frac{60}{9.81}a = 0$$

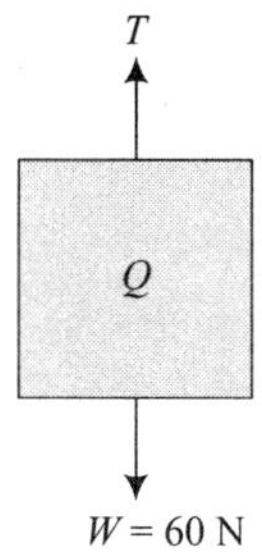

Fig. 8.17 FBD of Q

$$\therefore \qquad T = 60 + \frac{60}{9.81}a \qquad ...(1)$$

(2) Draw *FBD* of block P

Resultant vertical force

$$R_2 = P - 2T = 300 - 2T$$

As per D' Alembert's Principle

$$R_2 - m_2 a = 0$$

$$(300 - 2T) - \frac{300}{9.81}a = 0$$

$$T = \frac{300}{2} - \frac{300}{9.81 \times 2}a$$

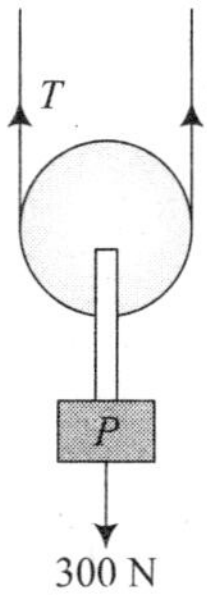

Fig. 8.18 *FBD of P*

Substituting the value of T from (1)

$$60 + \frac{60}{9.81}a = \frac{300}{2} - \frac{300}{9.81 \times 2}a$$

$$a = 3.68 \text{ m/s}^2$$

Acceleration of $Q = 2a = 6.54 \text{ m/s}^2$

Acceleration of $P = a = 3.68 \text{ m/s}^2$

$$T = 82.5 \text{ N}$$

Example 8.17: A body is weighing 196.2 N slides up a 30° inclined plane under the action of an applied force 300 N acting parallel to the inclined plane. The coefficient of friction is 0.2. The body moves from rest. Determine:

(a) Acceleration of the body.
(b) Distance travelled by the body in 4 seconds.
(c) Velocity of body after 4 seconds.
(d) Kinetic energy of the body after 4 seconds.
(e) Work done on the body in 4 seconds.
(f) Impulse applied in 4 seconds.

Solution:

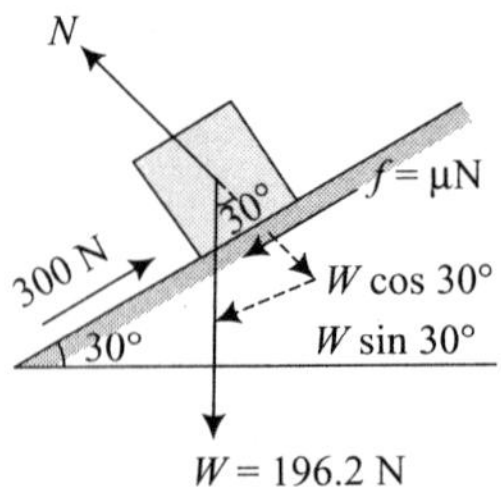

Fig. 8.19 Inclined plane

The resultant force parallel to inclined plane

$$R = P - W \sin 30° - f$$

$$= 300 - 196.2 \sin 30° - 0.2\,(196.2 \cos 30°)$$

$\therefore \quad N = W \cos 30°$

$\therefore \quad R = 300 - 98.1 - 33.98 = 167.92 \text{ N}$

(a) Applying D'Alembert's Principle

$$R - ma = 0$$

$$167.92 - \frac{196.2}{9.81}\,a = 0$$

$\therefore \quad a = 8.4 \text{ m/s}^2$

(b) $\quad S = ut + \frac{1}{2}at^2$

$$= 0 + \frac{1}{2} \times 8.4 \times 4^2 = 134.4 \text{ m}$$

(c) $v = u + at = 0 + 8.4 \times 4 = 33.6$ m/s

(d) K.E. $= \frac{1}{2}mv^2 = \frac{1}{2}\frac{196.2}{9.81}(33.6)^2 = 11.3$ kJ

(e) W.D. $= P \times S = 300 \times 134.3 = 40.3$ kJ

(f) Impulse $= (mv - mu) = \frac{196.2}{9.81}(33.6 - 0) = 672$ NS.

Example 8.18: A man weighing N newton entered a lift which moves with an acceleration of a m/s^2. Find the force exerted by the man on the floor of the lift when

(a) Lift is moving downwards

(b) Lift is moving upwards

Solution: (a) *Lift moving downwards*

The resultant vertical force

$$R = R_1 - W$$

Applying D'Alembert's principle

$$R + ma = 0$$

$$R_1 - W + \frac{W}{g}a = 0$$

$$R_1 = W\left(1 - \frac{a}{g}\right)$$

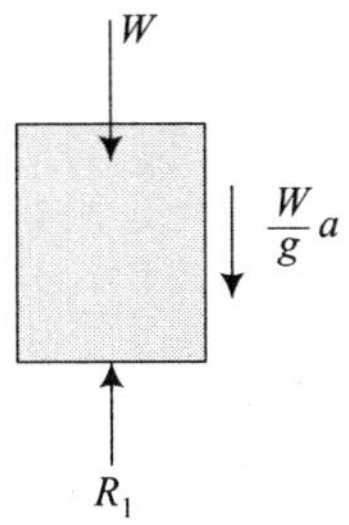

Fig. 8.20 Lift moving downwards

where, R_1 = Force exerted by man on left.

(b) *Lift moving upwards*

The resultant vertical force

$$R = R_2 - W$$

Applying D'Alembert's principle

$$R - ma = 0$$

$$R_2 - W - \frac{W}{g}a = 0$$

$$R_2 = W\left(1 + \frac{a}{g}\right)$$

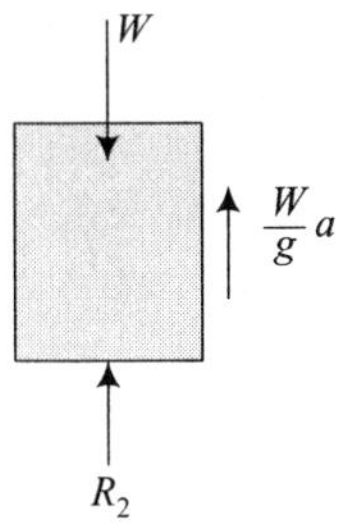

Fig. 8.21 Lift moving upwards

Example 8.19: Two bodies *A* and *B* are connected by a thread and move along a rough horizontal plane ($\mu = 0.3$) under the action of a force 400 N applied to the body *B* shown. Determine the acceleration of the two bodies and the tension in the thread, using D'Alembert's principle. [U.P.T.U. 2008-09]

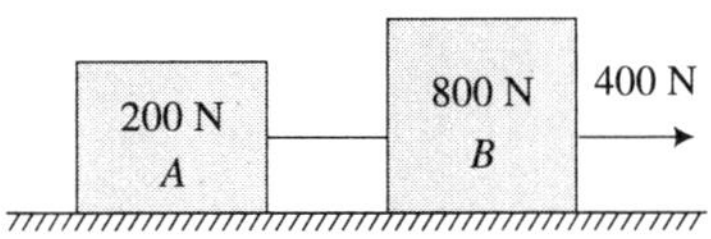

Solution:

(i) *Draw FBD of block B*

$$N_1 = W = 800 \text{ N}.$$

$$F_1 = \mu\, N_1 = 0.3 \times 800 = 240 \text{ N}.$$

Resultant horizontal force,

$$R_1 = P - T - f_1 = 400 - T - 240 = 160 - T$$

As per D'Alembert's principle,

$$R_1 - m_1\, a = 0$$

$$(160 - T) - \frac{800}{9.81}\, a = 0$$

$$T = 160 - \frac{800}{9.81}\, a \qquad \text{...(1)}$$

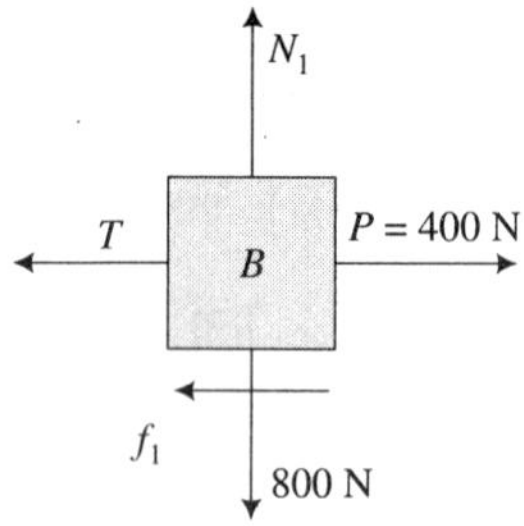

(ii) *Draw FBD of block A*

$$f_2 = \mu\, N_2 = 0.3 \times 200 = 60 \text{ N.}$$

Resultant horizontal force,

$$R_2 = T - f_2 = T - 60$$

As per D'Alembert's principle,

$$R_2 - m_2\, a = 0$$

$$(T - 60) - \frac{200}{9.81}\, a = 0$$

$$\therefore \qquad T = 60 + \frac{200}{9.81}\, a \qquad \text{...(2)}$$

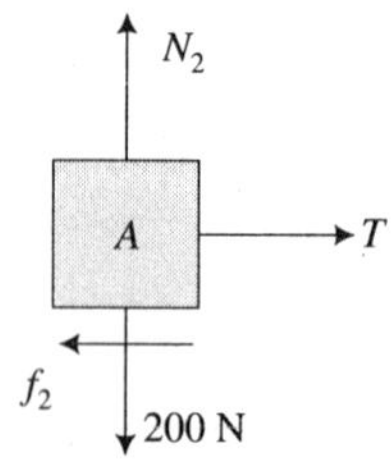

Equating T from equations (1) and (2)

$$160 - \frac{800}{9.81}a = 60 + \frac{200}{9.81}a$$

$$\left(\frac{800+200}{9.81}\right)a = 160 - 60 = 100$$

$$\therefore \quad a = \frac{100 \times 9.81}{1000}$$

$$= \mathbf{0.981\ m/s^2} \quad \textbf{Ans.}$$

$$\therefore \quad T = 60 + \frac{200}{9.81} \times 0.981$$

$$= \mathbf{80\ N} \quad \textbf{Ans.}$$

Example 8.20: Two weights P and Q are connected by the arrangement shown. Neglecting friction and inertia of the pulleys and cord, find the acceleration of the weight Q. Assume $P = 178$ N and $Q = 133.5$ N. [U.P.T.U. 2010-11]

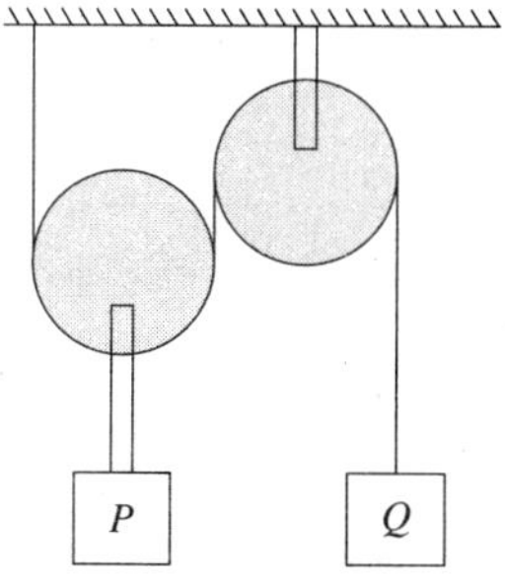

Solution:

(i) *Draw FBD of block Q*

Resultant vertical force,

$$R_1 = T - Q = T - 133.5$$

As per D'Alembert's principle,

$$R_1 - m_1 a_1 = 0$$

$$(T - 133.5) - \frac{133.5}{9.81}a_1 = 0$$

$$\therefore \quad T = 133.5 + \frac{133.5}{9.81}a_1 \qquad ...(1)$$

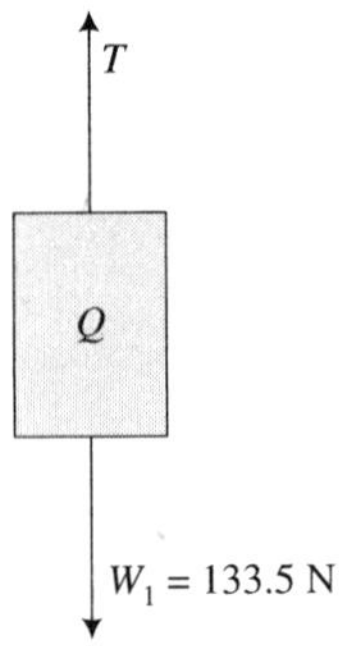

(ii) *Draw FBD of block P*

Resultant vertical force,

$$R_2 = W_2 - 2T = 178 - 2T$$

As per D'Alembert's principle,

$$R_2 - m_2 a_2 = 0$$

$$(178 - 2T) - \frac{178}{9.81} a_2 = 0$$

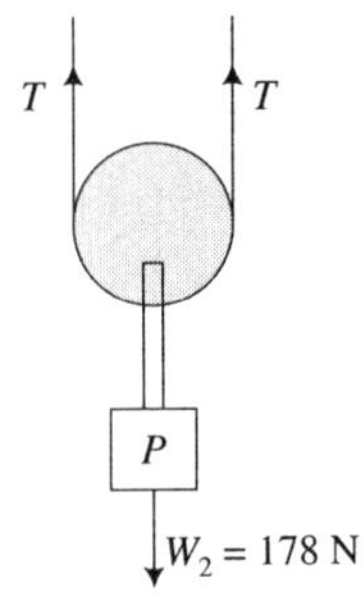

$$\therefore \quad T = \frac{178}{2} - \frac{178}{9.81 \times 2} a_2 \qquad ...(2)$$

From equations (1) and (2)

$$133.5 + \frac{133.5}{9.81} a_1 = \frac{178}{2} - \frac{178}{2 \times 9.81} a_2$$

Now $\quad a_1 = 2a_2$

$$\left(\frac{133.5}{9.81} + \frac{178}{9.81 \times 2}\right) a_2 = \frac{178}{2} - 133.5 = -44.5$$

$$\therefore \quad a_2 = -\frac{44.5 \times 9.81}{222.5} = -1.96 \text{ m/s}^2$$

$$\therefore \quad a_1 = 2a_2 = \mathbf{-3.92\ m/s^2} \quad \textbf{Ans.}$$

Example 8.21: A body of mass 25 kg resting on a horizontal table is connected by a string passing over a smooth pulley at the edge of the table to another body of mass 3.75 kg and hanging vertically as shown. Initially, the friction between 25 kg mass and the table is just sufficient to prevent the motion. If an additional mass of 1.25 kg is added to 3.75 kg mass, find the acceleration of the mass.

[U.P.T.U. 2008-09]

Solution:

(i) *Initial condition*

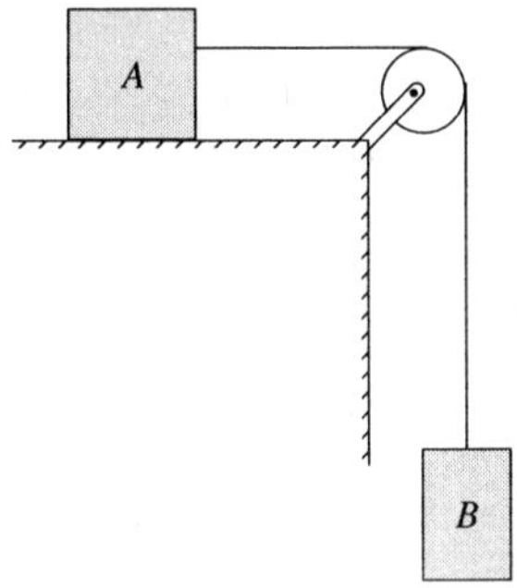

Draw FBD of block *A*

$$T = 3.75 \times 9.81 = 36.7875 \text{ N}$$

$$N = W = 25 \times 9.81 = 245.25 \text{ N}$$

$$f = T = 36.7875$$

$$\therefore \quad \mu N = f$$

$$\therefore \quad \mu = \frac{f}{N} = \frac{36.7875}{245.25} = 0.15$$

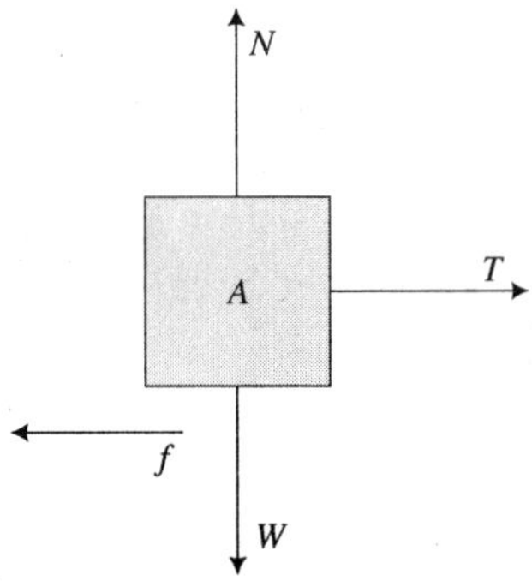

(ii) *Final condition*

$$T = (3.75 + 1.25)\ 9.81$$

$$= 49.05 \text{ N.}$$

The resultant horizontal force

$$R = T - f$$

$$= 49.05\ \mu\text{N}$$

$$= 49.05 - (0.15 \times 245.25)$$

$$= 12.2625 \text{ N.}$$

Applying D'Alembert's principle

$$R - ma = 0$$

$$12.2625 - 25\,a = 0$$

$$\therefore \quad a = \frac{12.2625}{25} = \mathbf{0.49\ m/s^2\ Ans}$$

Example 8.22: A cylinder shown is 70 cm in diameter and weighs 500 N. It is rotating about fixed axis *O* and has an angular velocity of 7 rad/s at the given instant. Using D'Alembert's principle, find the horizontal and vertical components of reaction at *O*. [U.P.T.U. 2009-10]

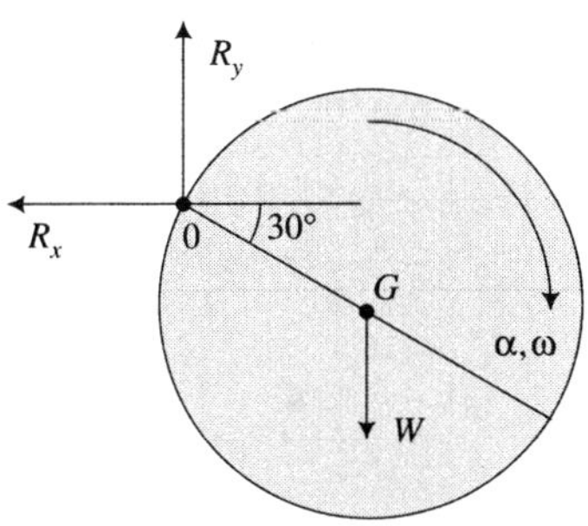

Solution:

As per D'Alembert's principle, a rigid body undergoing a rotary motion,

$$\sum M_O = I_o\,\alpha$$

$$W\,r\cos 30^\circ = (I_G + mh^2)\alpha$$

$$= \left(\frac{1}{2}mr^2 + mr^2\right)\alpha$$

$$= \frac{3}{2}mr^2\alpha$$

$$\therefore \quad \alpha = \frac{mgr\cos 30^\circ}{3/2mr^2} = \frac{2}{3}\frac{g}{r}\cos 30^\circ = \frac{2}{3}\times\frac{9.81}{0.35}\cos 30^\circ$$

$$= \mathbf{16.18\ rad/s^2}\quad \textbf{Ans.}$$

For translatory motion,

Resultant force $- ma = 0$

$$R_y - mg - ma = 0$$

$$R_y = mg + ma = 500 + \frac{500}{9.81}(r\alpha)$$

$$= 5\text{ N} + \frac{500}{9.81}(0.35\times 16.18) = 788.634\text{ N}$$

$$R_x = W\cos 30^\circ = 500\cos 30^\circ = \mathbf{433\ N}\quad \textbf{Ans.}$$

QUESTION BANK NO. 8

1. Define force. What are the main characteristics of a force? Explain with an example.
2. State and explain work energy principle. What are its applications?
3. Define Impulse, Momentum, Impulse Momentum Principle and Principle of Conservation of Momentum. What are the applications of these principles?
4. Explain the following:
   (a) D'Alembert's Principle
   (b) Equation of Dynamic Equilibrium
   (c) Applications of D'Alembert's Principle
5. Explain various forms of Mechanical Energy. What is conservation of energy?

QUICK REVIEW INFORMATION NO. 8

1. The principle of conservation of energy cannot be applied if a body is sliding down a rough inclined plane.
2. D'Alembert's principle is used to reduce kinematic problem to equivalent static problem.
3. A body of mass 2.5 kg is moving with a constant velocity of 5 m/s. In order to bring it to rest at a distance of 4 m, the work done is 31.25 Nm and the force required is 7.8 N.

Unit V: Stress and Strain Analysis

CHAPTER 9

Simple and Compound Stresses and Strains

9.1 INTRODUCTION

The knowledge of strength of materials is very essential for the design of machines and structures. It has been observed that when a force is applied to a material, it first deforms and then failure takes place. Different types of metallic and non-metallic materials are available for use. The ability of a material to resist failure under the action of external forces or loads is called the strength of material.

9.1.1 Strength or Internal Resistance of Material

If a bar AB is pulled by an external force P' at both ends, the bar will be elongated (though negligible and may not be visible by naked eye) and will be under threat of breakage say at C. For equilibrium of portion of bar AC, there should be a force

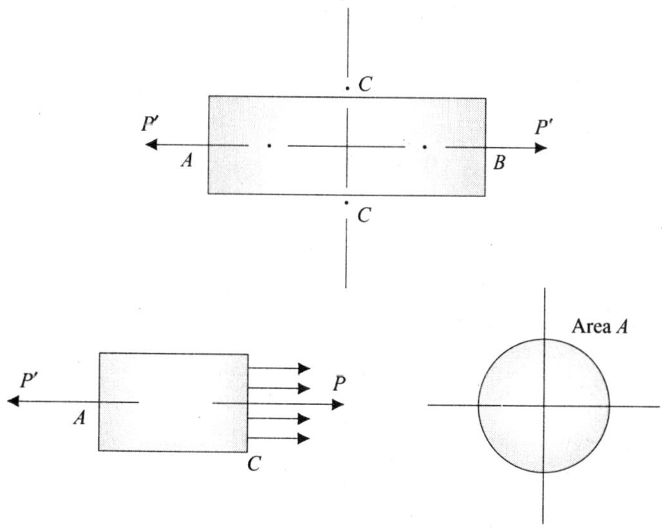

Fig. 9.1 Internal resistance

equal and opposite to force P' applied at section C. This force P is developed due to internal cohesion and attraction among molecules of the material. This is called internal resistance of the material and is measured as force per unit area.

$$P' = P = \sigma A$$

where, A = Area of cross-section of bar

σ = Internal resistance called stress.

If P developed is less than applied force P', failure will take place. The internal resistance (or strength) σ of the material depends upon the properties of the material and the total resistance, P depends upon σ and area of cross-section, A.

9.1.2 Elasticity

If external force, P' is removed, the internal resistance σ disappears and the bar springs back to its original length. This property of a material to spring back to its original size is called elasticity.

If there is no residual deformation (size and shape) left in the body after removal of external force, the material is called perfectly elastic. If there is some residual deformation left in the material after removal of force, it is called partially elastic. In most of the cases, elastic materials are used in the construction of machines and structures.

The strength and elasticity are exclusive properties of a material and depend upon the arrangement and distribution of molecules in the crystal lattice of the material.

9.2 LOADING OF AN ELASTIC MATERIAL

Whenever an elastic member of a machine or a structure is subjected to a load, the following effects are caused due to internal resistance of the material.

1. A stress is produced which is equal to force per unit area.

$$\sigma = \frac{P}{A} \text{ [N/mm}^2\text{]}$$

2. Deformation δL or strain $\frac{\delta L}{L}$. This depends upon the modulus of elasticity of the material, E.

$$\varepsilon = \frac{\sigma}{E}$$

3. Strain energy is stored within the material because of work done by the force in straining the material. The strain energy will depend upon the way the load is applied. For a gradually applied load,

$$\text{Strain energy} = \text{Work done}$$

$$U = \frac{1}{2}\text{stress} \times \text{strain}$$

$$= \frac{\sigma\varepsilon}{2}.$$

9.2.1 Stress

The external force, P can be measured in N, kN, MN or GN.

where, $\text{kN} = 10^3$ N (Kilo)

$\text{MN} = 10^6$ N (Mega)

$\text{GN} = 10^9$ N (Giga)

Similarly, depending upon magnitude of the force, the area of cross-section of the material may be measured as mm^2 or cm^2 or m^2.

Therefore, stress is measured as N/mm^2 or MN/m^2, etc.

where, $1 \text{ N/mm}^2 = 10^6 \text{ N/m}^2 = 1 \text{ MN/m}^2 = 1$ MPa.

and $1 \text{ N/m}^2 = 1$ Pascal = 1 Pa.

Types of Stress

The stress can be classified as simple stress or compound stress, normal (direct) stress and shear (tangential) stress and direct stress as tensile stress or compressive stress.

1. **Simple stress:** If the external load is applied on a body in one direction only, the stress developed is called simple stress.
2. **Compound stress:** If the external loads are applied in more than one direction, the stresses developed are called compound stresses.
3. **Normal stress (σ):** When a force acts longitudinally along the axis of the body, it produces a normal or direct stress which acts perpendicularly on the surface on which force is acting.
4. **Shear stress (τ):** When a force acts tangentially on the surface of the body, it produces shear stress parallel to the surface on which the force acts.
5. **Tensile stress:** If the length of the body increases due to the action of normal force or longitudinal force, the stress produced is called tensile stress and the force producing it is called tensile force or a pull.

$$\sigma_t = \frac{P}{A}$$

6. **Compressive stress:** If the length of the body decreases due to the action of the normal force or longitudinal force, the stress produced is called compressive stress and the force producing it is called compressive force or a push.

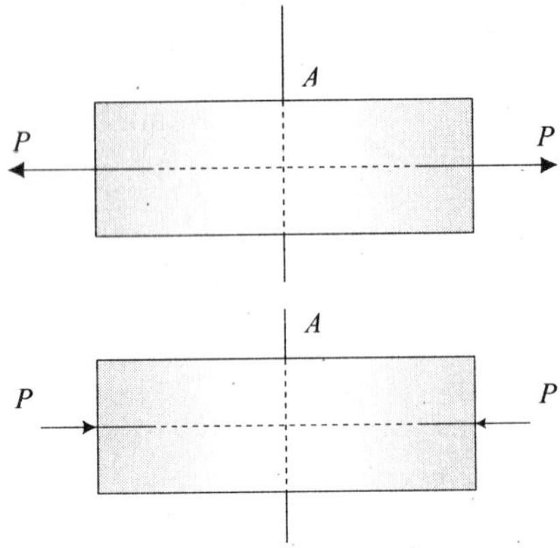

Fig. 9.2 Tensile and compressive stresses

9.2.2 Strain

There is a change in the length or dimension of the body subjected to external force or load. The ratio of change in length (δL) to the original length (L) of the material is called strain.

$$\varepsilon = \frac{\delta L}{L}$$

Tensile strain is the ratio of increase of length to its original length when subjected to tensile loading.

Compressive strain is the ratio of decrease in length to the original length of the material when subjected to compressive loading.

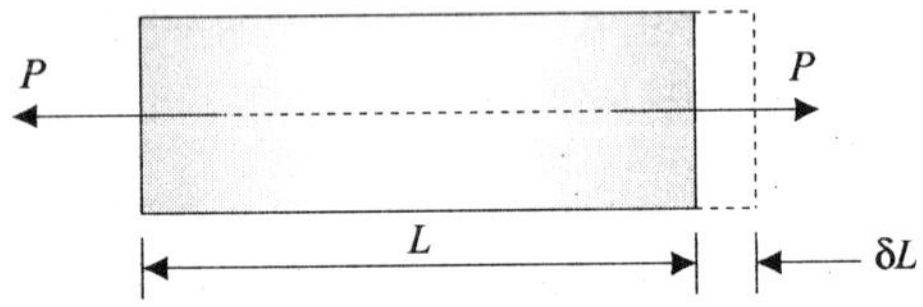

Fig. 9.3 Strain

9.2.3 Modulus of Elasticity

Hooke's law states that when a material is loaded within its elastic limit, the stress is proportional to strain. If a material is loaded within elastic limit, the body will return to its original size when load is removed.

$$\frac{\text{Stress}}{\text{Strain}} = \text{a constant} = E \qquad \text{(say)}$$

$$\therefore \qquad E = \frac{\sigma}{\varepsilon}$$

E is called modulus of elasticity or Young's modulus and has the same unit as stress. The value of *E* is more for ductile materials than for brittle materials. It is same for tensile and compressive loading.

Deformation of a body under external load

$$\text{Stress, } \sigma = \frac{P}{A}$$

$$\text{Strain, } \varepsilon = \frac{\delta L}{L}$$

Young's modulus,

$$E = \frac{\sigma}{\varepsilon}$$

$$\therefore \qquad \varepsilon = \frac{\sigma}{E} = \frac{P}{AE}$$

Now,
$$\varepsilon = \frac{\delta L}{L}$$

Deformation,

$$\delta L = \varepsilon L = \frac{\sigma L}{E} = \frac{PL}{AE}$$

The values of *E* for some important materials are given in Table 9.1.

Table 9.1 Young's modulus for engineering materials

Material	*Young's modulus (E) (N/mm²)*
1. Mild ctool	2.1×10^5
2. Cast iron	1.3×10^5
3. Aluminium	0.7×10^5
4. Copper	1.0×10^5
5. Timber	0.1×10^5

Young's modulus is a measure of stiffness of material or ductility of the material.

9.2.4 Strain Energy

Strain energy for a gradually applied load can be found out from the product of average force and deformation or area under the load-extension graph (shaded area).

$$U = \frac{P}{2}.\delta L$$

where,
$$P = \sigma.A$$

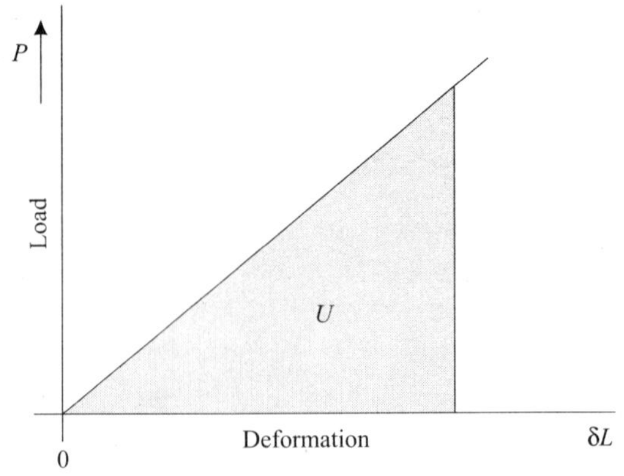

Fig. 9.4 Strain energy

and $$\delta L = \varepsilon L = \frac{\sigma L}{E}$$

$\therefore$ $$U = \frac{1}{2}.\sigma A.\frac{\sigma L}{E} = \frac{1}{2}\frac{\sigma^2}{E}.AL$$

$$= \frac{1}{2}\frac{\sigma^2}{E} \times \text{Volume.}$$

The strain energy per unit volume is called resilience.

$$\text{Resilience} = \frac{1}{2}\frac{\sigma^2}{E}$$

$$= \frac{1}{2}.\sigma.\frac{\sigma}{E} = \frac{1}{2}\sigma.\varepsilon$$

$\therefore$ $$\text{Resilience} = \frac{\text{stress} \times \text{strain}}{2}$$

Resilience is a measure of toughness of a material.

Example 9.1: Find the diameter of a steel wire to raise a load of 10 kN, if the minimum stress is not to exceed 100 MN/m^2.

Solution:

$$\text{Load, } P = 10 \text{ kN} = 10 \times 10^3 \text{ N.}$$

$$\text{Allowable stress, } \sigma = 100 \text{ MN/m}^2 = 100 \text{ N/mm}^2$$

$$\text{Area, } A = \frac{P}{\sigma} = \frac{10 \times 10^3}{100} = 100 \text{ mm}^2$$

Now, $A = \frac{\pi}{4} d^2.$

Diameter of wire,

$$d = \sqrt{\frac{4A}{\pi}} = \sqrt{\frac{4 \times 100}{\pi}} = \mathbf{11.28\ mm} \quad \textbf{Ans.}$$

Example 9.2: A steel wire 2 m long and 3 mm in diameter is extended by 0.75 mm due to a weight suspended from the wire. If the same weight is suspended from a brass wire, 2.5 m long and 2 mm in diameter, it is elongated by 5 mm. Determine the modulus of elasticity of brass, if that of steel is 2×10^5 N/mm^2.

Solution:

(i) *For steel wire*

$$L = 2 \text{ m} = 2000 \text{ mm}$$
$$d = 3 \text{ mm}$$
$$\delta L = 0.75 \text{ mm}$$
$$E = 2 \times 10^5 \text{ N/mm}^2$$

Area of wire,

$$A = \frac{\pi}{4} d^2 = \frac{\pi}{4} (3)^2$$
$$= 7.07 \text{ mm}^2$$

Strain, $\varepsilon = \frac{\delta L}{L} = \frac{0.75}{2000}$

$$= 0.000375$$

Stress, $\sigma = \varepsilon E = 0.000375 \times 2 \times 0^5$

$$= 75 \text{ N/mm}^2$$

Weight, $P = \sigma A = 75 \times 7.07$

$$= 530.\ 25 \text{ N}.$$

(ii) *For brass wire*

$$L = 2.5 \text{ m} = 2500 \text{ mm}$$
$$d = 2 \text{ mm}$$
$$\delta L = 5 \text{ mm}$$
$$P = 530.25 \text{ N}$$

Strain,

$$\varepsilon = \frac{\delta L}{L} = \frac{5}{2500}$$
$$= 2 \times 10^{-3}$$

Stress, $$\sigma = \frac{P}{A} = \frac{530.25}{\frac{\pi}{4}(2)^2}$$

$$= 168.78 \text{ N/mm}^2$$

Modulus of elasticity,

$$E = \frac{\sigma}{\varepsilon} = \frac{168.78}{2 \times 10^{-3}}$$

$$= \mathbf{0.8439 \times 10^5 \ N/mm^2} \quad \textbf{Ans.}$$

Example 9.3: A circular rod of diameter 20 mm and 500 mm long is subjected to a tensile load of 50 kN. The modulus of elasticity for steel may be taken as 200 kN/mm^2. Find the stress, strain and elongation of bar due to applied load.

Solutions:

$$d = 20 \text{ mm}$$

$$L = 500 \text{ mm}$$

$$P = 50 \text{ kN} = 50 \times 10^3 \text{ N.}$$

$$E = 200 \text{ kN/mm}^2 = 2 \times 10^5 \text{ N/mm}^2$$

Area of rod,

$$A = \frac{\pi}{4}(d)^2 = \frac{\pi}{4}(20)^2$$

$$= 314.16 \text{ mm}^2.$$

Stress, $$\sigma = \frac{P}{A} = \frac{50 \times 10^3}{314.16}$$

$$= \mathbf{159.15 \ N/mm^2} \quad \textbf{Ans.}$$

Strain, $$\varepsilon = \frac{\sigma}{E} = \frac{159.15}{2 \times 10^5}$$

$$= \mathbf{0.00079577} \quad \textbf{Ans.}$$

Elongation, $$\delta L = \varepsilon L = 0.00079577 \times 500$$

$$= \mathbf{0.39785 \ mm} \quad \textbf{Ans.}$$

Example 9.4: Calculate the strain energy stored in a bar 250 cm long, 5 cm wide and 4 cm thick, when it is subjected to a tensile load of 6 × 10^4 N. Take $E = 2 \times 10^5$ N/mm^2.

Solution:

$$L = 250 \text{ cm} = 2500 \text{ mm}$$

$$A = 5 \text{ cm} \times 4 \text{ cm} = 50 \text{ mm} \times 40 \text{ mm}$$

$$= 2000 \text{ mm}^2$$

$$P = 6 \times 10^4 \text{ N}$$

$$E = 2 \times 10^5 \text{ N/mm}^2$$

The extension,

$$\delta L = \frac{PL}{AE} = \frac{6 \times 10^4 \times 2500}{2000 \times 2 \times 10^5}$$

$$= 3.75 \text{ mm}$$

The strain energy,

$$U = \frac{P\delta L}{2} = \frac{6 \times 10^4 \times 3.75}{2}$$

$$= 112500 \text{ N-m}$$

$$= 112.5 \text{ kN-m}$$

$$= \mathbf{112.5\ kJ} \quad \textbf{Ans.}$$

Example 9.5: The piston of a steam engine is 300 mm in diameter and the piston rod is of 50 mm diameter. The steam pressure is 1 N/mm^2. Find the stress in the piston rod and elongation in a length of 800 mm. Take $E = 200$ GPa.

Solution:

$$D = 300 \text{ mm}$$

$$d = 50 \text{ mm}$$

$$p = 1 \text{ N/mm}^2$$

$$L = 800 \text{ mm}$$

$$E = 200 \text{ GPa} = 20 \times 10^9 \text{ Pa}$$

$$= 2 \times 10^5 \text{ N/mm}^2 .$$

The effective area of piston for tensile loading of rod,

$$A = \frac{\pi}{4}(D^2 - d^2) = \frac{\pi}{4}(300^2 - 50^2)$$

$$= 68722.34 \text{ mm}^2.$$

Force, $\quad P = pA = 1 \times 68722.34$

$$= 68722.34 \text{ N}$$

Area of piston rod,

$$A_r = \frac{\pi}{4} d^2 = \frac{\pi}{4}(50)^2$$

$$= 1963.5 \text{ mm}^2$$

Stress in piston rod,

$$\sigma = \frac{P}{A_r} = \frac{68722.34}{1963.5}$$

$$= \mathbf{35\ N/mm^2} \quad \textbf{Ans.}$$

Elongation of piston rod,

$$\delta L = \frac{PL}{A_r E} = \frac{68722.34 \times 800}{1963.5 \times 2 \times 10^5}$$

$$= \mathbf{0.14\ mm} \quad \textbf{Ans.}$$

Example 9.6: Three bars of equal length and having cross-sectional areas in the ratio 1:2:4 are all subjected to equal load. Compare their strain energy.

[U.P.T.U. I Sem. 2003–04]

Solution:

The strain energy,

$$U = \frac{P^2 L}{2AE}$$

$$A_1 = p$$

$$A_2 = 2A_1$$

$$A_3 = 4A_1$$

$$\therefore \quad U_1 = \frac{P^2 L}{2A_1 E}$$

$$U_2 = \frac{P^2 L}{2A_2 E} = \frac{P^2 L}{4A_1 E} = \frac{U_1}{2}$$

$$U_3 = \frac{P^2 L}{2A_3 E} = \frac{P^2 L}{8A_1 E} = \frac{U_1}{4}$$

$$\therefore \quad U_1 : U_2 : U_3 = 1 : \frac{1}{2} : \frac{1}{4} \quad \textbf{Ans.}$$

Example 9.7: A metallic rectangular rod 1.5 m long and 40 mm wide and 25 mm thick is subjected to an axial tensile load of 120 kN. The elongation of the rod is measured as 0.9 mm. Calculate the stress, strain and modulus of elasticity.

[U.P.T.U. II Sem. 2003–04]

Solution:

$$L = 1.5\ \text{m} = 1500\ \text{mm}$$

$$A = 40\ \text{mm} \times 25\ \text{mm} = 1000\ \text{mm}^2$$

$$P = 120\ \text{kN} = 120 \times 10^3\ \text{N}$$

$$\delta L = 0.9 \text{ mm}$$

Stress,
$$\sigma = \frac{P}{A} = \frac{120 \times 10^3}{1000}$$
$$= \mathbf{120\ N/mm^2} \quad \textbf{Ans.}$$

Strain,
$$\varepsilon = \frac{\delta L}{L} = \frac{0.9}{1500}$$
$$= 6 \times 10^4$$

Modulus of elasticity,
$$E = \frac{\sigma}{\varepsilon} = \frac{120}{6 \times 10^{-4}}$$
$$= \mathbf{2 \times 10^5\ N/mm^2} \quad \textbf{Ans.}$$

Example 9.8: A 1 m long steel rod of rectangular section 80 mm × 40 mm is subjected to an axial tensile load of 200 kN. Find the strain energy and maximum stress produced in it when the load is applied gradually. Take $E = 2 \times 10^5$ N/mm^2.

[U.P.T.U. II Sem. 2004–05]

Solution:

$$L = 1 \text{ m} = 1000 \text{ mm}$$
$$A = 80 \text{ mm} \times 40 \text{ mm} = 3200 \text{ mm}^2$$
$$P = 200 \text{ kN} = 200 \times 10^3 \text{ N}$$
$$E = 2 \times 10^5 \text{ N/mm}^2$$

Max. stress,
$$\sigma = \frac{P}{A} = \frac{200 \times 10^3}{3200} = 62.5 \text{ N/mm}^2$$

Strain energy,
$$U = \frac{P^2 L}{2AE}$$
$$= \frac{\left(200 \times 10^3\right)^2 \times 1000}{2 \times 3200 \times 2 \times 10^5}$$
$$= 31250 \text{ N-mm}$$
$$= 31.25 \text{ N-m}$$
$$= \mathbf{31.25\ J.} \quad \textbf{Ans.}$$

9.3 SHEAR STRESS

Stress produced by a force tangential to the surface of a body is called shear stress. The shear stress acts parallel to the surface on which the force acts. The corresponding strain is called shear strain.

A cube *ABCDEFG* of slide L is acted upon by a force P along the face *ABEF* tangentially. There is change in the shape of cube to $A'B'CDE'F'G$.

Shear stress is the force divided by surface area *ABEF*.

$$\tau = \frac{P}{L^2}.$$

Shear strain,

$$\gamma = \frac{BB'}{AB} = \frac{BB'}{L}$$

The ratio of shear stress to shear strain is called modulus of rigidity.

$$G = \frac{\tau}{\gamma}$$

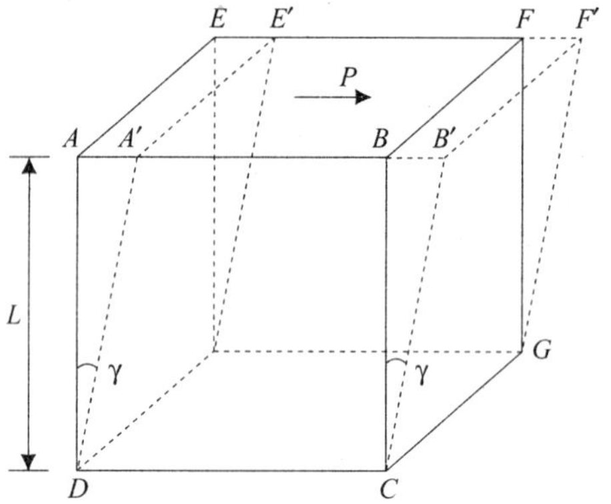

Fig. 9.5 Shear stress and strain

9.3.1 Complimentary Shear Stress

The block *ABCD* of unit thickness is subjected to a shear stress τ (clockwise) on faces *AB* and *CD*. Shear force on faces *AB* and *CD*.

$$= \tau \times AB \text{ or } \tau \times CD.$$

These parallel and equal forces form a couple.

The moment of couple

$$= \text{Force} \times \text{perpendicular distance}$$
$$= (\tau \times AB) \times AD \text{ or } (\tau \times CD) \times AD$$

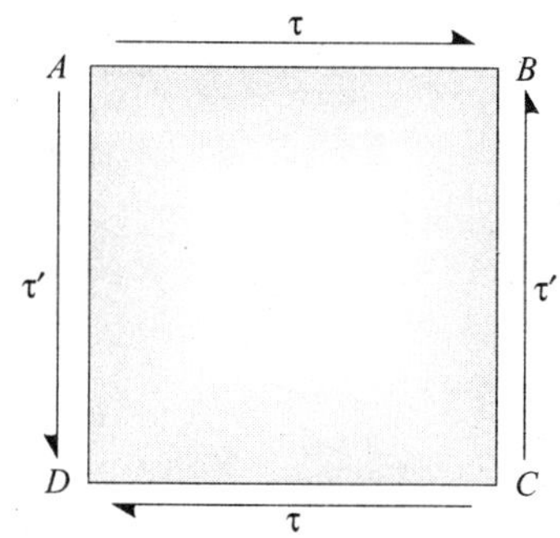

Fig. 9.6 Complimentary shear stress

In order to ensure equilibrium condition, a restoring couple equal and opposite to the above couple is generated.

Shear force acting on face *AD* or *BC*

$$= \tau' \times AD \text{ or } \tau' \times BC \qquad \text{(anticlockwise)}$$

These parallel forces will form the restoring couple.

The moment of the restoring couple

$$= (\tau' \times AD) \times AB \text{ or } (\tau' \times BC) \times AD$$

The moments of the two couples must be equal.

$\therefore \qquad (\tau \times AB) \times AD = (\tau' \times AD) \times AD$

or $\qquad \tau = \tau'.$

The stress τ' is called complimentary shear stress. Therefore, a set of shear stress across a plane is always accompanied by a set of balancing shear stress of the same intensity across the plane and normal to it.

9.3.2 Shear Strain Energy

Let the face *ABCD* of a cube is distorted to *A′B′CD* by a shear force *P*.

Shear strain, $\qquad \gamma = \dfrac{AA'}{AD} = \dfrac{AA'}{L}$

$$AA' = \gamma L$$

Shear stress, $\qquad \tau = \dfrac{P}{\text{Area of top face}} = \dfrac{P}{L^2}$

$\therefore \qquad P = \tau \times L^2$

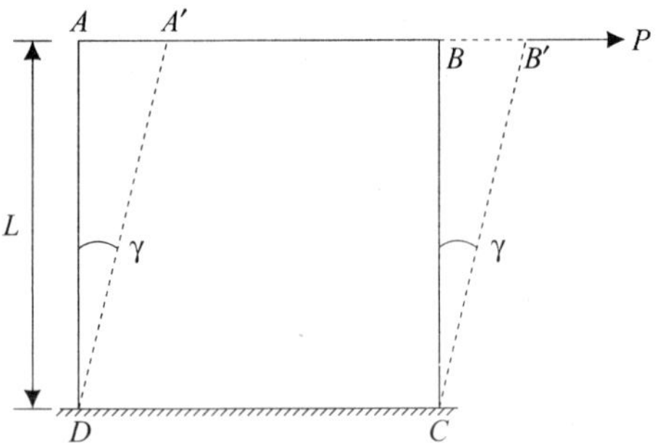

Fig. 9.7 Shear strain energy

Modulus of rigidity of the material of the cube,

$$G = \frac{\text{shear stress}}{\text{shear strain}} = \frac{\tau}{\gamma}$$

The strain energy stored in the cube due to shear stress is equal to work done by shear force, P. As the force is applied gradually, the average force $= \frac{P+o}{2} = \frac{P}{2}$.

Strain energy, $$U_s = \frac{P}{2} \times AA' = \left(\frac{\tau \times L^2}{2}\right) \times (\gamma \times L)$$

$$= \left(\frac{\tau}{2}L^2\right) \times \left(\frac{\tau}{G}.L\right)$$

$$= \frac{\tau^2}{2G} L^3$$

$$= \frac{\tau^2}{2G} \times \text{Volume}$$

Modulus of resilience is strain energy per unit volume

$$= \frac{\tau^2}{2G}$$

9.4 COMPARISON OF NORMAL AND SHEAR STRESS

The main characteristics of normal and shear stress are compared in Table 9.2.

Table 9.2 Comparison of normal and shear stress

Normal stress	*Shear stress*
1. Normal stress or direct stress is produced by longitudinal force.	1. Shear stress is produced by tangential force.
2. 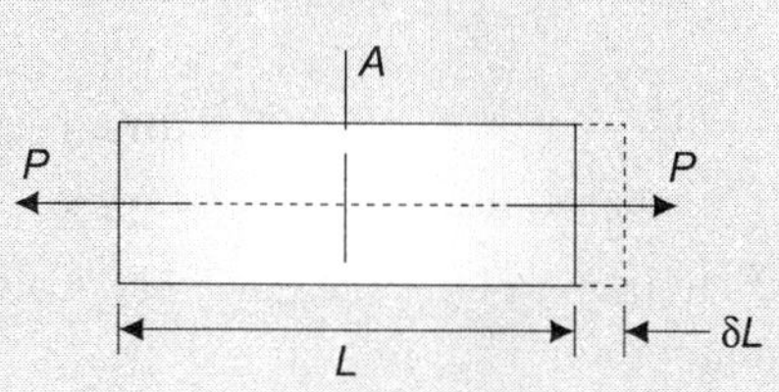 The normal stress is produced on a surface (A) perpendicular to the direction of force $\sigma = \frac{P}{A}$.	2. 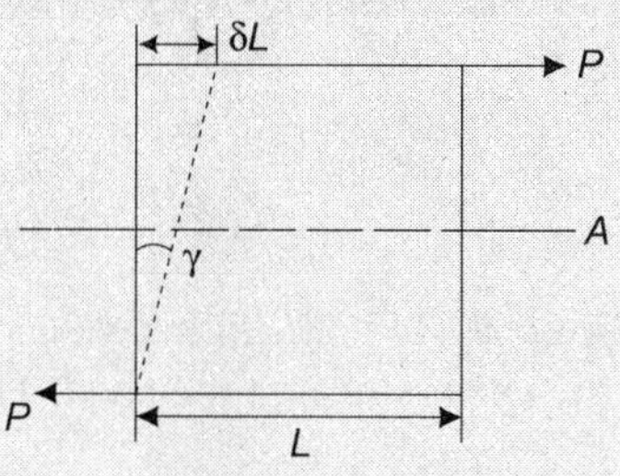 The shear stress is produced on a surface (A) parallel to the direction of force. $\tau = \frac{P}{A}$.
3. The normal strain is measured by the ratio of change of length and original length $\varepsilon = \frac{\delta L}{L}$.	3. The shear strain is measured by the ratio of deformation in the direction of force and original length. $\gamma = \frac{\delta L}{L}$.
4. The elastic constant is called Young's Modulus of Elasticity, $E = \frac{\sigma}{\varepsilon}$.	4. The elastic constant is called Modulus of Rigidity $G = \frac{\tau}{\gamma}$.
5. The strain energy due to tension or compression, $U = \frac{\sigma^2}{2E} \times \text{Volume}$.	5. The shear strain energy, $U_S = \frac{\tau^2}{2G} \times \text{Volume}$.
6. Resilience $= \frac{U}{V} = \frac{\sigma^2}{2E}$.	6. Resilience $= \frac{U_S}{V} = \frac{\tau^2}{2G}$.
7. *Examples* (i) Beams under pure bending. (ii) Bolts of flange of a pressure vessel.	7. *Examples* (i) Shafts under pure torsion (ii) Rivets of a truss.

Example 9.9: A steel punch can sustain a compressive stress of 800 N/mm^2. Find the least diameter of hole which can be punched through a steel plate 10 mm thick if its ultimate shear stress is 350 N/mm^2.

Solution:

Diameter of punch = diameter of hole = d.

The maximum load which can be borne by the punch,

$$P = \frac{\pi}{4} d^2 \sigma_c.$$

The area of the hole which can be sheared by the punch = πdt.

$\therefore$ Force required $= \pi dt\tau$.

Equating the two forces,

$$P = \frac{\pi}{4} d^2\sigma_c = \Pi dt\tau$$

$$\therefore \quad d = \frac{4t\tau}{\sigma_c} = \frac{4 \times 10 \times 350}{800} = \mathbf{17.5\ mm} \quad \textbf{Ans.}$$

Example 9.10: A single riveted lap joint is to transmit a force $P = 5$ kN. Calculate the shear stress in the rivet. The rivet diameter is 12.5 mm.

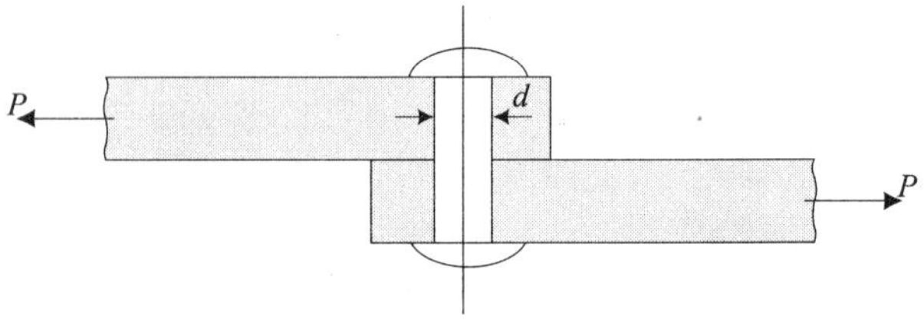

Solution:

$$\text{Load } P = 5 \text{ kN} = 5 \times 10^3 \text{ N.}$$

$$\text{Area of rivet, } A = \frac{\pi}{4} d^2$$

$$= \frac{\pi}{4} (12.5)^2 = 122.7 \text{ mm}^2$$

$$\therefore \quad \text{Shear stress, } \tau = \frac{P}{A} = \frac{5 \times 10^3}{122.7} = \mathbf{40.75\ N/mm^2} \quad \textbf{Ans.}$$

9.5 STRESS-STRAIN CURVE

Stress-strain curves are obtained experimentally by applying load gradually on a test specimen. Such curves are plotted from the data collected of stress and strain during testing on Universal Testing Machine. The test can be tensile test or compression test. The following material properties are obtained from stress-strain curve.

1. Type of material ductile or brittle. Structural steel can withstand sudden loads and are called ductile materials. Cast iron and non-ferrous metals cannot withstand sudden loads and break easily. These are called brittle materials. It can be known from the shape of stress-strain diagram whether the material tested was ductile or brittle.

2. The ultimate stress
3. The yield stress
4. % age elongation of specimen
5. % age reduction in area
6. Type of fracture
7. Area under the curve.

9.5.1 Stress-Strain Curve for Structural Steel (Ductile Material)

A typical stress-strain for structural steel is shown in Fig. 9.8. The salient points are:

o : Starting point
a : Elastic limit
b : Yield point
c : Lower yield point
d : Ultimate strength
e : Breaking point
I : Elastic region
II : Plastic region
III = I + II : Area under the curve.

The curve can be divided into the following parts:

o – a: The strain is proportional to stress. The material obeys Hooke's Law. Point 'a' is called elastic limit of proportionality. The slope of line $o – a$ is called modulus of elasticity. This is a measure of stiffness of material. Young's modulus, $E = \frac{\sigma}{\varepsilon}$ (N/mm^2).

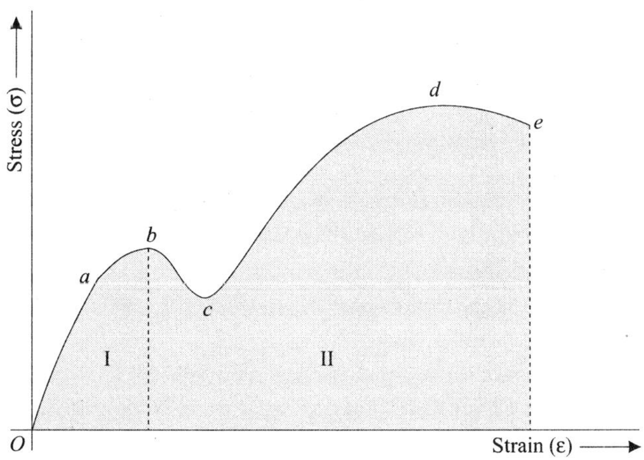

Fig. 9.8 Stress-strain diagram for ductile materials

a – b: The curve is not a straight line and material undergoes more rapid deformation. Hooke's Law is not obeyed although the material may still be elastic. The point '*b*' is called the yield point. The deformation is not fully recovered when the load is removed.

b – c: There is yielding of material and drop of load. The point '*c*' is called lower yield point.

c – d: The stress-strain curve continues to rise by increasing the load. The strain in portion *c-d* is nearly 100 times that in the portion *o-a*. There is a local neck formation in the material. The point '*d*' is called ultimate stress point and the ultimate strength of material is calculated at this point.

$$\text{Ultimate stress} = \frac{\text{Ultimate load}}{\text{Initial cross-sectional area}}.$$

d – e: The load falls off from maximum point and fracture takes place (at neck) at point '*e*' which is called breaking point.

9.5.2 Stress-Strain Curve for Cast Iron (Brittle Material)

A typical stress-strain curve for brittle materials is shown in Fig 9.9. The salient points on the curve are:

a : Limit of proportionality
b : Breaking point
c : Proof point
I . Elastic region
II : Plastic region
III = I + II: Area under the curve.

The curve can be divided into the following parts.

o – a: The strain is proportional to stress. Point '*a*' is called limit of proportionality. There is very little elongation as compared to ductile material. There is no yield point.

$$\text{Young's modulus, } E = \frac{\sigma}{\varepsilon}.$$

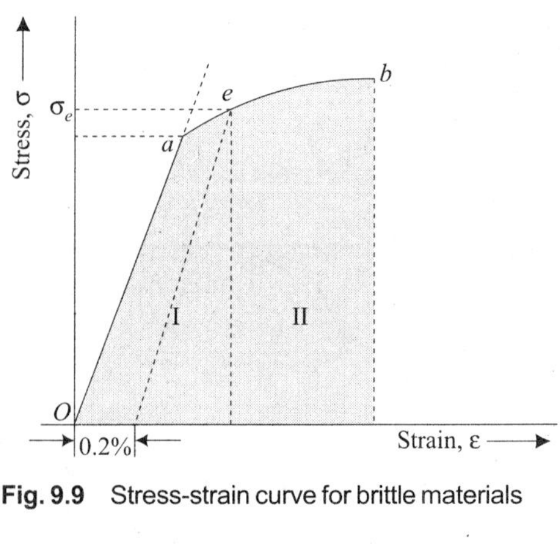

Fig. 9.9 Stress-strain curve for brittle materials

a – b: The curve is not a straight line. The material fractures at point '*b*' and is called breaking point.

Point *e*: In the design of machines and structures, the material of a member is not allowed to be stressed beyond yield point in order to avoid permanent set due to plastic deformation. In case of ductile materials, the yield point is well defined. But in case of brittle materials (cast iron, high carbon steel, non-ferrous metals), there is no defined yield point on the stress-strain diagram. In such a case, a proof stress (σ_e) is used to indicate the onset of plastic strain. A stress which produces a permanent strain of 0.2% is called the proof stress and is used for the design of machines and structures.

Strain energy stored in a material where it is stressed to proof point is called proof resilience (Ue)

$$Ue = \frac{1}{2}\frac{\sigma_e}{E} \times \text{Volume}$$

$$\text{Modulus of resilience} = \frac{Ue}{V} = \frac{1}{2}\frac{\sigma_e}{E}.$$

9.6 DUCTILE AND BRITTLE MATERIALS

The difference in the behaviour of ductile and brittle materials can be studied with the help of their stress-strain curves. Comparison of ductile and brittle material is given in Table 9.3.

Table 9.3 Comparison of ductile and brittle materials

Ductile Material	*Brittle Material*
1. It can absorb vibration and withstand sudden and dynamic loads.	1. It fails to withstand vibration, sudden and dynamic loads.
2. The stress-strain curve is shcwn in Fig. 9.8.	2. The stress-strain curve is shown in Fig. 9.9.
3. Slope of line *o–a* is large and therefore value of modulus of elasticity (*E*) is more.	3. Slope of line *o–a* is small and therefore value of modulus of elasticity (*E*) is low.
4. There is a well-defined yield point.	4. There is no well-defined yield point.
5. There is a local neck formation in the specimen.	5. There is no neck formation in the specimen.
6. The shape of the fracture is *cup and cone* type.	6. The shape of fracture is along a plane at 45° to the axis of the specimen.
7. The plastic region (II) is well extended as compared to elastic region (I).	7. The plastic region (II) is not much extended as compared to elastic region (I).
8. The area under the curve (III) is very large and has high resilience and modulus of resilience to adsorb vibration and impact loads.	8. The area under the curve (III) is small and has low resilience and modulus of resilience. Therefore, brittle material is unsuitable for impact loads.

Example 9.11: A load of 50 kN is suspended from a steel pipe of 50 mm external diameter. If the ultimate tensile strength is 500 N/mm^2 and factor of safety is 4, determine:

(i) internal diameter of pipe

(ii) elongation of pipe over a length of 200 mm if stressed to its maximum permissible value.

Take $E = 2 \times 10^5$ N/mm^2.

Solution:

Load, $P = 50 \text{ kN} = 50 \times 10^3$ N.

External diameter of pipe,

$$D = 50 \text{ mm}$$

$$L = 200 \text{ mm}$$

$$\sigma_{max} = 500 \text{ N/mm}^2$$

$$f.o.s. = 4.$$

$$E = 2 \times 10^5 \text{ N/mm}^2.$$

(i) Allowable stress, $\sigma_a = \dfrac{\sigma_{max}}{f.o.s.} = \dfrac{500}{4} = 125 \text{ N/mm}^2.$

Sectional area of pipe, $A = \dfrac{\pi}{4}(D^2 - d^2)$

$$\sigma_a = \frac{P}{A}$$

$$\therefore \quad A = \frac{P}{\sigma_a} = \frac{50 \times 10^3}{125}$$

Also $$A = \frac{\pi}{4}(50^2 - d^2) = \frac{50 \times 10^3}{125}$$

$$\therefore \quad d = \mathbf{44.62 \text{ mm}} \quad \textbf{Ans.}$$

$$A = \frac{\pi}{4}(50^2 - 44.62^2) = 399.8 \text{ mm}^2$$

(ii) Elongation, $\delta L = \dfrac{PL}{AE} = \dfrac{50 \times 10^3 \times 200}{399.8 \times 2 \times 10^5}$

$$= \mathbf{0.125 \text{ mm}} \quad \textbf{Ans.}$$

Example 9.12: A steel bar 20 mm in diameter, 200 mm long was tested to destruction. Following observations were recorded.

Load at elastic point = 60 kN

Load at yield point = 75 kN.

Max. load = 135 kN

Extension at elastic limit = 0.22 mm

Final length = 250 mm

Final diameter = 14 mm

Find: (i) Modulus of elasticity

(ii) Yield stress

(iii) Ultimate stress

(iv) % age elongation, and

(v) % age contraction of area.

Solution:

(i) Modulus of elasticity,

$$E = \frac{\sigma}{\varepsilon} = \frac{PL}{A\delta L} = \frac{60 \times 10^3 \times 200}{\frac{\pi}{4}(20)^2 \times 0.22}$$

$$= \mathbf{1.736 \times 10^5\ N/mm^2\ Ans.}$$

(ii) Yield stress,

$$\sigma_e = \frac{\text{Yield load}}{\text{Area}} = \frac{75 \times 10^3}{\frac{\pi}{4}(20)^2}$$

$$= \mathbf{238.73\ N/mm^2\ Ans.}$$

(iii) Ultimate stress,

$$\sigma_{max} = \frac{\text{Maximum load}}{\text{Original area}} = \frac{135 \times 10^3}{\frac{\pi}{4}(20)^2}$$

$$= \mathbf{429.72\ N/mm^2\quad Ans.}$$

(iv) % age elongation $= \dfrac{\text{Final length} - \text{Original length}}{\text{Original length}}$

$$= \frac{250 - 200}{200} \times 100 = \mathbf{25\%}\quad \mathbf{Ans.}$$

(v) % age contraction $= \dfrac{\text{Original area} - \text{Final area}}{\text{Original area}}$

$$= \frac{\frac{\pi}{4}[(20)^2 - (14)]^2}{\frac{\pi}{4}(20)^2} \times 100$$

$$= \mathbf{51\%} \quad \textbf{Ans.}$$

Example 9.13: During a tensile test on a mild steel specimen, 40 mm diameter and 200 mm long, the following data was obtained.

Extension at 40 kN load = 0.0304 mm

Yield load = 161 kN.

Length of specimen at fracture = 249 mm.

Determine modulus of elasticity, yield stress and % age elongation.

Solution:

(i) Modulus of elasticity,

$$E = \frac{\sigma}{\varepsilon} = \frac{(P/A)}{(\delta L/L)} = \frac{40/\frac{\pi}{4}(40)^2}{0.0304/200}$$

$$= \mathbf{209.4\ kN/mm^2} \quad \textbf{Ans.}$$

(ii) Yield stress,

$$\sigma_e = \frac{\text{Yield load}}{\text{Area}} = \frac{161 \times 10^3}{\frac{\pi}{4}(40)^2}$$

$$= \mathbf{128\ N/mm^2} \quad \textbf{Ans.}$$

(iii) % age elongation $= \dfrac{\text{Final length} - \text{Original length}}{\text{Original length}} \times 100$

$$= \left(\frac{249 - 200}{200}\right) \times 100 = \mathbf{24.5\%} \quad \textbf{Ans.}$$

Example 9.14: A bar 1.5 m long and 40 mm × 20 mm cross-section has elastic limit stress of 160 N/mm^2 and $E = 2 \times 10^5$ N/mm^2. Calculate the modulus of resilience and proof resilience.

Solution:

Volume of specimen,

$$V = 40 \times 20 \times 1500 = 12 \times 10^5 \text{ mm}^3$$

$$\sigma_e = 160 \text{ N/mm}^2$$

$$E = 2 \times 10^5 \text{ N/mm}^2.$$

(i) Proof resilience,

$$U_e = \frac{1}{2}\frac{\sigma_e^2}{E} \times V$$

$$= \frac{1}{2}\frac{(160)^2}{2\times10^5} \times 12 \times 10^5$$

$$= 7.68 \times 10^4 \text{ N.mm.}$$

(ii) Modulus of resilience,

$$= \frac{U_e}{V} = \frac{7.68\times10^4}{12\times10^5}$$

$$= \mathbf{0.064\ N/mm^2} \quad \textbf{Ans.}$$

Example 9.15: The following results were obtained in a tensile test on a rigid steel specimen of original diameter 20 mm and gauge length 50 mm. At the limit of proportionality, the load was 100 kN and extension was 0.5 mm. The specimen yielded at a load of 115 kN and the maximum load withstood was 200 kN. After rupture, the total length between gauge marks was found 6.67 cm and diameter of neck was 1.72 cm. Calculate Young's modulus, stress at limit of proportionality, the yield stress, ultimate tensile stress, percentage elongation and contraction.

Solution:

(i) Young's modulus,

$$E = \frac{\text{Stress at limit of proportionality}}{\text{Strain at limit of proportionality}}$$

$$= \frac{\dfrac{100\times10^3}{\dfrac{\pi}{4}(20)^2}}{\dfrac{0.5}{50}} = \mathbf{0.318 \times 10^5\ N/mm^2} \quad \textbf{Ans.}$$

(ii) Stress at limit of proportionality,

$$\sigma = \frac{100\times10^3}{\dfrac{\pi}{4}(20)^2} = \mathbf{318\ N/mm^2} \quad \textbf{Ans.}$$

(iii) Yield stress,

$$\sigma_e = \frac{115\times10^3}{\dfrac{\pi}{4}(20)^2} = \mathbf{366\ N/mm^2} \quad \textbf{Ans.}$$

(iv) Ultimate tensile stress,

$$\sigma_{max} = \frac{200 \times 10^3}{\frac{\pi}{4}(20)^2} = \mathbf{636.6\ N/mm^2} \quad \textbf{Ans.}$$

(v) % age elongation

$$= \frac{66.7 - 50}{50} \times 100$$

$$= \mathbf{33.4\%} \quad \textbf{Ans.}$$

(vi) % age contraction

$$= \frac{\frac{\Pi}{4}\left[(20)^2 - (17.2)^2\right]}{\frac{\pi}{4}(20)^2} \times 100$$

$$= \mathbf{26\%} \quad \textbf{Ans.}$$

9.7 ELASTIC CONSTANTS

The elastic constants are important properties of the material which influence the strains under different types of loading.

1. Modulus of Elasticity

Hooke's law staets that when a material is loaded within its elastic limits, the stress is proportional to strain,

$$\frac{\text{stress}}{\text{strain}} = \text{constant} = E$$

$$\therefore \quad E = \frac{\sigma}{\varepsilon}.$$

E is called modulus of elasticity or Young's modulus and determines the linear strain under linear loading. It is same for both tension and compression.

2. Modulus of Rigidity

The ratio of shear stress and shear strain is constant and is called modulus of rigidity.

$$G = \frac{\tau}{r}.$$

G is the material property which determines the strain due to shear stress.

3. Bulk Modulus

Whenever a member is subjected to a force or a system of forces, there is change in its dimensions in all the directions and, hence, there is change of volume.

The ratio of change in volume to original volume is called volumetric strain (ε_v) or dilatational strain.

$$\varepsilon_v = \frac{\delta V}{V}.$$

Bulk modulus is a material property which determines volumetric strain. The bulk modulus,

$$K = \frac{\text{Direct stress}}{\text{Volumetric strain}} = \frac{\sigma}{\varepsilon_v}.$$

4. Poisson's Ratio

The deformation of a bar per unit length in the direction of load or force is called linear or primary strain.

$$\text{Linear strain} = \frac{\delta L}{L}.$$

The linear strain is always accompanied by lateral or secondary strain. Lateral strain is the ratio of change of lateral dimension or diameter of a bar and the lateral dimension or diameter.

$$\text{Lateral strain} = \frac{\delta D}{D} \text{ or } \frac{\delta b}{b}$$

where, D is the diameter and b is the lateral dimension.

It is observed that within elastic limits, the lateral strain is proportional to linear strain.

$$\therefore \quad \frac{\text{Lateral strain}}{\text{Linear strain}} = \text{constant} = \frac{1}{m} = \nu$$

This constant, ν or $\frac{1}{m}$ is called Poisson's ratio.

$$\text{Lateral strain} = \nu\varepsilon \quad \text{or} \quad \frac{\varepsilon}{m}.$$

9.7.1 Relationship Between Elastic Constants

(a) Relation between E, K and ν

A cube is subjected to volumetric stress σ acting simultaneously along X, Y and Z directions.

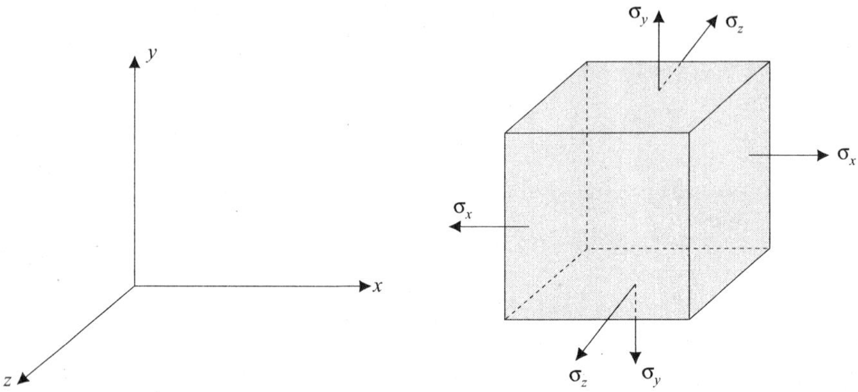

Fig. 9.10 Volumetric stress

$$\sigma_x = \sigma_y = \sigma_z = \sigma$$

Strain in x-direction,

$$\varepsilon_x = \frac{\sigma_x}{E} - \nu\frac{\sigma_y}{E} - \nu\frac{\sigma_z}{E} = \frac{\sigma}{E}(1-2\nu)$$

Strain in y-direction,

$$\varepsilon_y = \frac{\sigma_y}{E} - \nu\frac{\sigma_z}{E} - \nu\frac{\sigma_x}{E} = \frac{\sigma}{E}(1-2\nu)$$

Strain in z-direction,

$$\varepsilon_z = \frac{\sigma_z}{E} - \nu\frac{\sigma_x}{E} - \nu\frac{\sigma_y}{E} = \frac{\sigma}{E}(1-2\nu)$$

Volumetric strain,

$$\varepsilon_v = \varepsilon_x + \varepsilon_y + \varepsilon_z = \frac{3\sigma}{E}(1-2\nu)$$

Bulk modulus,

$$K = \frac{\sigma}{\varepsilon_v} = \frac{\sigma}{\frac{3\sigma}{E}(1-2\nu)} = \frac{E}{3(1-2\nu)}$$

$\therefore$ $$E = 3K(1-2\nu)$$

(b) Relation between E, G and ν

A cubic element is subjected to shearing force at the top face. The block experiences the following effects:

(i) Shearing stress, τ at face AB and CD.

(ii) Complimentary shear stress at faces AD and BC

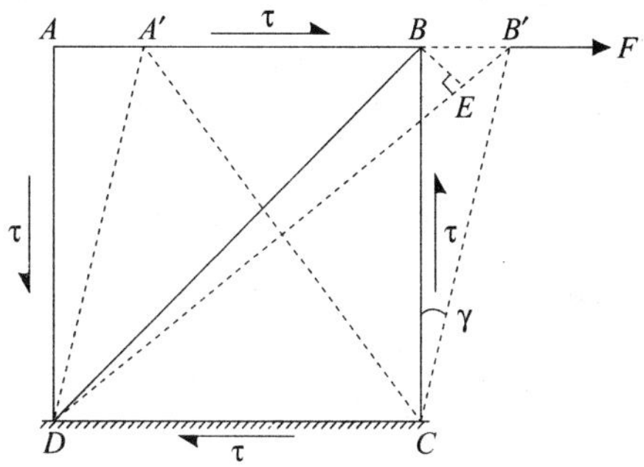

Fig. 9.11 Shearing stresses

(iii) Block is distorted as $A'B'CD$.

(iv) Diagonal DB elongates and diagonal AC is shortened. Therefore, longitudinal strain in diagonal DB,

$$= \frac{DB' - DB}{DB} = \frac{DB' - DE}{DB}$$

$$= \frac{EB'}{DB} \qquad \ldots(1)$$

BE is $\perp$ lar to DB'.

Let, $\angle DB'C = \angle DBC$ $\quad$ [$\because BB'$ is small]

$= 45°$.

$$\therefore \quad EB' = BB' \cos 45° = \frac{BB'}{\sqrt{2}}$$

$$\therefore \quad \text{Longitudinal strain} = \frac{EB'}{BD} = \frac{BB'}{\sqrt{2}} \times \frac{1}{\sqrt{2}BC} = \frac{BB'}{2BC}$$

$$= \frac{\tan r}{2} = \frac{\gamma}{2} \qquad \left[\because \gamma = \text{shear strain} = \frac{BB'}{BC}\right]$$

Now shear strain,

$$\gamma = \frac{\tau}{G}.$$

$$\therefore \text{ Longitudinal strain of diagonal } DB = \frac{\tau}{2G} \qquad \ldots(2)$$

Similarly, strain of diagonal DB = strain in DB due to tensile stress – strain in AC due to compresive stress.

$$= \frac{\tau}{E} - \left[-\nu\frac{\tau}{E}\right] = \frac{\tau}{E}(1 + \nu) \qquad \ldots(3)$$

From equations (2) and (3),

$$\frac{\tau}{2G} = \frac{\tau}{E}(1+\nu)$$

or $$E = 2G(1+\nu).$$

(c) Relation between E, G, and K

$$E = 3K(1-2\nu)$$

and $$E = 2G(1+\nu)$$

∴ $$E = 2G(1+\nu) = 3K(1-2\nu)$$

∴ $$\nu = \frac{E}{2G} - 1$$

$$E = 3K\left[1-2\left(\frac{E}{2G}-1\right)\right]$$

$$= 3K\left[1-\left(\frac{E}{2G}-2\right)\right]$$

or $$E = 3K\left[3-\frac{E}{G}\right] = 9K - \frac{3KE}{G}$$

∴ $$E + \frac{3KE}{G} = 9K$$

$$E\left[\frac{G+3K}{G}\right] = 9K$$

or $$E = \frac{9KG}{G+3K}$$

The relations among elastic constants can be summarised on follows:

$$E = 3K(1-2\nu) = 3K\left(1-\frac{2}{m}\right)$$

∴ $$K = \frac{mE}{3(m-2)}$$

$$E = 2G(1+\nu) = 2G\left(1+\frac{1}{m}\right)$$

∴ $$G = \frac{mE}{2(m+1)}$$

Example 9.16: A circular rod of 100 mm diameter and 500 mm long is subjected to a tensile force of 1000 kN. Determine the modulus of rigidity, bulk modulus and change in volume, if Poisson's ratio is 0.3 and Young's modulus is 2×10^5 N/mm^2.

Solution:

$$P = 100 \text{ kN} = 1000 \times 10^3 \text{ N.}$$

$$D = 100 \text{ mm}$$

$$L = 500 \text{ mm}$$

$$E = 2 \times 10^5 \text{ N/mm}^2$$

$$\nu = 0.3 = \frac{1}{m} \qquad [m = 3.33.]$$

(i) Modulus of rigidity.

$$G = \frac{mE}{2(m+1)} = \frac{3.33 \times 2 \times 10^5}{2(3.33+1)}$$

$$= \mathbf{0.78 \times 10^5 \text{ N/mm}} \quad \textbf{Ans.}$$

(ii) Bulk modulus,

$$K = \frac{mE}{2(m-2)} = \frac{3.33 \times 2 \times 10^5}{2(3.33-2)}$$

$$= \mathbf{1.67 \times 10^5 \text{ N/mm}^2} \quad \textbf{Ans.}$$

(iii) Volume,

$$V = \frac{\pi}{4} D^2 L = \frac{\pi}{4}(100)^2 \times 500$$

$$= 3926991 \text{ mm}^3.$$

$$\sigma = \frac{P}{A} = \frac{1000 \times 10^3}{\pi/4(100)^2} = 127.3 \text{ N/mm}^2.$$

$$K = \frac{\sigma}{\varepsilon_v} = \frac{\sigma V}{\delta V}$$

$$\therefore \qquad \delta V = \frac{\sigma V}{K} = \frac{127.3 \times 3926991}{1.67 \times 10^5}$$

$$= \mathbf{2994 \text{ mm}^2} \quad \textbf{Ans.}$$

Example 9.17: A bronze specimen has $E = 1.2 \times 10^5$ N/mm^2 and $G = 0.45 \times 10^5$ N/mm. Determine the Poisson's ratio of the material.

Solution:

$$G = \frac{mE}{2(m+1)}$$

$$\therefore \qquad 2mG + 2G = mE$$

$$m = \frac{-2G}{(2G - E)} = \frac{-2 \times 0.45 \times 10^5}{(2 \times .45 \times 10^5 - 1.2 \times 10^5)}$$

$$= \frac{0.9}{0.3} = 3$$

$$\therefore \qquad \nu = \frac{1}{m} = \frac{1}{3} = 0.33.$$

Example 9.18: Find the bulk modulus and lateral contraction of a round bar of 50 mm diameter and 2.5 m long when stretched to 2.5 mm. Take $E = 1.2 \times 10^5$ N/mm^2 and $\nu = 0.25$.

Solution:

Longitudinal strain,

$$\varepsilon_L = \frac{\delta L}{L} = \frac{2.5}{2500} = 0.001$$

Lateral strain,

$$\frac{\delta d}{d} = \nu\varepsilon_L = 0.25 \times 0.001$$

$$= 0.00025$$

$$\therefore \qquad \delta d = 0.00025 \times 50$$

$$= \mathbf{.0125\ mm} \quad \textbf{Ans.}$$

Bulk modulus,

$$K = \frac{mE}{3(m-2)} = \frac{4 \times 1.2 \times 10^5}{3(4-2)}$$

$$= 0.8 \times 10^5 \text{ N/mm}^2$$

Example 9.19: A steel bar 300 mm long, 50 mm wide and 12 mm thick is subjected to a tensile pull of 100 kN. If $E = 2 \times 10^5$ N/mm^2 and $\nu = 0.32$, determine the change in volume of the bar.

Solution:

Volume, $V = Lbt = 300 \times 50 \times 12$

$$= 180{,}000 \text{ mm}^3$$

Strain,
$$\varepsilon = \frac{\sigma}{E} = \frac{P/A}{E} = \frac{100\times10^3}{50\times12\times2\times10^5}$$
$$= 8.33 \times 10^{-4}$$

Volumetric strain,
$$\varepsilon_v = \frac{\delta V}{V} = \varepsilon\left[1-\frac{2}{m}\right]$$
$$\delta_V = 180{,}000 \times 8.33 \times 10^{-4}\,(1 - 2\times 0.32)$$
$$= 54\ \text{mm}^3.$$

Example 9.20: A 2 m long rectangular bar of 7.5 cm × 5 cm is subjected to an axial tensile load of 1000 kN. Bar gets elongated by 2 mm in length and decreases in width by 10×10^{-6} m. Determine the modulus of elasticity E and Poisson's ratio ν of the material of the bar.

$$L = 2\ \text{m}$$
$$b = 7.5\ \text{cm} = 7.5 \times 10^{-2}\ \text{m}$$
$$t = 5\ \text{cm} = 5 \times 10^{-2}\ \text{m}$$
$$P = 1000\ \text{kN}$$
$$\delta L = 2\ \text{mm} = 2 \times 10^{-3}\ \text{m}$$
$$\delta b = 10 \times 10^{-6}\ \text{m}.$$

The longitudinal stress,
$$\sigma = \frac{P}{A} = \frac{P}{b\times t} = \frac{1000}{7.5\times10^{-2}\times5\times10^{-2}}$$
$$= 26.67 \times 10^4\ \text{kN/m}^2$$
$$\varepsilon = \frac{\delta L}{L} = \frac{2\times10^{-3}}{2} = 10^{-3}$$
$$E = \frac{\sigma}{\varepsilon} = \frac{26.67\times10^4}{10^{-3}} = 26.67 \times 10^7\ \text{kN/m}^2$$

Lateral strain,
$$= \frac{\delta b}{b} = \frac{10\times10^{-6}}{7.5\times10^{-2}} = 1.333 \times 10^{-4}$$

∴ Poisson's ratio,
$$\nu = \frac{1.333\times10^{-4}}{10^{-3}} = 0.1333$$

Example 9.21: A bar of 25 mm diameter is subjected to a pull of 60 kN. The measured extension over a gauge length of 250 mm is 0.15 mm and change in diameter is 0.004 mm. Calculate the modulus of elasticity, modulus of rigidity and Poisson's ratio. [U.P.T.U. II Sem., 2005–06]

Solution:

$$d = 25 \text{ mm}$$
$$P = 60 \text{ kN}$$
$$L = 250 \text{ mm}$$
$$\delta L = 0.15 \text{ mm}$$
$$\delta d = 0.004 \text{ mm}$$

Stress, $$\sigma = \frac{P}{A} = \frac{60 \times 10^3}{\frac{\pi}{4}(25)^2}$$

$$= 122.23 \text{ N/mm}^2.$$

Strain, $$\varepsilon = \frac{\delta L}{L} = \frac{0.15}{250} = 6 \times 10^{-4}$$

Modulus of elasticity,

$$E = \frac{\sigma}{\varepsilon} = \frac{122.23}{6 \times 10^{-4}} = 203716.67 \text{ N/mm}^2$$

$$= \mathbf{2.037 \times 10^5 \text{ N/mm}^2} \quad \textbf{Ans.}$$

Lateral strain, $$= \frac{\delta d}{d} = \frac{0.004}{25}$$

$$= 1.6 \times 10^{-4}$$

Poisson's ratio, $$\nu = \frac{\text{Lateral strain}}{\text{Linear strain}} = \frac{1.6 \times 10^{-4}}{6 \times 10^{-4}}$$

$$= 0.2667$$

$$m = 3.75$$

Modulus of rigidity, $$G = \frac{mE}{2(m+1)}$$

$$= \frac{3.75 \times 2.037 \times 10^5}{2(3.75+1)}$$

$$= \mathbf{0.80 \times 10^5 \text{ N/mm}^2} \quad \textbf{Ans.}$$

Example 9.22: A metallic rectangular rod is 1.5 m long and 40 mm wide and 25 mm thick is subjected to an axial tensile load of 120 kN. The elongation of the rod is measured as 0.9 mm. Calculate the stress, strain and modulus of elasticity.

[U.P.T.U. II Sem., 2003–04]

Solution:

$$L = 1.5 \text{ m} = 1500 \text{ mm}$$
$$b = 40 \text{ mm}$$
$$t = 25 \text{ mm}$$
$$P = 120 \text{ kN} = 120 \times 10^3 \text{ N}$$
$$\delta L = 0.9 \text{ mm.}$$

Stress, $$\sigma = \frac{P}{A} = \frac{120 \times 10^3}{40 \times 25} = 120 \text{ N/mm}^2$$

Strain, $$\varepsilon = \frac{\delta L}{L} = \frac{0.9}{1500} = 6 \times 10^{-4}$$

Modulus of elasticity, $$E = \frac{\sigma}{\varepsilon} = \frac{120}{6 \times 10^{-4}} = \mathbf{2 \times 10^5 \text{ N/mm}^2}$$ **Ans.**

9.8 MEMBERS OF VARYING CROSS-SECTION

Members of varying cross-section are widely used in the construction of machines and structures. These members can be of the following shape:

1. Stepped bars.
2. Tapered bars.
3. Composite bars.

9.8.1 Stepped Bars

In stepped bars, the forces may be acting at some interior section of a body along its length, in addition to forces acting at the ends. The bar is divided in different lengths (sections) and forces for each section/length are worked out on the principle of equilibrium of each section. Free body diagram is drawn for each length and deformation is determined. The deformation for the entire length is worked out by summing the sectional deformations.

$$\delta L = \frac{PL}{AE} = \frac{1}{E}\left[\frac{P_1 L_1}{A_1} + \frac{P_2 L_2}{A_2} + \cdots\right]$$

The deformation of individual sections can be worked out as follows.

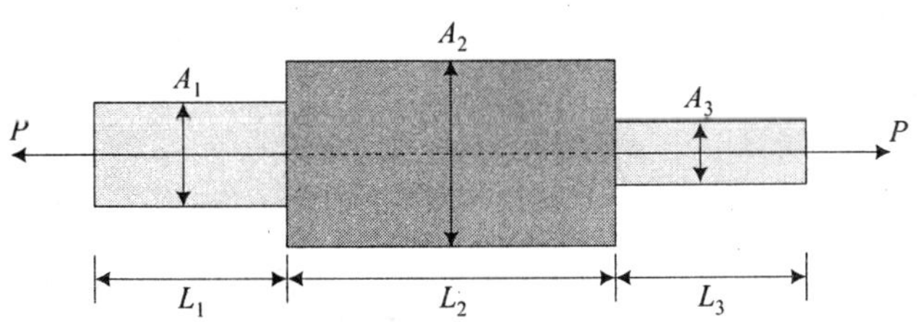

Fig. 9.12 Stepped bar

$$\delta L_1 = \frac{PL_1}{A_1 E}$$

$$\delta L_2 = \frac{PL_2}{A_2 E}$$

$$\delta L_3 = \frac{PL_3}{A_3 E}$$

Total elongation of the bar,

$$\delta L = \delta L_1 + \delta L_2 + \delta L_3$$

$$= \frac{P}{E}\left[\frac{L_1}{A_1} + \frac{L_2}{A_2} + \frac{L_3}{A_3}\right]$$

If the stepped bar is made from sections of different materials

$$\delta L = P\left[\frac{L_1}{A_1 E_1} + \frac{L_2}{A_2 E_2} + \frac{L_3}{A_3 E_3}\right]$$

where, E_1, E_2 and E_3 are moduli of elasticity of materials of different steps.

9.8.2 Tapered Bars

A tapered circular bar of big-end diameter D_1 and small-end diameter D_2 and length L is loaded with axial load P. It is required to find the extension of the bar. Consider a small strip of length dX at a distance of X from big-end.

Diameter of bar at this section, $D_x = D_1 - (D_1 - D_2)\dfrac{X}{L}$

$$= D_1 - mX$$

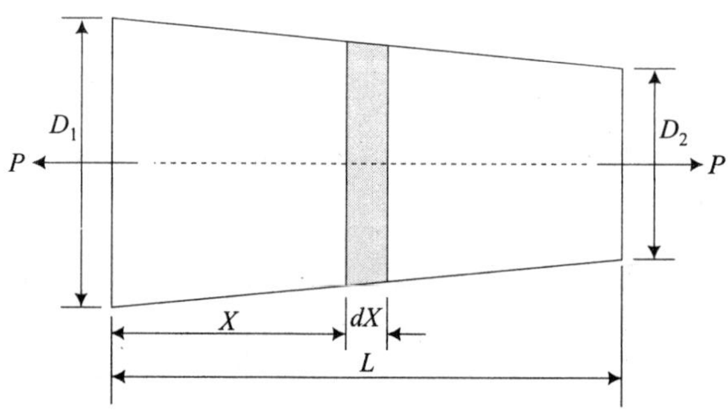

Fig. 9.13 Tapered bar

where, $$m = \frac{D_1 - D_2}{L}$$

Cross-sectional area of strip,

$$A_x = \frac{\pi}{4}(D_1 - mX)^2$$

Stress in the strip,

$$\sigma_x = \frac{P}{A_x} = \frac{P}{\frac{\pi}{4}(D_1 - mX)^2} = \frac{4P}{(D_1 - mX)^2}$$

Axial strain in the strip,

$$\varepsilon_x = \frac{\sigma_x}{E} = \frac{4P}{\pi(D_1 - mX)^2 E}$$

Extension of strip,

$$\delta L_x = \varepsilon_x \, dX = \frac{4PdX}{\pi(D_1 - mX)^2 E}$$

Total extension of bar

$$\delta L = \int_o^L \delta L_x = \int_o^L \frac{4PdX}{\pi(D_1 - mX)^2 E}$$

$$= \frac{4P}{\pi E}\int_o^L (D_1 - mX)^{-2} dx$$

$$= \frac{4P}{\pi E}\left[\frac{(D_1 - mX)^{-1}}{(-1)\times(-m)}\right]_o^L$$

$$= \frac{4P}{m\pi E}\left[\frac{1}{(D_1 - mX)}\right]_o^L$$

$$= \frac{4P}{\pi E\left(\frac{D_1 - D_2}{L}\right)}\left[\frac{1}{\left\{D_1 - \frac{D_1 - D_2}{L}\right\}L} - \frac{1}{D_1}\right]$$

$$\delta L = \frac{4P}{\pi E\left(\frac{D_1 - D_2}{L}\right)} \times \frac{(D_1 - D_2)}{D_1 D_2}$$

$$\delta L = \frac{4PL}{\pi E D_1 D_2}$$

Note: This is the relation for a circular taper bar. For other shapes of bar, proceed from first principles.

9.8.3 Composite Bars

Composite columns and ties are used by joining together two or more elements of different materials. These may be subjected to tensile or compressive loads axially. As the members are joined rigidly at both ends, the following principles apply.

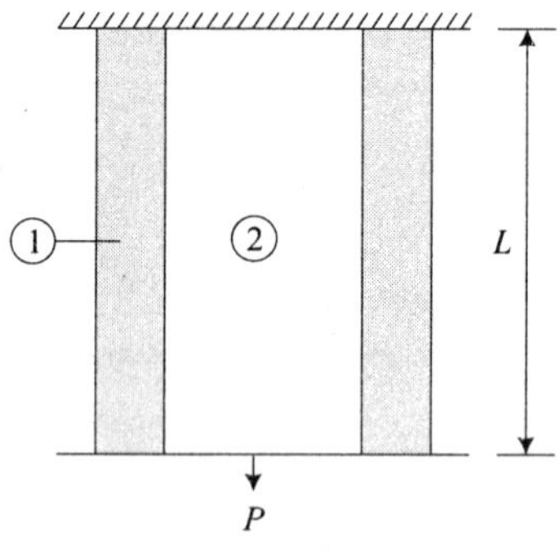

Fig. 9.14 Composite bar

1. The deformation (extension or contraction) of each member is same.

$$\delta L_1 = \delta L_2 = \ldots$$

2. All the elements of composite bar should have same strain as they have same length.

$$\frac{\delta L_1}{L} = \frac{\delta L_2}{L} = \ldots$$

$$\varepsilon_1 = \varepsilon_2 = \ldots$$

3. Total axial load must be the sum of loads carried by different members of the composite bar.

$$P = P_1 + P_2 + \ldots$$
$$= \sigma_1 A_1 + \sigma_2 A_2 + \ldots$$

$$\because \quad \varepsilon_1 = \varepsilon_2 = \ldots$$

$$\therefore \quad \frac{\sigma_1}{E_1} = \frac{\sigma_2}{E_2} = \ldots$$

$$\therefore \quad \frac{\sigma_1}{\sigma_2} = \frac{E_1}{E_2} = m_r$$

m_r is called the modular ratio.

Example 9.23: A round bar is subjected to an axial load of 100 kN. Calculate the diameter 'D' if the stress there is 100 MN/m^2. Find also the total elongation (E = 200 GPa).

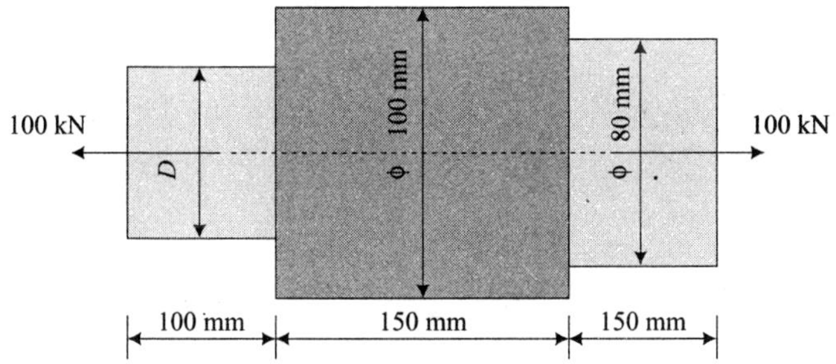

Solution:

Stress, $$\sigma = \frac{P}{A}$$

$$\therefore \quad A = \frac{P}{\sigma} = \frac{100 \times 10^3}{100} = 1000 \text{ mm}^2$$

$$D = \sqrt{\frac{4A}{\pi}} = \sqrt{\frac{4 \times 1000}{\pi}} = \mathbf{35.68\ mm} \quad \textbf{Ans.}$$

$$\delta L = \delta L_1 + \delta L_2 + \delta L_3$$

$$= \frac{P}{E}\left[\frac{L_1}{A_1} + \frac{L_2}{A_2} + \frac{L_3}{A_3}\right]$$

$$= \frac{100 \times 10^3}{2 \times 10^5} \times \frac{4}{\pi}\left[\frac{100}{(35.68)^2} + \frac{150}{(100)^2} + \frac{150}{(80)^2}\right]$$

$$= \frac{100 \times 10^3}{2 \times 10^5} \times \frac{4}{\pi}[0.07855 + 0.015 + 0.0234]$$

$$= \mathbf{0.0745\ mm} \quad \textbf{Ans.}$$

Example 9.24: A composite bar consists of aluminium section rigidly fastened between a bronze section and steel section as shown. Axial loads are applied at points indicated. Determine the stress in each section.

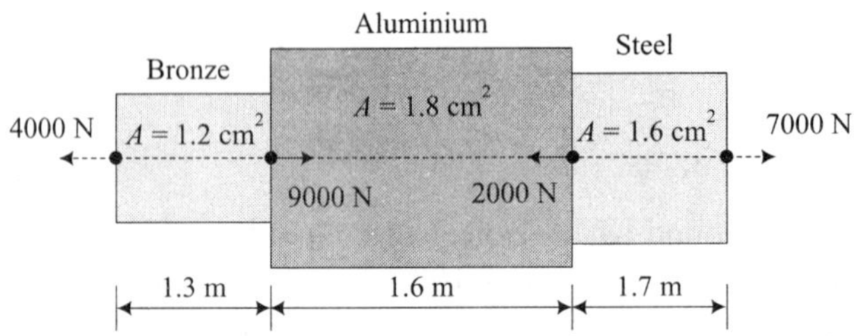

Solution: The forces in each step is found out from free body diagram.

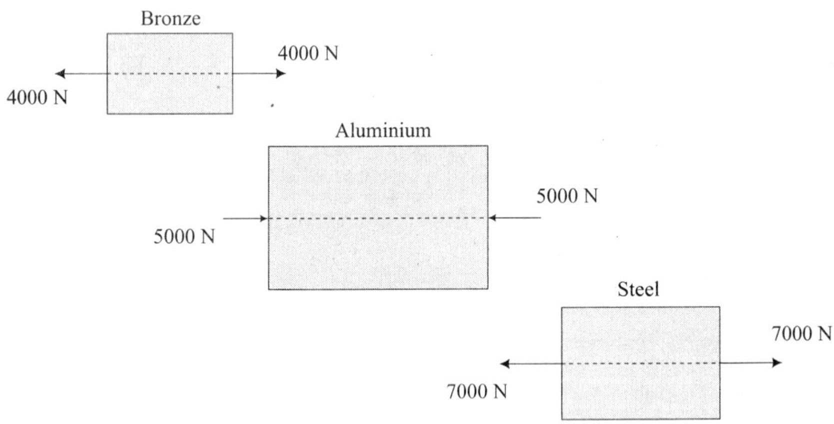

The bronze specimen is under a tensile load of 4000 N as applied at the end.

The steel specimen is under a tensile load of 7000 N as applied at the end.

Therefore, the central aluminium step will be under a compressive load of 5000 N (i.e., 9000 N – 4000 N) for equilibrium stress in bronze bar,

$$\sigma_b = \frac{P}{A} = \frac{4000}{1.2 \times 10^2} = 33.33 \text{ N/mm}^2 \text{(tensile)}$$

Stress in aluminium bar,

$$\sigma_a = \frac{5000}{1.8 \times 10^2} = 27.78 \text{ N/mm}^2 \text{ (compressive)}$$

Stress in steel bar,

$$\sigma_s = \frac{7000}{1.6 \times 10^2} = 43.75 \text{ N/mm}^2 \text{ (tensile)}$$

Example 9.25: A member *ABCD* is subjected to point loads P_1, P_2, P_3 and P_4 as shown. Calculate the force P_2 necessary for equilibrium, if $P_1 = 4.5$ kN, $P_3 = 45$ kN and $P_4 = 13$ kN. Determine the total elongation of the member, assuming modulus of elasticity to be 2×10^5 N/mm^2.

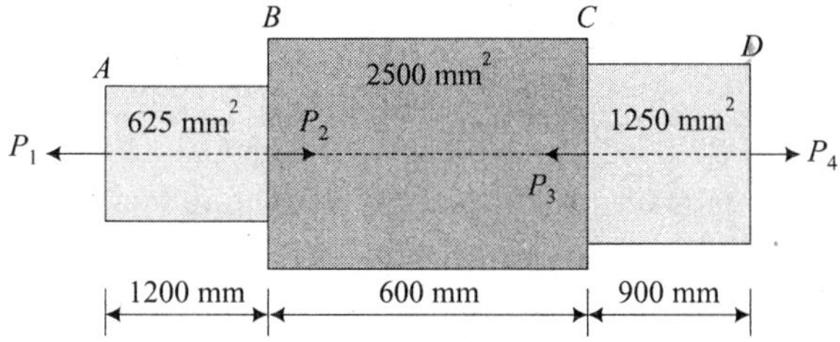

Solution:

For equilibrium of bar,

$$\sum_i (P_i)_{\text{left}} = \sum_i (P_i)_{\text{right}}$$

$$\therefore \quad P_1 + P_3 = P_2 + P_4$$

$$\therefore \quad P_2 = P_1 + P_3 - P_4 = 4.5 + 45 - 13$$

$$= \mathbf{36.5\ kN} \quad \textbf{Ans.}$$

The forces in each step are found out from free body diagram of individual step.

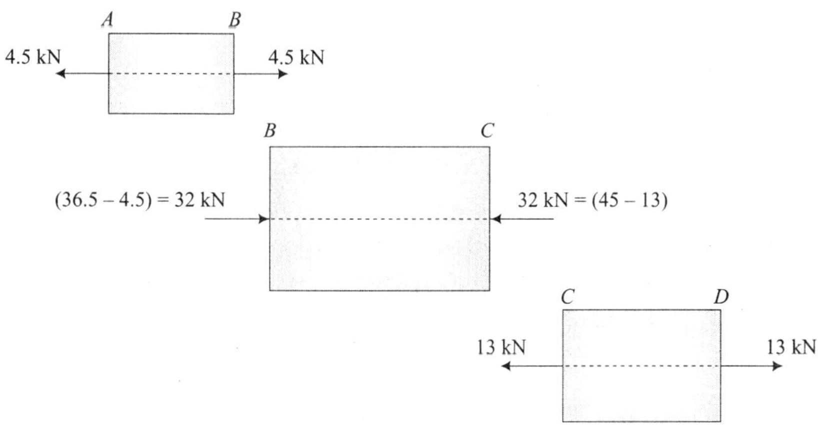

Step *AB* is under a tensile load of 4.5 kN as applied at end A.

Step *CD* is under a tensile load of 13 kN as applied at end *D*.

Step *BC* is under a compressive load of 32 kN (36.5 kN – 4.5 kN)

Total elongation of bar,

$$\sigma L = \sigma L_1 + \sigma L_2 + \sigma L_3 = \frac{1}{E}\left[\frac{P_1L_1}{A_1} + \frac{P_2L_2}{A_2} + \frac{P_3L_3}{A_3}\right]$$

$$= \frac{1}{2\times 10^5}\left[\frac{4500\times 1200}{625} - \frac{32000\times 600}{2500} + \frac{13000\times 900}{1250}\right]$$

$$= \mathbf{0.49\ mm} \quad \textbf{Ans.}$$

Example 9.26: A member *ABCD* is subjected to point loads P_1, P_2, P_3 and P_4 as shown. Calculate the force P_3 necessary for equilibrium, if P_1 = 120 kN, P_2 = 220 kN and P_4 = 160 kN. What will be the net change in the length of the bar, if E = 200 GN/m^2? [U.P.T.U. 2008-09]

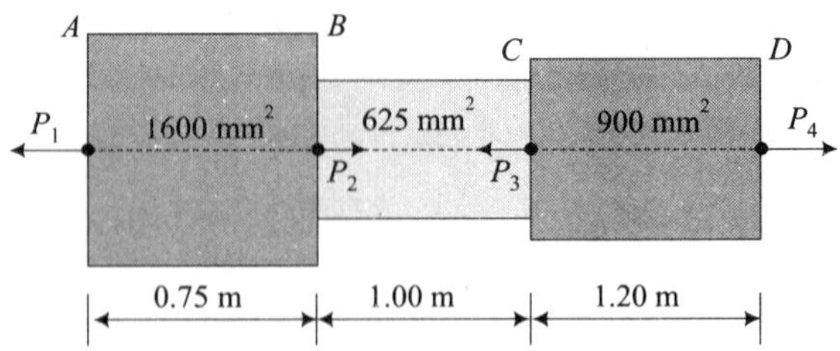

Solution:

For equilibrium of bar,

$$\sum_i (P_i)_{\text{left}} = \sum_i (P_i)_{\text{right}}$$

$$P_1 + P_3 = P_2 + P_4$$

$$\therefore \quad P_3 = P_2 + P_4 - P_1 = 220 + 160 - 120$$

$$= 260 \text{ kN}$$

E = 200 GN/m^2 = 200 × 10^3 MN/m^2 = 200 × 10^3 N/mm^2 = 200 kN/mm^2

The forces in each step can be found out from free-body diagram.

Step *AB* is under a tensile load of 120 kN as applied at end *A*.

Step *CD* is under a tensile load of 160 kN as applied at end *D*.

Step *BC* is under a compressive load of 100 kN (220 – 120 kN).

Total elongation,

$$\delta L = \delta L_1 + \delta L_2 + \delta L_3 = \frac{1}{E}\left[\frac{P_1L_1}{A_1} + \frac{P_2L_2}{A_2} + \frac{P_3L_3}{A_3}\right]$$

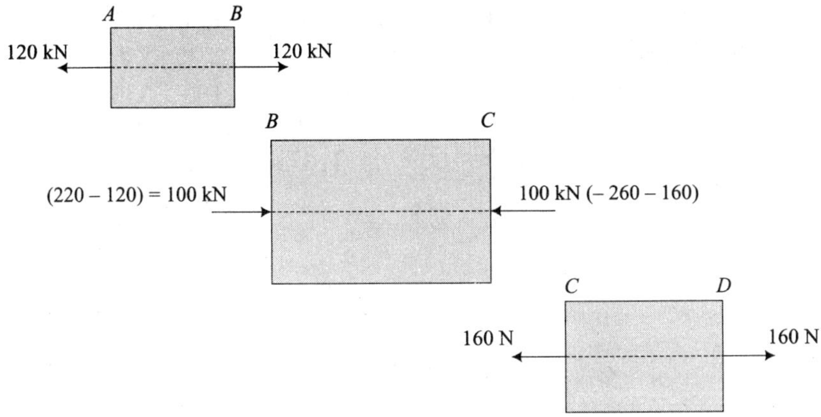

$$= \frac{1}{200}\left[\frac{120 \times 750}{1600} - \frac{100 \times 1000}{625} + \frac{160 \times 1200}{900}\right]$$

$$= \mathbf{0.548\ mm} \quad \textbf{Ans.}$$

Example 9.27: A steel bar is subjected to loads as shown. If Young's modulus for the bar material is 200 kN/mm^2, determine the change in the length of the bar. The bar is 200 mm in diameter. [U.P.T.U. I Sem., 2005–06]

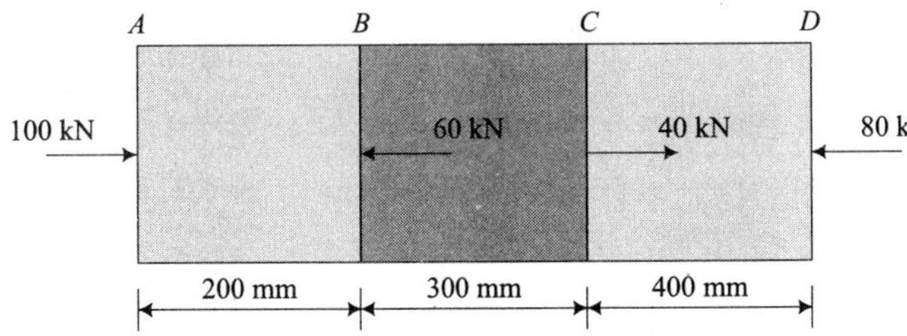

Solution: The forces in each step can be calculated from the free-body diagram.

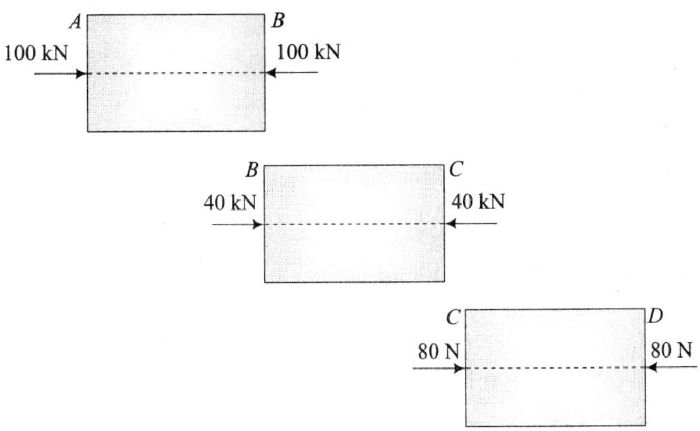

The step *AB* is under a compressive load of 100 kN as applied at end *A*.
The step *CD* is under a compressive load of 80 kN as applied at end *D*.
The step *BC* is under a compressive load of 40 kN (100 – 60 kN).

Area of bar, $A = \frac{\pi}{4}(200)^2 = 31416 \text{ mm}^2$

$E = 200 \text{ kN/mm}^2$

The total change in length of bar,

$$\delta L = \delta L_1 + \delta L_2 + \delta L_3$$

$$= \frac{1}{AE}[P_1 L_1 + P_2 L_2 + P_3 L_3]$$

$$= \frac{1}{31416 \times 200}[-100 \times 200 - 40 \times 300 - 80 \times 400]$$

$$= \frac{1}{31416 \times 200}[-20000 - 12000 - 32000]$$

= 0.01 mm (decrease) Ans.

Example 9.28: A tensile load of 50 kN is acting on a rod of diameter 50 mm and a length of 5 m. A bore of diameter 25 mm is made centrally in the rod at one end. To what length the rod should be bored, so that the total extension will be increased by 25% under the same tensile load. $E = 2 \times 10^5 \text{ N/mm}^2$.

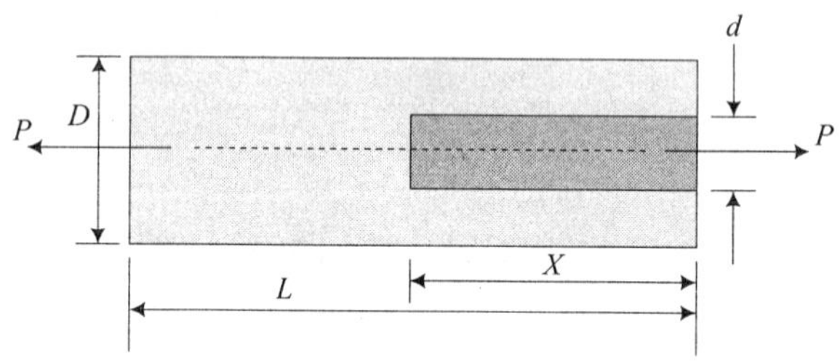

Solution:

$D = 50 \text{ mm}$

$d = 25 \text{ mm}$

$L = 5 \text{ m}$

$= 5000 \text{ mm}$

$P = 50 \text{ kN} = 50 \times 10^3 \text{ N}$

Extension of solid bar,

$$\sigma L_1 = \frac{PL}{AE} = \frac{50 \times 10^3 \times 5000}{\frac{\pi}{4}(50)^2 \times (2 \times 10^5)} = 0.6366 \text{ mm}$$

Extension of bar required after boring,

$$\delta L_2 = 0.6366 \times 1.25 = 0.7957 \text{ mm.}$$

Let the length of bore = X.

∴ Total extension of bar = Extension of solid portion + Extension of hollow portion.

$$\delta L_2 = \frac{P}{E}\left[\frac{L_1}{A_1} + \frac{L_2}{A_2}\right]$$

$$= \frac{P}{E}\left[\frac{X}{\frac{\pi}{4}(D^2 - d^2)} + \frac{(L - X)}{\frac{\pi}{4}D^2}\right]$$

$$= \frac{50 \times 10^3}{2 \times 10^5}\left[\frac{X}{\frac{\pi}{4}(50^2 - 25^2)} + \frac{(5000 - X)}{\frac{\pi}{4}50^2}\right] = 0.7957$$

∴ $X = 3748$ mm = **3.748 m** **Ans.**

Example 9.29: A steel rod 40 mm in diameter and 2 m long is subjected to a pull of 80 kN. To what length the rod should be bored centrally so that the total extension may increase by 20% under the same pull, the bore being 20 mm in diameter $E = 2 \times 10^5$ N/mm^2.

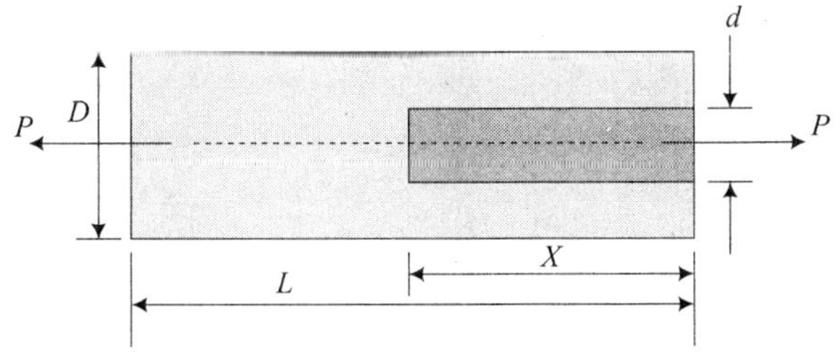

Solution:

$$D = 40 \text{ mm}$$

$$d = 20 \text{ mm}$$

$$L = 2 \text{ m} = 2000 \text{ mm}$$

$$P = 80 \text{ kN} = 80 \times 10^3 \text{ N}$$

$$E = 2 \times 10^5 \text{ N/mm}^2$$

Extension of rod before boring,

$$\delta L_1 = \frac{PL}{AE} = \frac{(80\times10^3)\times2000}{\left(\frac{\pi}{4}40^2\right)\times(2\times10^5)} = 0.6366 \text{ mm}$$

Let the length of bore = X mm.

Extension of rod required after boring = 1.2×0.6366

$$\delta L_2 = 0.764 \text{ mm.}$$

Extension of hollowed rod

= Extension of solid portion + Extension of hollow portion

$$\delta L_2 = \frac{P}{E}\left[\frac{L_1}{A_1}+\frac{L_2}{A_2}\right]$$

$$= \frac{80\times10^3}{2\times10^5}\left[\frac{X}{\frac{\pi}{4}(40^2-20^2)}+\frac{(2000-X)}{\frac{\pi}{4}(40^2)}\right] = 0.764$$

$\therefore$ **X = 1198.6 mm Ans.**

Example 9.29 (a): A round bar 40 cm long has 5 cm diameter for middle half of its length and a reduced diameter at the two ends (ends are equal in diameter and length-wise). Bar carries axial load of 10 kN. Find the diameter of end sections if the total allowable extension is 0.03 cm. $E = 200$ GN/m^2.

[U.P.T.U. II Sem. 2007-08]

Solution:

$P = 10$ kN

$\delta L = 0.03$ cm $= 0.03 \times 10^{-2}$ m

$L_1 = L_3 = 10$ cm $= 0.1$ m

$L_2 = 20$ cm $= 0.2$ m

$d_2 = 5$ cm $= 0.05$ m

$d_1 = d_3$

$E = 200$ GN/m^2 $= 200 \times 10^6$ kN/m^2

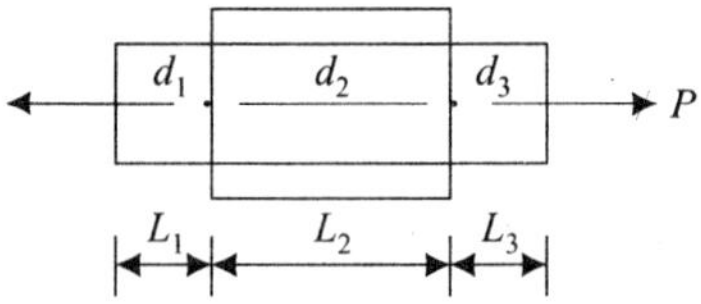

Using principle of superposition,

$$\delta L = \frac{P}{E}\left[\frac{L_1}{A_1}+\frac{L_2}{A_2}+\frac{L_3}{A_3}\right]$$

$$= \frac{4P}{\pi E}\left[\frac{2L_1}{d_1^2} + \frac{L_2}{d_1^2}\right]$$

$$= \frac{4 \times 10}{\pi \times 200 \times 10^6}\left[\frac{2 \times 0.1}{d_1^2} + \frac{0.2}{(0.05)^2}\right]$$

$$= \frac{10^{-6}}{5\pi}\left[\frac{0.2}{d_1^2} + 80\right]$$

$$\frac{0.2}{d_1^2} = \frac{0.03 \times 10^{-2}\ \pi \times 5}{10^{-6}} - 80 = 391.2389$$

$$\therefore \qquad d_1 = \sqrt{\frac{0.2}{391.2389}} = 0.0226 \text{ m} = \mathbf{2.26\ cm} \quad \textbf{Ans.}$$

Example 9.29 (b): Determine the stress in all three sections and total deformation of the steel rod shown. Cross-sectional area = 10 cm^2 and E = 200 GN/m^2.

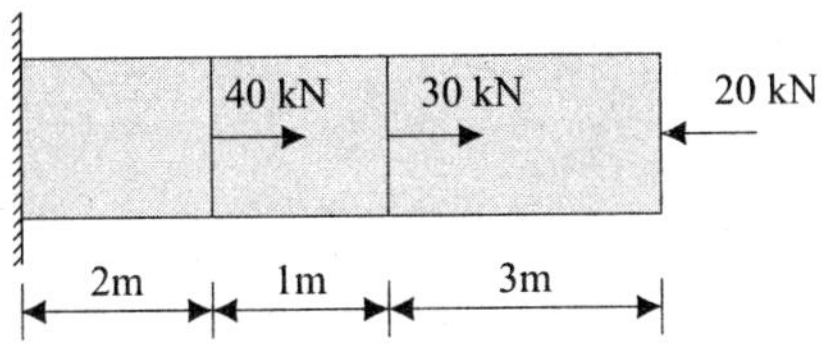

Solution:

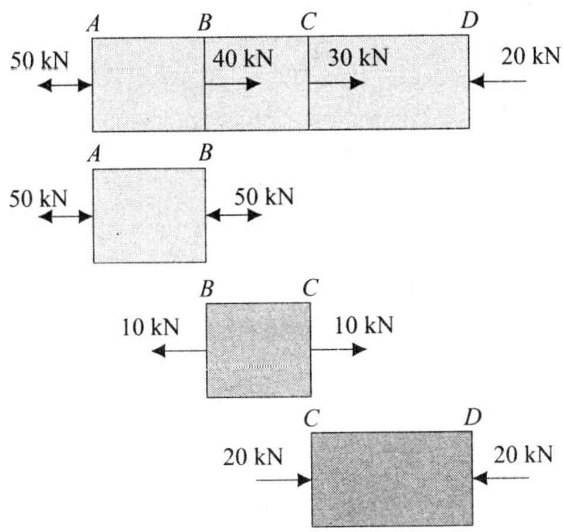

Area of cross-section, $A = 10 \text{ cm}^2 = 10 \times 10^{-4} \text{ m}^2$

$E = 200 \text{ GN/m}^2 = 200 \times 10^6 \text{ kN/m}^2$

(i) For equilibrium of steel rod, the reaction at A will be $40 + 30 - 20 = 50$ kN left side direction

(ii) Draw *FBD* of all the sections

Bar *CD* is under compression load of 20 kN.

Bar *AB* is under tensile load of 50 kN.

Bar *BC* is under tensile load of 10 kN.

(iii) Using principle of superposition,

$$\delta L_{AD} = \delta L_{AB} + \delta L_{BC} + \delta L_{CD} = \Sigma \frac{PL}{AE}$$

$$= \frac{1}{AE}[50 \times 2 + 10 \times 1 - 20 \times 3] = \frac{50}{10 \times 10^{-4} \times 200 \times 10^6}$$

$$= 2.5 \times 10^{-4} \text{ m} = \mathbf{0.25 \text{ mm}} \quad \textbf{Ans.}$$

Example 9.30: A vertical rod of 4 m length is rigidly fixed at upper end and carries an axial load of 50 kN force. Calculate the total extension of the bar if it tapers uniformly from a diameter of 50 mm at top to 30 mm at bottom. $E = 210 \text{ GN/m}^2$.

Solution:

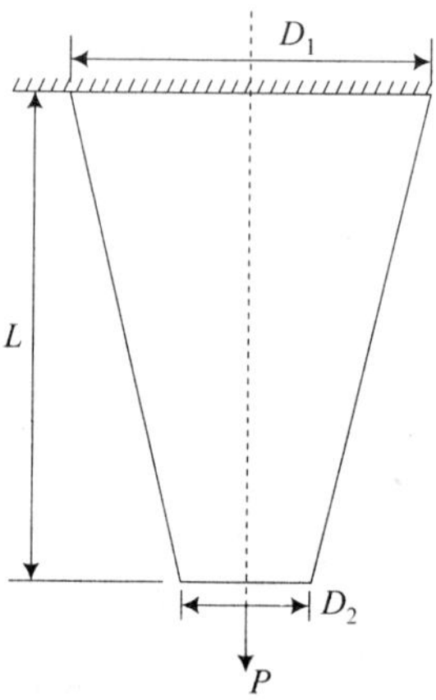

$P = 50 \text{ kN} = 50 \times 10^3 \text{ N}$

$L = 4 \text{ m} = 4000 \text{ mm}$

$E = 210 \text{ GN/m}^2 = 2.1 \times 10^5 \text{ N/mm}^2$

$$D_1 = 50 \text{ mm}$$
$$D_2 = 30 \text{ mm}$$

For a circular taper bar,

$$\delta L = \frac{4PL}{\pi E D_1 D_2} = \frac{4 \times (50 \times 10^3) \times 4000}{\pi \times (2.1 \times 10^5) \times 50 \times 30}$$

$$= \mathbf{0.8 \text{ mm}} \quad \textbf{Ans.}$$

Example 9.31: Two solid circular cross-section conical bars, one of titanium and the other of steel, are joined together as shown. The system is subjected to concentric axial tensile force of 500 kN together with an axisymmetric ring load applied at the junction of the bars having a horizontal resultant of 1000 kN. Determine the change in length of the system. For titanium, $E = 110$ GPa and for steel, $E = 200$ GPa.

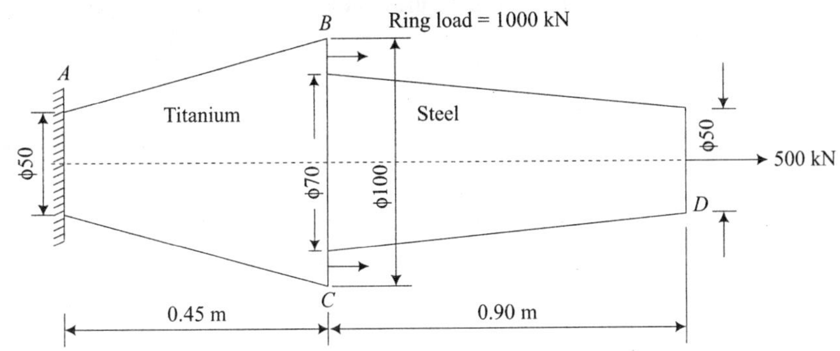

Solution:

For titanium cone AB,

$$P = 1000 + 500 = 1500 \text{ kN} = 1500 \times 10^3 \text{ N}$$
$$L = 0.45 \text{ m} = 450 \text{ mm}$$
$$D_1 = 100 \text{ mm}$$
$$D_2 = 50 \text{ mm}$$
$$E = 110 \text{ GPa} = 1.1 \times 10^5 \text{ N/mm}^2$$

$$\sigma L_1 = \frac{4PL}{\pi E D_1 D_2} = \frac{4 \times (1500 \times 10^3) \times 450}{\pi \times (1.1 \times 10^5) \times 100 \times 50} = 1.5626 \text{ mm}$$

For steel cone CD,

$$P = 500 \text{ kN} = 500 \times 10^3 \text{ N}$$

$$L = 0.90 \text{ mm} = 900 \text{ mm}$$
$$D_1 = 70 \text{ mm}$$
$$D_2 = 50 \text{ mm}$$
$$E = 2 \times 10^5 \text{ N/mm}^2$$

$$\delta L_2 = \frac{4PL}{\pi E D_1 D_2} = \frac{4 \times (50 \times 10^3) \times 900}{\pi \times (2 \times 10^5) \times 70 \times 50} = 0.8185 \text{ mm}$$

Total increase of length of the system,

$$\delta L = \delta L_1 + \delta L_2 = 1.5626 + 0.8185 \text{ mm}$$
$$= \mathbf{2.38\ mm} \quad \textbf{Ans.}$$

Example 9.32: A steel bar AB of uniform thickness 2 cm, tapers uniformly from 1.5 cm to 7.5 cm in a length of 50 cm. Determine the elongation of the bar, if an axial tensile force of 100 kN is applied on it. $E = 2 \times 10^5$ N/mm.

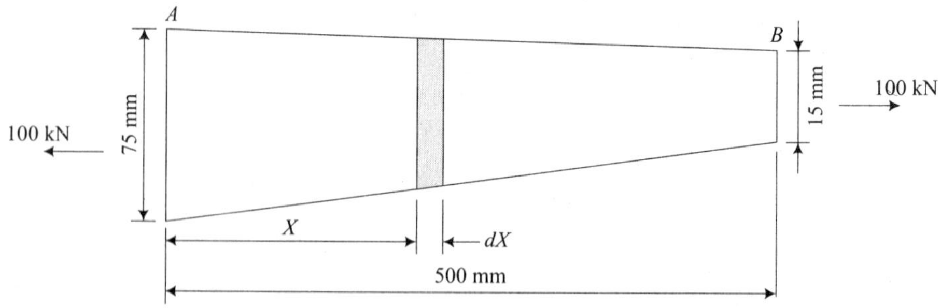

Solution:

Width of bar at a distance X from A,

$$b_X = b_A - \left(\frac{b_A - b_B}{L}\right) X = 75 - \left(\frac{75 - 15}{500}\right) X$$
$$= 75 - 0.12X$$

Area, $$A_X = 20\, b_X = 1500 - 2.4X$$

Stress, $$\sigma_X = \frac{P}{A_X} = \frac{100 \times 10^3}{(1500 - 2.4X)} = \frac{41667}{(625 - X)}$$

Strain, $$\varepsilon_X = \frac{\sigma_X}{E} = \frac{41667}{2 \times 10^5 (625 - X)} = \frac{1}{4.8(625 - X)}$$

$$\delta L_X = \varepsilon_X d_X = \frac{dX}{4.8(625 - X)}$$

Total elongation of bar.

$$\therefore \qquad \delta L = \int_0^{500} \frac{dX}{4.8(625-X)} = \frac{1}{4.8(-1)} \Big| \ln(625-X) \Big|_0^{500}$$

$$= -\frac{1}{4.8}(\ln 125 - \ln 625)$$

$$= \frac{1}{4.8}(\ln 625 - \ln 125) = \frac{1}{4.8} \ln \frac{625}{125}$$

$= \mathbf{0.335\ mm}$ **Ans.**

Example. 9.33: A beam weighing 500 N is held in horizontal position by three wires. The outer wires are of brass of 1.2 mm diameter and attached to each end of the beam. The central wire is of steel of 0.6 mm diameter and attached to the middle of the beam. The beam is rigid and the wires are of the same length and unstrained before the beam is attached. Determine the stress induced in each wire. The Young's modulus of brass is 80 GN/m^2 and that of steel is 200 GN/m^2.

Solution:

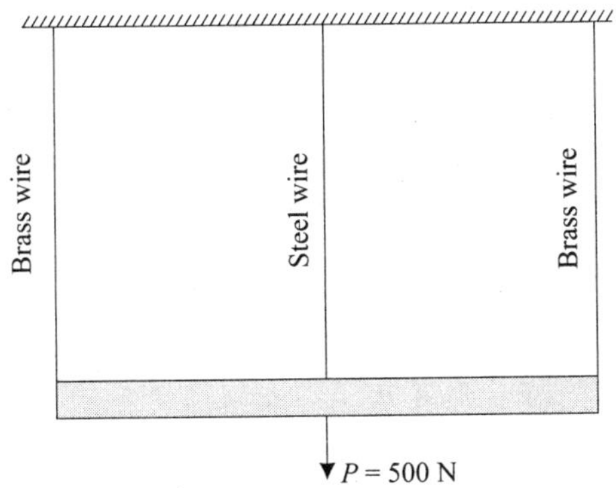

The total load,

$$P = 2P_b + P_S$$

where, P_b = load on brass wire

P_s = load on steel wire

$$\therefore \qquad P = 2\sigma_b A_b + \sigma_s A_s$$

Now, the wires have same strains.

$$\varepsilon_s = \varepsilon_b \text{ and } \delta L_s = \delta L_b$$

$$\therefore \qquad \frac{\sigma_s}{E_s} = \frac{\sigma_b}{E_b}$$

or,
$$\sigma_s = \frac{E_S}{E_b} \cdot \sigma_b = \frac{200}{80} \sigma_b = 2.5\sigma_b$$

$$500 = 2\sigma_b\left(\frac{\pi}{4} \times 1.2^2\right) + 2.5\,\sigma_b\left(\frac{\pi}{4} \times 0.6^2\right) = 2.9673\sigma_b$$

$\therefore$
$$\sigma_b \frac{500}{2.9673} = 168.5 \text{ N/mm}^2 \quad \text{Ans.}$$

$$\sigma_s = 2.5\sigma_b = 2.5 \times 168.5$$
$$= \mathbf{421.25 \ N/mm^2} \quad \textbf{Ans.}$$

Example 9.34: A brass tube of 60 mm outside diameter completely encloses a steel bar of 40 mm diameter. The composite system measures 300 mm in length and carries an axial thrust which induces a stress of 50 N/mm^2 in brass tube. Determine:

(a) Stress developed in steel bar.

(b) Magnitude of compressive force and change in length of composite bar. (E_s = 210 GPa, E_b = 105 GPa)

Solution:

Area of steel bar, $A_s = \frac{\pi}{4}(40^2) = 12566 \text{ mm}^2.$

Area of brass tube, $A_b = \frac{\pi}{4}(60^2 - 40^2) = 1570.8 \text{ mm}^2.$

(a) Total load = Load carried by steel bar + load carried by brass tube.

$$P = P_S + P_b$$

The brass tube and steel rod have same strain

$$\varepsilon_b = \varepsilon_s$$

$\therefore$
$$\frac{\sigma_b}{E_b} = \frac{\sigma_s}{E_s}$$

or
$$\sigma_s = \frac{E_s}{E_b} . \sigma_b = \frac{210}{105} \times 50 = \mathbf{100 \ N/mm^2} \quad \textbf{Ans.}$$

(b) Again,
$$P = P_b + P_s$$
$$= \sigma_b A_b + \sigma_s A_s$$
$$= 50 \times 1570.8 + 100 \times 12566$$
$$= 204200 \text{ N} = 204.2 \text{ kN}.$$

Extension of tube = Extension of rod

$$\delta L_b = \delta L_s$$

$$= \frac{\sigma_b L_b}{E_b} = \frac{P_b}{A_b} \cdot \frac{L_b}{E_b}$$

$$= \frac{50 \times 300}{1.05 \times 10^5}$$

$$= \mathbf{0.14285\ mm} \quad \textbf{Ans.}$$

9.9 EXTENSION OF BAR UNDER OWN WEIGHT

9.9.1 Bar of Uniform Section

A bar of cross-sectional area A and length L hanging vertical will extend under its own weight.

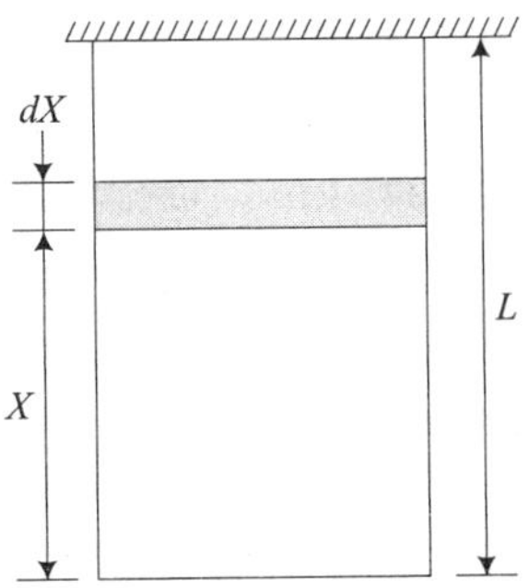

Fig. 9.15 Uniform bar under own weight

The weight of the bar of length X will be:

$$P - \rho g A X$$

This load will extend from dX by say δX.

$$\therefore \quad \delta X = \frac{PdX}{AE} = \frac{\rho g A X}{AE} dx = \frac{\rho g}{E} \times dx.$$

Total change in length of the bar under its own weight can be found out by integrating the above equation.

$$\therefore \quad \delta L = \int_o^L \frac{\rho g}{E} \times dX = \frac{\rho g}{E} \left| \frac{X^2}{2} \right|_0^L$$

$$= \frac{\rho g}{E} \cdot \frac{L^2}{2}.$$

If $$W = \rho g A L = \text{Total weight of bar}$$

$$\therefore \quad \delta L = \frac{W}{AL} \cdot \frac{L^2}{2E} = \frac{WL}{2AE}.$$

Therefore, extension of bar under its own weight is equal to extension of bar if effective load is half of its weight applied at the bar end.

Example 9.35: Determine the maximum length of mild steel bar of uniform cross-section that can be suspended vertically if the maximum stress is not to exceed 250 MN/m^2. Given, $g = 9.81$ m/sec^2 and density of steel, $\rho = 7.85 \times 10^3$ kg/m^3.

Solution: The maximum tensile force at the suspension point due to weight of rod,

$$P = \rho g A L = (7.85 \times 10^3 \times 9.81) AL (N)$$

$$\sigma_{max} = \frac{P}{A} = \frac{(7.85 \times 10^3 \times 9.81) AL}{A}$$

$$= 77 \times 10^3 \text{L N/m}^2$$

$$\therefore \quad 77 \times 10^3 \text{ L} = 250 \times 10^6$$

$$L = \mathbf{3.247\ km} \quad \textbf{Ans.}$$

Example 9.36: An electrical copper wire ($E = 1 \times 10^5$ N mm^2), 40 m long has cross-sectional area of 80 mm^2 and weighs 0.6 N/m run. If the wire is suspended vertically, calculate:

(a) the elongation of wire due to self-weight.

(b) the total elongation of wire when a weight of 200 N is attached to its lower end, and

(e) the maximum weight this wire can support at its lower end if the limiting value of stress is 65 N/mm^2.

Solutions:

(a) Elongation of wire due to self-weight,

$$\delta L = \frac{WL}{2AE} = \frac{(wL)L}{2AE} = \frac{(0.6 \times 40) \times 40 \times 10^3}{2 \times 80 \times (1 \times 10^5)}$$

$$= \mathbf{0.06\ mm} \quad \textbf{Ans.}$$

(b) Elongation of wire due to external load,

$$\delta L_1 = \frac{PL}{AE} = \frac{200 \times (40 \times 10^3)}{80 \times (1 \times 10^5)}$$

$$= 1.00 \text{ mm.}$$

Total extension of wire $= \delta L + \delta L_1 = 0.01 + 1.0$

$$= \mathbf{1.00\ mm} \quad \textbf{Ans.}$$

(c) Stress due to self-weight,

$$\sigma = \frac{W/2}{A} = \frac{wL^2}{2A} = \frac{0.6 \times 40 \times (40 \times 10)}{2 \times 80}$$

$$= 0.15 \text{ N/mm}^2$$

The balance stress, $\sigma_1 = 650 - 0.15 = 64.85 \text{ N/mm}^2$

Maximum weight which can be supported by wire

$$P = \sigma_1 A = 64.85 \times 80 = \mathbf{5188\ N} \quad \textbf{Ans.}$$

Example 9.36(a): A rectangular bar of uniform cross-section 4 cm × 2.5 cm and of length 2.2 m is hanging vertically from a rigid support. It is subjected to axial tensile loading of 10 kN. If density of steel is 8000 kg/m^3 and E = 200 GN/m^2, find the maximum stress and elongation of bar. [U.P.T.U. 2009-10]

Solution:

(i) Elongation of a bar due to external load

$$\delta L_1 = \frac{PL}{AE} = \frac{10 \times 2.2}{(4 \times 2.5) \times 10^{-4} \times 200 \times 10^6} = 1.1 \times 10^{-5} \text{ m}$$

$$= 0.011 \text{ mm}$$

The stress due to external loading

$$\sigma_1 = \frac{P}{A} = \frac{10}{(4 \times 2.5) \times 10^{-4}} = 10^4 \text{ kN/m}^2$$

(ii) The weight of the rod,

$$W = (4 \times 2.5) \times 10^{-4} \times 2.2 \times 8000 \times 9.81$$

$$= 172.656 \text{ N}$$

Elongation of bar due to self weight

$$\delta L_2 = \frac{WL}{2AE} = \frac{172.656 \times 10^{-3} \times 2.2}{2 \times (4 \times 2.5) \times 10^{-4} \times 200 \times 10^6}$$

$$= 0.095 \times 10^{-5} \text{ m}$$

$$= 0.00095 \text{ mm}$$

The stress due to self weight

$$\sigma_2 = \frac{w/2}{A} = \frac{172.656 \times 10^{-3} \times \frac{1}{2}}{(4 \times 2.5) \times 10^{-4}} = 86.328 \text{ kN/m}^2$$

Total elongation of rod,

$$\delta L = \delta L_1 + \delta L_2 = 0.011 + 0.00095$$

$$= \mathbf{0.01195\ mm} \quad \textbf{Ans.}$$

Total stress in the rod,

$$\sigma = \sigma_1 + \sigma_2 \quad = 1000 + 86.328$$

$$= \mathbf{10086.328\ kN/m^2} \quad \textbf{Ans.}$$

9.10 STRAIN ENERGY

When an external force acts on an elastic material it deforms and internal resistance is developed in the material due to molecular cohesion. The work done by the interval resistance is stored as strain energy within the elastic limits. This strain energy is called the resilience.

The maximum strain energy absorbed in this material when loaded upto elastic limit is called *proof resilience.*

Resilience, $$U = \frac{\sigma^2}{2E} \times \text{volume}$$

Proof resilience, $$U_e = \frac{\sigma_e^{\ 2}}{2E} \times \text{volume.}$$

where, σ = stress developed when the material is loaded within the elastic limit.

σ_e = proof stress developed when the material is loaded upto the elastic limit.

The proof resilience (maximum strain energy at elastic limit) per unit volume of material is called *modulus of resilience.*

Modulus of resilience, $$\frac{Ue}{V} = \frac{\sigma_e^{\ 2}}{2E}.$$

9.10.1 Strain Energy due to Gradually Applied Load

During gradually applied load the load starts from zero and reaches upto a value P linearly. The work done in straining the material will be given by the shaded area under the load-extension curve.

The strain energy,

$$U = \frac{P}{2}\delta L$$

But, $$P = \sigma A$$

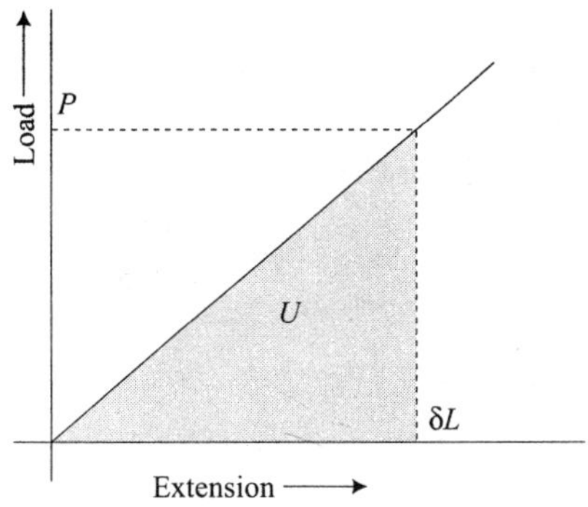

Fig. 9.16 Strain energy due to gradually applied load

and
$$\delta L = \varepsilon L = \frac{\sigma L}{E}$$

∴
$$U = \frac{1}{2}\sigma A \times \frac{\sigma L}{E}$$

$$= \frac{1}{2}\frac{\sigma^2}{E}AL = \frac{1}{2}\frac{\sigma^2}{E} \times \text{volume}$$

Modulus of resilience $= \frac{U}{V} = \frac{1}{2}\frac{\sigma^2}{E} = \frac{1}{2}\sigma.\frac{\sigma}{E} = \frac{\sigma.\varepsilon}{2}$

$$= \frac{\text{Stress} \times \text{Strain}}{2}$$

9.10.2 Strain Energy due to Suddenly Applied Load

The load is applied suddenly and it remains constant during extension of the material. The area of load-extension curve will be a rectangle.

Strain energy,

$$U = P.\delta L$$

$$\sigma = \frac{2P}{A} \text{ for suddenly applied load.}$$

The strain energy,

$$U = P.\left(\frac{PL}{AE}\right)$$

$$= \left(\frac{P}{A}\right)^2 . \frac{AL}{E} = \frac{\sigma^2}{E}.\text{volume}$$

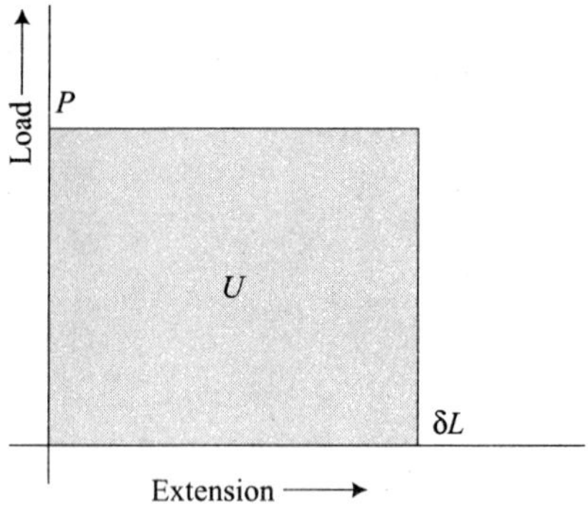

Fig. 9.17 Strain energy due to suddenly applied load

$$\text{Modulus of resilience,} = \frac{U}{V} = \frac{\sigma^2}{E}.$$

9.10.3 Strain Energy due to Impact Loading

When a weight P is dropped on a member from a height h, the falling weight causes impact loading. The strain energy is equal to the loss of potential energy of the weight.

$$U = P(h + \delta L)$$

where, h = Free height of fall of weight

δL = Extension due to loading

$$\therefore \quad U = Ph + P.\delta L$$

$$= Ph + \sigma^2 \frac{AL}{E}$$

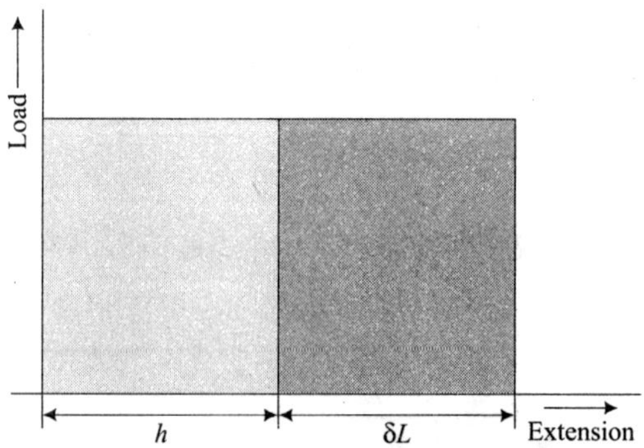

Fig. 9.18 Strain energy due to impact loading

Example 9.37: A steel bar of rectangular section 50 mm × 30 mm and length 1.5 m is subjected to a gradually applied load of 150 kN. Find the strain energy stored in the bar. If the elastic limit of material is 150 kN/mm^2, determine the proof resilience and modulus of resilience. Take $E = 2 \times 10^5$ N/mm^2.

Solution:

(a)
$$\sigma = \frac{P}{A} = \frac{150 \times 10^3}{50 \times 30} = 100 \text{ N/mm}^2$$

$$V = 50 \times 30 \times 1500 = 225 \times 10^4 \text{ mm}^2$$

$$\therefore \quad U = \frac{\sigma^2}{2E} V = \frac{(100)^2}{2 \times \left(20 \times 10^5\right)} \times (225 \times 10^4)$$

$$= 56250 \text{ N-mm.}$$

(b) Proof resilience
$$= U_e = \frac{\sigma_e{}^2}{2E} . V$$

$$= \frac{(150)^2}{2 \times \left(2 \times 10^5\right)} \times (225 \times 10^4)$$

$$= 144000 \text{ N-mm.}$$

(c) Modulus of resilience $= \dfrac{U_e}{V} = \dfrac{144000}{225 \times 10^4} = 0.064 \text{ N/mm}^2$

Example 9.38: A steel rod of 30 mm diameter and 2 m long has a pull of 50 kN suddenly applied to it. Calculate:

(a) Maximum instantaneous stress induced and the instantaneous elongation, and

(b) Strain energy stored and modulus of resilience.

Solution:

(a)
$$\sigma = \frac{2P}{A} = \frac{2 \times 50 \times 10^3}{\frac{\pi}{4}(30)^2}$$

$$= 141.54 \text{ N/mm}^2$$

$$\sigma_L = \frac{\sigma L}{E} = \frac{141.5 \times 2000}{(2 \times 10^5)} = 1.415 \text{ mm}$$

(b) Volume, $V = \frac{\pi}{4} D^2 \; L = \frac{\pi}{4} (30)^2 \times 2000$

$$= 14.13 \times 10^5 \text{ mm}^3$$

$$U = \frac{\sigma^2}{2E} . V = \frac{(141.54)^2}{2 \times (2 \times 10^5)} \times 14.13 \times 10^5$$

$$= 70768.6 \text{ N-mm}$$

Modulus of resilience,

$$\frac{U}{V} = \frac{70768.6}{14.13 \times 10^5} = \mathbf{0.05008 \text{ N/mm}^2}$$

9.11 COMPOUND STRESS AND STRAIN

The tensile and compressive stresses act on planes normal to the direction of external tensile and compressive forces. The shear stresses are induced on planes parallel to the direction of external shear forces. On planes inclined to applied external forces, both normal and shear stresses are induced. The combination of normal and shear stress is called compound stress. In most of the situations, the machine and structure members are subjected to two or more stresses simultaneously. These members have to be designed and analysed for compound stresses and strains.

9.11.1 State of Plane Stress

A cubic element *ABCD* within a strained material, when isolated, is found to be subjected to normal stress σ_x, σ_y and σ_z and shear stresses parallel to planes. Such a system is called three-dimensional stress system.

However, if on planes parallel to plane *ABCD*, there is no stress (σ_z = o), the cubic element is said to be under 2-*D* stress system or plane stress condition.

The following stresses will be acting on the isolated element:

1. The parallel planes *AD* and *BC* are under tensile stress, σ_x and shear stress, τ_{xy}.

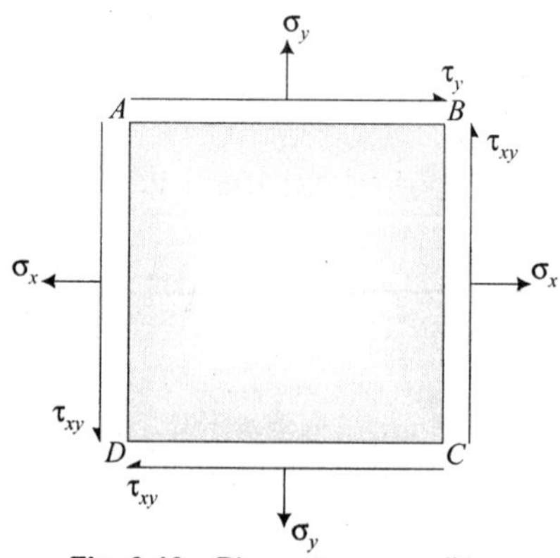

Fig. 9.19 Plane stress condition

2. The parallel planes *AB* and *CD* are under tensile stress, σ_y and shear stress (or complimentary shear stress), τ_{xy}.

9.11.2 Principal Planes

At any point within a stressed body, no matter how complex the state of stress may be, there always exist three mutually perpendicular planes on each of which the resultant stress is a normal stress (shear stress being zero). These mutually perpendicular planes are called principal planes.

In Fig. 9.20, the isolated element *ABCD* has been rotated anticlockwise through an angle θ to new position *A′B′CD′*. The stresses acting in the new position are normal stresses σ_1 and σ_2 and the shear stresses on all faces are absent. Then these mutually perpendicular planes, *B′C* and *A′B′* are called *principal planes*.

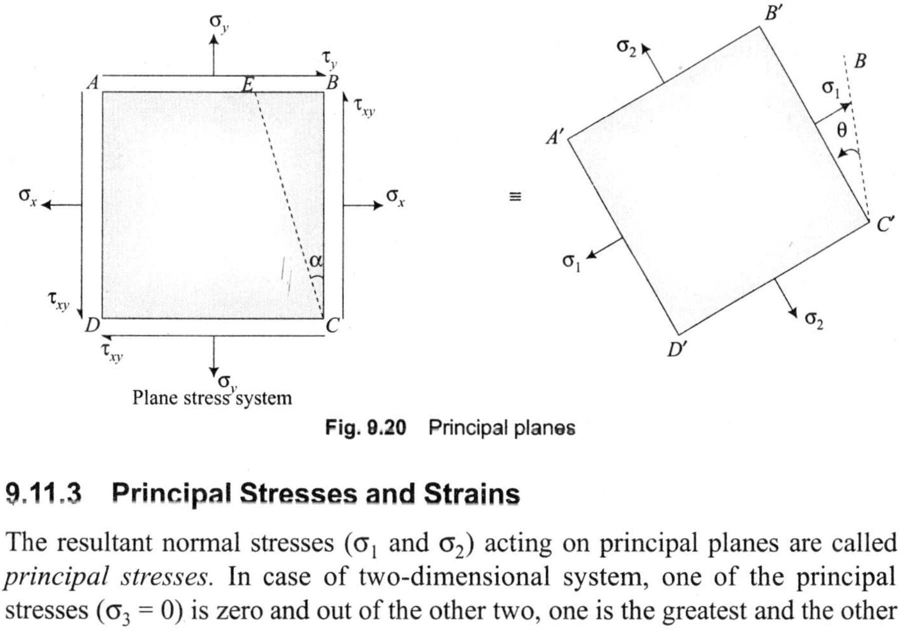

Fig. 9.20 Principal planes

9.11.3 Principal Stresses and Strains

The resultant normal stresses (σ_1 and σ_2) acting on principal planes are called *principal stresses.* In case of two-dimensional system, one of the principal stresses ($\sigma_3 = 0$) is zero and out of the other two, one is the greatest and the other is the least stress. The greatest stress is called the *major principal stress* and the least is called the *minor principal stress.*

The plane carrying the major principal stress is called the *major principal plane.* The plane carrying the minimum or minor principal stress is called minor *principal plane.*

In Fig. 9.20, if σ_1 is the major principal stress and σ_2 the minor principal stress, then plane *B′C* is the major principal plane and plane *A′B′* is the minor principal plane.

9.12 DETERMINATION OF PRINCIPAL STRESSES AND STRAINS

There are two methods for determination of principal stresses and strains.

1. Analytical method by the use of mathematical expression.
2. Graphical method by the use of Mohr's circle.

(a) The principal stresses can be calculated from the following general expresson.

The principal stress,

$$\sigma_p = \frac{1}{2}(\sigma_x + \sigma_y) \pm \frac{1}{2}\sqrt{\left(\sigma_x - \sigma_y\right)^2 + 4\tau_{xy}^{\ 2}}$$

Major principal stress,

$$\sigma_1 = \frac{1}{2}(\sigma_x + \sigma_y) + \frac{1}{2}\sqrt{\left(\sigma_x - \sigma_y\right)^2 + 4\tau_{xy}^{\ 2}}$$

Minor principal stress

$$\sigma_2 = \frac{1}{2}(\sigma_x + \sigma_y) - \frac{1}{2}\sqrt{\left(\sigma_x - \sigma_y\right)^2 + 4\tau_{xy}^{\ 2}}$$

where, σ_x = Normal stress along x-axis.

σ_y = Normal stress along y-axis.

τ_{xy} = Shear stress acting parallel to planes.

Sign Convention

1. Tensile stress is positive.
2. Compressive stress is negative.
3. Clockwise shear stress is positive.
4. Anticlockwise shear stress is negative.

(b) Maximum shear stress can be calculated as:

$$\tau_{max} = \frac{1}{2}\sqrt{\left(\sigma_x - \sigma_y\right)^2 + 4\tau_{xy}^{\ 2}}$$

$$= \frac{1}{2}(\sigma_1 - \sigma_2).$$

(c) Principal planes

If θ_1 is the angle between plane of major principal stress σ_1 with the plane of σ_x and θ_2 is the angle of plane of minor principal stress σ_2 with the plane of σ_x, then

$$\tan 2\theta = \frac{2_{\tau \times y}}{(\sigma_x - \sigma_y)}$$

$$\tan\theta_1 = \frac{\sigma_1 - \sigma_x}{\tau_{xy}}$$

$$\tan\theta_2 = \frac{\sigma_2 - \sigma_x}{\tau_{xy}}$$

Maximum shear stress occurs on planes at 45° to the principal planes.

(a) Principal strains.

Principal strains are the greatest and least direct strains in a material subjected to complex stresses. These strains are produced in the direction of the principal stresses.

The major principal strain,

$$\varepsilon_1 = \frac{\sigma_1}{E} - \frac{\sigma_2}{mE} = \frac{\sigma_1}{E} - \frac{v\sigma_2}{E}$$

The minor principal strain,

$$\varepsilon_2 = \frac{\sigma_2}{E} - \frac{v\sigma_1}{E}.$$

σ_1 and σ_2 are the major and minor principal stresses and v or $\frac{1}{m}$ is the Poisson's ratio.

Example 9.39: In an elastic material, the direct stresses of 120 MN/m^2 and 90 MN/m^2 are applied at a certain point on planes at right angles to each other in tension and compression respectively. Estimate the shear stress to which material could be subjected, if the maximum principal stress is 150 MN/m^2.

Also find the magnitude of other principal stress and its inclination to 120 MM/m^2. [U.P.T.U. II Sem., 2000–01]

Solution:

$$\sigma_x = 120 \text{ MN/m}^2 \text{ (T)}$$
$$\sigma_y = -90 \text{ MN/m}^2 \text{ (C)}$$
$$\sigma_1 = 150 \text{ MN/m}^2$$

Now,
$$\sigma_1 = \frac{\sigma_x + \sigma_y}{2} + \sqrt{\left(\frac{\sigma_x - \sigma_y}{2}\right)^2 + \tau_{xy}^2}$$

$$150 = \frac{120 - 90}{2} + \sqrt{\left(\frac{120 + 90}{2}\right)^2 + \tau_{xy}^2}$$

$\therefore$
$$\tau_{xy} = 84.85 \text{ MN/m}^2.$$

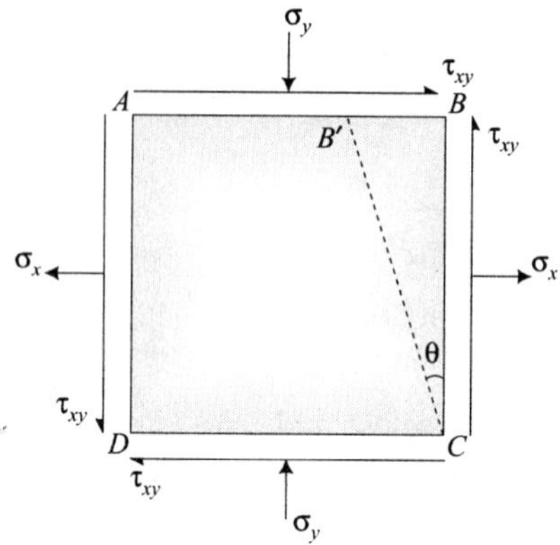

$$\therefore \qquad \sigma_2 = \frac{\sigma_x + \sigma_y}{2} - \sqrt{\left(\frac{\sigma_x - \sigma_y}{2}\right)^2 + \tau_{xy}^2}$$

$$= \frac{120 - 90}{2} - \sqrt{\left(\frac{120 + 90}{2}\right)^2 + (84.85)^2}$$

$$= -120 \text{ N/mm}^2$$

$$= 120 \text{ N/mm}^2 \text{ (C)}.$$

Angle of minor principal plane,

$$\tan \theta_2 = \frac{\sigma_2 - \sigma_x}{\tau_{xy}} = \frac{-120 - 120}{84.85}$$

$$\theta_2 = -70.5° \text{ or } 19.5°$$

Angle of plane of σ_2 with plane of σ_x i.e., angle of σ_2 with σ_x.

Example 9.40: A load-carrying member is subjected to the following stress conditions:

Tensile stress, $\sigma_x = 400$ MPa

Tensile stress, $\sigma_y = 300$ MPa

Shear stress, $\tau_{xy} = 200$ MPa clockwise

Obtain:

(i) Principal stresses and their planes

(ii) The maximum shear stress and its plane. [U.P.T.U. II Sem., 2001–02]

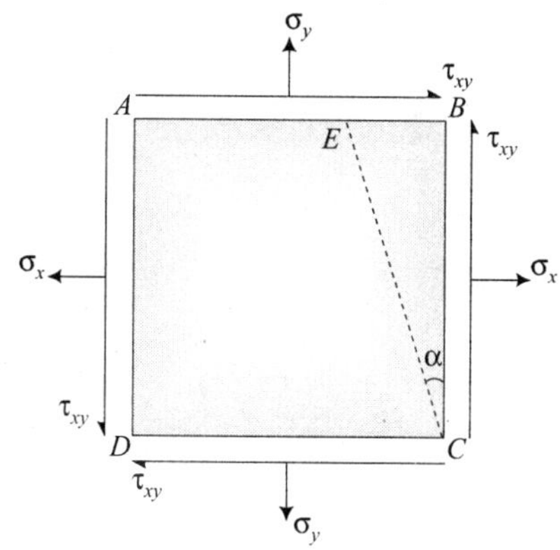

Solution:

(1) Principal stresses

$$\sigma_1 = \frac{\sigma_x + \sigma_y}{2} + \sqrt{\left(\frac{\sigma_x - \sigma_y}{2}\right)^2 + \tau_{xy}^2}$$

$$= \frac{400 - 300}{2} + \sqrt{\left(\frac{400 + 300}{2}\right)^2 + 200^2}$$

$$= 50 + 403$$

$$= 453 \text{ MN/ m}^2 \text{ (T)}$$

$$\sigma_2 = 50 - 403$$

$$= 353 \text{ MN/m}^2 \text{ (C)}.$$

$$\tan \theta_1 = \frac{\sigma_1 - \sigma_x}{\tau_{xy}} = \frac{453 - 400}{200}$$

∴ $\theta_1 = 14.84°$ Angle of plane of σ_1 with plane of σ_x (i.e. *BC*)

$$\tan \theta_2 = \frac{\sigma_2 - \sigma_x}{\tau_{xy}} = \frac{353 - 400}{200}$$

∴ $\theta_2 = -13.2°$ or $90 + 14.84° = 104.84°$. Angle of plane of σ_2 will plane of σ_x.

$$\tau_{max} = \sqrt{\frac{(\sigma_x - \sigma_y)^2}{2} + 4\tau_{xy}^2} = 403 \text{ MN/m}^2$$

Maximum shear stress marks an angle of 45° with principal plane.

Example 9.41: Find the principal stresses for the state of stress given below:

[U.P.T.U. II Sem., 2001–02]

Solution:

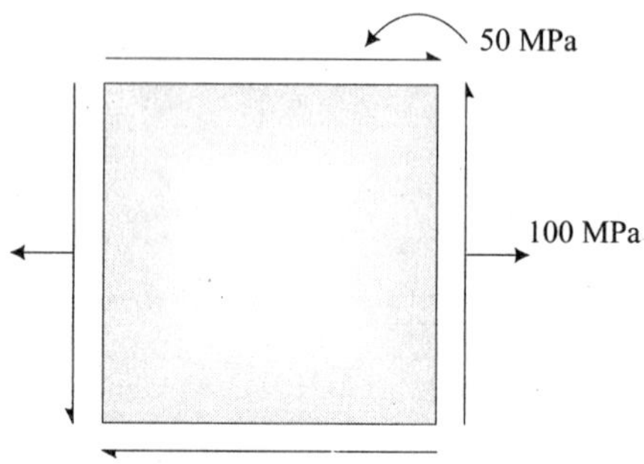

$$\sigma_x = 100 \text{ MPa}$$

$$\sigma_y = 0$$

$$\tau_{xy} = -50 \text{ MPa}$$

$$\sigma_1 = \frac{\sigma_x + \sigma_y}{2} + \sqrt{\left(\frac{\sigma_x - \sigma_y}{2}\right)^2 + \tau_{xy}^2}$$

$$= \frac{100 + 0}{2} + \sqrt{\left(\frac{100 - 0}{2}\right) + (-50)^2}$$

$$= 50 + 70.7$$

$$= 112.7 \text{ MPa (T)}$$

$$\sigma_2 = \frac{\sigma_x + \sigma_y}{2} - \sqrt{\left(\frac{\sigma_x - \sigma_y}{2}\right)^2 + \tau_{xy}^2}$$

$$= 50 - 70.7$$

$$= -20.7 \text{ MPa.}$$

$$\tan \theta_1 = \frac{\sigma_1 - \sigma_x}{\tau_{xy}} = \frac{70.7 - 100}{-50}$$

$$\theta_1 = 30.36°$$

$$\tan \theta_2 = \frac{\sigma_1 - \sigma_x}{\tau_{xy}} = \frac{-20.7 - 100}{-50}$$

$$\theta_2 = 67.5°$$

Example 9.42: The state of stress at a point in a loaded component is found to be as given below:

$$\sigma_x = 50 \text{ GN/m}^2$$
$$\sigma_y = 150 \text{ GN/m}^2$$
$$\tau_{xy} = 100 \text{ GN/m}^2$$

Determine the principal stresses and shearing stresses. Find the orientation of the planes on which they act. [U.P.T.U. II Sem., 2003–04)

Solution:

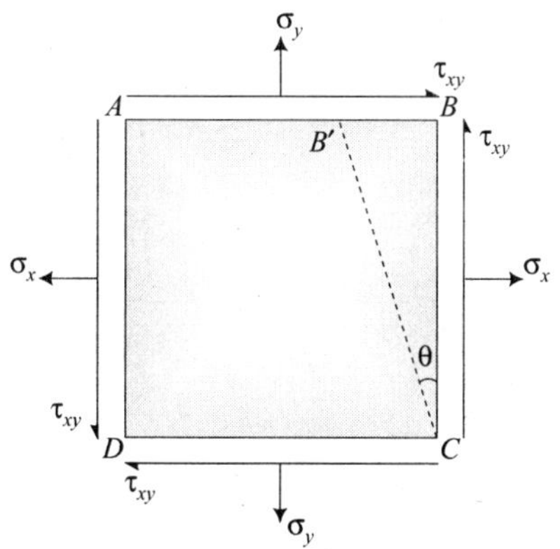

$$\sigma_1 = \frac{\sigma_x + \sigma_y}{2} + \sqrt{\left(\frac{\sigma_x - \sigma_y}{2}\right)^2 + \tau_{xy}^2}$$

$$= \frac{50 + 150}{2} + \sqrt{\left(\frac{50 - 150}{2}\right) + 100^2}$$

$$= 100 + 111.8$$

$$= 211.8 \text{ GN/m}^2 \text{ (T)}$$

$$\sigma_2 = \frac{\sigma_x + \sigma_y}{2} - \sqrt{\left(\frac{\sigma_x - \sigma_y}{2}\right)^2 + \tau_{xy}^2}$$

$$= 100 - 111.8 = -11.8 \text{ GN/m}^2 \text{ (C)}$$

$$\tau_{max} = \pm \sqrt{\left(\frac{\sigma_x - \sigma_y}{2}\right)^2 + \tau_{xy}^2}$$

$$= \pm\, 111.8 \text{ GN/m}^2$$

$$\tan\theta_1 = \frac{\sigma_1 - \sigma_x}{\tau_{xy}} = \frac{211.8 - 50}{100}$$

$$\theta_1 = 58.28°$$

$$\tan\theta_2 = \frac{\sigma_2 - \sigma_x}{\tau_{xy}} = \frac{-11.8 - 50}{100}$$

$$\theta_2 = -31.7° \text{ or } 58.28 + 90 = 148.28°$$

τ_{max} marks an angle of 45° with the principal stress or 45° + 58.28 = 103.28°.

Example 9.43: A rectangular element is subjected to a plane stress system as shown. Determine the principal stresses. [U.P.T.U. II Sem., 2003–04]

Solution:

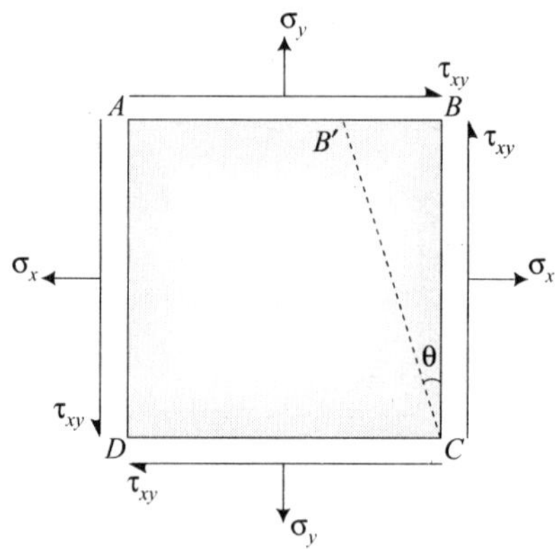

$$\sigma_x = 16 \text{ kN/m}^2$$

$$\sigma_y = 6 \text{ kN/m}^2$$

$$\tau_{xy} = 4 \text{ kN/m}^2$$

$$\sigma_1 = \frac{\sigma_x + \sigma_y}{2} + \sqrt{\left(\frac{\sigma_x - \sigma_y}{2}\right)^2 + \tau_{xy}^2}$$

$$= \frac{16+6}{2} + \sqrt{\left(\frac{16-6}{2}\right)^2 + 4^2}$$

$$= 11 + 6.4$$

$$= 17.4 \text{ kN/m}^2 \text{ (T)}$$

$$\sigma_2 = \frac{\sigma_x + \sigma_y}{2} - \sqrt{\left(\frac{\sigma_x - \sigma_y}{2}\right)^2 + \tau_{xy}^2}$$

$$= 11 - 6.4$$
$$= 4.6 \text{ kN/m}^2 \text{ (T)}$$

$$\tan \theta_1 = \frac{\sigma_1 - \sigma_x}{\tau_{xy}} = \frac{17.4 - 16}{4} = 0.35°$$

$$\theta = 19.3°$$

$$\tan \theta_2 = \frac{\sigma_2 - \sigma_x}{\tau_{xy}} = \frac{4.6 - 16}{4} = -2.85$$

$$\theta_2 = -70.67°.$$

Example 9.44: In an elastic material the direct stresses of 100 MN/m^2 and 80 MN/m^2 are applied at certain point on planes at right angle to each other in tension and compression respectively. Estimate the shear stress to which material can be subjected, if the maximum principal stress is 130 MN/m^2. Also find the magnitude of the other principal stress and its inclination to 100 MN/m^2 stress.

[U.P.T.U. I Sem., 2005–06]

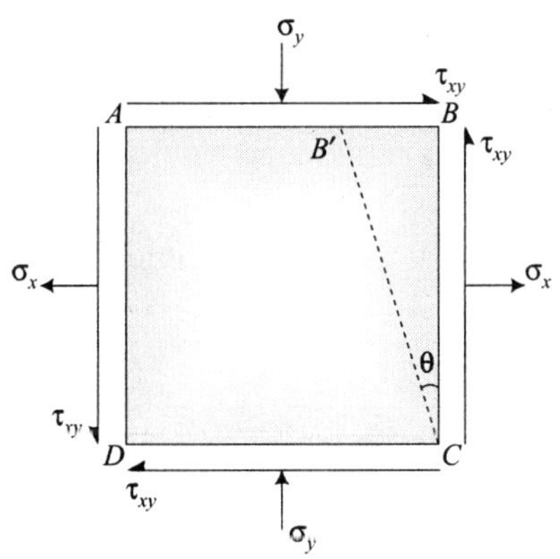

Solution:

$$\sigma_x = 100 \text{ MN/m}^2$$
$$\sigma_y = -80 \text{ MN/m}^2$$
$$\sigma_1 = 130 \text{ MN/m}^2.$$

$$\sigma_1 = \frac{\sigma_x + \sigma_y}{2} + \sqrt{\left(\frac{\sigma_x - \sigma_y}{2}\right)^2 + \tau_{xy}^2}$$

$$130 = \frac{100 - 80}{2} + \sqrt{\left(\frac{100 + 80}{2}\right)^2 + \tau_{xy}^2}$$

$$90^2 + \tau_{xy}^2 = 120^2$$

$$\tau_{xy} = \sqrt{120^2 - 90^2} = \mathbf{79.4\ MN/m^2} \quad \textbf{Ans.}$$

$$\sigma_2 = \frac{\sigma_x + \sigma_y}{2} - \sqrt{\left(\frac{\sigma_x - \sigma_y}{2}\right)^2 + \tau_{xy}^2}$$

$$= \frac{100 - 80}{2} - \sqrt{\left(\frac{100 + 80}{2}\right)^2 + (79.4)^2}$$

$$= 20 - 120 = -100 \text{ MN/m}^2 \text{ (compressive)}$$

$$\tan\theta_2 = \frac{\sigma_2 - \sigma_x}{\tau_{xy}} = \frac{-100 - 100}{79.4} = -2.51889$$

$$\theta_2 = -68.3° \text{ or } 21.65°.$$

Example 9.45: At a point in a beam, the bending stress is 50 MPa tensile and shear stress is 20 MPa. Determine:

(i) The principal stresses and maximum shear stress.

(ii) The tensile stress which acting alone would produce the same maximum shear stress as in (i) above.

(iii) The shear stress which acting alone would produce the same principal stress as in (i) above.

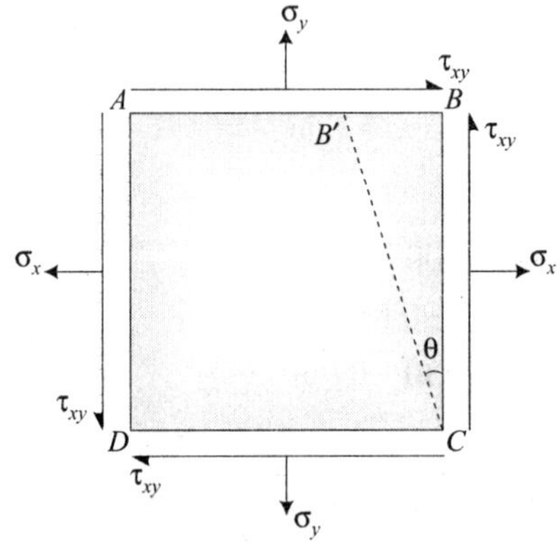

Solution:

(i)

$$\sigma_x = \sigma_b = \text{MPa}$$

$$\sigma_y = 0$$

$$\tau_{xy} = 20 \text{ MPa}$$

$$\sigma_1 = \frac{\sigma_x + \sigma_y}{2} + \sqrt{\left(\frac{\sigma_x - \sigma_y}{2}\right)^2 + \tau_{xy}^2}$$

$$= \frac{50-0}{2} + \sqrt{\left(\frac{50-0}{2}\right)^2 + 20^2}$$

$$= 25 + 64 = 89 \text{ MPa.}$$

$$\sigma_2 = \frac{\sigma_x + \sigma_y}{2} - \sqrt{\left(\frac{\sigma_x - \sigma_y}{2}\right)^2 + \tau_{xy}^2}$$

$$= 25 – 64 = – 39 \text{ MPa.}$$

$$\tan \theta_1 = \frac{\sigma_1 - \sigma_x}{\tau_{xy}} = \frac{89-56}{20} = 1.95$$

$\theta_1 = 62.85°$ i.e., angle of plane of σ_1 with plane of σ_x.

$\theta_2 = \theta_1 + 90° = 62.85 + 90° = 152.85$, i.e., angle of plane of σ_2 with plane of σ_x.

$$\tau_{max} = \frac{\sigma_1 - \sigma_2}{2} = \frac{89+39}{2} = 64 \text{ MPa}$$

τ_{max} acts on a plane at 45° to the principal plane.

$$\theta_s = 62.85° + 45° = 107.85°.$$

(ii) If $\tau_{max} = 64$ MPa, $\sigma_y = 0$ and $\tau_{xy} = 0$

$$\tau_{max} = \pm\sqrt{\left(\frac{\sigma_x - \sigma_y}{2}\right)^2 + \tau_{xy}^2}$$

$$= \frac{\sigma_x}{2} = 64$$

$\therefore$ $\sigma_x = 128$ MPa

(iii) If $\sigma_x = \sigma_y = 0$

$\sigma_1 = 89$ MPa

$$\sigma_1 = \frac{\sigma_x + \sigma_y}{2} + \sqrt{\left(\frac{\sigma_x - \sigma_y}{2}\right)^2 + \tau_{xy}^2}$$

$$= \tau_{xy}$$

$\therefore$ $\tau_{xy} = 89$ MPa

Example 9.46: The state of stress at a given point in a material component is shown. The loading on the component is increased by a factor of k. Determine the maximum value of k if the material can withstand maximum normal and shear stress of 300 MPa and 200 MPa respectively.

Solution: The new values of stresses are

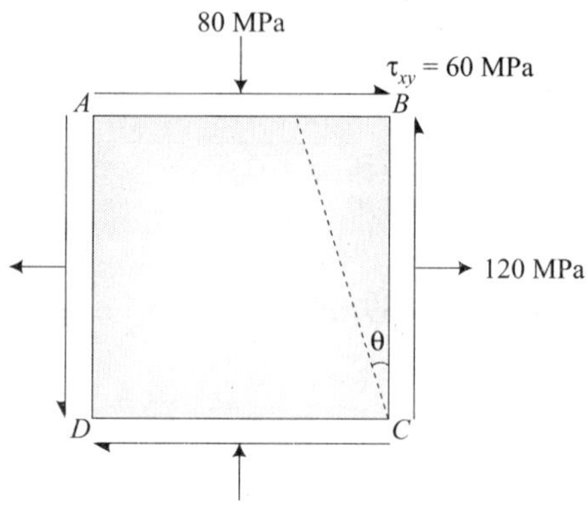

$$\sigma_x = 120\,k \text{ MPa}$$
$$\sigma_y = -\,80\,k \text{ MPa}$$
$$\tau_{xy} = 60\,k \text{ MPa.}$$

The maximum normal stress,

$$\sigma_1 = \frac{\sigma_x + \sigma_y}{2} + \sqrt{\left(\frac{\sigma_x - \sigma_y}{2}\right)^2 + \tau_{xy}^2}$$

$$\sigma_1 = \frac{120k - 80k}{2} + \sqrt{\left(\frac{12k + 80k}{2}\right)^2 + (60k)^2}$$

$$= 20\,k + 116.6\,k$$

$$= 136.62\,k = 300 \text{ MPa}$$

$$\therefore \quad k = \frac{300}{136.62} = 2.196$$

Maximum shear stress,

$$\tau_{max} = \sqrt{\left(\frac{\sigma_x - \sigma_y}{2}\right)^2 + \tau_{xy}^2}$$

$$= \sqrt{\left(\frac{12k + 80k}{2}\right)^2 + (60k)^2}$$

$$\tau_{max} = 116.6\,k = 200 \text{ MPa}$$

$$\therefore \quad k = \frac{200}{116.6} = 1.715$$

Maximum permissible value of k is **1.715** **Ans.**

9.13 OBLIQUE STRESSES

In a plane stress system, the direct and shear stresses acting on two mutually perpendicular planes of a stressed material can be transformed to another plane inclined with the planes of the system. The stresses on inclined plane are called oblique stresses.

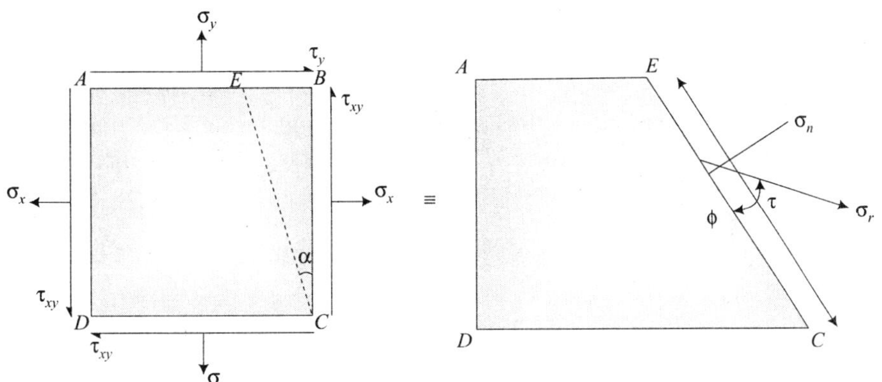

Fig. 9.21 Oblique stresses

Oblique stresses consisting of a normal stress σ_n and shear stress τ are acting on oblique plane *CE* which is at an angle α with plane *BC* in the anticlockwise direction.

$$\sigma_n = \frac{\sigma_x + \sigma_y}{2} + \frac{\sigma_x - \sigma_y}{2} \cos 2\alpha + \tau_{xy} \sin 2\alpha$$

$$\tau = \frac{\sigma_x - \sigma_y}{2} \sin 2\alpha - \tau_{xy} \cos 2\alpha.$$

The resultant stress,

$$\sigma_r = \sqrt{\sigma_n^2 + \tau^2}$$

The inclination of resultant stress with the oblique plane can be calculated as:

$$\tan \phi = \frac{\sigma_n}{\tau}.$$

The oblique stresses can also be found out graphically by the use of Mohr's circle.

Note: 1. If σ_y is compressive stress,

$$\sigma_n = \frac{\sigma_x - \sigma_y}{2} + \frac{\sigma_x + \sigma_y}{2}\cos 2\alpha + \tau_{xy}\sin 2\alpha$$

$$\tau = \frac{\sigma_x + \sigma_y}{2}\sin 2\alpha - \tau_{xy}\cos 2\alpha.$$

2. If shear stress is absent,

$$\tau_{xy} = 0$$

$$\sigma_n = \frac{\sigma_x + \sigma_y}{2} + \frac{\sigma_x - \sigma_y}{2}\cos 2\alpha$$

$$\tau = \frac{\sigma_x + \sigma_y}{2}\sin 2\alpha.$$

Example 9.47: The principal stresses at a point in a plane are 200 N/mm^2 (tensile) and 100 N/mm^2 (compressive). Determine the magnitude and direction of the resultant stress on a plane making an angle of 60° with the axis of major principal stress.

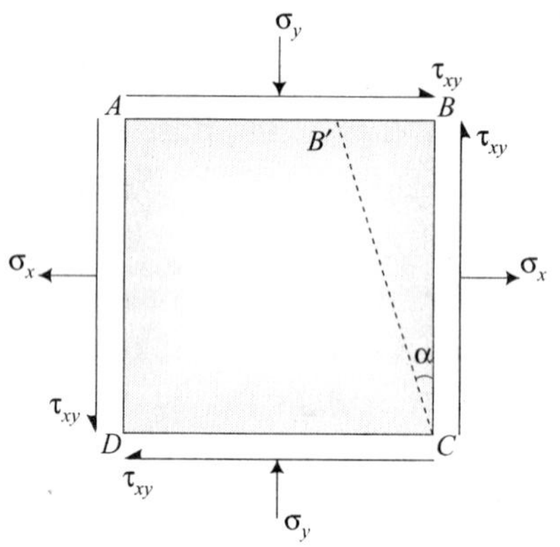

Solution:

$$\sigma_x = 200 \text{ N/mm}^2$$

$$\sigma_y = -100 \text{ N/mm}^2$$

$$\alpha = 90° - 60° = 30°$$

$$\sigma_n = \frac{\sigma_x + \sigma_y}{2} + \frac{\sigma_x - \sigma_y}{2}\cos 2\alpha + \tau_{xy}\sin 2\alpha$$

$$= \frac{200 - 100}{2} + \frac{200 + 100}{2}\cos 60° - 0$$

$$= 125 \text{ N/mm}^2.$$

$$\tau = \frac{\sigma_x - \sigma_y}{2} \sin 2\alpha - \tau_{xy} \cos 2\alpha$$

$$= \frac{200 + 100}{2} \cos 60°$$

$$= 129.9 \text{ N/mm}^2$$

The resultant stress,

$$\sigma_r = \sqrt{\sigma_n^2 + \tau^2}$$

$$= \sqrt{(125)^2 + (129.9)^2} = 180.27 \text{ N/mm}^2.$$

$$\tan \phi = \frac{\sigma_n}{\tau} = \frac{125}{129.9}$$

$\phi =$ 43.9° i.e., the inclination of σ_r with oblique plane.

Example 9.48: At a point in a bracket, the stresses on two mutually perpendicular planes are 350 N/mm² and 150 N/mm² both tensile in nature. The shear stress across these planes is 90 N/mm². Find the magnitude and direction of the resultant stress on a plane making an angle 40° with the plane of first stress. Also find the normal and tangential stress on the plane.

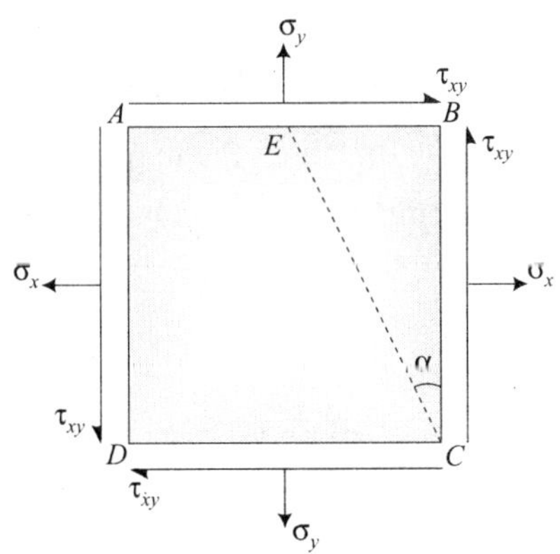

Solution:

$\sigma_x = 350 \text{ N/mm}^2$

$\sigma_y = 150 \text{ N/mm}^2$

$\tau_{xy} = 90 \text{ N/mm}^2$

$\alpha = 40°.$

Normal stress,

$$\sigma_n = \frac{\sigma_x + \sigma_y}{2} + \frac{\sigma_x - \sigma_y}{2} \cos 2\alpha + \tau_{xy} \sin 2\alpha$$

$$= \frac{350+150}{2} + \frac{350-150}{2} \cos 80° + 90 \sin 80°$$
$$= 250 + 17.365 + 88.632$$
$$= 366 \text{ N/mm}^2.$$

Tangential stress,

$$\tau = \frac{\sigma_x - \sigma_y}{2} \sin 2\alpha - \tau_{xy} \cos 2\alpha$$
$$= \frac{350-150}{2} \sin 80° - 90 \cos 80°$$
$$= 82.85 \text{ N/mm}^2$$

Resultant stress,

$$\sigma_r = \sqrt{\sigma_n^2 + \tau^2} = \sqrt{366^2 + 82.85^2}$$
$$= 375.28 \text{ N/mm}^2$$

If ϕ is the angle of resultant with the oblique plane,

$$\tan \phi = \frac{\sigma_n}{\tau} = \frac{366}{82.85}$$
$$\phi = 77°.$$

Example 9.49: The principal stresses at a point in a body are 30 and 50 MN/m^2, both tensile. Determine by calculations or by Mohr's circle, the following on a plane inclined at 40° to the plane on which the major principal stress acts:

(i) Normal and tangential compartments of stress

(ii) Magnitude of resultant stress

(iii) Angle of obliquity. [U.P.T.U. II Sem., 2002–03]

Solution:

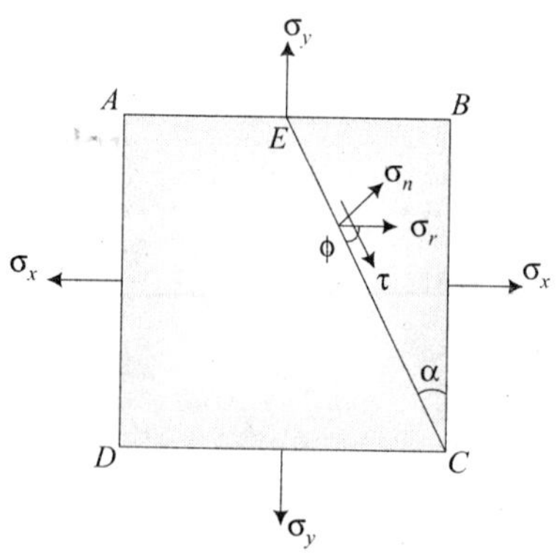

$$\sigma_x = 30 \text{ MN/m}^2$$
$$\sigma_y = 50 \text{ MN/m}^2$$
$$\tau_{xy} = 0$$
$$\alpha = 40°$$

$$\sigma_n = \frac{\sigma_s + \sigma_y}{2} + \frac{\sigma_x + \sigma_y}{2}\cos 2\alpha + \tau_{xy}\sin 2\alpha$$

$$= \frac{30 + 50}{2} + \frac{30 - 50}{2}\cos 80° - 0$$

$$= 40 - 3.47$$

$$= 36.\ 527 \text{ MN/m}^2.$$

$$\tau = \frac{\sigma_x - \sigma_y}{2}\sin 2\alpha - \tau_{xy}\cos 2\alpha$$

$$= \frac{30 - 50}{2}\sin 80° - 0$$

$$= -9.848 \text{ MN/m}^2$$

$$\tan\phi = \frac{\sigma_n}{\tau} = \frac{36.527}{-9.848}$$

$$\phi = -74.9° \text{ or } 105.1°.$$

9.14 MOHR'S CIRCLE

A graphical method was developed by Otto Mohr in 1882 to determine the following:

1. Oblique stresses, i.e., normal and shear stresses on any inclined plane through a point in the plane stress system.
2. Principal stresses and their planes.
3. Principal strains.

9.14.1 Construction of Mohr's Stress Circle

Sign Convention:

Tensile stress = +ve

Compressive stress = –ve

Clockwise shear stress = + ve

Anticlockwise shear stress = – ve.

1. Select a suitable scale on a graph paper.

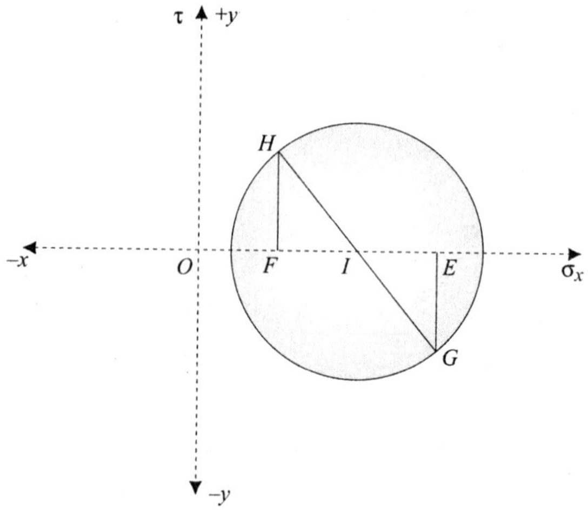

Fig. 9.22 Mohr's stress circle

1. Select a suitable scale on a graph paper.
2. Plot normal stresses on x-axis and shear stress on y-axis.
3. Cut $OE = \sigma_x$ and $OF = \sigma_y$.
4. Draw EG and FH perpendicutar to x-axis and cut $EG = -\tau_{xy}$ and $FH = +\tau_{xy}$.
5. Join GH intersecting x-axis in I.
6. With I as centre and HG as diameter, draw a circle. This circle is called Mohr's stress circle. The coordinates of I are $\dfrac{\sigma_x + \sigma_y}{2}$ and diameter of circle is $\sqrt{\left(\dfrac{\sigma_x - \sigma_y}{2}\right)^2 + \tau_{xy}^2}$

9.14.2 Use of Mohr's Circle for Finding Principal Stresses

(a) Convention

1. All normal stresses (σ_x, σ_y, σ_1, σ_2) are shown on x-axis.
2. All shear stresses (τ_{xy}, τ_{max}) are shown on y-axis.
3. All angles of planes are taken in anticlockwise direction from plane BC of stress σ_x.
4. Tensile stresses are +ve. Compressive stresses are –ve.
5. Shear stresses in clockwise direction are +ve and in anticlockwise direction are –ve.

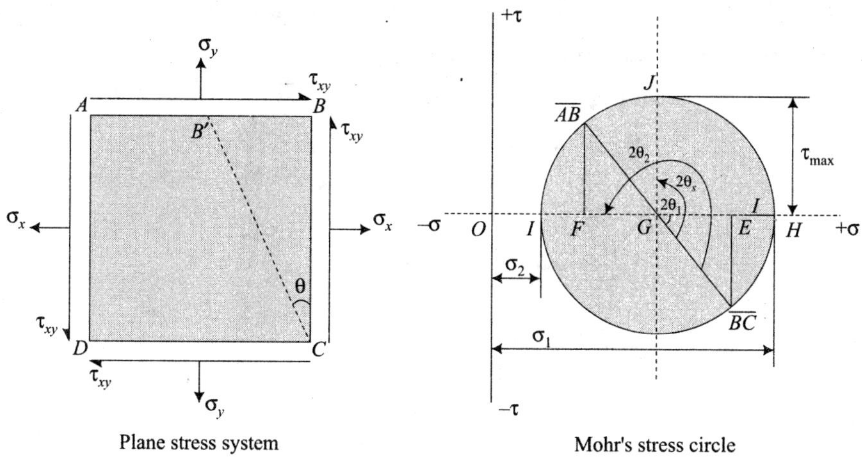

Fig. 9.23 Principal stresses by Mohr's circle

(b) Construction

Draw or cut to scale:

1. $OE = \sigma_x$ along x-axis
2. $OF = \sigma_y$ along x-axis
3. $F\,\overline{AB} = +\tau_{xy}$ and perpendicular to x-axis
4. $E\,\overline{BC} = -\tau_{xy}$ and perpendicular to x-axis
5. Join $\overline{AB}$ and $\overline{BC}$ intersecting x-axis in G. G is the centre of Mohr's stress circle and has coordinates $\left(\dfrac{\sigma_x + \sigma_y}{2}, 0\right)$
6. Draw a circle with G as centre and $\overline{AB} - \overline{BC}$ as diameter. The radius of Mohr's circle is $\dfrac{\sigma_1 - \sigma_2}{2}$ or τ_{max}. The circle is the Mohr's stress circle.

(c) Measurements

Measure the following and multiply by scale selected.

1. $OH = \sigma_1$, major principal stress
2. $OI = \sigma_2$, minor principal stress.
3. $GJ = \tau_{max}$, maximum shear stress.
4. Measure $\angle \overline{BC}\,GH$ as $2\theta_1$.
   θ_1 is the angle of major principal plane $(G - \overline{BC})$ with plane BC.
5. Measure $\angle \overline{BC} - G - I$ as $2\theta_2$.
   θ_2 is the angle of minor principal plane $(\overline{AB} - G)$ with plane BC.
6. Measure $\angle \overline{BC} - GJ$ as $2\theta_s$.
   θ_s is the angle of plane of maximum shear stress with plane BC.

9.14.3 Mohr's Stress Circle for Different Loadings

(a) Uniaxial stress system

Uniaxial stress system is shown in Fig. 9.24.

$$\sigma_y = 0 \text{ and } \tau_{xy} = 0$$

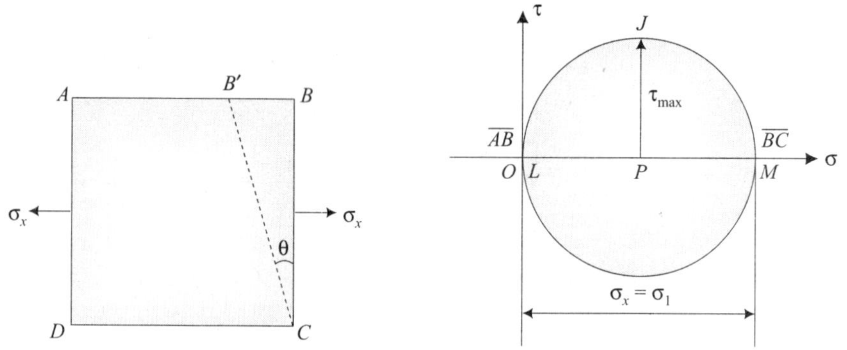

Fig. 9.24 Uniaxial stress system

1. Draw $OM = \sigma_x$ and $OL = \sigma_y = 0$.
2. Take $LP = PM$.
3. With P as centre and LM as diameter draw Mohr's stress circle.
4. $\overline{BC}$ coincides with M and $\overline{AB}$ coincides with L as $\tau_{xy} = 0$.
5. $\sigma_1 = \sigma_x = OM$
6. $\sigma_2 = \sigma_y = OL = 0$.
7. $\tau_{max} = JP = \dfrac{OM}{2} = \dfrac{\sigma_x}{2}$.
8. $\theta_1 = \dfrac{1}{2}\angle MP\overline{BC} = 0°$
9. $\theta_2 = \dfrac{1}{2}\angle MOL = \dfrac{180°}{2} = 90°$.
10 $\theta_5 = \dfrac{1}{2}\angle MPJ = \dfrac{90°}{2} = 45°$.

(b) Biaxial stress system

Biaxial stress system is shown in Fig. 9.25. $\tau_{xy} = 0$.

1. Draw $OM = \sigma_x$ and $OL = \sigma_y$
2. Take P as bisector of LM.
3. With P as centre and LM as diameter, draw Mohr's stress circle.
4. As $\tau_{xy} = 0$, $\overline{BC}$ coincides with M and $\overline{AB}$ coincides with L.
5. Measure $OM = \sigma_1 = \sigma_x$.

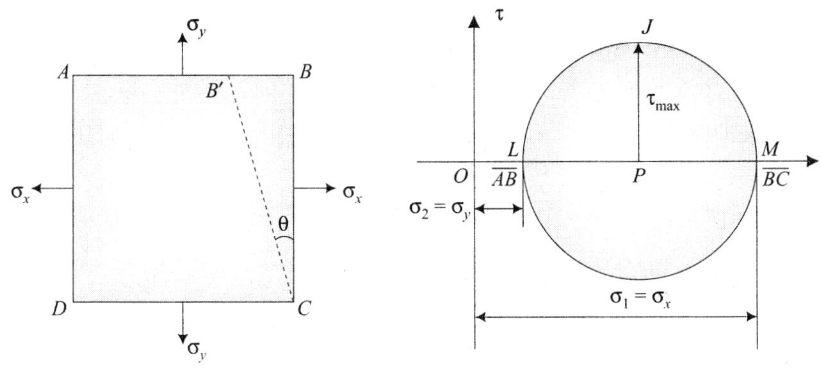

Fig. 9.25 Biaxial stress system

6. Measure $OL = \sigma_2 = \sigma_y$

7. Measure $JP = \tau_{max} = \dfrac{\sigma_x - \sigma_y}{2} = \dfrac{\sigma_1 - \sigma_2}{2}$.

8. $\theta_1 = \dfrac{1}{2} \angle \overline{BC}\text{-}P\text{-}M = 0.$

9. $\theta_2 = \dfrac{1}{2} \angle \overline{BC}\text{-}P\text{-}L = \dfrac{180°}{2} = 90°$

10. $\theta_s = \dfrac{1}{2} \angle \overline{BC}\text{-}P\text{-}J = \dfrac{90°}{2} = 45°.$

(c) Biaxial stress system with equal tension and compression

The stress system is shown in Fig. 9.26. $\sigma_x = \sigma_y$ and $\tau_{xy} = 0$.

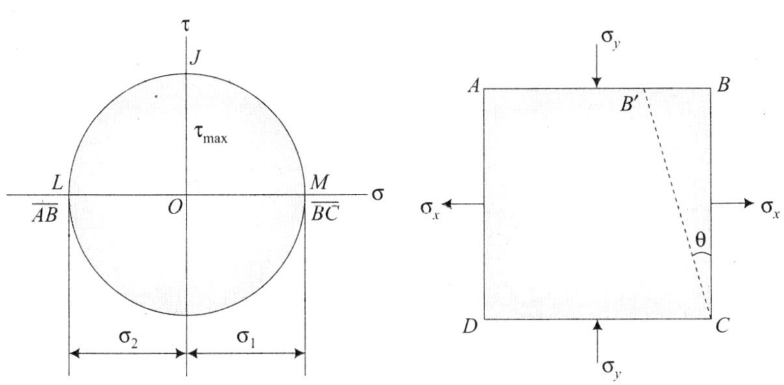

Fig. 9.26 Biaxial stress system with equal tension and compression

1. Draw $OM = \sigma_x$ centre and $OL = \sigma_y$.
2. With O as centre and LM as diameter, draw Mohr's stress circle.
3. $\overline{BC}$ coincides with M and $\overline{AB}$ coincides with L as $\tau_{xy} = 0$.
4. Measure $OM = \sigma_x = \sigma_1$ (tensile)

   $OL = \sigma_y = \sigma_2$ (compression)

   $OJ = \tau_{max} = \sigma_1 = \sigma_2$

5. $\theta_1 = \frac{1}{2} \angle MO\overline{BC} = 0°$.

   $\theta_2 = \frac{1}{2} \angle MOL = \frac{180°}{2} = 90°$

   $\theta_s = \frac{1}{2} \angle MOJ = \frac{90°}{2} = 45°$.

(d) Pure shear

The pure shear stress system is shown in Fig. 9.27. $\sigma_x = 0$ and $\sigma_y = 0$.

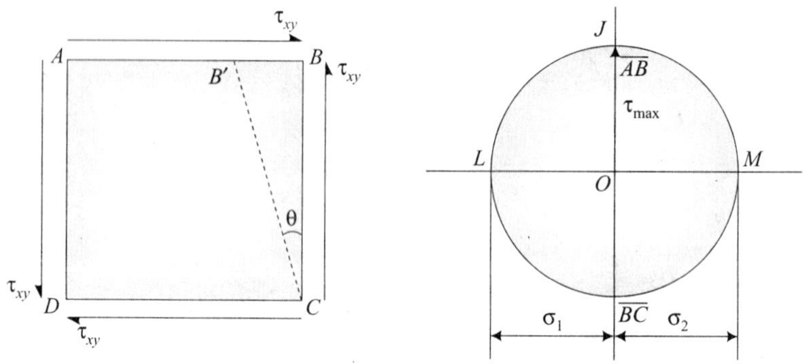

Fig. 9.27 Pure shear

1. Draw $OJ = \tau_{xy}$.
2. With O as centre and OJ as radius draw Mohr's stress circle cutting x-axis in L and M.
3. Measure $OL = \sigma_2 = \tau_{xy}$ $\qquad OJ = \tau_{max} = \tau_{xy}$

   $OM = \sigma_1 = \tau_{xy}$

4. Measure $\theta_1 = \frac{1}{2} \angle \overline{BC}\, OM = \frac{90°}{2} = 45°$.

   $\theta_2 = \frac{1}{2} \angle \overline{BC}\, OL = \frac{270°}{2} = 135°$

   $\theta_s = \frac{1}{2} \angle \overline{BC}\, OJ = \frac{180°}{2} = 90°$.

Example 9.50: The stresses on two mutually perpendicular planes ar 40 N/mm^2 (tensile) and 20 N/mm^2 (tensile). The shear stress across these planes is 10 N/mm^2. Find out principal stresses and their orientation using Mohr's circle. Also find the magnitude and direction of maximum shear stress.

Solution:

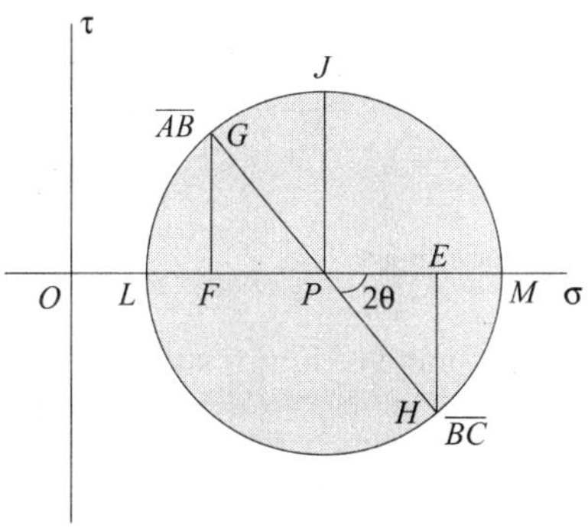

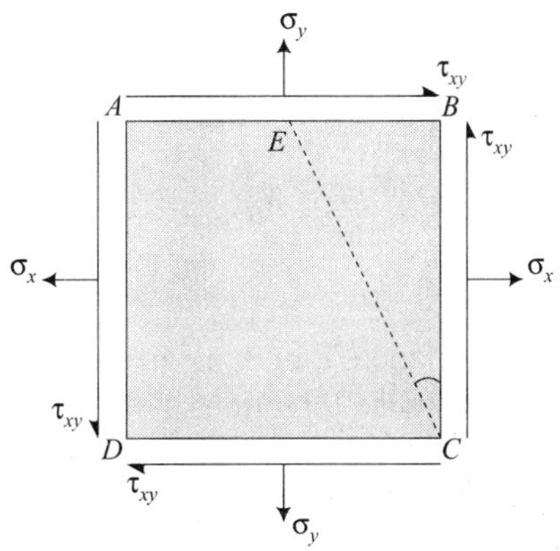

Draw Mohr's circle to a suitable scale.

1. Cut OE = 40 N/mm^2 and OF = 20 N/mm^2 on x-axis.
2. Draw FG perpendicular to x-axis and cut FG = 10 N/mm^2.
3. Draw EH perpendicular to x-axis and cut EN = –10 N/mm^2.
4. Joins GH intersecting x-axis in P.
5. With P as centre and PG as radius draw the Mohr's stress circle.
6. Measure and multiply by the selected scale.

   $OM = \sigma_1$

   $OL = \sigma_2$

   $PJ = \tau_{max}$

$$\theta_1 = \frac{1}{2} \angle EPH$$

$$\theta_2 = \frac{1}{2} \angle HPF \text{ or } \theta_1 + 90°$$

$$\theta_s = \frac{1}{2} \angle HPJ \text{ or } \theta_1 + 45°$$

9.14.4 Use of Mohr's Circle for Oblique Stresses

Mohr's circle is a powerful tool for finding oblique stress. Cube $ABCD$ represents a plane stress system. It is required to find normal, shear and resultant stress on an

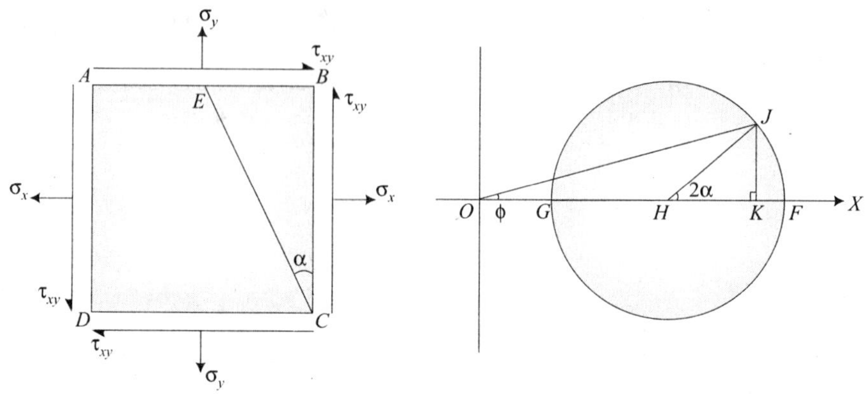

Fig. 9.28 Mohr's circle for oblique stress

oblique plane *CE* making an angle α with *BC*, i.e., plane of major stress σ_x in the anti-clockwise direction.

(a) Construction

1. Draw *OX* horizontally.
2. Cut $OF = \sigma_x$ and $OG = \sigma_y$ to a selected scale.
3. Bisect *GF* so that $GH = HF$.
4. With *H* as centre, draw a circle with *HF* as radius.
5. Draw *HJ* so that $\angle FHJ = 2\alpha$.
6. Draw *JK* perpendicular to *HF*.
7. Join *OJ*.

(b) Measurements

Measure the following and multiply by scale.

1. $OK = \sigma_n$
2. $JK = \tau$
3. $OJ = \sigma_r$
4. $\angle KOJ = \phi$, i.e., angle of obliquity.

Note: If σ_x is tensile and σ_y is a compressive stress, the Mohr's circle can be drawn as shown in Fig. 9.29.

1. Draw *XOX* horizontally.
2. Cut $OF = \sigma_x$ and $OG = \sigma_y$ on left-hand side to scale.
3. Bisect *GF* in *H*.
4. With *H* as centre and *HF* as radius, draw a circle.
5. Draw the line *HJ* so that $\angle FHJ = 2\alpha$.
6. Draw line *JK* perpendicular on *XOX*.
7. Join *OJ*.

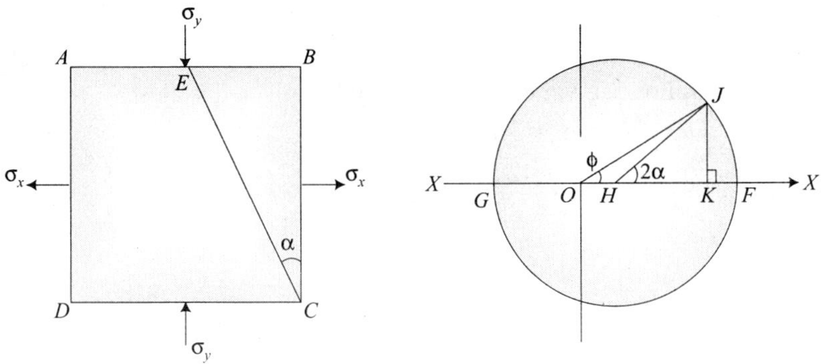

Fig. 9.29 Mohr's circle for unlike stresses

Measure the following and multiply by scale.

1. $OK = \sigma_n$
2. $JK = \tau$
3. $OJ = \sigma_r$
4. $\angle KOJ = \phi$, i.e., angle of obliquity.

Example 9.51: A uniform steel bar of 2 cm × 2 cm area of cross-section is subjected to an axial pull of 4000 kg. Calculate the intensity of normal stress, shear stress on and resultant stress a plane, normal to which is inclined at 30° to the axis of the bar. Solve the problem graphically by drawing Mohr's circle.

[U.P.T.U. I Sem., 2002–03]

Solution:

$$\sigma_x = \frac{P}{A} = \frac{4000 \times 9.81}{20 \times 20} = 98 \text{ N/mm}^2$$

$$\alpha = 30°$$

$$\sigma_n = \sigma_x \cos^2\alpha = 98 \cos^2 30° = 73.5 \text{ N/mm}^2$$

$$\tau = \sigma_x \cos\alpha \sin\alpha = 98 \cos 30° \sin 30° = 42.435 \text{ N/mm}^2$$

$$\sigma_r = \sqrt{\sigma_n^2 + \tau^2} = \sqrt{73.5^2 + 42.435^2}$$

$$= 84.87 \text{ N/mm}^2.$$

Graphical Solution

1. Draw a horizontal line *X–O–X*.
2. Cut $OF = \sigma_x = 98$ N/mm^2 to a suitable scale.
3. Bisect the line *OF* in *H*.
4. Draw a circle with *H* as centre and *HF* as radius.
5. Draw a line *HJ* so that $\angle FHJ = 2\alpha = 60°$.

6. Draw JK perpendicular to XOX.
7. Join OJ.

Measure the following values and multiply by selected scales:

$$OK = \sigma_n = 73.5 \text{ N/mm}^2$$
$$JK = \tau = 42.5 \text{ N/mm}^2$$
$$OJ = \sigma_r = 85 \text{ N/mm}^2.$$

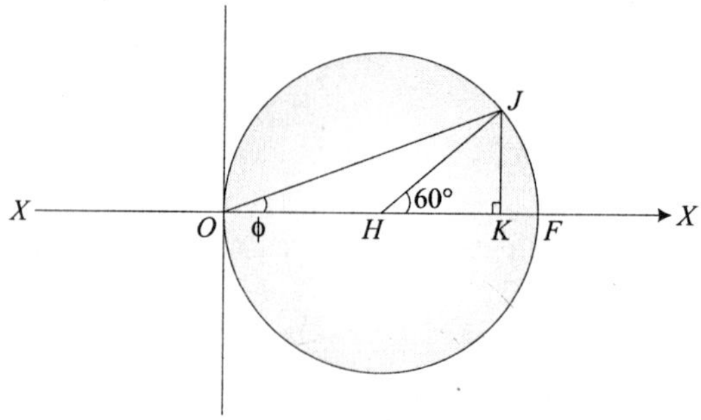

QUESTION BANK NO. 9

1. Discuss the effect of loading of an elastic material.
2. How do you differentiate between normal and shear stress?
3. Draw stress-strain diagram for structural steel and cast iron and briefly explain the various salient points.
4. With the help of suitable stress-strain diagram, explain the difference in the behaviour of ductile and brittle materials.
5. Explain the following elastic constants and establish relation between them:
   (i) Modulus of Elasticity
   (ii) Modulus of Rigidity
   (iii) Bulk Modulus
   (iv) Poisson's Ratio.
6. (a) Define resilience, proof resilience and modulus of resilience.
   (b) What is the value of strain energy for gradually applied load, sudden load and impact load?
7. Describe briefly:
   (i) State of plane stress
   (ii) Principal planes
   (iii) Principal stresses and strains.
8. Describe the construction of Mohr's stress circle for a plane stress system. Explain how is it used to find:

(i) Principal stresses and their planes.
(ii) Maximum shear stress and its plane.

9. Draw Mohr's stress circle for the following loadings:
   (i) Uniaxial stress
   (ii) Biaxial stress
   (iii) Equal tension and compression
   (iv) Pure shear.
10. Explain clearly Mohr's circle method of finding oblique stresses in a rectangular element subjected to normal stresses σ_x, σ_y and shear stress τ_{xy}.
11. Explain the following:
    (i) Complementary shear stress
    (ii) Angle of obliquity
    (iii) Shear strain energy.
12. Derive the following expression for elastic constants.

$$K = \frac{E}{3(1-2\nu)}$$

13. Define Poisson's Ratio. Prove that its value lies between zero and half.
14. What is strain energy? Derive its value in terms of strain and E.

QUICK REVIEW INFORMATION NO. 9

1. In a tensile test on mild steel specimen, the yield stress is less than the ultimate stress and breaking stress is less than the ultimate stress if all the stresses are calculated on the basis of initial cross-sectional area of specimen.
2. Maximum possible value (theoretical) for the Poisson's ratio is 0.5 and actual value for the cork is approximately 0.4.

CHAPTER 10

Bending and Torsion

10.1 INTRODUCTION

Beam is a structural member to support external loads at right angle to its axis. The load applied on a beam gets transferred to its supports. Any section of the beam experiences the following:

1. Shear force that tries to shear off the section. This force is resisted by the beam and shear stress is induced.
2. Bending moment which tries to bend it. Consequently, longitudinal or bending stresses are induced in the cross-section of the beam.

The bending moment and shear force vary from point to point along the length of the beam. However, in a large number of practical cases, the bending moment is maximum when the shear force is zero or changes the sign and passes through zero value. Therefore, it is desired to design the beam for maximum bending moment.

If a length of a beam is subjected to a constant bending moment and no shear force, it is said to be in *pure bending* or simple bending. The bending stress for that part of the beam remains constant.

10.2 THEORY OF SIMPLE BENDING

When a beam is subjected to simple bending moment M, the beam will bend with some radius of curvature R. The curvature can be taken as circular.

The maximum compression develops at the topmost layer of beam and maximum tension develops at the bottom-most layer in a hogging beam.

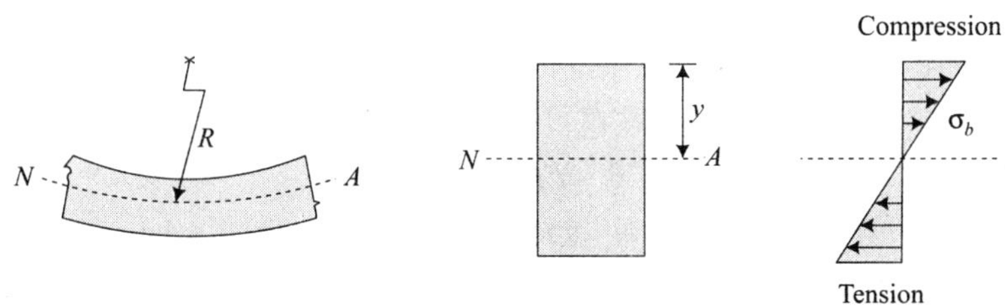

Fig. 10.1 Stress diagram in simple bending

The neutral axis passing through the centroid of the beam section is neither in tension nor compression and value of stress is zero. The magnitude of bending stress, σ_b and bending strain, ε_b at any depth of beam are proportional to its distance from neutral layer of the beam. The neutral layer is called neutral axis. The stress variation along the beam depth is linear and can be represented by triangles.

10.2.1 Assumptions in the Theory of Simple Bending

The following assumptions are made in the theory of simple bending:

1. The material of the beam is homogeneous and isotropic. The material properties are same in all directions.
2. The beam is initially straight and of uniform section throughout the length.
3. The radius of curvature of bending is very large in comparison to cross-sectional dimensions.
4. The stresses developed are within elastic limit of the material.
5. The material is elastic and obeys Hooke's law.
6. The value of Young's modulus of elasticity, E is same in tension and compression.
7. The loads are perpendicular to longitudinal axis of the beam.
8. Only longitudinal stresses are present in the beam and there is no lateral stresses.
9. The transverse section of the beam remains plane before and after bending.
10. Each layer of beam is free to expand (elongate) or contract (shorten) independently of the layer below or above it. There is no friction or shear force between the layers.

10.2.2 Bending Stress

The neutral axis of a beam is bent under the action of bending moment or a couple M, and forms an arc of the circle of radius R.

A beam section is shown before and after bending in Fig. 10.2. Consider a small length δx of a beam subjected to bending moment M.

Original length of neutral axis $NL = \delta x$

Changed length of neutral axis $N'L' = R\theta$

As there is no change in the length of neutral axis,

$$NL = N'L'$$

$$\therefore \quad \delta x = R\theta = EF$$

where, EF is a layer parallel to NA and at a distance y. This layer will change to $E'F'$ after bending,

$$\therefore \quad E'F' = (R + y)\,\theta$$

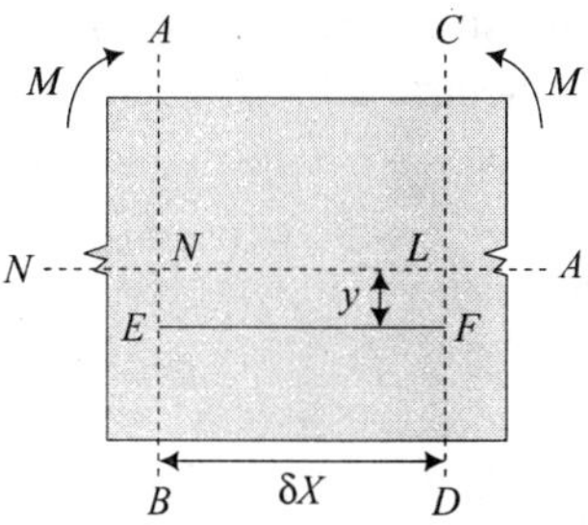

Beam before bending

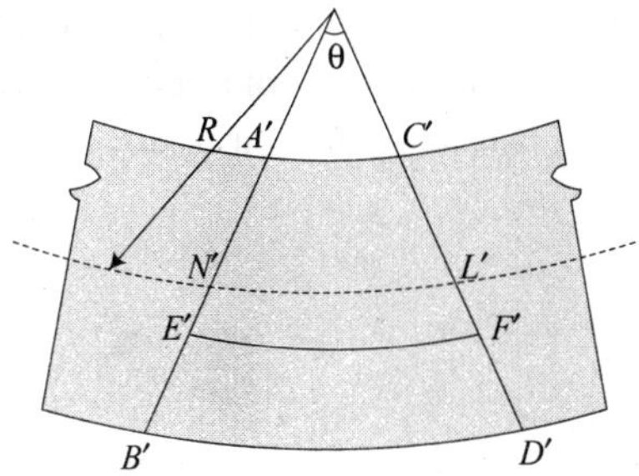

Beam after bending

Fig. 10.2 Bending stress

Change in the length of layer EF,

$$\delta EF = E'F' - EF$$

$$= (R + y)\theta - R\theta = y\theta.$$

$$\text{Strain} = \frac{\text{Change in length}}{\text{Original length}}$$

$$\varepsilon = \frac{Y\theta}{R\theta} = \frac{Y}{R}$$

$$\text{Modulus of elasticity} = \frac{\text{stress}}{\text{strain}}$$

$\therefore$

$$E = \frac{\sigma_b}{\varepsilon} = \frac{\sigma_b}{Y/R}$$

$\therefore$ Bending stress,

$$\sigma_b = \frac{YE}{R}$$

or,

$$\frac{\sigma_b}{Y} = \frac{E}{R}$$

Since, $\frac{E}{R}$ is a constant for a beam, bending stress, σ_b is proportional to its distance from neutral axis and varies linearly.

10.3 BENDING EQUATION

For pure sagging moment on a beam, all layers above neutral axis undergo compressive stress varying from zero at neutral axis and maximum at top layer.

Layers below neutral axis will be under tension with stress varying from zero at neutral axis to maximum at bottom-most layer. The neutral axis is the centroidal axis and can be found out by the centroid of the section.

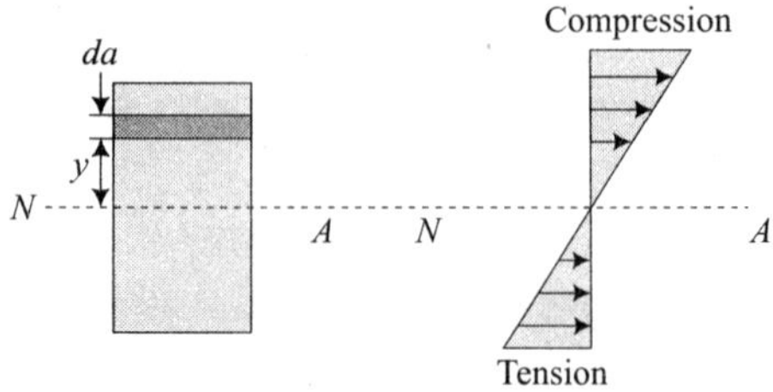

Fig. 10.3 Stress distribution

The section of a rectangular beam is shown in Fig. 10.3.

The force acting on an elementary area, *da*

$$= \sigma_b \times da$$

But, $$\sigma_b = \frac{E}{R} \times y$$

∴ $$\text{Force} = \left(\frac{E}{R} \times y\right) da$$

The bending moment will be resisted by an equal moment to ensure equilibrium. This is called moment of resistance and is calculated as product of force and its distance from neutral axis.

Moment of resistance of force acting on elementary area about *N-A*.

$$\delta MR = \left(\frac{E}{R} \times y\right) da \times y.$$

The total moment of resistance of beam section can be found out by integration.

Total moment of resistance of the beam section,

$$MR = \int \delta MR = \int \left(\frac{E}{R} \times y\right) . da \times y$$

$$= \frac{E}{R} \int y^2 da.$$

Now, $$MR = M$$

where, M = bending moment on the beam section

∴ $$M = MR = \frac{E}{R} \int y^2 da$$

But moment of inertia (M.O.I.) of beam section about N-A,

$$I = \int y^2 da$$

$\therefore$
$$M = \frac{E}{R} I$$

But,
$$\frac{\sigma_b}{y} = \frac{E}{R}$$

$\therefore$
$$M = \frac{\sigma_b}{y} . I$$

$\therefore$
$$\frac{M}{I} = \frac{E}{R} = \frac{\sigma_b}{y}$$

This is called *bending equation* based on theory of bending.

Example 10.1: A steel bar 10 cm wide and 8 mm thick is subjected to bending moment. The radius of neutral surface is 100 cm. Determine the maximum and minimum bending stresses in the beam. [U.P.T.U. II Sem., 2001–02]

Solution:

$b = 10$ cm $= 100$ mm

$d = 8$ mm

$R = 100$ cm $= 1000$ mm

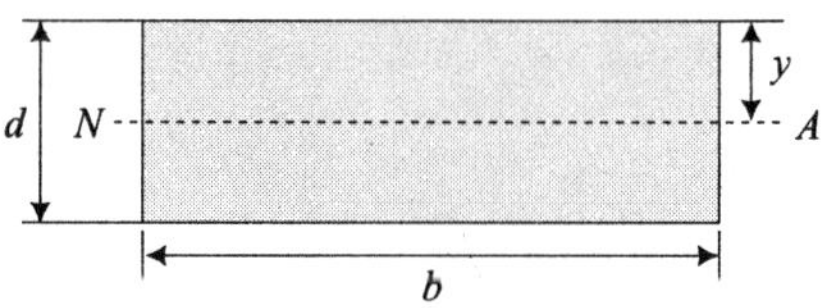

Assume, $E - 210$ kN/mm^2.

From bending equation,

$$\frac{\sigma_b}{y} = \frac{E}{R}$$

$\therefore$
$$\sigma_b = \frac{Ey}{R} = \frac{210 \times 10^3 \times 4}{1000} = 840 \text{ N/mm}^2$$

The maximum stress is 840 N/mm^2 at the outer fibres. Minimum tress is zero at neutral axis.

Example 10.2: Determine the least radius to which a steel bar 100 mm wide and 10 mm thick can be bent if the maximum allowable bending stress is 500 N/mm^2. Take $E = 2 \times 10^5$ N/mm^2.

Solution:

$$b = 100 \text{ mm}$$
$$d = 10 \text{ mm}$$
$$\sigma_b = 500 \text{ N/mm}^2$$
$$E = 2 \times 10^5 \text{ N/mm}^2$$

From bending equation,

$$\frac{\sigma_b}{y} = \frac{E}{R}$$

$$\therefore \quad R = \frac{Ey}{\sigma_b} = \frac{2 \times 10^5}{500} \times \left(\frac{10}{2}\right) = 2000 \text{ mm.}$$

Radius at the inner surface of bar $= R - y$
$= 2000 - 5$
$= 1995$ mm

10.4 SECTION MODULUS

The moment of inertia of an area about any axis is the product of the area and square of the perpendicular distance of its centre of gravity from the axis. The moment of inertia is also called second moment of area.

Moment of inertia of area A about x-axis,

$$I_{xx} = A(\bar{y})^2$$

where, A = Area of section

$\bar{y}$ = Distance of *c.g.* of area from x-axis.

Similarly, moment of inertia of area A about y-axis

$$I_{yy} = A(\bar{X})^2$$

where, $\bar{X}$ = Distance of *c.g.* of area from y-axis. Section modulus is the ratio of moment of inertia of a beam section about neutral axis to the distance of extreme fibre from neutral axis.

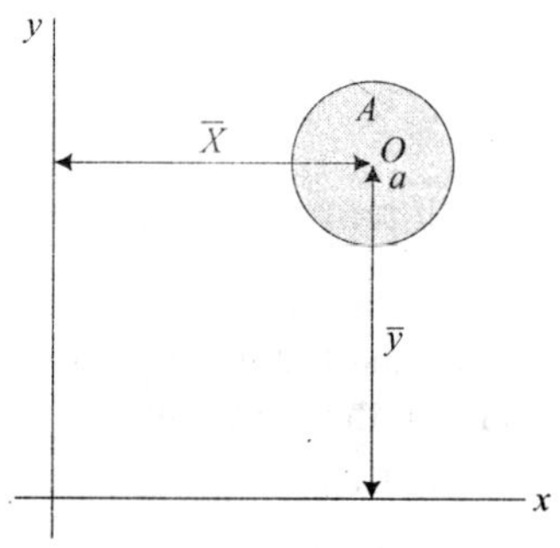

Fig. 10.4 Moment of inertia

Section modulus, $Z = \dfrac{I}{y_{max}}$

where, I = M.O.I. of section about neutral axis.

y_{max} = Distance of extreme fibre of beam from $N-A$.

From equation of bending,

$$\frac{M}{I} = \frac{\sigma_b}{y}$$

$$\therefore \quad \sigma_b = \frac{My}{I} = \frac{M}{I/y} = \frac{M}{Z}$$

The strength of the beam section depends upon the section modulus to resist bending moment. The section modulus of standard section is available in hand books. The section modulus of common section used for beams is calculated below.

10.4.1 Rectangular Section

$$Z = \frac{I}{y_{max}}$$

But, $$I = \frac{bd^3}{12}$$

and $$y_{max} = \frac{d}{2}$$

$$\therefore \quad Z = \frac{bd^3/12}{d/2} = \frac{bd^2}{6}.$$

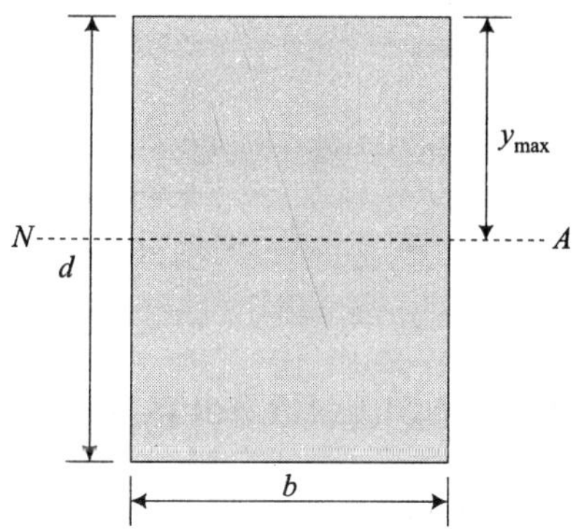

Fig. 10.5 Rectangular section

$$M = \sigma Z = \frac{\sigma}{6} bd^2.$$

Refer Fig. 10.5 for dimensions of section.

10.4.2 Hollow Rectangular Section

The dimensions of a hollow rectangular section are shown in Fig. 10.6.

$$I = \frac{BD^3}{12} - \frac{bd^3}{12} = \frac{(BD^3 - bd^3)}{12}$$

$$y_{max} = D/2$$

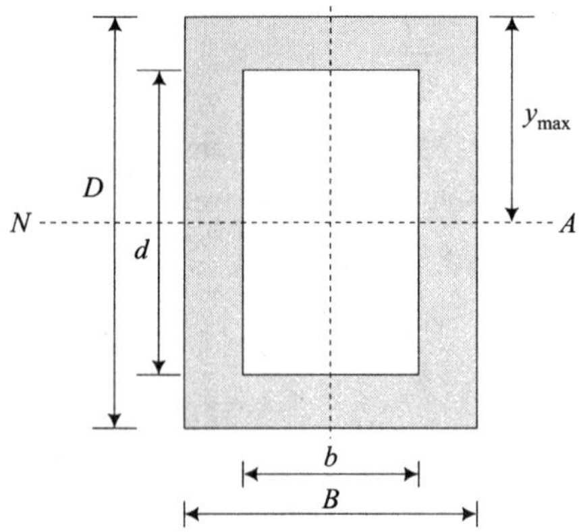

Fig. 10.6 Hollow rectangular section

$$\therefore \quad Z = \frac{I}{y_{max}} = \frac{(BD^3 - bd^3) \times 2}{D \times 12}$$

$$= \frac{BD^3 - bd^3}{6D}$$

Moment of resistance

$$M = \sigma Z = \sigma \left[\frac{BD^3 - bd^3}{6D} \right]$$

10.4.3 Circular Section

M.O.I. about $N - A$,

$$I = \frac{\Pi D^4}{64}$$

$$y_{max} = D/2$$

$\therefore$ Section modulus,

$$Z = \frac{I}{y_{max}} = \frac{\Pi D^4/64}{D/2}$$

$$= \frac{\Pi D^3}{32}.$$

Moment of resistance,

$$M = \sigma Z = \sigma Z \frac{\Pi D^3}{32}$$

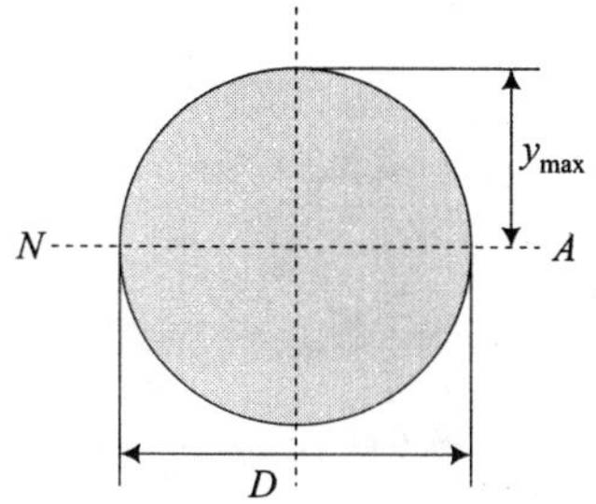

Fig. 10.7 Circular section

10.4.4 Hollow Circular Section

The dimensions are shown in Fig. 10.8.

M.O.I. of section about $N–A$,

$$I = \frac{\Pi}{64}(D^4 - d^4).$$

$$y_{max} = D/2$$

∴ Section modulus,

$$Z = \frac{I}{y_{max}} = \frac{\Pi}{64}(D^4 - d^4) \times \frac{2}{D}$$

$$= \frac{\Pi}{32}\left(\frac{D^4 - d^4}{D}\right)$$

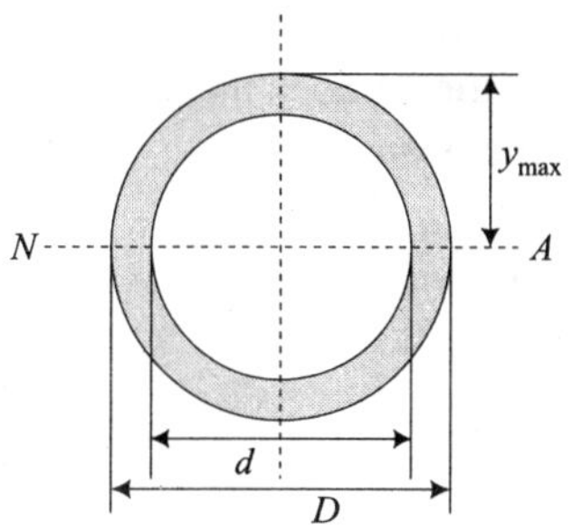

Fig. 10.8 Hollow circular section

Moment of resistance,

$$M = \sigma Z = \frac{\Pi\sigma}{32}\left(\frac{D^4 - d^4}{D}\right)$$

10.4.5 I-Section

The dimensions of a symmetrical I-section are shown in Fig. 10.9.

M.O.I. about $N - A$ = Area of rectangle $(B \times D)$ – Area of rectangle $[(B - b) \times d]$

$$I = \frac{BD^3}{12} - \frac{(B-b)d^3}{12}$$

$$y_{max} = D/2$$

$$\therefore \qquad Z = \frac{I}{y_{max}} = \frac{\frac{BD^3}{12} - \frac{(B-b)d^3}{12}}{D/2}$$

$$= \frac{BD^3 - (B-b)d^3}{6D}$$

Moment of resistance, $M = \sigma Z$.

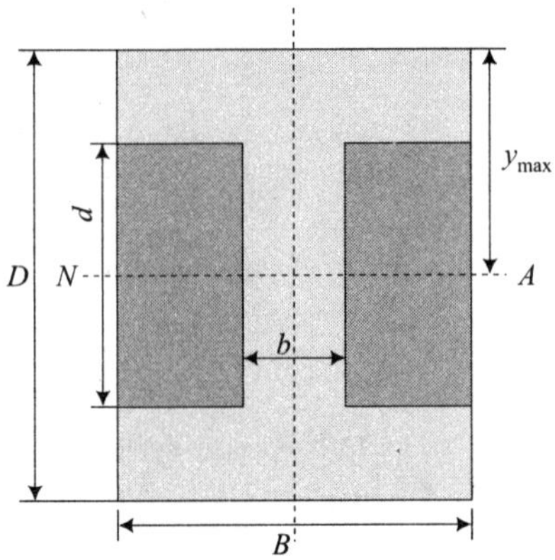

Fig. 10.9 I-section

Example 10.3: A wooden beam of rectangular section is subjected to a bending moment of 5 kN-m. If the depth of the section is to be twice the breadth and stress in the wood is not to exceed 60 N/cm^2, find the dimensions of the cross-section of the beam. [U.P.T.U. I Sem., 2005–06]

Solution:

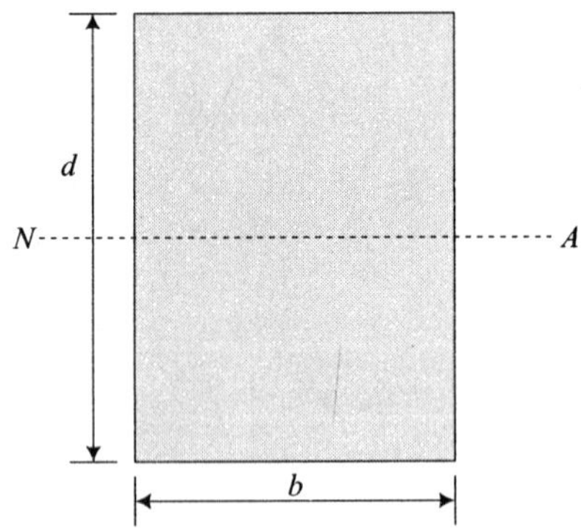

$$M = 5 \text{ kN-m} = 5 \times 10^3 \text{ Nm}$$
$$d = 2b$$
$$\sigma_b = 60 \text{ N/cm}^2 = 6 \times 10^4 \text{ N/m}^2$$
$$M = \sigma_b Z$$

$$\therefore \quad Z = \frac{M}{\sigma_b} = \frac{bd^2}{6}$$

$$\therefore \quad \frac{d^3}{3} = \frac{M}{\sigma_b}$$

$$d = \sqrt[3]{\frac{3M}{\sigma_b}} = \sqrt[3]{\frac{3 \times 5 \times 10^3}{60 \times 10^4}} = 1.357 \text{ m}$$

$$\therefore \quad b = d/2 = 0.6786 \text{ m}$$

The size of beam = **0.6786 m × 1.357 m** **Ans.**

Example 10.4: A rectangular beam with depth 150 mm and width 100 mm is subjected to a maximum bending moment of 300 kN-m. Calculate the maximum stress in the beam. [U.P.T.U. II Sem., 2003–04]

Solution:

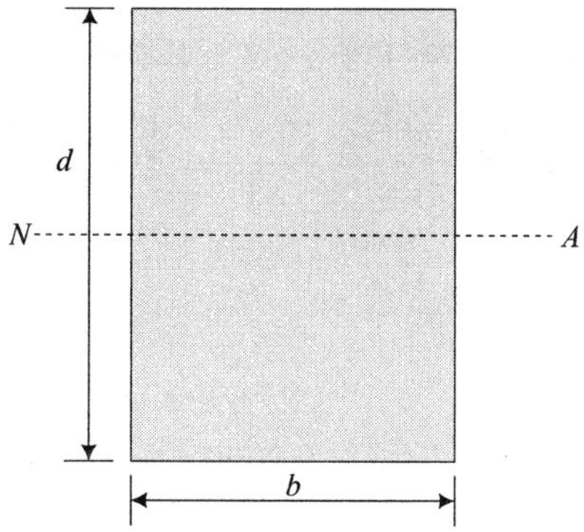

$$d = 150 \text{ mm} = 0.15 \text{ m}$$
$$b = 100 \text{ mm} = 0.10 \text{ m}$$
$$M = 300 \text{ kN-m}$$

Bending stress,

$$\sigma_b = \frac{M}{Z} = \frac{300}{bd^2/6}$$
$$= \frac{300 \times 6}{0.1 \times 0.15^2}$$
$$= 8 \times 10^5 \text{ kN/m}^2$$
$$= 80 \text{ kN/cm}^2$$

Example 10.5: A rectangular beam of 200 mm in width and 400 mm in depth is simply supported over a span of 4 m and carried a distributed load of 10 kN/m. Determine the maximum bending stress in the beam.

Solution:

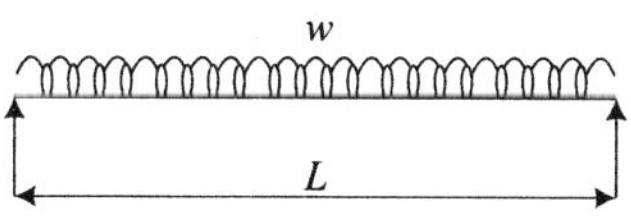

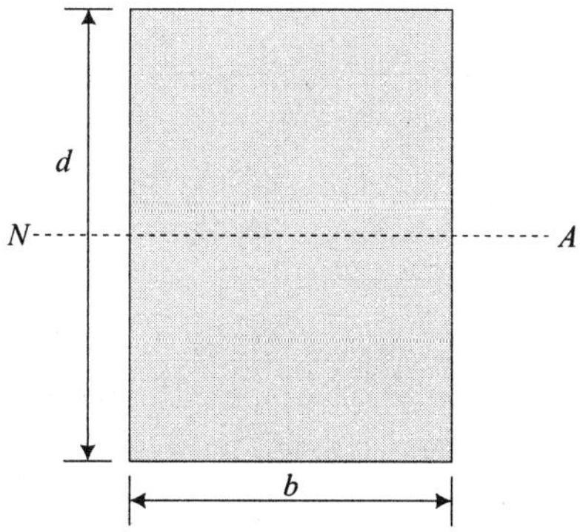

$$b = 200 \text{ mm} = 0.2 \text{ m}$$
$$d = 400 \text{ mm} = 0.4 \text{ m}$$
$$L = 4 \text{ m}$$
$$w = 10 \text{ kN/m}$$

Max. bending moment for a simply supported beam carrying *UDLw*,

$$M = \frac{wL^2}{8} = \frac{10 \times 4^2}{8} \quad 20 \text{ kNm}$$

Max. bending stress,

$$\sigma_b = \frac{M}{Z}$$

$$Z = \frac{bd^2}{6} = \frac{0.2 \times 0.4^2}{6} = 5.33 \times 10^{-3} \text{ m}^3.$$

$$\therefore \quad \sigma_b = \frac{20}{5.33 \times 10^{-3}} = 3750 \text{ kN/m}^2$$

Example 10.6: A rectangular beam of cross-section (300×200) mm^2 is simply supported over a span of 5 m. What uniformly distributed load per metre the beam may carry in two options:

(i) when height is 300 mm

(ii) when height is 200 mm

The bending stress is not to exceed 130 N/mm^2.

Solution:

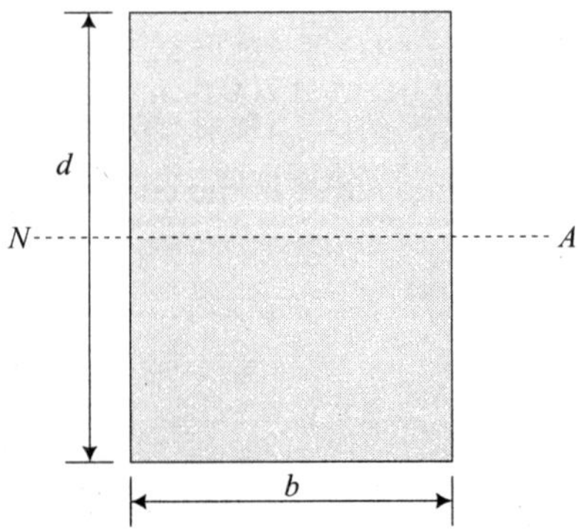

$$M = \frac{wL^2}{8} = \frac{25}{8} w \text{ N-m}$$

$$\sigma_b = 130 \text{ N/mm}^2 = 130 \times 10^6 \text{ N/m}^2$$

(i)
$$b = 200 \text{ mm} = 0.2 \text{ m}$$
$$d = 300 \text{ mm} = 0.3 \text{ m}$$
$$Z = \frac{bd^2}{6} = \frac{0.2 \times 0.3^2}{6} = 3 \times 10^{-3} \text{ m}^3$$
$$\sigma_b = \frac{M}{Z}$$
$$\therefore \quad M = \sigma_b Z$$
$$= \frac{25}{8} w = (130 \times 10^6) \times (3 \times 10^{-3})$$
$$\therefore \quad w = (130 \times 10^6) \times (3 \times 10^{-3}) \times \frac{8}{25} = 124.8 \times 10^3 \text{ kN/m.}$$

(ii)
$$b = 0.3 \text{ m}$$
$$d = 0.2 \text{ m}$$
$$Z = \frac{0.3 \times 0.2^2}{6} = 2 \times 10^{-3} \text{ m}^3$$
$$w = (130 \times 10^6) \times (2 \times 10^{-3}) \times \frac{8}{25} = 83.2 \times 10^3 \text{ kN/m.}$$

Example 10.7: A beam made of C.I. having a circular section of 50 mm external diameter and 25 mm internal diameter is supported at two points 4 m apart. The beam carries a concentrated load of 100 N at the centre. Find the maximum bending stress induced in the beam. [U.P.T.U. I Sem., 2001–02]

Solution:

Bending moment is maximum at the centre of beam span.

$$M = \frac{WL}{4} \quad \frac{100 \times 4000}{4} = 10^5 \text{ N-mm}$$
$$Z = \frac{\Pi}{32}\left(\frac{D^4 - d^4}{D}\right) = \frac{\Pi}{32}\left|\frac{50^4 - 25^4}{50}\right| = 0.011506 \times 10^6 \text{ mm}^3$$
$$\sigma_b = \frac{M}{Z} = \frac{10^5}{0.011506 \times 10^6}$$
$$= \mathbf{8.69 \text{ N/mm}^2} \quad \textbf{Ans.}$$

Example 10.8: A hollow circular bar having outside diameter twice the inside diameter is used as a beam subjected to a bending moment of 50 kN-m. Determine the inside diameter of the bar if the allowable bending stress is limited to 100 MN/m^2.

Solution:

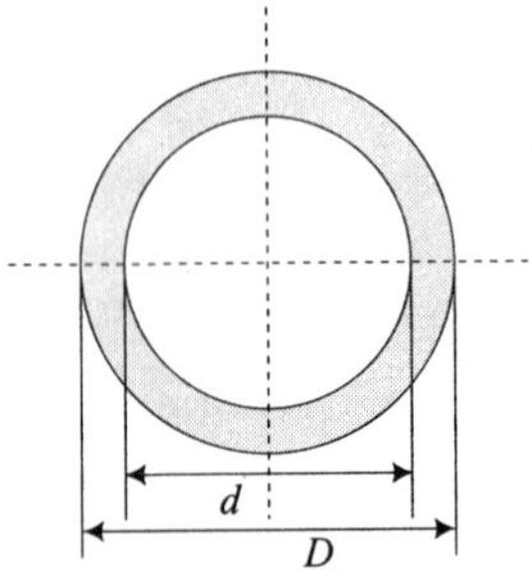

$$M = 50 \text{ kN-m},$$

$$\sigma_b = 100 \text{ MN/m}^2 = 100 \times 10^3 \text{ kN/m}^2$$

$$D = 2\,d.$$

$$Z = \frac{\Pi}{32}\left(\frac{D^4 - d^4}{D}\right)$$

$$= \frac{\Pi}{32}\left|\frac{(2d)^4 - d^4}{2d}\right|$$

$$= \frac{15\Pi d^3}{64}.$$

$$\sigma_b = \frac{M}{Z}.$$

$$Z = \frac{M}{\sigma_b} = \frac{50}{100 \times 10^3}$$

$$\therefore \quad d^3 = \frac{64}{15\Pi} \times \frac{50}{100 \times 10^3}$$

$$d = 0.08812 \text{ m} = \mathbf{88.12 \text{ mm}} \quad \textbf{Ans.}$$

Example 10.9: A rectangular beam 300 mm deep has $I = 8.5 \times 10^6$ mm^4. The beam is simply supported and has a span of 5 m. Calculate *UDL* for a bending stress of 130 N/mm^2.

Solution:

$$I = 8.5 \times 10^6 \text{ mm}^4$$

$$y = \frac{300}{2} = 150 \text{ mm}$$

$$\sigma_b = 130 \text{ N/mm}^2$$

$$M = \frac{wL^2}{8} = \frac{\sigma_b I}{y}$$

$$\therefore \quad w = \frac{8\sigma_b I}{yL^2} = \frac{8 \times 130 \times 8.5 \times 10^6}{150 \times (5000)^2} = 2.357 \text{ N/mm}$$

$$= 2.36 \text{ N/mm} = \mathbf{2.36 \text{ kN/m}}. \quad \textbf{Ans.}$$

Example 10.10: Find the dimension of the strongest rectangular beam that can be cut out of a log of 250 mm diameter.

Solution:

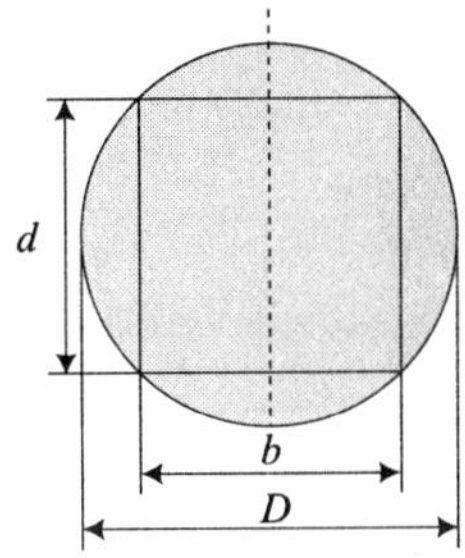

Log diameter, $D = 250$ mm

$$D^2 = b^2 + d^2$$

$$\therefore \quad d^2 = B^2 - b^2$$

Now,

$$Z = \frac{bd^2}{6} = \frac{b}{6}(D^2 - b^2)$$

$$= \frac{1}{6}(bD^2 - b^3)$$

For strongest beam, Z should be maximum.

$$\therefore \quad \frac{dZ}{db} = 0$$

$$\frac{d}{db}\left[\frac{1}{6}bD^2 - \frac{b^3}{6}\right] = 0.$$

$$\therefore \quad \frac{D^2}{6} - \frac{3b^2}{6} = 0$$

$$b = \frac{D}{\sqrt{3}} = \frac{250}{\sqrt{3}} = 144 \text{ mm}$$

$$d = \sqrt{D^2 - b^2} = \sqrt{250^2 - 144^2} = 204 \text{ mm}$$

The size of the strongest beam = 204 mm × 144 mm.

Example 10.11: Three beams have the same length, the same allowable stress and the same bending moment. The cross-sections of the beam are a square, a rectangle with depth twice the width and a circle. Determine the ratios of the weights of circular and rectangle beams with respect to that of square beam.

[U.P.T.U. I Sem., 2000–01]

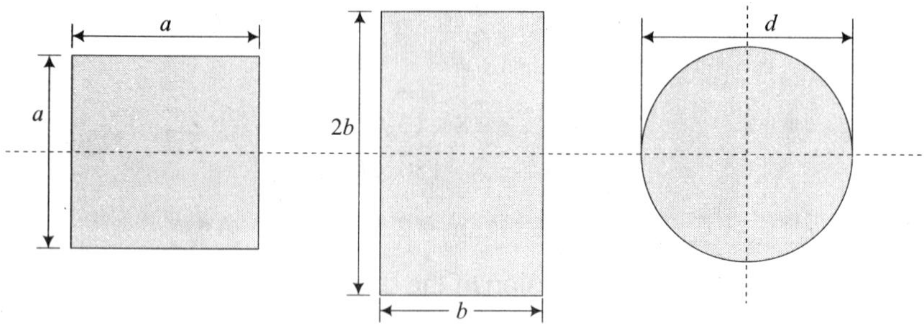

(1) Square section (2) Rectangular section (3) Circular section

Solution: The dimensional sketches of the beam are shown.

The three beams have same bending moment and same allowable stress.

$$\sigma_b = \frac{M}{Z}$$

$$\therefore \quad Z_1 = Z_2 = Z_3$$

$$\frac{a \times a^2}{6} = \frac{b(2b)^2}{6} = \frac{\Pi}{32} d^3$$

$$\frac{a^3}{6} = \frac{2b^3}{3} = \frac{\pi}{32} d^3$$

$$4b^3 = a^3$$

$$b = 0.63a$$

$$\frac{\pi d^3}{32} = \frac{a^3}{6}$$

$$d = \sqrt[3]{\frac{32a^3}{\pi \times 6}} = 1.19a.$$

$$\frac{\text{Weight of rectangular beam}}{\text{Weight of square beam}} = \frac{\rho g A_2 L}{\rho g A_1 L} = \frac{A_2}{A_1} = \frac{2b^2}{a^2}$$

$$= \frac{2(0.63a)^2}{a^2} = 0.71$$

$$\frac{\text{Weight of circular beam}}{\text{Weight of square beam}} = \frac{\rho g A_3 L}{\rho g A_1 L} = \frac{A_3}{A_1} = \frac{\frac{\pi}{4} d^2}{a^2}$$

$$= \frac{\pi}{4} \frac{(1.19a)^2}{a^2} = 1.12$$

Example 10.12: For a given stress, compare the moment of resistance of a beam of square section when placed (i) with two sides horizontal, and (ii) with its diagonal horizontal.

Solution: The moment of resistance is proportional to section modulus because $M = \sigma Z$.

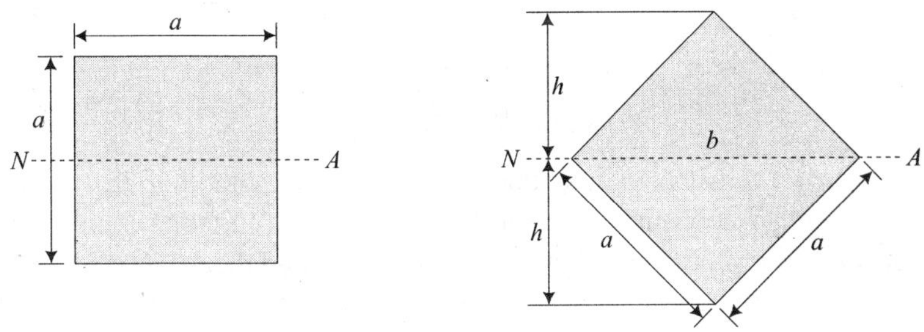

Position (i) Position (ii)

Case 1 $$Z_1 = \frac{bd^2}{6} = \frac{a \times a^2}{6} = \frac{a^3}{6}.$$

Case 2 $$Z_2 = 2 \times \frac{bh^3}{12} \times \frac{1}{h} = \frac{bh^2}{6}$$

where, $b = \sqrt{a^2 + a^2} = \sqrt{2}\,.\,\text{a}$

$$h = \frac{b}{2} = \frac{\sqrt{2}a}{2} = \frac{a}{\sqrt{2}}$$

$$Z_2 = \frac{\sqrt{2}\,a.}{6} \left(\frac{a}{\sqrt{2}}\right)^2 = \frac{a^3}{6\sqrt{2}}$$

$$\therefore \quad \frac{M_1}{M_2} = \frac{Z_1}{Z_2} = \frac{(a^3/6)}{(a^3/6\sqrt{2})} = \sqrt{2} = \mathbf{1.414\ Ans.}$$

Example 10.13: Calculate the maximum compressive stress in a beam of a rectangular section 80 × 120 mm carrying an *UDL* of 40 kN/m over a span of 1 m. There is additional compressive force of 5 kN.

Solution:

Direct compressive stress,

$$\sigma_d = \frac{P_c}{A} = \frac{5000}{80 \times 120} = 0.52 \text{ N/mm}^2.$$

Compressive stress due to bending,

$$\sigma_b = \frac{M}{Z} = \frac{(wL^2/8)}{\frac{bd^2}{6}} = \frac{\left(\frac{40 \times 10^3 \times 1}{8}\right) \times 10^3}{\left(\frac{80 \times 120^2}{6}\right)}$$

$$= 26 \text{ N/mm}^2$$

∴ Maximum fibre stress,

$$\sigma_{max} = \sigma_b + \sigma_d = 26 + 0.52$$
$$= 26.52 \text{ N/mm}^2.$$

Example 10.14: A water mains 500 mm external diameter and 25 mm thick is full of water and is freely supported for 20 m span. Determine maximum bending stress induced in the pipe metal if the weight of water and that of pipe is taken into account. Take specific weight of water and steel as10 kN/m^3 and 75 kN/m^3.

Solution: The water mains acts as a simply supported beam carrying *UDL* consisting of weight of water and weight of steel pipe.

$D = 500$ mm $= 0.5$ m,

$d = 500 - 2 \times 25 = 450$ mm $= 0.45$ m.

Weight of steel pipe per metre length $= \frac{\Pi}{4}(D^2 - d^2)\gamma_s L$

$$= \frac{\Pi}{4}(0.5^2 \,.\, 0.45^2) \times 75 \times 1$$

$$= 2.8 \text{ kN/m}.$$

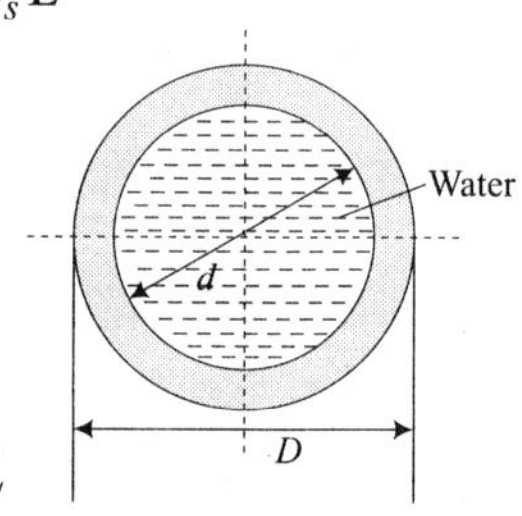

Weight of water per metre length of pipe $= \frac{\Pi}{4} d^2 \gamma_W L$

$$= \frac{\Pi}{4} \times (0.45)^2 \times 10 \times 1 = 1.59 \text{ kN/m}$$

Total $\quad UDL = 2.8 + 1.59 = 4.39$ kN/m.

Bending moment, $M = \frac{wL^2}{8} = \frac{4.39 \times 20^2}{8} = 219.5$ kN-m

Moment of inertia, $I = \frac{\Pi}{64}(D^2 - d^4)$

$$= \frac{\Pi}{64}(0.5^4 - 0.45^4)$$

$$= 1.055 \times 10^{-3} \text{ m}^4$$

Bending stress, $\sigma_b = \frac{MY}{I} = \frac{219.5 \times 0.5/2}{1.055 \times 10^{-3}}$

$$= 52014 \text{ kN/m}^2$$

$$= \mathbf{52.014 \text{ N/mm}^2} \quad \textbf{Ans.}$$

Example 10.15: A cantilever with a constant breadth of 100 mm has a span of 2.5 m. It carries a uniformly distributed load of 20 kN/m. Determine the depth of the section at the middle of the length of the cantilever and also at the fixed end if the stress remains the same throughout and is equal to 120 MN/m^2.

Solution:

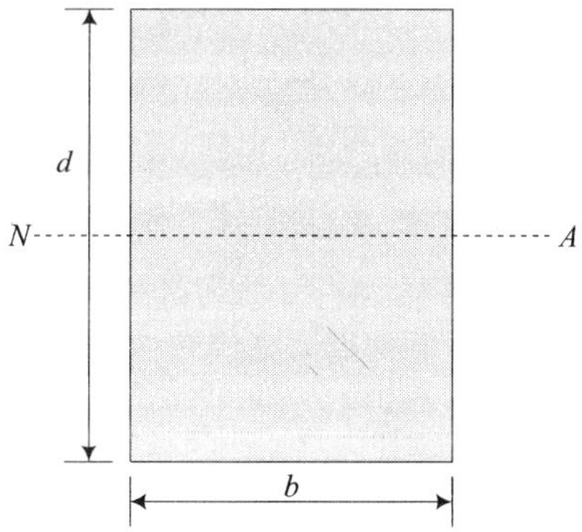

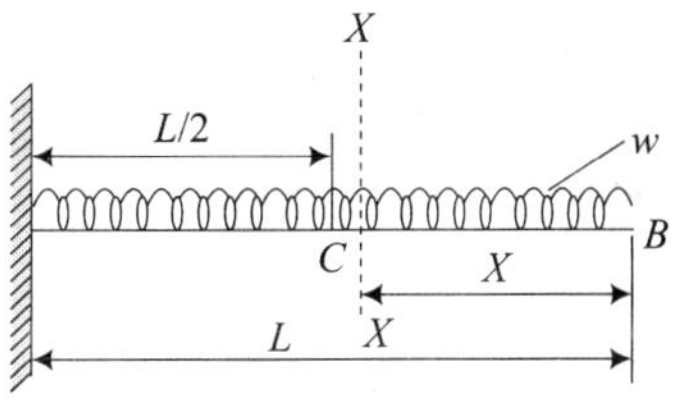

$$w = 20 \text{ kN/m}$$

$$L = 2.5 \text{ m}$$

$$b = 100 \text{ mm} = 0.1 \text{ m}$$

$$\sigma_b = 120 \text{ MN/m}^2$$

$$= 120 \times 10^3 \text{ kN/m}^2$$

Bending moment at section $X - X$,

$$M_x = \frac{wx^2}{2} = \frac{20}{2}x^2 = 10x^2 \text{ kN-m.}$$

$$Z = \frac{M}{\sigma_b}$$

$$\frac{bd^2}{6} = \frac{10x^2}{120 \times 10^3}$$

$$\therefore \qquad d = \sqrt{\frac{6 \times 10 \times x^2}{0.1 \times 120 \times 10^3}} = 0.0707x$$

Depth at fixed end,

$$x = 2.5 \text{ m}$$

$$\therefore \qquad d = 0.176776 \text{ m} = 176.777 \text{ mm}$$

Depth at mid-point, $x = 1.25$ m

$$\therefore \qquad d = 0.0707 \times 1.25 = 0.088 \text{ m} = \mathbf{88\ mm} \quad \textbf{Ans.}$$

Example 10.16: A simply supported beam 1 m long and 20 mm × 20 mm in cross-section fails when a central load of 600 N is applied to it. What density of UDL would cause failure of a cantilever beam 2 m long and 40 mm wide × 80 mm deep made of same material?

Solution: *Simply supported beam,*

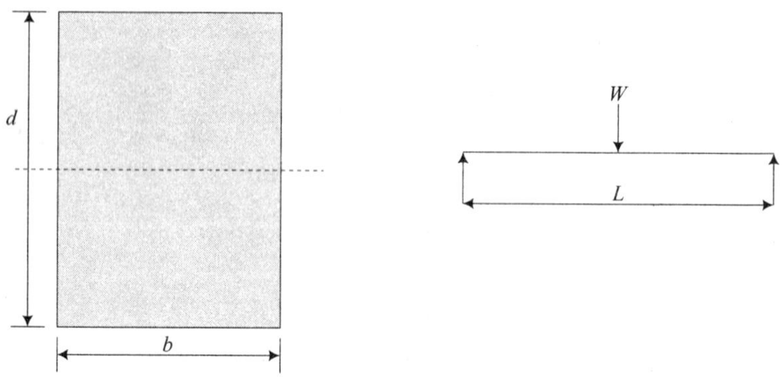

$$L = 1 \text{ m}$$

$$b = 20 \text{ mm} = 0.02 \text{ m}$$

$$d = 0.02 \text{ m}$$

$$W = 600 \text{ N.}$$

$$Z = \frac{bd^2}{6} = \frac{(0.02)^3}{6}$$

$$= 1.333 \times 10^{-6} \text{ mm}^3$$

$$M = \frac{WL}{4} = \frac{600 \times 1}{4} = 150 \text{ N-m}$$

$$\sigma_b = \frac{M}{Z} = \frac{150}{1.333 \times 10^{-6}} = 112.5 \times 10^6 \text{ N/m}^2$$

$$= \mathbf{112.5 \text{ N/mm}^2} \quad \textbf{Ans.}$$

Example 10.17: Determine the longest span of a simply supported beam to carry a UDL of 6 kN/m without exceeding a bending stress of 120 MN/m². The depth and moment of inertia of the symmetrical *I*-section are 20 cm and 2640 cm⁴ respectively.

Solution:

$$w = 6 \text{ kN/m}$$

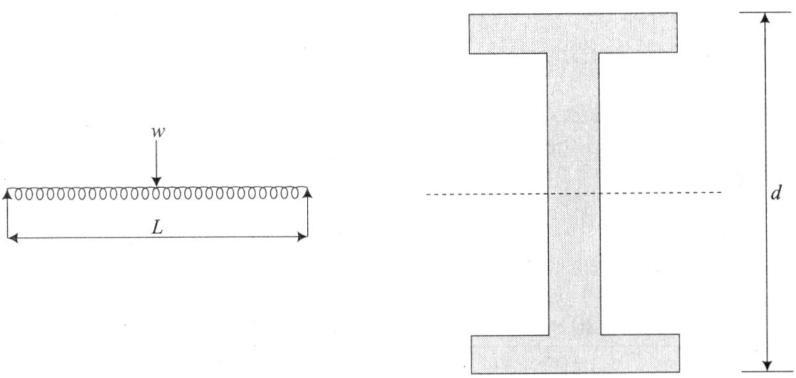

$$\sigma_b = 120 \text{ MN/ m}^2 = 120 \times 10^3 \text{ kN/m}^2$$

$$d = 20 \text{ cm} = 0.2 \text{ m}$$

$$I = 2640 \text{ cm}^4 = 2640 \times 10^{-16} \text{ m}^4.$$

$$Z = \frac{I}{y} = \frac{2640 \times 10^{-16}}{0.2/2} = 2.64 \times 10^{-12} \text{ m}^3$$

$$M = \sigma_b Z = \frac{wL^2}{8}$$

$$L = \sqrt{\frac{\sigma_b Z \times 8}{w}} = \sqrt{\frac{120 \times 10^3 \times 2.64 \times 10^{-12} \times 8}{6}} = 6.5 \text{ m.}$$

Example 10.18: The cross-section of a cast iron beam is shown and is simply supported at the ends and carries a uniformly distributed load of 20 kN/m. If the span of the beam is 3 m, determine the maximum tensile and compressive bending stress in the beam.

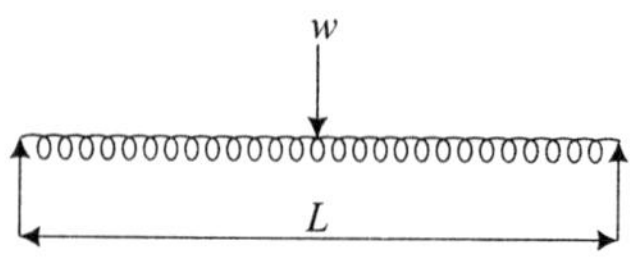

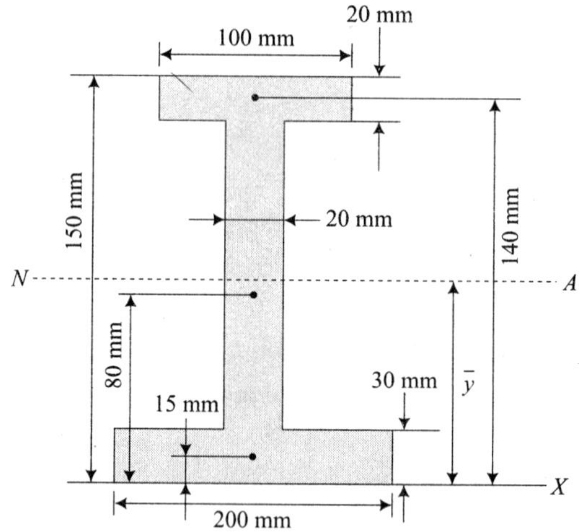

Solution:

1. In order to find the height of *c.g.* from edge $X - X$, take moments of area about $X - X$,

$$\bar{y} = \frac{\Sigma AY}{\Sigma A}$$

$$= \frac{(100 \times 20 \times 140) + (20 \times 100 \times 80) + (200 \times 30 \times 15)}{(100 \times 20) + (100 \times 20) + (200 \times 30)}$$

$$= \frac{280{,}0000 + 160{,}000 + 90{,}000}{2{,}000 + 2{,}000 + 6{,}000} = \frac{530{,}000}{10{,}000} = 53 \text{ mm}$$

Moment of inertia of total section about N–A

$$I = \Sigma I_G + \Sigma Ah^2 = \Sigma(I_G + Ah^2)$$

$$= \frac{100 \times 20^3}{12} + 100 \times 20 \times (53 - 140)^2 + 100 \times 20 \times (53 - 80)^2 + \frac{20 \times 100^3}{12}$$

$$+ 200 \times 30 \times (53 - 15)^2 + \frac{200 \times 30^3}{12}$$

$$= 66{,}666.667 + 15.14 \times 10^6 + 1.45 \times 10^6 + 1.667 \times 10^6 + 8.66 \times 10^6 + 0.45 \times 10^6$$

$$= 0.06667 \times 10^6 + 1.1667 \times 10^6 + 0.45 \times 10^6 + 15.14 \times 10^6 + 1.45 \times 10^6 + 8.66 \times 10^6$$

$$= 2.18367 \times 10^6 + 2525 \times 10^6 = 27.43 \times 10^6 \text{ mm}^4$$

$$M = \frac{wL^2}{8} = \frac{20 \times (3000)^2}{8} = 22.5 \times 10^6 \text{ kN-mm}$$

$$\sigma_c = \frac{MYc}{I} = \frac{22.5 \times 10^6 \times (150 - 53)}{27.43 \times 10^6} = 79.566 \text{ kN/mm}^2 \qquad \text{(at top)}$$

$$\sigma_t = \frac{MYt}{I} = \frac{22.5 \times 10^6 \times 53}{27.43 \times 10^6} = 43.474 \text{ kN/mm}^2 \qquad \text{(at bottom)}$$

Example 10.18 (a): Determine the dimensions of a rectangular simply supported steel beam 5 m long to carry an UDL of 10 kN/m, if the maximum permissible bending stress is 1000 N/cm^2. The depth of the beam is 1.5 times its width. [U.P.T.U. II Sem., 2007–08]

Solution:

$$L = 5 \text{ m}$$

$$w = 10 \text{ kN/m}$$

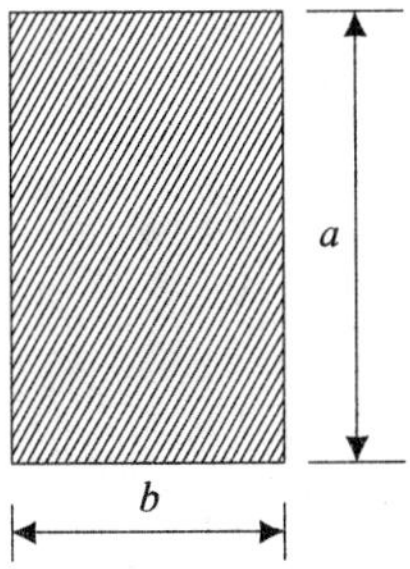

$$\therefore \quad M = \frac{wL^2}{8} = \frac{10 \times 5^2}{8} = 31.25 \text{ kN-m}$$

$$d = 1.5\ b$$

$$Z = \frac{bd^2}{6} = \frac{b(1.5b)^2}{6}$$

$$= 0.375\ b^3 \text{ m}^3$$

$$\sigma_b = 1000 \text{ N/cm}^2 = 10000 \text{ kN/m}^2$$

$$= \frac{M}{Z}$$

$$\therefore \quad Z = \frac{M}{\sigma_b}$$

$$0.375\ b^3 = \frac{M}{\sigma_b}$$

$$\therefore \quad b = \sqrt[3]{\frac{M}{0.375\,\sigma_b}} = \sqrt[3]{\frac{31.25}{0.375 \times 10{,}000}}$$

$b = 0.2027 \text{ m} =$ **20.27 cm**

$d = 1.5 \text{ b} =$ **30.4 cm** **Ans.**

Example 10.18 (b): Determine the dimensions of a simply supported rectangular steel beam 6 m long to carry a brick wall 250 mm thick and 3 m high, if the brick work weighs 19.2 kN/m^2 and maximum permissible bending stress is 800 N/cm^2. The depth of beam is 3/2 times its width. [U.P.T.U. II Sem., 2006–07]

Solution:

The beam length $L = 6$ m

The beam depth $d = \frac{3}{2}\ b$

Total volume of wall $= \frac{250}{1000} \times 3 \times 6 = 4.5\ \text{m}^3$

Total weight of wall $= 19.2 \times 4.5 = 86.4$ kN

$$\text{Udl, } w = \frac{86.4}{6} = 14.4 \text{ kN/m}$$

$$= 14.4 \times 10^3 \text{ N/m}$$

The maximum bending moment,

$$M = \frac{wL^2}{8} = \frac{14.4 \times 10^3 \times 6^2}{8} = 64800 \text{ N-m}$$

The permissible bending stress,

$$\sigma_b = 800 \text{ N/cm}^2 = 800 \times 10^4 \text{ N/m}^2$$

$$Z = \frac{M}{\sigma_b} = \frac{64800}{800 \times 10^4} = 8.1 \times 10^{-3}\ \text{m}^3 = 8.1 \times 10^6\ \text{mm}^3$$

$$Z = \frac{bd^2}{6} = \frac{b(1.5b)^2}{6} = 0.375\ b^3 = 8.1 \times 10^6$$

$$\therefore \qquad b = \sqrt[3]{\frac{8.1 \times 10^6}{0.375}} = 278.5 \text{ mm}$$

$$d = 1.5\ b = 417.7 \text{ m}$$

The beam size = 417.7 mm × 278.5 mm **Ans.**

Example 10.18(c): A 300 mm deep rectangular beam is simply supported over a span of 6 m. What uniformly distributed load per metre the beam can carry if the bending stress is not to exceed 110 N/mm^2? Take $I = 8.5 \times 10^6$ mm^4.

[U.P.T.U. 2009-10]

Solution:

$$I = 8.5 \times 10^6 \text{ mm}^4$$

$$Y = \frac{300}{2} = 150 \text{ mm}$$

$$\sigma = 110 \text{ N/mm}^2$$

The bending equation:

$$\frac{M}{I} = \frac{\sigma}{y}$$

$$\therefore \quad M = \frac{\sigma I}{y} = \frac{110 \times 8.5 \times 10^6}{150} = 6.23 \times 10^6 \text{ N. mm}$$

$$= 6.23 \times 10^3 \text{ N.m}$$

For a uniformly distributed load,

$$M = \frac{wL^2}{8}$$

$$\therefore \quad w = \frac{8M}{L^2} = \frac{8 \times 6.23 \times 10^3}{(6)^2} = \mathbf{1384.4\ N/m} \quad \textbf{Ans.}$$

Example 10.18(d): A steel beam of hollow square section of 60 mm outer side and 50 mm inner side is simply supported on a span of 4 m. Find the maximum concentrated load the beam can carry at the middle of the span if the bending stress is not to exceed 120 N/mm^2. [U.P.T.U. 2008-09]

Solution:

$$Z = \frac{BD^2}{6} - \frac{bd^2}{6} = \frac{B^3 - b^3}{6}$$

$$= \frac{(0.06)^3 - (0.05)^3}{6} = 0.15 \times 10^{-4} \text{ m}^3$$

$$M = \sigma z = 120 \times 10^6 \times 0.15 \times 10^{-4} = 1820 \text{ N-m}$$

For a simply supported beam

$$M = \frac{WL}{4}$$

$$\therefore \quad W = \frac{4M}{L} = \frac{4 \times 1820}{4} = \mathbf{1820\ N} \quad \textbf{Ans.}$$

Example 10.18(e): A beam of *I*-section is 250 mm deep. The flanges are 15 mm thick, 100 mm wide while the web is 8 mm thick. Compare the flexural strength of this beam section with a rectangular section of the same material and area whose width is two-thirds of depth. [U.P.T.U. 2009-10]

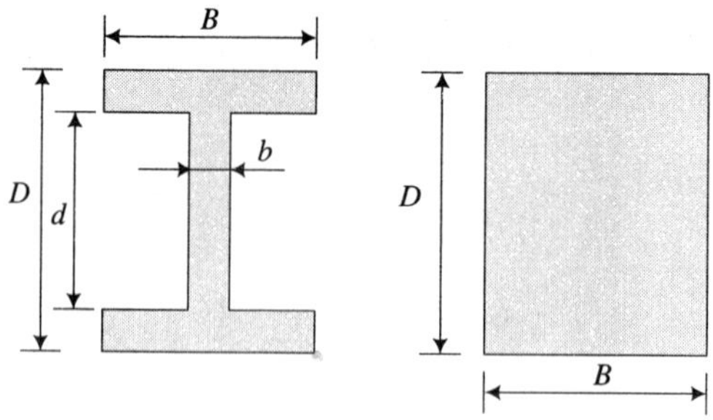

Solution:

(i) *I- section*

$B = 100$ mm

$D = 250$ mm

$d = 250 - 2 \times 15 = 220$ mm

$b = 8$ mm

Area $= 250 \times 100 - 220 \times 92$

$= 4760 \text{ mm}^2$

$$I = \frac{BD^3}{12} - \frac{(B-b)d^3}{12}$$

$$= \frac{100 \times (250)^3}{12} - \frac{92 \times (220)^3}{12}$$

$= 0.4857 \times 10^8 \text{ mm}^4$

$$Z_1 = \frac{I}{Y} = \frac{0.4857 \times 10^8}{125} = 3.886 \times 10^5 \text{ mm}^3$$

(ii) *Rectangular- section*

$$\text{Area} = BD = \frac{2}{3} D^2 = 4760$$

$$\therefore \quad D = \sqrt{\frac{4760 \times 3}{2}} = 84.5 \text{ mm}$$

$$B = \frac{2}{3} D = 56.33 \text{ mm}$$

$$Z_2 = \frac{BD^2}{6} = \frac{56.33 \times (84.5)^2}{6} = 6.7 \times 10^4 \text{ mm}^3$$

The flexural strength of a beam section is proportional to section modulus as the beams have the same material and length.

$$\text{The ratio of flexural strength} = \frac{Z_1}{Z_2} = \frac{3.886 \times 10^5}{6.7 \times 10^4} = \mathbf{5.8} \quad \textbf{Ans.}$$

Example 10.18(f): A cast iron test beam 20 mm × 20 mm in section and 1 m long and supported at the ends fails when a central load of 640 N is applied. What uniformly distributed load will break a cantilever of the same material 50 mm wide, 100 mm deep and 2 m long? [U.P.T.U. 2009-10]

Solution:

(i) *Simply supported beam*

$$Z = \frac{BD^2}{6} = \frac{20 \times (20)^2}{6} = 1333.33 \text{ mm}^3$$

$$M = \frac{WL}{4} = \frac{640 \times 1000}{4} = 1.6 \times 10^5 \text{ N-mm}$$

The bending stress

$$\sigma = \frac{M}{Z} = \frac{1.6 \times 10^5}{1333.33} = 120 \text{ N/m}^2$$

(ii) *Cantilever*

$$Z = \frac{BD^2}{6} = \frac{50 \times (100)^2}{6} = 8.33 \times 10^4 \text{ mm}^3$$

$$M = \sigma z = 120 \times 8.33 \times 10^4 = 9.996 \times 10^6 \text{ N mm}$$

$$= 9.996 \times 10^3$$

For a cantilever under *UDL*,

$$M = \frac{wL^2}{2}$$

$$\therefore \quad w = \frac{2M}{L^2} = \frac{2 \times 9.996 \times 10^3}{(2)^2} = 4.998 \times 10^3 \text{ N/m}$$

$$= \mathbf{4.998 \text{ kN/m}} \quad \textbf{Ans.}$$

Example 10.18(g): A simply supported beam, 2 cm wide by 4 cm high and 1.5 m long is subjected to a concentrated load of 2 kN (perpendicular to beam) at a point 0.5 m from one of the supports. Determine:

(i) The maximum fibre stress, and

(ii) The stress in a fibre located 1 cm from the top of the beam at mid-span.

[U.P.T.U. 2008-09]

Solution:

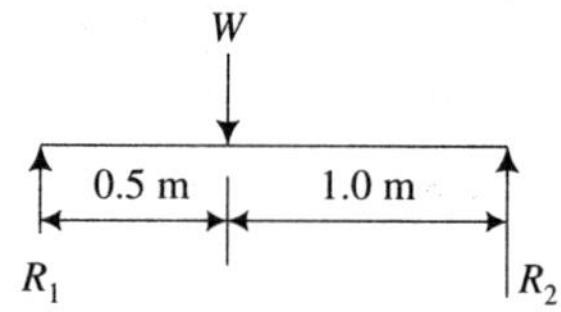

(i) $R_1 + R_2 = W$

$$R_1 = \frac{W \times 1}{1.5} = \frac{2 \times 1}{1.5} = \frac{4}{3} \text{ kN}$$

$$R_2 = \frac{2}{3} \text{ kN}$$

The maximum bending is under the load,

$$\therefore \qquad M_{max} = R_1 \times 0.5 = \frac{4}{3} \times 0.5$$

$$= \frac{2}{3} \text{ kNm} = \frac{2}{3} \times 10^2 \text{ kN-cm}$$

$$Z = \frac{BD^2}{6} = \frac{2 \times (4)^2}{6} = \frac{16}{3} \text{ cm}^3$$

$$\sigma_{max} = \frac{M_{max}}{Z} = \frac{2/3 \times 10^2}{16/3} = \frac{100}{8} = 12.5 \text{ kN/cm}^2$$

(ii) *Bending moment at mid-span,*

$$M = R_2 \times 0.75$$

$$= \frac{2}{3} \times \frac{3}{4}$$

$$= 0.5 \text{ kN-m}$$

$$= 0.5 \times 10^2 \text{ kN-cm}$$

$$\sigma_1 = \frac{M}{Z} = \frac{0.5 \times 10^2}{16/3} = 9.375 \text{ kN/cm}^2$$

Bending stress in a fibre 1 cm from top

$$\frac{\sigma_1}{\sigma_2} = \frac{y_1}{y_2}$$

$$\therefore \qquad \sigma_2 = \frac{\sigma_1 y_2}{y_1} = \frac{9.375 \times 1}{2}$$

$$= \mathbf{4.669 \text{ kN/cm}^2} \textbf{ Ans.}$$

Example 10.18(h): Determine the dimensions of a simply supported rectangular steel beam 6 m long to carry a brick wall 250 mm thick and 3 m high, if the brick weighs 20 kN/m^3 and maximum permissible bending stress is 800 N/cm^2. The depth of the beam is 1.50 times its width. [UP.T.U. 2010-11]

Solution:

Total volume of brick wall

$$= \frac{250}{1000} \times 3 \times 6 = 4.5 \text{ m}^3$$

Total weight of the wall

$$= 20 \times 4.5 = 90 \text{ kN.}$$

Weight per metre length,

$$w = \frac{90}{6} = 15 \text{ kN/m}$$

Maximum bending moment,

$$M = \frac{wL^2}{8} = \frac{15 \times 10^3 \times 6^2}{8}$$

$$= 67.5 \times 10^3 \text{ N-m}$$

$$\sigma = 800 \text{ N/cm}^2 = 800 \times 10^4 \text{ N/m}^2$$

$$Z = \frac{M}{\sigma} = \frac{67.5 \times 10^3}{800 \times 10^4} = 8.4375 \times 10^{-3} \text{ m}^3$$

$$= 8.4375 \times 10^6 \text{ mm}^3$$

$$Z = \frac{BD^2}{6} = \frac{B(1.5B)^2}{6} - 0.375\ B^3$$

$$B = \sqrt[3]{\frac{Z}{0.375}} = \sqrt[3]{\frac{8.4375 \times 10^6}{0.375}}$$

$$= 282.3 \text{ mm}$$

$$D = 1.5\ B = 423.47 \text{ mm}$$

The beam size $= \mathbf{423.47\ mm \times 282.3\ mm}$ **Ans.**

10.5 TORSION OF SHAFTS OF CIRCULAR SECTION

Solid and hollow circular shafts are used in industry to transmit power from prime movers to various types of machines. Other shapes of shafts are seldom used. The shafts are subjected to the following loads.

1. Torsion or twisting moment or torque.
2. Bending moment due to transverse loads or weights.

3. Axial thrust due to longitudinal loads.
4. Combination of above three loads.

A shaft is under *pure torsion*, if it is not associated with bending moment or axial force. If a shaft is under pure torsion, its cross-section experiences the following effects,

1. Shear stress or torsional shear stress due to torque.
2. Twisting or torsional shear strain measured as angle of twist,
3. Strain energy.

Therefore, shafts are designed so that shear stress induced and angle of twist produced are within allowable limits.

10.6 POWER TRANSMISSION

The shaft rotated by a motor or a prime mover (engine, turbine, etc.) has rotational energy. This energy can be expressed by shaft work.

Shaft work,

$$W = \int_1^2 T\, d\theta$$

where, T = Torque or turning moment applied to the shaft (Nm)

$d\theta$ = Angular displacement of shaft

Shaft power,

$$P = \int_1^2 T\frac{d\theta}{dt} = T\omega$$

where, ω = Angular velocity of shaft

or $P = 2\Pi NT$ (W)

where, N = shaft speed (rps)

$$P = \frac{2\Pi NT}{60 \times 1000} \text{ (kW)}$$

where, N = shaft speed (rpm).

The rotating shaft possesses kinetic energy which can be expressed as:

$$KE = \frac{1}{2} I\,\omega^2$$

where, I = mass moment of inertia

$= mk^2$

m = mass of rotating shaft (kg)

k = radius of gyration (m)

The work done by a rotating shaft is equal to change of kinetic energy.

$$W_{1-2} = KE_2 - KE_1$$

1 stands for initial and 2 for final condition.

10.7 ASSUMPTIONS IN PURE TORSION OF CIRCULAR SHAFTS

Certain assumptions are made to consider pure torsion in circular shafts. These assumptions are made to develop the torsion equation.

1. Material of the shaft is homogeneous and isotropic. The material is assumed free of impurities and mechanical properties are same in all directions.
2. The shaft is perfectly elastic and obeys Hooke's law.
3. The stress does not exceed the limit of proportionality.
4. The value of Young's modulus of elasticity is same in tension and compression.
5. The shaft is subjected to pure twisting moment or torque acting in a plane perpendicular to the axis of the shaft.
6. The twist is uniform along the total length of the shaft.
7. The shaft remains circular in section before and after application of torque.
8. The plane cross-sections of shafts remain plane before and after application of torque.
9. All diameters of the cross-section of the shaft remain straight without change in their magnitude before and after twisting.
10. The relative rotation between any two cross-sections of the shaft is proportional to the distance between them.
11. There is no change in the length of the shaft between any two normal cross-sections after application of torque.

10.8 TORSIONAL SHEAR STRESS

When a shaft is transmitting power, it behaves as if a force F acts tangentially on the shaft of diameter D at one end and is fixed at the other end. The twisting moment can be represented by a couple $F \times l$ as shown.

The torque, $T = F \times l.$

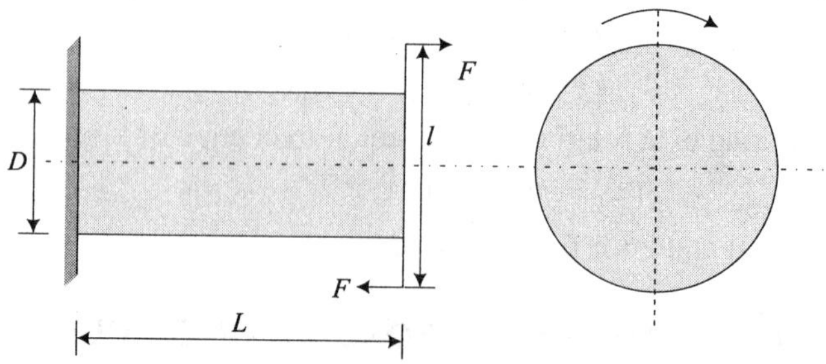

Fig. 10.10 Torque

Torque can be expressed in terms of shaft diameter, the force being assumed acting on the surface of the shaft. Then,

$$T = F \times D.$$

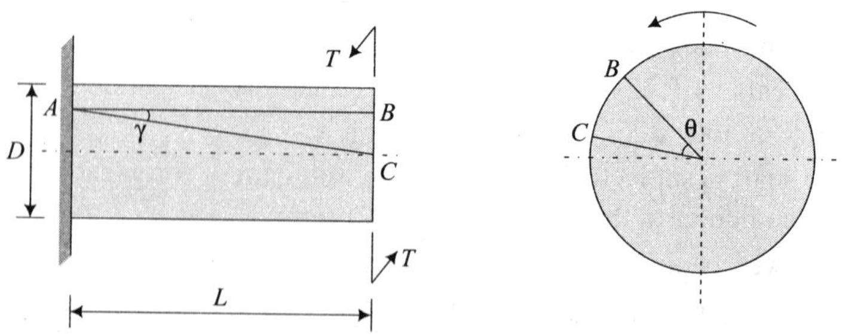

Fig. 10.11 Torsion of circular shaft

A solid circular shaft of diameter D and length L is subjected to a torque T at one end and fixed at the other end (by a couple equal to torque). A straight line AB parallel to shaft axis on the surface of the shaft will take the form of helix AC after twisting. Assume γ is the angle of shear strain on the surface.

In the front view,

$$BC = L\gamma$$

or
$$\gamma = \frac{BC}{L}$$

But shear strain,

$$\gamma = \frac{\tau}{G}$$

where, τ = shear stress

G = modulus of rigidity.

or $$\tau = G\gamma = G \times \frac{BC}{L}$$

In the end view of the shafts,

$$BC = \frac{D}{2}.\theta.$$

where, θ is called the angle of twist.

$$\therefore \quad \tau = G.\frac{D/2\theta}{L} = \left(\frac{G\theta}{L}\right)\frac{D}{2} = \left(\frac{G\theta}{L}\right)R$$

$$= kR$$

where, $R = D/2$ = radius of shaft

$$k = \frac{G\theta}{L} = \text{constant for a given shaft.}$$

Therefore, shear stress is proportional to the radius of the shaft and varies linearly with the radius.

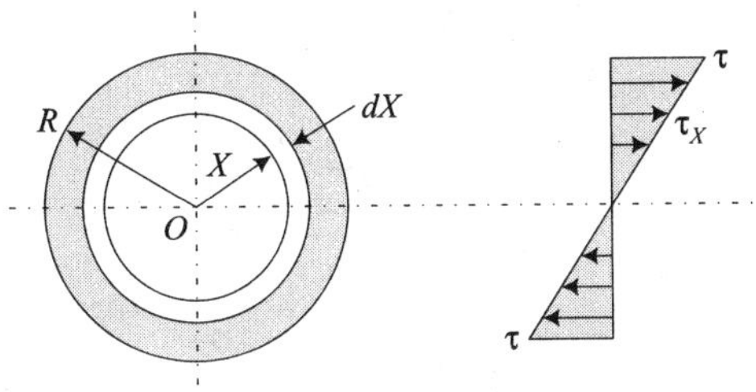

Fig. 10.12 Shear stress distribution

At any radius X,

$$\frac{\tau_x}{\tau} = \frac{X}{R}$$

$$\therefore \quad \tau_x = \frac{X}{R}.\tau$$

But, $$\frac{\tau}{R} = k = \frac{G\theta}{L}$$

Consider an elementary ring of the shaft at radius X and thickness dX.

The turning force on the ring,

$$F_X = \tau_X.dA = \tau_x \times 2\pi X dX$$

Moment of turning of this force about shaft axis

Moment of turning of this force about shaft axis

$$dT = 2\pi X dX\, \tau_x . X = 2\pi X^2 dX\, \tau_x$$

$$= 2\pi X^2 dX \frac{X}{R}\tau = 2\pi X^3 dX \frac{\tau}{R}$$

Total turning moment of resistance,

$$TR = \int dT = \int_O^R 2\pi X^3 \times \frac{\tau}{R} dX$$

$$= \frac{2\pi\tau}{R}\int_O^R X^3 dX = \frac{2\pi\tau}{R}\left|\frac{X^4}{4}\right|_o^R$$

$$= \frac{2\pi\tau}{R}\frac{R^4}{4} = \frac{\pi R^3}{2}\tau = \frac{\pi}{16}D^3\tau$$

But applied torque, T = Total turning moment of resistance,

$$TR = T = \frac{\pi}{16}D^3\tau.$$

$\therefore$

$$T = \frac{\tau}{(D/2)} \cdot \frac{\pi D^4}{32} = \frac{\tau}{D/2} \cdot J$$

where,

$$J = \frac{\pi D^4}{32}$$

J is the polar moment of inertia of the shaft cross-section

$\therefore$

$$\frac{T}{J} = \frac{\tau}{R} = \frac{G\theta}{L}.$$

This is called the *torsion equation* for the circular shaft.

10.9 TORSION EQUATION FOR HOLLOW CIRCULAR SHAFT

Figure 10.13 shows a hollow circular shaft with outside radius R and inside radius r, and subjected to a torque T. The turning moment of resistance on elementary ring,

$$dT = \tau_X . 2\pi X . dX . X$$

The total turning moment resistance shaft cross-section

$$T = \int dT = \int_r^R 2\pi X^3 dX \frac{\tau}{R}$$

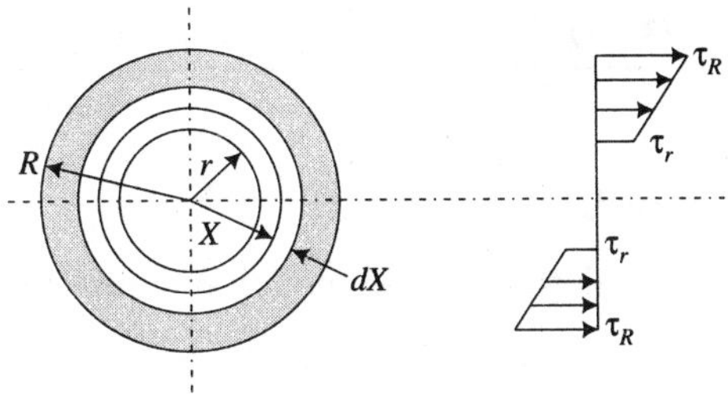

Fig. 10.13 Shear stress distribution for hollow circular shaft

$$T = \frac{2\pi\tau}{R}\int_r^R X^3 dX = \frac{2\pi\tau}{R}\left|\frac{X^4}{4}\right|_r^R$$

$$= \frac{\pi}{2}.\frac{\tau}{R}.(R^4 - r^4)$$

$$= \frac{\pi}{16}\tau\left[\frac{D^4 - d^4}{D}\right]$$

But $$J = \frac{\pi}{32}(D^4 - d^4) = \frac{\pi}{2}(R^4 - r^4) \text{ (for a hollow shaft)}$$

∴ $$T = \frac{\tau}{R}.J.$$

or $$\frac{T}{J} = \frac{\tau}{R} = \frac{G\theta}{L}.$$

The torsion equation equally holds good for solid and hollow circular shafts.

10.10 POLAR SECTION MODULUS

Polar section modulus is the ratio of polar moment of inertia and outer radius of the shaft. It is analogous to section modulus in case of beam subjected to bending. The polar section modulus is a measure of strength and rigidity of a shaft against a twisting moment or torque. Polar moment of inertia is the moment of inertia about z-axis or longitudinal axis of the shaft.

$$J = I_{zz} = I_{xx} + I_{yy}$$

Polar section modulus,

$$Z_p = \frac{J}{R} = \frac{\frac{\pi}{32}\times D^4}{D/2} = \frac{\pi}{16}D^3 \quad \text{for solid circular shaft.}$$

$$Z_p = \frac{\pi/32(D^4 - d^4)}{D/2} = \frac{\pi}{16}\left(\frac{D^4 - d^4}{D}\right) \quad \text{for hollow circular shaft.}$$

10.10.1 Strength of Circular Shaft

Strength of a shaft may be defined as the maximum torque which can be applied to the shaft without exceeding allowable shear stress and angle of twist.

(a) Solid shaft

The torque transmitted by a shaft,

$$T = \tau Z_p = \frac{\pi}{16} D^3 \tau.$$

The maximum torque which can be transmitted,

$$T_{max} = \frac{\pi}{16} D^3 \tau_{max} \qquad \text{...(1)}$$

where, τ_{max} = maximum allowable shear stress.

Also,

$$\frac{T}{J} = \frac{G\theta}{L}$$

$\therefore$

$$T = \frac{G\theta}{L} . J = \frac{G\theta}{L} . \frac{\pi}{32} D^4$$

Similarly,

$$T_{max} = \frac{\pi}{32} \frac{GD^4}{L} . \theta_{max.} \qquad \text{...(2)}$$

where, θ_{max} = Maximum allowable angle of twist (rad)

The strength of the shaft is the maximum value of T_{max} as obtained in equations (1) and (2).

(b) Hollow shaft

The maximum torque which can be transmitted by a hollow shaft,

$$T_{max} = Zp\, \tau_{max} = \frac{\pi}{16}\left(\frac{D^4 - d^4}{D}\right) \tau_{max} \qquad \text{...(1)}$$

where, D = External shaft diameter

d = Internal shaft diameter

τ_{max} = Maximum allowable shear stress

Also,

$$T_{max} = \frac{G\theta_{max}}{L} \quad J = \frac{G\theta_{max}}{L} . \frac{\pi}{32} (D^4 - d^4) \qquad \text{...(2)}$$

The strength of the shaft is the maximum value of T_{max} as obtained from equations (1) and (2).

10.10.2 Torsional Rigidity

The torsion equation gives

$$\frac{T}{J} = \frac{G\theta}{L}$$

$$\therefore \qquad \theta = \frac{TL}{GJ}$$

The values of G, J and L are constant for a given shaft.

∴ Angle of twist θ is directly proportional to the torque T.

Now,
$$\frac{\theta}{L} = \frac{T}{GJ}$$

or
$$\frac{T}{\theta/L} = GJ = \text{torsional rigidity.}$$

The torque divided by the angle of twist per unit length of shaft is called *torsional rigidity.*

The parameter GJ is analogous to the quantity EI. The quantity EI is called flexural rigidity in case of beam subjected to bending.

10.10.3 Torsional Stiffness

The torque per unit angle of twist is called torsional stiffness of the shaft.

$$\text{Torsional stiffness} = \frac{T}{\theta} \text{ (N-m/rad).}$$

10.11 STRAIN ENERGY DUE TO TORSION

Energy is stored in the shaft due to angular distortion θ. This is also called *torsional resilience* or torsional energy.

Torsional energy = Work done by the torque
= Average torque × Angular twist

$$U = \frac{1}{2} T\theta$$

From torsion equation,

$$\frac{T}{J} = \frac{\tau}{R} = \frac{2\tau}{D}$$

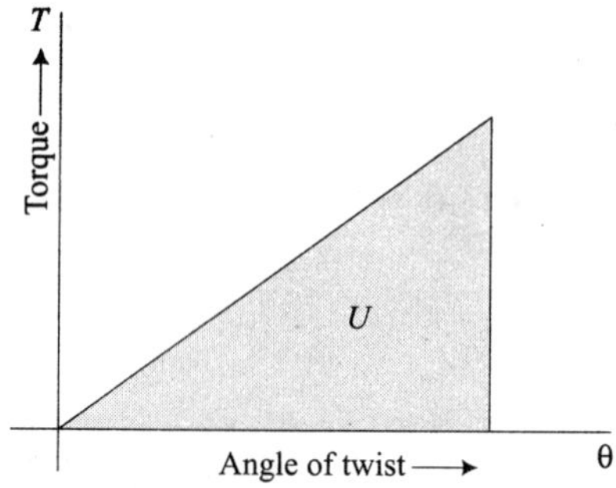

Fig. 10.14 Torsional energy

$$\therefore \quad T = \frac{\tau J}{R}.$$

Also, $$\frac{T}{J} = \frac{G\theta}{L} = \frac{\tau}{R}$$

$$\therefore \quad \theta = \frac{\tau L}{GR}$$

$$\therefore \quad U = \frac{1}{2}\left(\frac{\tau J}{R}\right)\left(\frac{\tau L}{GR}\right)$$

$$= \frac{1}{2}\frac{\tau^2}{G} \cdot \frac{\pi R^4}{2R^2} \cdot L$$

$$= \frac{1}{4} \times \frac{\tau^2}{G}(\pi R^2 L) = \frac{1}{4}\frac{\tau^2}{G} \text{ (Volume)}$$

Modulus of resilience $= \frac{U}{V} = \frac{1}{4}\frac{\tau^2}{G}$ for solid shaft.

For a hollow shaft,

$$T = J.\frac{2\tau}{D} = \frac{\pi}{32}(D^2 - d^2)\frac{2\tau}{D}$$

$$= \frac{\pi}{16}\tau\left(\frac{D^4 - d^4}{D}\right)$$

$$\therefore \quad \theta = \frac{2\tau}{D} \cdot \frac{L}{G}$$

$$\therefore \qquad U = \frac{1}{2} \times \frac{\pi}{16} \tau \left(\frac{D^4 - d^4}{D} \right) . \frac{2\tau}{D} . \frac{L}{G}$$

$$= \frac{\tau^2}{4} \left[\left(\frac{D^2 + d^2}{D^2} \right) \left(\frac{D^2 - d^2}{4} \right) \right] . \pi L$$

$$= \frac{\tau^2}{4G} \left(\frac{D^2 + d^2}{D^2} \right) \quad \text{(Volume shaft)}$$

Modulus of resilience = Torsional energy for unit volume

$$\frac{U}{V} = \frac{\tau^2}{4G} . \left(\frac{D^2 + d^2}{D^2} \right)$$

Example 10.19: Find the power transmitted by a solid circular shaft of steel of 50 mm diameter at 120 rpm, if the permissible shear stress is 62.5 N/mm^2.

Solution:

$$D = 50 \text{ mm}$$
$$N = 120 \text{ rpm}$$
$$\tau = 62.5 \text{ N/mm}^2$$

Torque,

$$T = \frac{\pi}{16} D^3 \tau$$

$$= \frac{\pi}{16} (50)^3 \times 62.5$$

$$= 1533980 \text{ N-mm}$$

$$= 1533.98 \text{ N-m}$$

Power,

$$P = \frac{2\pi NT}{60 \times 1000} = \frac{2\pi \times 120 \times 1533.98}{60 \times 1000}$$

$$= 19.276 \text{ kW.}$$

Example 10.20: If the maximum torque transmitted by a solid shaft exceeds the mean torque by 30% in each revolution, find a suitable shaft diameter to transmitted 75 kW power at 200 rpm. Take allowable shear stress as 70 N/mm^2.

Solution:

$$T_{max} = 1.3\ T_{mean}.$$
$$P = 75 \text{ kW}$$
$$N = 200 \text{ rpm}$$
$$\tau_{max} = 70 \text{ N/mm}^2$$

Power, $P = \dfrac{2\pi N\, T_{mean}}{60 \times 1000}$

$\therefore$ $T_{mean} = \dfrac{60{,}000P}{2\pi N} = \dfrac{60{,}000 \times 75}{2\pi \times 200}$

$= 3581$ N-m,

$T_{max} = 1.3 \times 3581$

$= 4655.3$ N-m.

Now, $T_{max} = \dfrac{\pi}{16} D^3 \tau_{max}$

$$D = \sqrt[3]{\frac{T_{max} \times 16}{\pi \times \tau_{max}}} = \sqrt[3]{\frac{4655.3 \times 16 \times 10^3}{\pi \times 70}}$$

= 69.7 mm Ans.

Example 10.21: Find the torque which a shaft of 25 cm diameter can safely transmit, if stress is not to exceed 460 N/mm^2.

Solution: $D = 25$ cm $= 250$ mm

$\tau = 460$ N/mm^2

Now torque,

$$T = \frac{\pi}{16} D^3 \tau$$

$$= \frac{\pi}{16} (250)^3 \times 460 \text{ N-mm}$$

$$= \frac{\pi}{16} (250)^3 \times 460 \times 10^{-3} \text{ N-m}$$

= 1.6 × 10^6 N-m Ans.

Example 10.22: Find the diameter of a shaft to transmit a torque of 3×10^4 N-m if allowable shear stress is 50 N/mm^2.

Solution: $T = 3 \times 10^4$ N-m $= 3 \times 10^4 \times 10^3$ N-mm

$\tau = 50$ N/mm^2

Now, $T = \dfrac{\pi}{16} D^3 \tau$

$\therefore$ $D = \sqrt[3]{\dfrac{16T}{\pi\tau}} = \sqrt[3]{\dfrac{16 \times 10^4 \times 10^3}{\pi \times 50}}$

= 145 mm Ans.

Example 10.23: A hollow shaft with outer diameter 25 cm and inner diameter 15 cm has to transmit maximum torque. Find the value of torque if allowable shear stress is 40 N/mm^2.

Solution:

$$D = 25 \text{ cm} = 250 \text{ mm}$$

$$d = 15 \text{ cm} = 150 \text{ mm}$$

$$\tau = 40 \text{ N/mm}^2$$

Now,

$$T = \frac{\pi}{16}\left(\frac{D^4 - d^4}{D}\right)\tau$$

$$= \frac{\pi}{16}\left(\frac{250^4 - 150^4}{150}\right) \times 40$$

$$= 0.1066 \times 10^9 \text{ N-mm}$$

$$= \mathbf{106.6 \text{ kN-m}} \quad \textbf{Ans.}$$

Example 10.24: For one propeller drive shaft, compute the torsional shear stress when it is transmitting a torque of 1.76 kN-m. The shaft is a hollow tube having an outside diameter of 60 mm and an inside diameter of 40 mm. Find the stress at both the outer and inner surfaces. [U.P.T.U. II Sem., 2001–02]

Solution:

$$T = 1.76 \text{ kN-m} = 1.76 \times 10^6 \text{ N/mm}$$

$$D = 60 \text{ mm}$$

$$d = 40 \text{ mm}$$

$$T = \frac{\pi}{16}\left(\frac{D^4 - d^4}{D}\right)\tau$$

$$\tau = \frac{16DT}{\pi(D^4 - d^4)} \quad \frac{16 \times 60 \times 1.76 \times 10^6}{\pi(60^4 - 40^4)}$$

$$= 51.7 \text{ N/mm}^2.$$

This is the stress at outer radius.

The stress varies linearly with the radius.

The stress at the inner radius of shaft,

$$\tau_i = \tau_o \frac{d}{D} = 51.7 \times \frac{40}{60}$$

$$= \mathbf{34.48 \text{ N/mm}^2} \quad \textbf{Ans.}$$

Example 10.25: Show that for a given maximum shear stress, the minimum diameter required for solid circular shaft to transmit P(kW) at N(rpm) can be expressed as $D = \text{constant} \times \left(\frac{P}{N}\right)^{\frac{1}{3}}$.

Solution:

$$P = \frac{2\pi NT}{60 \times 1000} \text{ [kW]}$$

$$T = \frac{60{,}000 P \times 10^3}{2\pi N} \text{ N-mm}$$

$$= 3 \times 10^7 \frac{P}{\pi N} \text{ N-mm}$$

But,
$$T = \frac{\pi}{16} D^3 \tau$$

$\therefore$
$$D^3 = \frac{16T}{\pi\tau} = \frac{16 \times 3 \times 10^7 P}{\pi\tau \times \pi N}$$

$\therefore$
$$D = k\left(\frac{P}{N}\right)^{\frac{1}{3}}$$

where,
$$k = \frac{16 \times 3 \times 10^7}{\pi^2 \tau} = \text{constant}$$

Example 10.26: What external and internal diameters are required for a hollow shaft to transmit 50 kW of power at 300 rpm if the shear stress is limited to 100 MN/m^2? Take outside diameter to be twice of inside diameter.

Solution:

$$P = 50 \text{ kW}$$
$$N = 300 \text{ rpm}$$
$$\tau = 100 \text{ MN/m}^2 = 100 \text{ N/mm}^2$$
$$D = 2d.$$

$$P = \frac{2\pi NT}{60 \times 1000}$$

$$\therefore \qquad T = \frac{60,000P}{2\pi N} = \frac{60,000 \times 50}{2\pi \times 300} \text{ N-m}$$

$$= \frac{60,000 \times 50 \times 10^3}{2\pi \times 300} \text{ N-mm.}$$

Also,

$$T = \frac{\pi}{16}\left(\frac{D^4 - d^4}{D}\right)\tau$$

$$= \frac{\pi}{16}\left[\frac{D^4 - \left(\frac{D}{2}\right)^4}{D}\right]\tau$$

$$= \frac{\pi}{16} \times \frac{15}{16} D^3\tau$$

$$\therefore \qquad D = \sqrt[3]{\frac{256}{\pi \times 15} \times \frac{T}{\tau}} = \sqrt[3]{\frac{256}{15\pi} \times \frac{60,000 \times 50 \times 10^3}{2\pi \times 300 \times 100}}$$

$$= 44.2 \text{ mm}$$

$$d = \frac{D}{2} = \frac{44.2}{2} = \mathbf{22.1\ mm} \quad \textbf{Ans.}$$

Example 10.27: Determine the torque required to give a twist of 10 rounds to a steel wire which is of 0.60 mm diameter and 30 m length. $G = 80$ GPa.

Solution:

$$\theta = 2\pi \times 10 \text{ rad.}$$

$$D = 0.6 \text{ mm}$$

$$L = 30 \text{ m} = 30 \times 10^3 \text{ mm}$$

$$G = 80 \text{ GPa} = 80 \times 10^3 \text{ N/mm}^2.$$

From torsion equation,

$$\frac{T}{J} = \frac{G\theta}{L}$$

$$T = \frac{G\theta J}{L} = \frac{80 \times 10^3 \times 2\pi \times 10}{30 \times 10^3} \times .\left(\frac{\pi}{32} \times 0.6^4\right)$$

$$= \mathbf{2.1297\ N\text{-}mm} \quad \textbf{Ans.}$$

Example 10.28: Calculate the diameter of a circular shaft to transmit 75 kW at 200 rpm. Allowable shear stress is restricted to 50 MN/m^2 and twist 1° in 2 m shaft length. $G = 100$ GN/m^2.

Solution:

$$P = 75 \text{ kW}$$

$$N = 200 \text{ rpm}$$

$$\tau = 50 \text{ MN/m}^2 = 50 \text{ N/mm}^2$$

$$\theta = 1^\circ \times \frac{\pi}{180} = \frac{\pi}{180} \text{ rad}$$

$$L = 2 \text{ m} = 2000 \text{ mm}$$

$$G = 100 \text{ GN/m}^2 = 100 \times 10^3 \text{ N/mm}^2$$

$$P = \frac{2\pi NT}{60,000}$$

$$T = \frac{60,000}{2\pi N} = \frac{60,000 \times 75}{2\pi \times 200}$$

$$= 3581 \text{ N-m.} = 3581 \times 10^3 \text{ N-mm.}$$

(a) *Using shear stress as design basis*

$$T = \frac{\pi}{16} D^3 \tau$$

$$D = \sqrt[3]{\frac{16T}{\pi\tau}} = \sqrt[3]{\frac{358 \times 10^3 \times 16}{\pi \times 50}}$$

$$= 71.45 \text{ mm.}$$

(b) *Using twist as design basic*

$$\frac{T}{J} = \frac{G\theta}{L}$$

$$\therefore \quad J = \frac{TL}{G\theta}$$

$$\frac{\pi}{32} D^4 = \frac{TL}{G\theta}$$

$$\therefore \quad D = \sqrt[4]{\frac{32 \times T \times L}{\pi \times G \times \theta}} = \sqrt[4]{\frac{32 \times 3581 \times 10^3 \times 2000}{\pi \times 10^5 \times \frac{\pi}{180}}}$$

$$D = 80.4 \text{ mm}$$

Select $D = 80.4$ mm, i.e, the larger value.

Example 10.29: It is required to transmit 112.5 kW at 200 rpm by a solid circular shaft. Select a suitable diameter of the shaft if allowable twist is 1° in a length of 3 m and allowable shear stress is 75 N/mm^2. $G = 0.082 \times 10^6$ N/mm^2.

Solution:

$$P = 112.5 \text{ kW}$$

$$N = 200 \text{ rpm}$$

$$\theta = 1° = \frac{\pi}{180} \text{ rad}$$

$$L = 3 \text{ m} = 3000 \text{ mm.}$$

$$\tau = 75 \text{ N/mm}^2$$

$$G = 0.086 \times 10^6 \text{ N/mm}^2$$

$$P = \frac{2\pi NT}{60 \times 1000}$$

$$T = \frac{60{,}000\, P}{2\pi N} = \frac{60{,}000 \times 112.5}{2\pi \times 200} = 5.37 \text{ N-m.}$$

$$= 5.37 \times 10^3 \text{ N-mm}$$

(a) *Using twist as design basis*

$$\frac{T}{J} = \frac{G\theta}{L}$$

$$J = \frac{TL}{G\theta}$$

$$\frac{\pi}{32} D^4 = \frac{TL}{G\theta}$$

$$D = \sqrt[4]{\frac{TL}{G\theta} \times \frac{32}{\pi}} = \sqrt[4]{\frac{5.37 \times 10^3 \times 3000 \times 32}{0.08 \times 10^6 \times \frac{\pi}{180} \times \pi}}$$

$$= 103.4 \text{ mm}$$

(b) *Using shear stress as design basis*

$$T = \frac{\pi}{16} D^3 \tau$$

$$D = \sqrt[3]{\frac{16T}{\pi\tau}} = \sqrt[3]{\frac{16 \times 5.37 \times 10^3}{\pi \times 75}}$$

$$= 71.47 \text{ mm.}$$

Select $D = 103.4$ mm, i.e, the larger value.

Example 10.30: A circular steel shaft 25 mm diameter is subjected to a torque of 500 N-m. Determine:

(a) The maximum shear stress developed in the shaft.

(b) The shear stress at a distance 10 mm from the centre of the shaft.

(c) The angular twist over 1 m length of the shaft. Take $G = 80$ GPa.

Solution:

$$D = 25 \text{ mm}$$

$$R = 12.5 \text{ mm}$$

$$T = 500 \text{ N-m} = 500 \times 10^3 \text{ N-mm}$$

(a) *Maximum shear stress at the surface of shaft*

$$T = \frac{\pi}{16} D^3 \tau$$

$$\therefore \quad \tau = \frac{16T}{\pi D^3} = \frac{16 \times 500 \times 10^3}{\pi \times (25)^3} = 163 \text{ N/mm}^2$$

(b) *Shear stress at* $r = 10$ mm

$$\tau_r = \frac{\tau}{R} \times r = \frac{163}{12.5} \times 10 = 130.4 \text{ N/mm}^2$$

(c)

$$\frac{T}{J} = \frac{G\theta}{L}$$

$$\theta = \frac{TL}{GJ} = \frac{500 \times 10^3 \times 1000}{\left(80 \times 10^3\right) \times \frac{\pi}{32}(25)^4} = 0.163 \text{ rad}$$

$$= 0.163 \times \frac{180}{\pi} = 9.344°.$$

Example 10.31: A hollow shaft is to transmit a uniform torque of 30 kN-m. The total angular twist in 2.5 m length of the shaft is not to exceed 2 degrees and the allowable shear stress is 85 MPa. Determine the external and internal diameters if $G = 80$ GPa.

Solution:

$$T = 30 \text{ kN-m} = 30 \times 10^6 \text{ N-mm}$$

$$L = 2.5 \text{ m} = 2500 \text{ mm}$$

$$\theta = 2° = 2 \times \frac{\pi}{180} \text{ rad}$$

$$\tau = 85 \text{ MPa} = 85 \text{ N/mm}^2$$

$$G = 80 \text{ GPa} = 80 \times 10^3 \text{ N/mm.}$$

(a) *Using twist as design basis*

$$\frac{T}{J} = \frac{G\theta}{L} \qquad \text{(Torsion Equation)}$$

$$J = \frac{TL}{G\theta}$$

$$\frac{\pi}{32}(D^4 - d^4) = \frac{TL}{G\theta}.$$

$$(D^4 - d^4) = \frac{32TL}{\pi G\theta} = \frac{32 \times 30 \times 10^6 \times 2500 \times 180}{\pi \times 80 \times 10^3 \times 2\pi}$$

$$= 273.567 \times 10^6 \text{ mm.} \qquad ...(1)$$

(b) *Using shear stress as design basis*

$$T = \frac{\pi}{16}\left(\frac{D^4 - d^4}{D}\right)\tau$$

$$\therefore \quad \left(\frac{D^4 - d^4}{D}\right) = \frac{16T}{\pi\tau} = \frac{16 \times 30 \times 10^6}{\pi \times 85} = 1.798 \times 10^6 \qquad ...(2)$$

From eqns. (1) and (2),

$$1.798 \times 10^6\, D = 273.567 \times 10^6$$

$$\therefore \quad D = \frac{273.567 \times 10^6}{1.798 \times 10^6} = 152.3 \text{ mm}$$

$$(D^4 - d^4) = 273.567 \times 10^6$$

$$\therefore \quad d^4 = D^4 - 273.567 \times 10^6$$

$$= (152.3)^4 - 273.567 \times 10^6$$

$$= 264.\,453 \times 10^6$$

$$\therefore \quad d = 127.5 \text{ mm.}$$

Example 10.32: The diameter of a shaft is 20 cm. Find the safe maximum torque which can be transmitted by the shaft if the permissible stress in the shaft material be 4000 N/cm^2 and permissible angle of twist is 0.2 degree per metre length. Take $G = 8 \times 10^6$ N/cm^2. If the shaft rotates at 320 rpm, what maximum power can be transmitted by the shaft? (U.P.T.U. I Sem., 2005–06)

Solution: $D = 20$ cm $= 200$ mm

$\tau = 4000$ N/cm^2 $= 40$ N/mm^2.

$$\theta = 0.2° = 0.2 \times \frac{\pi}{180} \text{ rad.}$$

$$L = 1000 \text{ mm.}$$

$$G = 8 \times 10^6 \text{ N/cm}^2 = 8 \times 10^4 \text{ N/mm}^2$$

(a) *Using shear stress as design basis*

$$T = \frac{\pi}{16} D^3 \tau$$

$$= \frac{\pi}{16} (200)^3 \times 40 = 62831853 \text{ N-mm}$$

$$= 62.83 \text{ kN-m.}$$

(b) *Using twist as design basis*

$$\frac{T}{J} = \frac{G\theta}{L}$$

$$T = \frac{G\theta J}{L} = \frac{8 \times 10^4 \times 0.2}{1000} \times \frac{\pi}{180} \times \frac{\pi (200)^4}{32}$$

$$= 43864908 \text{ N-mm}$$

$$= 43.865 \text{ kN-m}$$

The safe maximum torque is 43.865 kN-m, the lower value.

Power, $$P = \frac{2\pi NT}{60 \times 1000}$$

$$= \frac{2\pi \times 320 \times 43.865}{60}$$

$$= \mathbf{1470 \text{ kW}} \quad \textbf{Ans.}$$

Example 10.33: A solid circular shaft is to transmit 160 kW at 180 rpm. What will be the suitable diameter of this shaft if the permissible stress in the shaft material should not exceed 2×10^6 Pa and twist per unit length should not exceed 2°? Take $G = 200$ GPa. [U.P.T.U. II Sem., 2005–06]

Solution: $$P = 160 \text{ kW}$$

$$N = 180 \text{ rpm}$$

$$\tau = 2 \times 10^6 \text{ Pa}$$

$$\theta = 2° = 2 \times \frac{\pi}{180} \text{ rad.}$$

$$L = 1 \text{ m}$$

$$G = 200 \text{ GPa} = 200 \times 10^9 \text{ Pa.}$$

$$P = \frac{2\pi NT}{60 \times 1000}$$

$$\therefore \quad T = \frac{60{,}000\, P}{2\pi N} = \frac{60{,}000 \times 160}{2\pi \times 180}$$

$$= 8488.26 \text{ N-m}$$

(a) *Using shear stress as design basis*

$$T = \frac{\pi}{16} D^3 \tau$$

$$\therefore \quad D = \sqrt[3]{\frac{16T}{\pi\tau}}$$

$$= \sqrt[3]{\frac{16 \times 8488.26}{\pi \times 2 \times 10^6}}$$

$$= 0.27856 \text{ m}$$

$$= 278.56 \text{ mm}$$

(b) *Using twist as design basis*

$$\frac{T}{J} = \frac{G\theta}{L}$$

$$J = \frac{TL}{G\theta}$$

$$\frac{\pi}{32} D^4 = \frac{TL}{G\theta}$$

$$D = \sqrt[4]{\frac{32TL}{\pi G\theta}} = \sqrt[4]{\frac{32 \times 8488.26 \times 1 \times 180}{\pi \times 200\, 10^9 \times 2 \times \pi}}$$

$$= 0235.25 \text{ m}$$

$$= 235.25 \text{ mm.}$$

Select $D = 278.56$ mm. i.e., the larger value.

Example 10.34: A hollow shaft of 3 m length is subjected to a torque such that the maximum shear stress produced is 75 MN/m^2. The external and internal diameters of the shaft are 150 mm and 100 mm respectively. Find the shear stress at the inside surface and angle of twist. Take $G = 75$ GN/m^2.

[U.P.T.U. II Sem., 2003–04]

Solution:

$$L = 3 \text{ m}$$

$$\tau = 75 \text{ MN/m}^2$$

$$D = 150 \text{ mm} = 0.15 \text{ m}$$

$$d = 100 \text{ mm} = 0.10 \text{ m}$$

$$G = 75 \text{ GN/m}^2 = 75 \times 10^3 \text{ MN/m}^2.$$

$$T = \frac{\pi}{16}\left(\frac{D^4 - d^4}{D}\right)\tau$$

$$= \frac{\pi}{16}\left(\frac{0.15^4 - 0.1^4}{0.15}\right) \times 75$$

$$= 0.03988 \text{ MN-m.}$$

$$\frac{T}{J} = \frac{G\theta}{L}$$

$$\therefore \quad \theta = \frac{TL}{GJ} = \frac{0.03988 \times 3 \times 32}{10^3 \times 75 \times \pi(0.15^4 - 0.1^4)} = 0.04 \text{ rad}$$

$$= 0.04 \times \frac{180}{\pi} = 2.3°.$$

The shear stress varies linearly with radius

$$\therefore \quad \frac{\tau_i}{\tau_o} = \frac{d}{D}$$

$$\therefore \quad \tau_i = \frac{d}{D} \times \tau_o = \frac{100}{150} \times 75 = \mathbf{50\ MN/m^2} \quad \textbf{Ans.}$$

Example 10.35: A solid circular shaft is required to transmit 200 kW at 100 rpm. Determine the diameter of the shaft if permissible shear stress is 60 N/mm^2. Calculate the energy stored per metre length of the shaft. $G = 1 \times 10^5$ N/mm^2.

Solution:

$$P = 200 \text{ kW}$$

$$N = 100 \text{ rpm}$$

$$\tau = 60 \text{ N/mm}^2$$

$$G = 1 \times 10^5 \text{ N/mm}^2$$

Now,

$$P = \frac{2\pi NT}{60 \times 1000}$$

$$\therefore \quad T = \frac{60{,}000P}{2\pi N} = \frac{60{,}000 \times 200}{2\pi \times 100}$$

$$T = 19108 \text{ N-m}$$
$$= 19108 \times 10^3 \text{ N-mm.}$$

$$T = \frac{\pi}{16} D^3\tau$$

$$\therefore \quad D = \sqrt[3]{\frac{16T}{\pi\tau}} = \sqrt[3]{\frac{16 \times 19108 \times 10^3}{\pi \times 60}}$$

$$= 116.95 \text{ mm}$$

Strain energy, $$U = \frac{T^2}{4G} \times V$$

$$\therefore \quad U = \frac{60^2}{4 \times (1 \times 10^5)} \left[\frac{\pi}{4}(116.95)^2 \times 1000\right]$$

$$= 96630 \text{ N-mm} = 96.63 \text{ N-m}$$

= 96.63 J. Ans.

Example 10.36: A solid shaft of 200 mm diameter is to be replaced by a hollow shaft of the same material, weight and length. Find the external and internal diameter of the hollow shaft if its strain energy is 1.15 is higher than that of the solid shaft when transmitting torque at the same maximum shear stress.

Solution: Both shafts have same material, same length and same weight.

Weight of solid shaft $= \rho A_s L$

Weight of hollow shaft $= \rho A_h L$

$$\therefore \quad A_s = A_h$$

$$\frac{\pi}{4}(200)^2 = \frac{\pi}{4}(D^2 - d^2)$$

$$D^2 - d^2 = 40{,}000$$

Again strain energy stored by solid shaft

$$U_s = \frac{\tau^2}{4G} \text{ Volume} = \frac{\tau^2}{4G} \cdot \frac{\pi}{4} D^2 L$$

Strain energy stored by hollow shaft,

$$U_h = \frac{\tau^2}{4G} \cdot \frac{\pi}{4}(D^2 - d^2)L \times \left(\frac{D^2 + d^2}{D^2}\right)$$

Now, $$U_h = 1.15U_s$$

Now, the two shafts must have same volume.

$$\therefore \qquad D^2 + d^2 = 1.15D^2$$

$$D^2 = 6.67\, d^2$$

But,
$$5.67\, d^2 = 40{,}000$$

$$d = \sqrt{\frac{40,000}{5.67}} = 84 \text{ m}$$

$$D = \sqrt{6.67} \times 84 = 216.94 \text{ mm}$$

Example 10.37:

(a) Show that for a hollow shaft of external and internal radius r_o and r_i respectively, the torsional strain energy is given by

$$\frac{\tau^4}{4G}\left[\frac{r_0^2 + r_i^2}{r_0^2}\right] \times \text{volume of shaft.}$$

(b) A hollow shaft having the external diameter 1.5 times the internal diameter is subjected to pure torque. If it attains a maximum shear stress τ_{max}, show that strain energy per unit volume is $\frac{13}{36G}\tau^2_{max}$.

Solution: (a) Strain energy,

$$U = \frac{1}{2} T\theta$$

From torsion equation,

$$\frac{T}{J} = \frac{\tau}{R} = \frac{G\theta}{L}$$

$$\theta = \frac{\tau L}{GR} = \frac{\tau L}{Gr_0}$$

and
$$T = \frac{\tau J}{R} = \frac{\tau J}{r_0}$$

$$\therefore \qquad U = \frac{1}{2}\frac{\tau J}{r_0} \times \frac{\tau L}{Gr_0} = \frac{\tau^2}{2G} \times \frac{JL}{r_0^2}$$

$$= \frac{\tau^2}{4G} \cdot \frac{\pi}{2} \frac{(r_0^4 + r_i^4)}{r_0^2} . L$$

$$= \frac{\tau^4}{4G} \frac{(r_0^2 + r_i^2)}{r_0^2} \times \pi (r_0^2 - r_i^2) L$$

$$U = \frac{\tau^4}{4G} \frac{(r_0^2 + r_i^2)}{r_0^2} \times \text{Vol. of shaft.}$$

(b) Strain energy per unit volume,

$$\frac{U}{V} = \frac{\tau_{max}^2}{4G} \cdot \left(\frac{d_0^2 + d_i^2}{d_0^2} \right)$$

$$= \frac{\tau_{max}^2}{4G} \times \frac{d_0^2 + \left(\frac{2}{3} d_0 \right)^2}{d_0^2}$$

$$= \frac{13}{36G} \tau_{max}^2$$

10.12 COMPARISON OF SOLID AND HOLLOW SHAFTS

In industry power is transmitted both by solid as well as hollow shafts. For same power or torque transmission, hollow shafts are lighter in weight due to unfavourable stress distribution across the cross-section of a solid shaft.

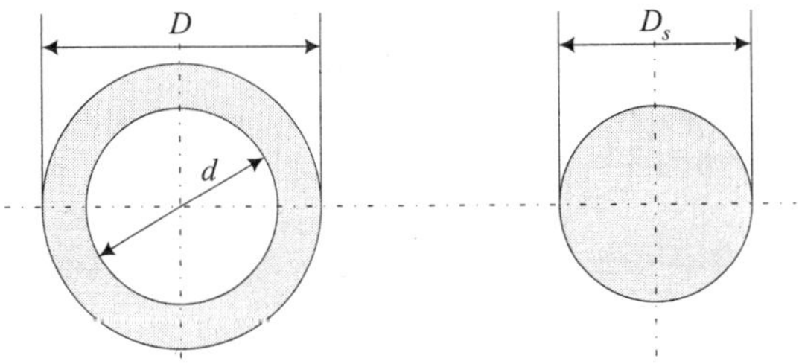

Fig. 10.15 Solid and hollow shafts

1. Geometry. If diameter of solid shaft is D_s and external and internal diameters of hollow shaft are D and d respectively:

Area of cross-section of solid shaft $= \frac{\pi}{4} D_s^2$

Area of cross-section of hollow shaft $= \frac{\pi}{4} (D^2 - d^2)$

Polar moment of inertia of solid shaft, $J = \frac{\pi}{32} D_s^4$

Polar section modulus of solid shaft, $Z_p = \frac{\pi}{16} D_s^3$

Polar moment of inertia of hollow shaft, $J = \frac{\pi}{32}(D^4 - d^4)$

Polar section modulus of hollow shaft, $Zp = \frac{\pi}{16}\left(\frac{D^4 - d^4}{D}\right)$

2. Weight. The weight of a shaft, $W = \rho g V$ [kN]

where,
ρ = density of material [kg/m^3]
g = acceleration due to gravity
= 9.81 m/s^2
V = volume of the shaft [m^3]

For a solid shaft, $V = \frac{\pi}{4} D_s^2\, L$ [m^3]

For a hollow shaft, $V = \frac{\pi}{4}(D^2 - d^2)L$ [m^3]

For same material and length, the volume of solid and hollow shaft must be same.

$$\therefore \quad D_s^2 = (D^2 - d^2).$$

3. Strength. The torque transmission by a shaft

$$T = Zp\tau$$

For a solid shaft,

$$T_s = \frac{\pi}{16} D_s^3 \tau$$

For a hollow shaft,

$$T_h = \frac{\pi}{16}\left(\frac{D^4 - d^4}{D}\right)\tau$$

$$\therefore \quad \frac{T_s}{T_h} = \frac{D_s^3 \times D}{(D^4 - d^4)}, \quad \text{if } \tau \text{ is same.}$$

4. Rigidity. Torsional rigidity, $\frac{T}{\theta} = \frac{GJ}{L}$

$$\frac{\text{Torsional rigidity of solid shaft}}{\text{Torsional rigidity of hollow shaft}} = \frac{GJ_s/L}{GJ_h/L}$$

For same material and shaft length,

$$\frac{\text{Torsional rigidity of solid shaft}}{\text{Torsional rigidity of hollow shaft}} = \frac{J_s}{J_4}$$

$$= \frac{\frac{\pi}{32}.D_S^4}{\frac{\pi}{32}(D^4 - d^4)} = \frac{D_S^4}{(D^4 - d^4)}$$

5. Angle of twist. From torsion equation:

$$\frac{T}{J} = \frac{\tau}{R} = \frac{G\theta}{L}$$

$$\therefore \qquad \theta = \frac{\tau L}{GR}$$

For same material and length,

$$\frac{\theta_s}{\theta_h} = \frac{D_h}{D_s}.$$

Example 10.38: Compare the weights of a solid shaft and hollow shaft of same material, same length, same torque and same allowable shear stress. The internal diameter of hollow shaft is 2/3 of its outer diameter.[U.P.T.U. II Sem., 2003–04]

Solution:

(a) *Solid shaft*

Torque, $$T = \frac{\pi}{16} D_s^3 \tau$$

Weight, $$W_s = \frac{\pi}{4} D_s^2 L\rho g$$

(b) *Hollow shaft*

Torque, $$T = \frac{\pi}{16}\left(\frac{D^4 - d^4}{D}\right)\tau$$

$$= \frac{\pi}{16}\left[\frac{D^4 - \left(\frac{2D}{3}\right)^4}{D}\right]\tau$$

$$= \frac{\pi}{16} \times 0.8D^3 \tau$$

Weight, $$W_h = \frac{\pi}{4}(D^2 - d^2)L\rho g$$

$$= \frac{\pi}{4}\left[D^2 - \left(\frac{2D}{3}\right)^2\right]L\rho g$$

$$= \frac{\pi}{4} \times 0.55\, D^2 L\rho g$$

(c) *Comparison*

For same torque,

$$\frac{\pi}{16} D_s^3 \tau = \frac{\pi}{16} \times 0.8\, D^3 \tau$$

$$\therefore \quad \frac{D_s}{D} = 0.928.$$

$$\frac{\text{Weight of solid shaft}}{\text{Weight of hollow shaft}} = \frac{W_s}{W_h} = \frac{\pi/4 D_s^2 L\rho g}{\frac{\pi}{4} \times 0.55 D^2 L\rho g}$$

$$= \frac{(0.928D)}{0.55D} = \mathbf{1.566} \quad \textbf{Ans.}$$

Example 10.39: A solid round shaft is replaced by a hollow shaft, the external diameter of which is $1\frac{1}{4}$ times the internal diameter. Allowing the same intensity of torsional stress in each, compare the weight of the solid shaft with that of hollow shaft.

Solution: The torque transmitted by solid shaft,

$$T = \frac{\pi}{16} D_s^3 \tau$$

The torque transmitted by hollow shaft,

$$T = \frac{\pi}{16} \frac{(D^4 - d^4)}{D} \tau.$$

For same torque and allowable shear stress

$$\frac{\pi}{16} D_s^3 \tau = \frac{\pi}{16} \frac{(D^4 - d^4)}{D} \tau.$$

$$D_s^3 = \frac{\left[D^4 - \left(\frac{D}{1.25}\right)^4\right]}{D}$$

$$\therefore \qquad \frac{D_s}{D} = 0.84$$

$$\frac{\text{Weight of solid shaft}}{\text{Weight of hollow shaft}} = \frac{\frac{\pi}{4} D_s^2 \rho g L}{\frac{\pi}{4}(D^2 - d^2)\rho g L}$$

$$= \frac{\frac{\pi}{4}(0.84D)^2}{\frac{\pi}{4}\left[D^2 - \left(\frac{D}{1.25}\right)^2\right]} = \mathbf{1.96} \quad \textbf{Ans.}$$

Example 10.40: Two shafts of the same material and same length are subjected to same torque. The first shaft is of a solid circular section and the second shaft is of a hollow circular section with internal diameter 0.6 times the external diameter. Compare the weight of the two shafts if the maximum shear stress developed in each of them is the same.

Solution:

Torque, $T = \frac{\pi}{16} D^3 \tau$

For same strength,

$$Z_s = Z_h$$

$$\frac{\pi}{16} D_s^3 = \frac{\pi}{16}\left(\frac{D^4 - d^4}{D}\right) = \frac{\pi}{16}\frac{[D^4 - (0.6D)^4]}{D} = \frac{\pi}{16} 0.87D^3$$

$$\therefore \qquad \frac{D_s}{D} = \sqrt[3]{0.87} = 0.955$$

The ratio of weights,

$$\frac{W_h}{W_s} = \frac{\frac{\pi}{4}[D^2 - (0.6D)^2]\rho g L}{\frac{\pi}{4} D_s^2 \rho g L} = \frac{D^2 - (0.6D)^2}{(0.87D)^2}$$

$$= \mathbf{0.475} \quad \textbf{Ans.}$$

Example 10.41: A solid circular shaft transmits 75 kW power at 180 rpm. Calculate the shaft diameter if the twist in the shaft is not to exceed 1 degree in 2 m length and shear stress is limited to 50 MN/m^2. Take modulus of rigidity, $G = 100$ GN/m^2. [U.P.T.U. II Sem., 2006–07]

Solution:

$$P = 75 \text{ kW}$$

$$N = 180 \text{ rpm}$$

$$\tau = 50 \text{ MN/m}^2 = 50 \times 10^6 \text{ Pa}$$

$$\theta = 1° = \frac{\pi}{180} \text{ rad.}$$

$$L = 2 \text{ m}$$

$$G = 100 \text{ GN/m}^2 = 100 \times 10^9 \text{ Pa}$$

$$P = \frac{2\pi\, NT}{60 \times 1000}$$

$$\therefore \quad T = \frac{60{,}000\, p}{2\pi\, N} = \frac{60{,}000 \times 75}{2\pi \times 180} = 3978.87 \text{ N-m}$$

(a) *Using shear stress as design basis.*

$$\text{T} = \frac{\pi}{16}\text{D}^3\tau$$

$$\therefore \quad D = \sqrt[3]{\frac{16\,T}{\pi\,\tau}} = \sqrt[3]{\frac{16 \times 3978.87}{\pi \times 50 \times 10^6}} = 0.074 \text{ m} = 74 \text{ mm}$$

(b) *Using twist as design basis*

$$\frac{T}{J} = \frac{G\theta}{L}$$

$$J = \frac{TL}{G\theta}$$

$$\frac{\pi}{32} D^4 = \frac{TL}{G\theta}$$

$$\therefore \quad D = \sqrt[4]{\frac{32\,TL}{\pi\, G\theta}} = \sqrt[4]{\frac{32 \times 3978.87 \times 2 \times 180}{\pi \times 100 \times 10^9 \times \pi}}$$

$$= 0.08255 \text{ m} = 82.55 \text{ mm}$$

Selection ***D* = 82.55 mm**, i.e., the larger value. **Ans.**

Example 10.42: Design a circular solid shaft to transmit 80 kW power at 200 rpm, if the twist in the shaft is not to exceed 2° in 3 m length of the shaft and maximum shear stress is limited to 70 MN/m^2. Take modulus of rigidity, $G = 90$ GN/m^2. [U.P.T.U. II Sem., 2008–09]

Solution:

$$P = 80 \text{ kW}$$

$$N = 200 \text{ rpm}$$

$$\theta = 2° = \frac{2\pi}{180} \text{ rad}$$

$$L = 3 \text{ m}$$

$$\tau = 70 \text{ MN/m}^2 = 70 \times 10^3 \text{ kN/m}^2$$

$$G = 90 \text{ GN/m}^2 = 90 \times 10^6 \text{ kN/m}^2$$

$$P = \frac{2\pi NT}{60 \times 1000}$$

$$\therefore \quad T = \frac{60{,}000\, P}{2\pi\, N} = \frac{60{,}000 \times 80}{2\pi \times 200} = 3.82 \text{ KN-m}$$

(a) *Using shear stress as design basis*

$$T = \frac{\pi}{16} D^3 \tau$$

$$D = \sqrt[3]{\frac{16\,T}{\pi\,\tau}} = \sqrt[3]{\frac{16 \times 3.82}{\pi \times 70 \times 10^3}} = 0.0652 \text{ m} = 62.2 \text{ mm}$$

(b) *Using twist as design basis*

$$\frac{T}{J} = \frac{G\theta}{L}$$

$$J = \frac{TL}{G\theta}$$

$$\frac{\pi}{32} D^4 = \frac{TL}{G\theta}$$

$$\therefore \quad D = \sqrt[4]{\frac{TL}{G\theta} \times \frac{32}{\pi}} = \sqrt[4]{\frac{3.82 \times 3 \times 32 \times 180}{90 \times 10^6 \times 2 \times \pi}}$$

$$= 0.1039 \text{ m} = 103.9 \text{ mm}$$

Select $D = 103.9$ mm, i.e., the larger value.

Example 10.43: Calculate the minimum diameter of a solid steel shaft which is not allowed to twist more than 3° in a 6 m length when subjected to a torque of 12 kNm. Also calculate the maximum shear stress developed. Take $G = 83$ GPa.

[U.P.T.U. 2008-09]

Solution:

$$G = 83 \text{ GPa} = 83 \times 10^6 \text{ kN/m}^2$$

$$\theta = 3° = 3 \times \frac{\pi}{180} = 0.05236 \text{ rad}$$

$$L = 6 \text{ m}$$

$$T = 12 \text{ kNm}$$

$$\frac{T}{J} = \frac{G\theta}{L}$$

$$J = \frac{TL}{G\theta}$$

$$\frac{\pi}{32}D^4 = \frac{TL}{G\theta}$$

$$\therefore \quad D = \sqrt[4]{\frac{TL}{G\theta} \times \frac{32}{\pi}} = \sqrt[4]{\frac{12 \times 6 \times 32}{83 \times 10^6 \times 0.05236 \times \pi}}$$

$$= 0.113976 \text{ m} = 113.976 \text{ mm}$$

$$\approx \mathbf{114\ mm} \quad \textbf{Ans.}$$

Now

$$T = \frac{\pi}{16}D^3\tau$$

$$\therefore \quad \tau = \frac{16T}{\pi D^3} = \frac{16 \times 12}{\pi \times (0.113976)^3}$$

$$= 41277 \text{ kN/m}^2 = \mathbf{41.277\ MPa} \quad \textbf{Ans.}$$

Example 10.44: The average torque transmitted by a shaft is 2255 N/m. The maximum torque is 146% of the average torque. If the allowable shear stress in the shaft material is 45 N/mm², determine the suitable diameter of the shaft.

[U.P.T.U. 2009-10]

Solution:

$$T_{av} = 2255 \text{ Nm}$$

$$T_{\max} = 1.46\, T_{av} = 1.46 \times 2255$$

$$= 3292.3 \text{ Nm} = 3.29 \times 10^6 \text{ Nmm}$$

$$\tau = 45 \text{ N/mm}^2$$

$$T_{\max} = \frac{\pi}{16} D^3 \tau$$

$$D = \sqrt[3]{\frac{16\, T \max}{\pi \tau}} = \sqrt[3]{\frac{16 \times 3.29 \times 10^6}{\pi \times 45}} = 71.94 \text{ mm}$$

$$\approx \textbf{72 mm} \quad \textbf{Ans.}$$

Example 10.45: A solid shaft is subjected to a maximum torque of 15 MN-cm. Determine the diameter of the shaft, if the allowable shear stress and twist are limited to 1 kN/cm^2 and 1° respectively for 210 cm length. G = 8 MN/cm^2.

[U.P.T.U. 2009-10]

Solution:

$$T = 15 \text{ MN} - \text{cm} = 15 \times 10^3 \text{ kN-cm}$$

$$\tau = 1 \text{ kN/cm}^2$$

$$\theta = 1° = 1 \times \frac{\pi}{180} = \frac{\pi}{180} \text{ rad}$$

$$G = 8 \text{ MN/cm}^2 = 8 \times 10^3 \text{ kN/cm}^2$$

$$L = 210 \text{ cm}$$

(i) *Using shear stress as design basis*

$$T = \frac{\pi}{16} D^3 \tau$$

$$D = \sqrt[3]{\frac{16T}{\pi \tau}} = \sqrt[3]{\frac{16 \times 15 \times 10^3}{\pi \times 1}} = 42.43 \text{ cm}$$

(ii) *Using twist as design basis*

$$\frac{T}{J} = \frac{G\theta}{L}$$

$$J = \frac{TL}{G\theta}$$

$$\frac{\pi}{32} D^4 = \frac{TL}{G\theta}$$

$$\therefore \quad D = \sqrt[4]{\frac{TL}{G\theta} \times \frac{32}{\pi}} = \sqrt[4]{\frac{15 \times 10^3 \times 210 \times 180 \times 32}{8 \times 10^3 \times \pi \times \pi}}$$

$$= 21.8945 \text{ cm}$$

Take D = 42.43 cm, the greater value

Example 10.46: Determine the suitable value for inside and outside diameters of hollow shaft whose interval diameter is 0.6 times its external diameter. The shaft transmits 220 kW at 200 r.p.m. The allowable shear stress is limited to 75 MPa, and angle of twist is limited to 1° per metre.

The modulus of rigidity for shaft material is 80 kN/mm^2.

[U.P.T.U. 2010-11]

Solution:

$$P = \frac{2\pi NT}{60 \times 1000}$$

$$\therefore \quad T = \frac{60{,}000P}{2\pi N} = \frac{60{,}000 \times 220}{2\pi \times 200} = 10504.23 \text{ Nm}$$

also
$$T = \frac{\pi}{16}\left(\frac{D^4 - d^4}{D}\right)\tau$$

$$T = \frac{\pi}{16}\left|\frac{D^4 - (0.6D)^6}{D}\right|\tau = 0.171\ D^3\ \tau$$

$$\therefore \quad D = \sqrt[3]{\frac{10504.23}{0.171 \times 75}} = 9.356 \text{ mm}$$

$$\frac{T}{J} = \frac{G\theta}{L}$$

$$J = \frac{TL}{G\theta}$$

$$\frac{\pi}{32}(D^4 - d^4) = \frac{TL}{G\theta}$$

$$D^4 - (0.6D)^4 = \frac{TL}{G\theta} \times \frac{32}{\pi}$$

$$0.87\ D^4 = \frac{TL}{G\theta} \times \frac{32}{\pi}$$

$$\therefore \quad D = \sqrt[4]{\frac{TL}{G\theta} \times \frac{32}{\pi \times 0.87}}$$

$$= \sqrt[4]{\frac{10504.23 \times 1000 \times 180 \times 32}{80 \times 10^3 \times \pi \times \pi \times 0.87}} = 17.22 \text{ mm}$$

Take $D = 17.22$ mm, the greater value

Example 10.47: In a tensile test, a test piece 25 mm in diameter, 200 mm gauge length stretched 0.0975 mm under a pull of 50,000 N. In a torsion test, the same rod twisted 0.025 radian over a length of 200 mm, when a torque of 400 Nm was applied. Evaluate Poisson's ratio and the three elastic moduli for the material.

[U.P.T.U. 2008-09]

Solution:

$$\sigma = \frac{\rho}{A} = \frac{50{,}000}{\frac{\pi}{4}(25)^2} = 101.86 \text{ N/mm}^2$$

$$\varepsilon = \frac{\delta L}{L} = \frac{0.0975}{200} = 4.875 \times 10^{-4}$$

Modulus of elasticity

$$E = \frac{\sigma}{\varepsilon} = \frac{101.86}{4.875 \times 10^{-4}} = 2.089 \times 10^5 \text{ N/mm}^2$$

$$\frac{T}{J} = \frac{G\theta}{L}$$

$$\therefore \quad G = \frac{TL}{J\theta} = \frac{400 \times 10^3 \times 200}{\frac{\pi}{32} \times (2.5)^4 \times 0.025}$$

$$= 0.8344 \times 10^5 \text{ N/mm}^2$$

$$E = 2G\,(1 + \mu)$$

$\therefore$ Poisson's ratio

$$\mu = \frac{E}{2G} - 1 = \frac{2.089 \times 10^5}{2 \times 0.8344 \times 10^5} - 1$$

$= \mathbf{0.2518}$ **Ans.**

$$m = \frac{1}{\mu} = 3.97$$

The bulk modulus

$$K = \frac{mE}{3(m-2)} = \frac{3.97 \times 2.089 \times 10^5}{3(3.97 - 2)}$$

$$= 2.85 \times 10^5 \text{ N/mm}^2$$

10.13 COMPARISON OF BENDING AND TORSION

The various parameters of bending of beams and torsion of shafts are compared in Table 10.1.

Table 10.1 Comparison of bending and torsion.

Sl. No.	Parameter	Bending	Torsion
1.	Loading	Bending of beam under bending moment	Torsion of shafts under torque
2.	Governing Equation	Bending equation $\frac{M}{I}=\frac{\sigma}{Y}=\frac{E}{R}$	Torsion equation $\frac{T}{J}=\frac{\tau}{R}=\frac{G\theta}{L}$
3.	Stress	$\sigma=\frac{M}{Z}$. Z = section modulus	$\tau=\frac{T}{Z_p}$, Zp = polar section modulus.
4.	Section Modulus	$Z=\frac{I}{y}$ Rectangle, $Z=\frac{bd^2}{6}$ Circle, $Z=\frac{\pi d^3}{32}$	$Z_p=\frac{J}{R}$ Solid shaft $Z_p=\frac{\pi D^3}{16}$ Hollow shaft, $Z_p=\frac{\pi(D^4-d^4)}{16D}$
5.	Rigidity	Flexural rigidity, $EI = MR$	Tosional rigidity, $GJ=\frac{T}{\theta/L}$
6.	Combined Loading	Bending + direct load $\sigma=\frac{P}{A}\pm\frac{My}{I}$ $=\frac{P}{A}\pm\frac{Pey}{I}$	Torsion + bending $M_e=\frac{1}{2}\left[M+\sqrt{M^2+T^2}\right]$ $T_e=\left[\sqrt{M^2+T^2}\right]$
7.	Strain Energy	$U=\frac{M^2L}{2EI}$	$\frac{U}{V}=\frac{\tau^2}{4G}$, $U=\frac{\tau^2L}{2GJ}$

QUESTION BANK NO. 10

1. What is pure bending? Derive the relation between bending moment and radius of curvature of the beam under pure bending.
2. What are the assumptions made in the theory of pure bending?
3. Define and explain the following:

(i) Neutral surface and neutral axis.
(ii) Moment of resistance in bending.
(iii) Section modulus
(iv) Flexural rigidity

4. Prove that bending stress changes with height.
5. Plot bending stress curve for the following beam sections.
   (i) Rectangular beam
   (ii) *I*-section beam.
6. Why *I*-section beam is preferred over a rectangular beam?
7. Explain the following terms:
   (i) Pure torsion
   (ii) Polar section modulus
   (iii) Torsional rigidity
   (iv) Torsional stiffness.
8. What do you understand by pure torsion? What are its assumptions? Derive torsion equation, $\frac{T}{J}=\frac{\tau}{R}=\frac{G\theta}{L}$ where symbols are having their usual meanings.
9. What is shear stress distribution for a shaft under pure torsion if the shaft cross-section is:
   (i) Solid
   (ii) Hollow.
10. Show that the strain energy stored in a solid shaft for unit volume is $0.25\ \tau^2/G$, where τ is the shear stress at the outer radius of the shaft and G is the modulus of rigidity.

QUICK REVIEW INFORMATION NO. 10

1. In case of pure bending, the radius of curvature of neutral axis is circular and bending stress varies linearly with height of beam.
2. The bending stress at the neutral axis (surface) is zero and at the top layer is maximum.
3. The bending equation is $\frac{M}{1}=\frac{J}{y}=\frac{E}{R}$, where all symbols have their usual meaning.
4. For same cross-sectional area, an *I*-beam has more flexural strength than a rectangular beam.
5. The section modulus for different beam sections are as follows:

Rectangular section $= \dfrac{BD^2}{6}$

Hollow rectangular section $= \dfrac{BD^3 - bd^3}{6D}$

Circular section $= \dfrac{\pi}{32} D^3$

Hollow circular section $= \dfrac{\pi}{32}\left[\dfrac{D^4 - d^4}{D}\right]$

Symmetrical *I*-section $= \dfrac{BD^3 - (B-b)d^3}{6D}$

6. The shear stress induced in a shaft due to applied torque is zero at the centre and maximum at the circumference.
7. The torque equation is $\dfrac{T}{J} = \dfrac{\tau}{R} = \dfrac{G\theta}{L}$, where all symbols have their usual meaning.
8. The mathematical expressions for the following terms are:

   Flexural rigidity $= EI$

   Torsional rigidity $= GT$

   Section modulus $= I/y$

   Torsional stiffness $= T/\theta$.
9. The polar section modulus for different shaft sections are:

   Solid shaft $= \dfrac{\pi}{16} D^3$

   Hollow shaft $= \dfrac{\pi}{16}\left[\dfrac{D^4 - d^4}{D}\right]$
10. For same power transmitted, the weight of a hollow shaft is less than that of a solid shaft.
11. For same cross-sectional area, a hollow shaft has larger polar moment of inertia as compared to solid shaft.
12. For same material, weight and length, a hollow shaft is stronger as compared to a solid shaft.
13. As compared to a solid shaft of outer diameter D, a hollow shaft of same outer diameter will have larger angle of twist per unit length when subjected to same torque.

ANNEXURES

Annexure I: Introduction to SI Units

Annexure II: Section Modulus

ANNEXURE I: INTRODUCTION TO SI UNITS

International system of Units (SI units) have seven fundamental units and two supplementary units, which cover the entire field of science and engineering. These basic units are given in Table 1.1.

Table I.1 SI System: Basic Units

S. No.	*Physical Quantity*	*Unit*	*Symbol*
	Fundamental Units		
1.	Length (L)	Metre	m
2.	Mass (M)	Kilogram	kg
3.	Amount of substance (n)	Mole	mol
4.	Time (t)	Second	s
5.	Temperature (T)	Kelvin	K
6.	Electric current (I)	Ampere	A
7.	Luminous intensity (Iv)	Candela	cd
	Supplementary Units		
1.	Plane angle (α, β, θ, ϕ)	Radian	rad
2.	Solid angle (Ω)	Steradian	sr

The dimensions of all other quantities are derived from the basic units. Some of the derived units are given in Table 1.2.

Table I.2 SI System: Derived Units

S. No.	*Quantity*	*Unit*	*Symbol*	*Alternative unit*	*In basic units*
1	2	3	4	5	6
1.	Force, Weight (F, W)	newton	N	–	kg m/s^2
2.	Work, Energy, Enthalpy (W, E, H)	joule	J	N-m	kg m^2/s^2
3.	Power (P)	watt	W	J/s	kg m^2/s^2
4.	Pressure (p)	pascal	Pa	N/m^2	kg/m^2
5.	Gas constant (R)	-	J/kg K	-	m^2/Ks^2
6.	Molar mass or Molecular mass (M)	-	kg/mol	-	kg/mol
7.	Specific heat (c)	-	J/kg K	-	m/s^{2}K
8.	Thermal conductivity (k)	-	w/mK	-	kg m/s^{2}K
9.	Mass density (ρ)	-	kg/m^3	-	kg/m^3
10.	Linear velocity (V)	-	m/s	-	m/s
11.	Linear acceleration (a)	-	m/s^2	-	m/s^2
12.	Angular velocity (ω)	-	rad/s	-	rad/s
13.	Angular acceleration (α)	-	rad/s^2	-	rad/s^2
14.	Dynamic viscosity (μ)	-	N-s/m^2	-	kg/sm
15.	Kinematic viscosity (μ)	-	m^2/s	-	m^2/s
16.	Frequency (f)	hertz	Hz	-	s^{-1}

It is often convenient and desirable to use multiples of various units, the standard list of which is given in Table 1.3.

Table I.3 SI System: Standard Multipliers

Factor	*Prefix*	*Abbreviation*
10^{12}	tera	T
10^{9}	giga	G
10^{6}	mega	M
10^{3}	kilo	k
10^{-3}	milli	m
10^{-6}	micro	m
10^{-9}	nano	n
10^{-12}	pico	P

Rules for SI Units

There are certain styles and usage of SI units.

1. For large numerals use groups of three without a coma. Grouping should start both sides from decimal. No grouping is done if figures are only 4.

Correct	*Incorrect*
234 869.739 45	23 4869.7 394 5
40 000	40000
3456.72	3456.7 2
0.123 457	0.123457

2. A dash should be used to separate units that are multiplied together.

Correct	*Incorrect*
N-m	Nm or mN which stand for millinewton

3. Plurals are never used with symbols or full names. For example, metre or metres are written as *m*.
4. Full names of units or prefixes shall be written in upright roman, lower case letters. No capitals are used even if unit is named after a person.

Correct	*Incorrect*
kelvin	Kelvin
newton	Newton
joule	Joule
watt	Watt
volt	Volt

5. All symbols are written in small letters except those derived from the proper names.

Correct	*Incorrect*
m s kg W J N K	M S Kg w j n k

6. No abbreviation should be used for units. Write either as a symbol or full name.

Correct	*Incorrect*
N or newton kg	newt kgram

7. No full stop or a punctuation mark be used.

Correct	*Incorrect*
3 m 100.25 kg	3m, 100.25 kg.

8. If a prefix is used with a symbol no space be left in between.

Correct	*Incorrect*
MJ	M J

9. Not more than one prefix be used. For 10^{-9} m.

Correct	*Incorrect*
nm	m μ m

10. Power to base unit applies to whole and not to base alone.

Correct	*Incorrect*
$km^2 = (km)^2 = 10^6\ m^2$	$km^2 \neq k(m)^2 = 10^3\ m^2$

11. Leave a space between numeral and its unit.

Correct	*Incorrect*
64 kW	64kW

12. A solidus or oblique may be used or a negative power may be used.

Correct	
$\frac{m}{s}$, m/s or ms^{-1}	

ANNEXURE II: SECTION MODULUS

Section modulus is the ratio of moment of inertia of a beam section about neutral axis (passing through the centre of gravity of the section) to the distance of extreme fibre from neutral axis.

Mathematically, section modulus,

$$Z = \frac{I}{y_{max}}$$

where, I = M.O.I. about neutral axis

y_{max} = Distance of the outermost layer of beam from neutral axis

The bending equation for a beam is

$$\frac{M}{I} = \frac{\sigma}{y} = \frac{E}{R}$$

The stress will be maximum where y is maximum.

$$\therefore \quad \frac{M}{I} = \frac{\sigma_{max}}{y_{max}}$$

$$\therefore \quad M = \sigma_{max} . \frac{I}{y_{max}} = \sigma_{max} . Z$$

where,
$$Z = \frac{I}{y_{max}}$$

The moment of resistance of a section depends upon section modulus. Therefore, section modulus represents the strength of the section against bending moment.

I. SECTION MODULUS FOR COMMON BEAM SECTIONS

1. Rectangular Section

If b is the width and d the depth of a rectangular section, the M.O.I. about neutral axis passing through *c.g.* is given by

$$I = \frac{bd^3}{12}$$

$$y_{max} = \frac{d}{2}$$

$\therefore$ Section modulus,

$$Z = \frac{I}{y_{max}} = \frac{bd^3}{12 \times \left(\frac{d}{2}\right)} = \frac{bd^2}{6}.$$

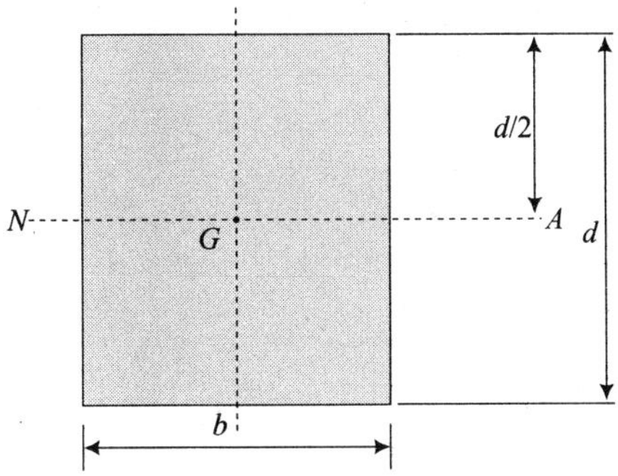

Fig. II.1 Rectangular section

2. Hollow Rectangular Section

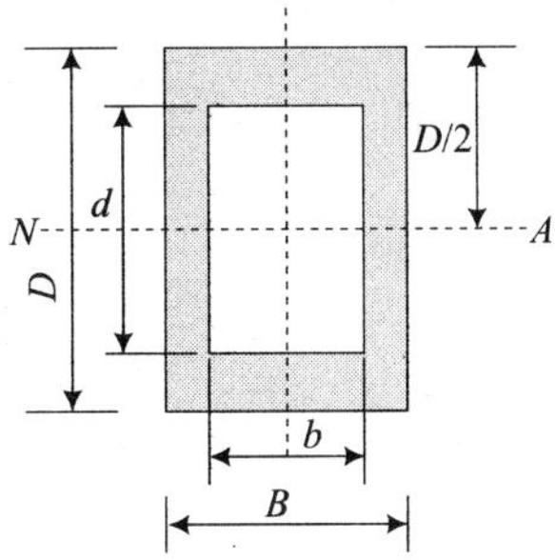

Fig. II.2 Hollow rectangular section

$$\text{M.O.I.} = \frac{BD^3}{12} - \frac{bd^3}{12}$$

$$= \frac{BD^3 - bd^3}{12}$$

$$y_{\max} = \frac{D}{2}$$

$$\therefore \quad Z = \frac{I}{y_{\max}}$$

$$= \frac{[BD^3 - bd^3]}{12 \times \frac{D}{2}}$$

$$= \left[\frac{BD^3 - bd^3}{6D}\right]$$

3. Circular Section

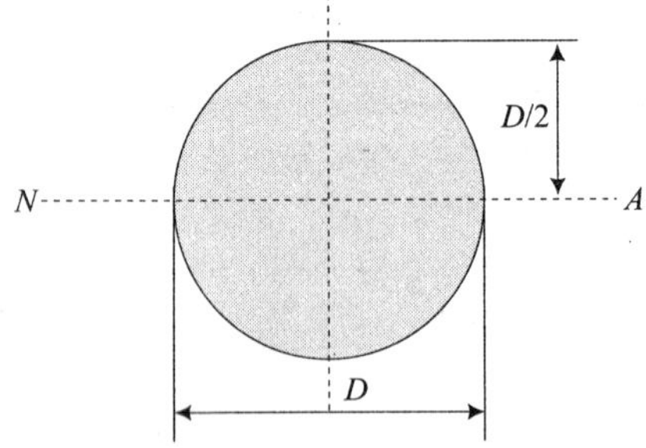

Fig. II.3 Circular section

M.O.I. of a circular section of diameter D.

$$I = \frac{\pi}{64} D^4$$

$$y_{\max} = \frac{D}{2}$$

$$\therefore \qquad Z = \frac{I}{y_{\max}} = \frac{\frac{\pi}{64} D^4}{\frac{D}{2}}$$

$$= \frac{\pi}{32} D^3.$$

4. Hollow Circular Section

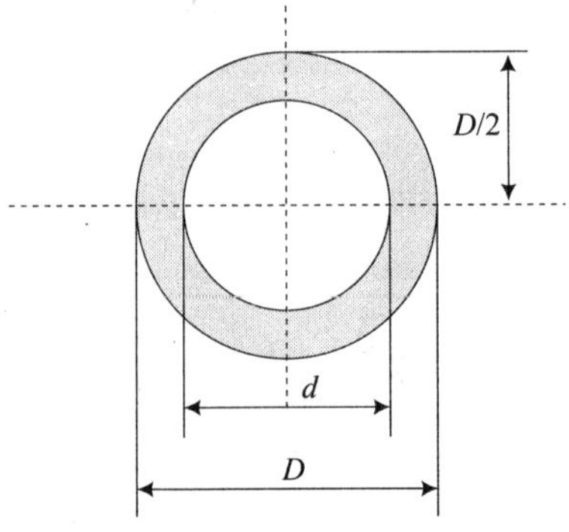

Fig. II.4 Hollow circular section

M.O.I. of a hollow circular section,

$$I = \frac{\pi}{64}[D^4 - d^4]$$

$$y_{max} = \frac{D}{2}$$

$$\therefore \quad Z = \frac{I}{y_{max}} = \frac{\frac{\pi}{64}[D^4 - d^4]}{\frac{D}{2}}$$

$$= \frac{\pi}{32}\left[\frac{D^4 - d^4}{D}\right]$$

II. SECTION MODULUS FOR SYMMETRICAL SECTIONS

5. I-Section

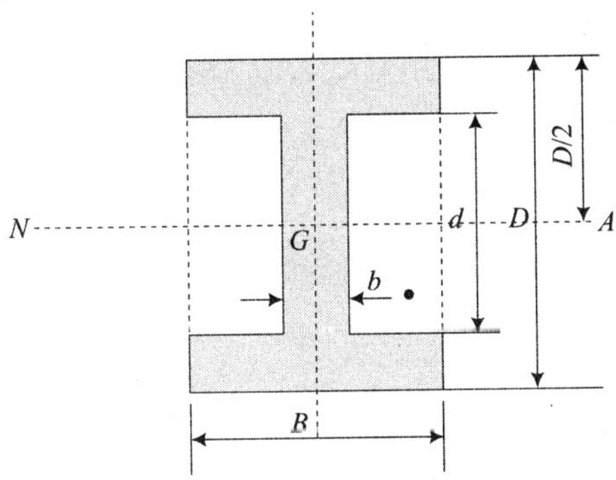

Fig. II.5 I-section

M.O.I. of I-section

$$I = \frac{BD^3}{12} - \frac{(B-b)d^3}{12}$$

$$y_{max} = \frac{D}{2}$$

$$Z = \frac{[BD^3 - (B-b)d^3]}{12 \times \frac{D}{2}}$$

$$= \frac{BD^3 - (B-b)d^3}{6D}$$

III. SECTION MODULUS FOR UNSYMMETRICAL SECTIONS

6. Unsymmetrical I-Section

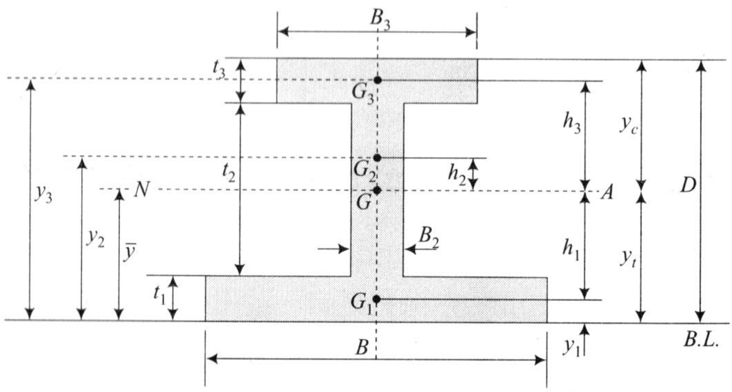

Fig. II.6 Unsymmetrical I-section

The dimensions of unsymmetrical I-section are as follows:

Bottom flange

Width = B_1

Thickness = t_1

Distance between G_1 (*c.g.* of flange section) and G (*c.g.* of total section) = h_1

Distance between G_1 and bottom layer = y_1

Area $\quad A_1 = B_1 t_1$

Web

Width = B_2

Depth = t_2

Distance between G_2 (*c.g.* of web section) and G (*N-A*) = h_2

Distance between G_2 and bottom layer of section = y_2

Area of web, $\quad A_2 = B_2 t_2$.

Top flange

Width = B_3

Thickness = t_3

Distance between G_3 (*c.g.* of top flange section) and G (N-A) = h_3

Distance between G_3 and bottom layer of section = y_3

Area = $A_3 = B_3 t_3$

y_t = Distance of extreme bottom layer to N-A

y_c = Distance of extreme top layer to N-A

$\bar{y}$ = Distance of G from bottom layer

$$\bar{y} = \frac{A_1 y_1 + A_2 y_2 + A_3 y_3}{(A_1 + A_2 + A_3)}$$

M.O.I. of complete section about N-A,

$$I = \left(\frac{B_1 t_1^3}{12} + B_1 t_1 h_1^2\right) + \left(\frac{B_2 t_2^3}{12} + B_2 t_2 h_2^2\right) + \left(\frac{B_3 t_3^3}{12} + B_3 t_3 h_3^2\right)$$

$$y_t = \bar{y}$$

$$y_c = (D - \bar{y}) = (D - y_t)$$

where, D = depth of I-section.

Maximum compressive stress in extreme top layer,

$$\sigma_c = \frac{M y_c}{I}$$

Maximum tensile stress in extreme bottom layer,

$$\sigma_t = \frac{M y_t}{I}$$

7. T-section

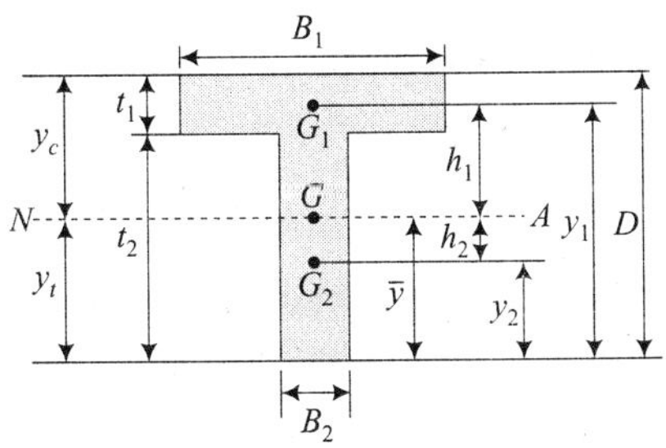

Fig. II.7 *T*-section

The dimensions of T-section are:

Top flange

Width = B_1

Thickness = t_1

Area, $A_1 = B_1 t_1$

Distance of G_1 (*c.g.* of flange) form bottom line = y_1

Distance of G_1 form N-A = h_1

Web

Width = B_2

Thickness = t_2

Area $A_2 = B_2 t_2$

Distance of G_2 (*c.g.* of web) form bottom line = y_2

Distance of G_2 form N-A = h_2

Section

Total depth = D

Distance of G (N-A) from bottom line = $\bar{y}$

Distance of top layer form N-A = y_c

Distance of bottom layer form N-A = y_t

$$\bar{y} = \frac{A_1 y_1 + A_2 y_2}{(A_1 + A_2)}$$

$$I = \left(\frac{B_1 t_1^3}{12} + B_1 t_1 h_1^2\right) + \frac{B_2 t_2^3}{12} + (B_2 t_2 h_2^2)$$

$$y_t = \bar{y}$$

$$y_c = (D - \bar{y})$$

Maximum compressive stress in extreme top layer,

$$\sigma_c = \frac{M y_c}{I}$$

Maximum tensile stress in extreme bottom layer,

$$\sigma_t = \frac{M y_t}{I}$$

IV. POLAR SECTION MODULUS

Polar section modulus or polar modulus is the ratio of polar moment of inertia of a shaft section and its outer radius. Polar section modulus represents the strength of a section against twisting or turning moment.

The torsion formula gives

$$\frac{T}{J} = \frac{\tau}{R} = \frac{G\theta}{L}$$

where, J = Polar moment of inertia,

$$\therefore \qquad T = \frac{\tau J}{R} = \tau Z_p$$

where, $Z_p = \frac{J}{R}$ = Polar section modulus.

V. POLAR SECTION MODULUS FOR COMMON SHAFT SECTIONS

1. Solid Shaft

Polar moment of inertia,

$$J = \frac{\pi}{32} D^4$$

Polar section modulus,

$$Z_p = \frac{J}{R} = \frac{\frac{\pi}{32} D^4}{\frac{D}{2}}$$

$$= \frac{\pi}{16} D^3$$

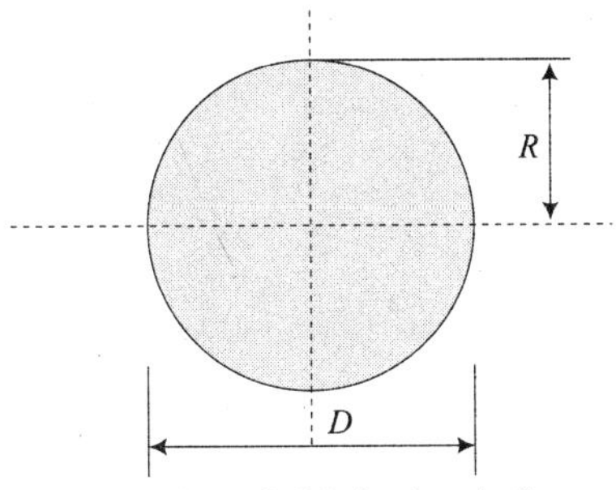

Fig. II.8 Solid circular shaft

2. Hollow Shaft

Polar moment of inertia,

$$J = \frac{\pi}{32}(D^4 - d^4)$$

Polar section modulus,

$$Z_p = \frac{J}{R} = \frac{\frac{\pi}{32}(D^4 - d^4)}{\frac{D}{2}}$$

$$= \frac{\pi}{16}\left[\frac{D^4 - d^4}{D}\right].$$

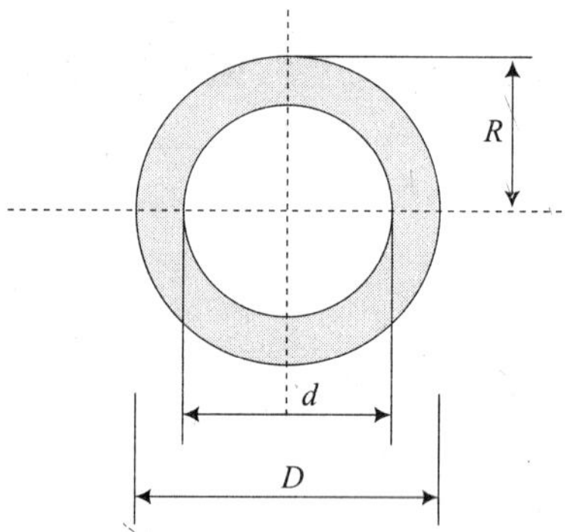

Fig. II.9 Hollow circular shaft

Index